Target Sites of Fungicide Action

Editor

Wolfram Köller
Assistant Professor
Department of Plant Pathology
Cornell University
New York State Agricultural Experiment Station
Geneva, New York

CRC Press
Taylor & Francis Group
Boca Raton London New York

CRC Press is an imprint of the
Taylor & Francis Group, an informa business

First published 1992 by CRC Press
Taylor & Francis Group
6000 Broken Sound Parkway NW, Suite 300
Boca Raton, FL 33487-2742

Reissued 2018 by CRC Press

Library of Congress Cataloging-in-Publication Data

Target sites of fungicide action/editor, Wolfram D. Köller,
p. cm.
Includes bibliographical references and index.
ISBN 0-8493-6855-3
1. Phytopathogenic fungi--Control. 2. Fungicides--Physiological effect. 3. Phytopathogenic fungi--Physiology. I. Koller, Wolfram D.
SB733.T37 1991
632'.952--dc20 91-22650

A Library of Congress record exists under LC control number: 91022650

Publisher's Note
The publisher has gone to great lengths to ensure the quality of this reprint but points out that some imperfections in the original copies may be apparent.

Disclaimer
The publisher has made every effort to trace copyright holders and welcomes correspondence from those they have been unable to contact.

ISBN 13: 978-1-315-89798-1 (hbk)
ISBN 13: 978-1-351-07708-8 (ebk)

Visit the Taylor & Francis Web site at http://www.taylorandfrancis.com and the
CRC Press Web site at http://www.crcpress.com

PREFACE

Fungi are not only the most important class of plant pathogenic organisms, but they are also the causal agents of animal and human mycoses. Fungal diseases of crops can be devastating, and disease symptoms have been described since farming became the major source of our food and fiber supply.[1] Although the incidence of human diseases caused by fungi is less than that of bacterial and viral infections, they do occur and can be life-threatening.[2] Major weapons in the combat of fungal diseases have been, in both cases, chemicals with antifungal activities. Interestingly, two different names have emerged in the past. The term "fungicide" describes antifungal compounds developed for agriculture, whereas "antifungal agent", "antifungal drug" and "antimycotic" are more commonly used in medicine. From a historic point of view, this distinction is not surprising. The first generation of agricultural fungicides had no application in medicine, and the first antifungal drugs introduced in the late 1950s, amphotericin B and 5-fluorocytosine, were useless for farmers. Over the last two decades, however, the distinction between the two classes of antifungal compounds has become less pronounced. In particular the discovery and development of azoles with their broad field of application in agriculture and medicine has triggered increasing cooperation between the mycological disciplines concerned with either control of plant diseases or antifungal chemotherapy. While the recognition of mutual scientific interests was on the rise, the future perspectives for fungicides and antifungal drugs began to drift apart. Although the common goal remained the same, agricultural fungicides and medical drugs are viewed differently. What has happened?

It had been the progress in chemistry and chemical engineering in the first half of the 20th century that made the reliable control of insects, weeds and plant diseases a reality. The initial acceptance of early organic pesticides was overwhelming, and it is worth remembering that P. Müller, the inventor of DDT, was honored with the Nobel Prize in 1948. This positive perception has vanished and given place to a global awareness of pesticide threats to our environment. Undoubtedly, the chemical control of plant diseases, although still irreplaceable, has become part of our current "chemophobia". The issue has become increasingly emotional, and public concerns are not always justified from a scientific point of view. Moreover, the tremendous progress in safety and application rates made over the last four decades is not always fully recognized and appreciated. Regardless of many existing misperceptions, however, some of the fungicide-related problems are real, and scientists engaged in fungicide research will have to answer a question of considerable importance: Was the development of chemical fungicides a dead-end street with low expectations for the future, or will it be feasible to apply our scientific progress to the development of novel and effective chemical control agents with extremely specific targets and thus without side effects on non-target organisms and the environment?

Many scientists at the forefront of plant disease control appear to have answered this question. They have dismissed chemical measures from their arsenal of future strategies, and the term "innovative" has become a synonym for "non-chemical". For example, future prospects for chemical control are no longer part of a recent book on innovative approaches to plant disease control.[3] Expectations are high of finding widespread and durable solutions by recombinant DNA technology, be it through the engineering of disease-resistant crop cultivars, or through the construction of greatly improved biocontrol agents. Undoubtedly, the exciting potential of biotechnology will have beneficial impact in the future. But will it make chemical control entirely obsolete, and will it always lead to entirely "nontoxic" means of plant disease control? Zehr[4] concluded in his recent reflections on innovation in plant disease control: "We might even find that new chemical pesticides carefully regulated by government agencies and applied by informed users may be among the safest of the mechanisms available to control many plant diseases and pests." It certainly would be unwise to abandon innovative research on novel fungicides before safe and lasting "non-chemical" alternatives are in sight. Most likely, the breeding of disease-resistant cultivars, the development of biocontrol agents, and new chemical fungicides will co-exist in the future as they have in the past.

In contrast to chemical fungicides, the public demand for novel antifungal drugs has increased over the last decade. The incidence of fungal infections is on the rise since an increasing number of immunocompromised patients became prone to opportunistic and sometimes life-threatening fungal infections.[5,6] The treatment of deep-seated and systemic mycoses is difficult with the currently prescribed arsenal of antimycotic drugs, and a number of new antimycotics are under clinical development.[7] Mandatory for the treatment of many opportunistic fungal infections is the availability of highly specific antifungal drugs without side effects on the patient. Progress toward high levels of fungal specificity has already been made with the group of systemic fungicides developed for agriculture, and newer developmental antifungal drugs are becoming increasingly similar to these site-specific fungicides, in particular within the group of sterol biosynthesis inhibitors.

In general, the systemic treatment of any patient, be it plant, animal or human, is only possible with inhibitors of fungal-specific targets. Specificity, however, also has its shortcomings. As the experience with specific agricultural fungicides has shown, these inhibitors are more prone to the development of resistance. Although fungal resistance has not been a serious problem for the current antimycotics, this might change if immunocompromised patients with mycotic diseases have to be maintained under long-term treatment. The identification of suitable and novel antifungal targets, the modification of already existing inhibitors, and the prediction of inherent risks of resistance will require profound data on the molecular mode of inhibitor action. The contributions in this book are a comprehensive summary of our current knowledge, but they are also meant to raise questions and to stimulate future research.

The first two chapters are dedicated to the carboxamides and hydroxypyrimidines. They represent groups of relatively old systemic fungicides with a remarkable level of specificity and selectivity. The benzimidazoles described in Chapter 3 with their broad spectrum of plant disease control have been severely affected by the development of resistance. The more recent cloning and sequencing of the target gene has not only provided insight into the molecular specifics of resistance, but it also explained the mode of action of compounds with exclusive activity against resistant genotypes. Edlich and Lyr (Chapter 4) describe the mode of action concepts that emerged only recently for a group of rather unrelated fungicides, the dicarboximides and the aromatic hydrocarbons. The class of specific fungicides available for the control of oomycetes is described in depth in Chapter 5. All antifungal inhibitors covered in the first five chapters are exclusively used in agriculture. This restriction is not always apparent from a mode of action point of view, and the better understanding of molecular principles might lead to closely related inhibitors with activities against a broader or different spectrum of fungal pathogens. The melanin biosynthesis inhibitors (Chapter 6) are different from all other antifungal agents and represent a novel and desirable trait in fungal disease control. They are neither fungitoxic nor fungistatic but interfere with the penetration of fungal pathogens into their plant hosts and, thus, are the first representatives with an indirect mode of action. The inhibitors of fungal sterol biosynthesis described in Chapters 7, 8 and 9 comprise the currently most successful group of antifungal compounds. Several different enzymes of the sterol biosynthetic pathway are already antifungal targets, and interesting inhibitor lead structures for additional enzymes are promising. Moreover, the sterol biosynthesis inhibitors have crossed the border of disciplines and are successfully used in both agriculture and medicine. Several specific fungicides with known modes of action did not warrant a separate chapter. They are described, however, in Chapter 10, which is dedicated to the role of molecular studies in fungicide research. Finally, the question is asked in Chapter 11 whether it might become feasible to use industrial enzymes as disease control agents.

The contributions as a whole are a mirror of the tremendous progress made over the last two decades. This progress becomes immediately evident from the contents of a monograph dedicated to the same subject, written by Lukens in 1971.[8] There are no reasons to believe that progress will slow down in the future. Hopefully, this book will stimulate ideas and encourage research in the field of antifungal inhibitors.

Wolfram Köller
Geneva, New York

1. **Agrios, G. N.,** *Plant Pathology,* 3rd ed., Academic Press, San Diego, 1988.
2. **Rippon, J. R.,** *Medical Mycology,* 2nd ed., W. B. Saunders, Philadelphia, 1982.
3. **Chet, I.,** *Innovative Approaches to Plant Disease Control,* John Wiley, New York, 1987.
4. **Zehr, E. I.,** Reflections on innovation in plant disease control, *Plant Dis.,* 73, 372, 1989.
5. **Bodey, G. P.,** Fungal infections in cancer patients, *Ann. N.Y. Acad. Sci.,* 544, 431, 1988.
6. **Spencer, P. M. and Jackson, G. G.,** Fungal and mycobacterial infections in patients infected with the human immunodeficiency virus, *J. Antimicr. Chemother.,* 23 (Suppl. A), 107, 1989.
7. **Saksena, A. K., Girijavallabhan, V. M., Cooper, A. B., and Loebenberg, D.,** Recent advances in antifungal agents, *Annu. Rep. Med. Chem.,* 24, 111, 1989.
8. **Lukens, R. J.,** *Chemistry of Fungicidal Action,* Springer-Verlag, New York, 1971.

EDITOR

Wolfram Köller is assistant professor of plant pathology at the Geneva Campus of Cornell University, where he joined the faculty in 1986. He was trained in chemistry and biochemistry and received his Ph.D. in chemistry from the Philipps University of Marburg in 1979. He spent two postdoctoral years as a recipient of a DFG fellowship at the Institute of Biological Chemistry of Washington State University in Pullman. Thereafter he became the head of a biochemistry laboratory in fungicide research at Bayer in Leverkusen.

His research interest is focused on the biochemistry of plant disease control. He has authored more than 60 scientific contributions and is co-author of a textbook on undergraduate chemistry training. He has given more than 30 invited lectures and seminars. His current appointments are on the editorial board of *Phytopathology* and on several national and international committees in pesticide resistance.

CONTRIBUTORS

Pierre Benveniste, D.S.
Professor
Institut de Biologie Moléculaire des Plantes
Institut de Botanique
Université Louis Pasteur
Strasbourg, France

Amanda Davis, B.Ag.Sci.
Research Scholar
Department of Biochemistry
University of Melbourne
Parkville, Victoria, Australia

Wilfried Edlich, Dr.
Research Scientist
Insitute of Plant Protection Research Kleinmachnow
Kleinmachnow, Germany

S. G. Georgopoulos, Ph.D.
Professor
Department of Plant Pathology
Agricultural University of Athens
Athens, Greece

Bruce Grant, Ph.D.
Reader in Biochemistry
Department of Biochemistry
University of Melbourne
Parkville, Victoria, Australia

Julia M. Griffith, Ph.D.
Research Fellow
Department of Biochemistry
University of Melbourne
Parkville, Victoria, Australia

D. W. Hollomon, Dr.
Department of Agricultural Sciences
University of Bristol
AFRC Institute of Arable Crops Research
Long Ashton Research Station
Bristol, United Kingdom

Hideo Ishii, M.Agr.
Laboratory Head
Division of Plant Protection
Fruit Tree Research Station, MAFF
Tsukuba, Ibaraki, Japan

Paul A. J. Janssen, Dr.
President and Director of Research
Janssen Pharmaceutica
Beerse, Belgium

Wolfram Köller, Ph.D.
Assistant Professor
Cornell University
Department of Plant Pathology
New York State Agricultural Experiment Station
Geneva, New York

Yasuyuki Kubo, Ph.D.
Assistant Professor
Laboratory of Plant Pathology
Faculty of Agriculture
Kyoto University
Kyoto, Japan

Horst Lyr, Dr.
Professor and Department Director
Institute of Plant Protection Research Kleinmachnow
Kleinmachnow, Germany

A. J. Poulose, Ph.D.
Research Director
Exploratory Research
Genencor International
South San Francisco, California

Alain Rahier, D.S.
Directeur de Recherche
Centre National de la Recherche Scientifique
Institut de Biologie Moléculaire des Plantes
Strasbourg, France

Hugo Vanden Bossche
Department Head
Department of Comparative Biochemistry
Janssen Research Foundation
Beerse, Belgium

G. A. White, Ph.D.
Research Scientist
Agriculture Canada
Research Centre
London, Ontario, Canada

Isamu Yamaguchi, Ph.D.
Senior Scientist and Professor
Microbial Toxicology Laboratory
The Institute of Physical and Chemical Research (RIKEN)
Wako-shi, Saitama, Japan

TABLE OF CONTENTS

Chapter 1

TARGET SITES OF CARBOXAMIDES

G. A. White and S. G. Georgopoulos

TABLE OF CONTENTS

I. INTRODUCTION

It appears that the initial carboxamide shown to be fungicidal was salicylanilide (2-hydroxybenzanilide), which was described nearly 60 years ago by Fargher et al.[1] as a selective, protectant fungicide. However, it was not until 1966 that von Schmeling and Kulka reported the first truly systemic carboxamide, i.e., carboxin (5,6-dihydro-2-methyl-1,4-oxathiin-3-carboxanilide) and its sulfone analogue, oxycarboxin.[2] Carboxamides since then became of considerable interest for plant disease control, and subsequently an array of compounds with different planar and nonplanar ring systems were synthesized and tested for antifungal properties. Woodcock,[3] and more recently Kuhn[4] and Kulka and von Schmeling,[5] have reviewed these heterocyclic carboxamide structures in some detail. Representative structures are shown in Figure 1.

Carboxin and oxycarboxin act as systemic fungicides for disease control displaying apoplastic movement as shown by experiments with cereals, beans, and other plants. Thus, foliage diseases can be controlled by treating the seeds or the soil with fungicide. Cytological studies have revealed the details of protective and curative effects of oxycarboxin against the bean rust fungus, *Uromyces phaseoli*.[6,7] When bean plants were root-treated with oxycarboxin, germination of uredospores proceeded normally, but the germ tubes were shorter than those of uredospores germinating on untreated plants. Formation of appressoria and intercellular hyphae was not hindered, but no haustoria could be detected. When treatment was postponed until the chlorotic spot stage, no further symptoms developed and the original spots became less evident. Microscopic examinations showed disrupted haustoria, encased by material of plant origin. It is interesting that the mitochondria of such haustoria had become swollen with dilated cristae. Also, wall appositions were formed within the plant cell wall, indicating that plant defense mechanisms might contribute to the curative action of oxycarboxin. Hirooka et al.[8] found that 3′-isopropoxy-2-trifluoromethylbenzanilide (Flutolanil) was absorbed and translocated internally through rice plants giving protectant control of sheath blight *(Rhizoctonia solani)* by retardation of fungal growth and penetration of infection cushions. Severely collapsed hyphae and infection cushions were observed.

Carboxamide fungicides have generally been found most active against Basidiomycete fungi, e.g., rusts, smuts, and bunts of cereals and damping off caused by *R. solani*.[2,9-11] The 2′-phenyl analogue of carboxin (F 427) has a broader spectrum which shows preference for specific taxonomic groups of Basidiomycetes, Porosporae, and *Aspergillus* spp.[9] It has also been noted that 2,5-dimethylfuran-3-carboxanilide is effective against some *Helminthosporium* and *Fusarium* spp.[12] *In vitro* it has been possible, by structural modification, to extend the fungitoxic spectrum of carboxamides to include some Ascomycete and Phycomycete fungi where, for example, a number of thiophene carboxamides were active spore germination inhibitors, particularly toward *Phytophthora infestans* and *Verticillium dahliae*.[13] *In vivo* experiments with late blight *(P. infestans)* on tomato plants showed that several thiophene carboxamides, e.g., 3′-*n*-butyl-3-methylthiophene-2-carboxanilide, gave satisfactory protectant activity.[13]

II. MECHANISM OF ANTIFUNGAL ACTIVITY

The first indication that carboxamides inhibit respiration was obtained in experiments with intact cells by Mathre[14] and Ragsdale and Sisler.[15] Glucose oxidation by *R. solani* mycelium in phosphate buffer was blocked to about 60% by a growth-inhibiting concentration of carboxin. With *Ustilago maydis,* inhibition of glucose oxidation was less severe and declined with sporidial age. However, acetate, pyruvate, and succinate oxidation was much more sensitive to carboxin than was glucose oxidation.[15] When teliospores of *U. nuda* or mycelia of *R. solani* were incubated with [U-^{14}C] glucose or [U-^{14}C] acetate, carboxin caused

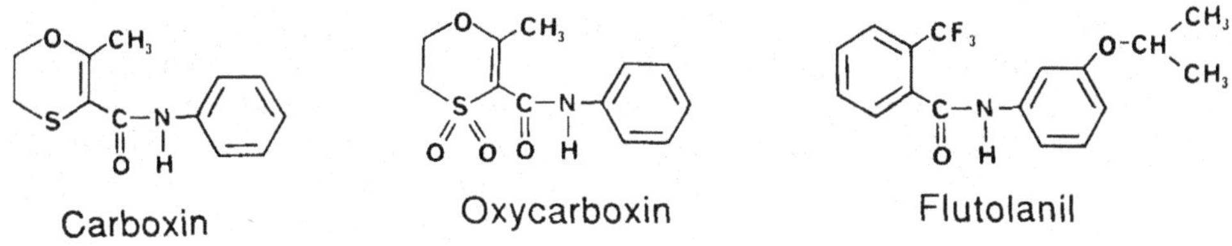

FIGURE 1. Chemical structures of carboxamides.

a decreased incorporation of label into citrate, malate, and fumarate but an increased incorporation into succinate.[14] This was the primary indication that carboxamide fungicides might affect succinate metabolism. Another early observation that carboxin affects respiration and therefore the mitochondrial electron transport system was made by Georgopoulos and Sisler.[16] These workers found that carboxin selects for mutants of *U. maydis* which lack the alternative antimycin A-insensitive electron transport pathway.

Of considerable interest during these studies was the rather marked difference in carboxin sensitivity on acetate substrate as compared to glucose.[15] Growth tests made with three organisms showed a tenfold increase in toxicity when acetate replaced glucose as the carbon source, and this was considered to indicate that ATP generated in glycolysis and possibly during glucose oxidation is less affected by the fungicide than is ATP formed during acetate metabolism. Yeast cells utilizing glucose anaerobically were indeed appreciably more tolerant to carboxin than those aerobically cultured on acetate. In fungi unable to grow anaerobically, growth inhibition on glucose, though less pronounced than on acetate, is quite considerable and cannot be accounted for by the small amount, if any, of inhibition of glucose oxidation by carboxin. This was explainable if carboxin inhibited some reaction in the citric acid cycle. Since glucose was not oxidized via this cycle in the presence of carboxin, it was suggested by Mathre[14] and Ragsdale and Sisler[15] that a lack of cycle intermediates and/or limited availability of ATP was responsible for growth inhibition.

Following studies with whole cells, the identification of the succinate dehydrogenase complex (SDC,[1] complex II, succinate-ubiquinone reductase) of the respiratory electron transport system as the site of action of carboxin and related carboxamides was established by experiments with mitochondrial preparations. Succinate oxidation by mitochondria of *U. maydis* was very sensitive to carboxin with a K_i of 0.32 μM, while NADH oxidation was not inhibited even at a carboxin concentration of 150 μM.[17] The oxidation of succinate was also inhibited in mitochondria from rat liver and pinto bean, although to a lesser extent than in mitochondria from the sensitive fungus. Particulate preparations from *U. maydis* and *Cryptococcus laurentii*, exhibiting succinate dehydrogenase activity, were also used by White[18] and White and Thorn[19] to study the effect of carboxin and a number of carboxin analogues. The 1,4-oxathiin carboxamides inhibited the reduction of DCIP dye by succinate and this inhibition was sufficiently high to account for both the accumulation of succinate and inhibition of cell growth. It was concluded that interference of succinate oxidation constitutes the primary action of these fungicides and the suggestion was made that the site of action could be on succinate dehydrogenase or on an electron carrier component proximal to the dehydrogenase.[18] That the site of action of carboxin lies between succinate and Coenzyme Q was supported by the experiments of Ulrich and Mathre[20] with mitochondria from *U. maydis* and *Saccharomyces cerevisiae*. These authors noted that succinate dehydrogenase, assayed with PMS and DCIP, was inhibited when intact mitochondria but not the solubilized enzyme was assayed. The same observation was made by Lyr et al.[21] working with *Trametes versicolor*.

The conclusions arising form the initial biochemical investigations on the mode of carboxin action were strongly supported by genetic studies done by Georgopoulos et al.[22,23]

and Gunatilleke et al.[24] In *U. maydis,* mutants (oxathiin-resistant; *oxr*), recognized by selection on carboxin-containing medium, were also found to be cross-resistant to thiazole carboxamides. The dosage-response curves for inhibition of succinoxidase activity, as measured by the reduction of ferricyanide by mitochondrial preparations, were in good agreement with those for growth and for oxidation of tricarboxylic cycle acids by whole cells.[22] These results indicated that resistance to carboxin results from a mutational modification of the SDC. Mitochondrial preparations from mutant cells could also be identified in the absence of carboxin from the instability of the succinate dehydrogenase activity. Enzyme assays showed that mitochondria prepared from a heterozygous diploid strain of *U. maydis* displayed intermediate sensitivity to carboxin and probably contained a mixture of carboxin-sensitive and carboxin-resistant SDCs. Georgopoulos et al.[23] obtained results which supported this hypothesis. When mitochondria from the heterozygous strain were "aged" by incubation at 30°C to eliminate the portion of succinate-DCIP reductase activity contributed by the mutant haploid, the activity left exhibited a carboxin sensitivity similar to that of mitochondria from the sensitive wild-type strain.

In *Aspergillus nidulans,*[24] among three freely recombining mutant genes for carboxin resistance, all reduced carboxin inhibition of succinate oxidase, succinate-cytochrome c reductase, and succinate dehydrogenase in mitochondrial fractions. Also, there was good correlation between the degree of resistance *in vitro* and *in vivo,* indicating that in this organism, the ability of carboxin to inhibit mycelial growth is a direct consequence of the inhibition of cellular succinate oxidation.

III. MUTATION MODIFICATION OF CARBOXAMIDE SENSITIVITY OF FUNGAL SDCs

Fungal mutants with modified sensitivity to carboxamides can be obtained easily in the laboratory. Resistance problems in the field have been encountered in the control of chrysanthemum rust *(Puccinia horiana),* carnation rust *(Uromyces caryophylinus),* and loose smut of barley *(Ustilago nuda).*[25-27]

The first mutation recognized by selection on medium containing carboxin was the *ants* (antimycin A sensitivity) mutation in *U. maydis.*[16] Approximately 90% of the cell colonies arising after incubation of UV-irradiated sporidia on a glucose medium containing 40 μM carboxin were of the *ants* type and the frequency at which such mutants were isolated was 250 per 10^6 survivors of the UV treatment.[28] On the carboxin-agar medium, *ants* colonies were generally small and growth measurements in liquid culture with glucose as the carbon source showed a slightly lower carboxin sensitivity of *ants* mutants as compared to wild-type strains.[29] A much clearer distinction can be made on medium containing antimycin A which permits growth of wild-type but not mutant cells.

Respiration of wild-type sporidia of *U. maydis,* as with cells from a variety of other organisms,[30] is insensitive to antimycin A and to inhibitors of cytochrome oxidase due to the operation of an alternate electron transport pathway. It was observed, by Georgopoulos and Sisler,[16] that O_2 uptake by the wild-type strain of *U. maydis* is stimulated by antimycin A, cyanide, and azide, while respiration of *ants* mutants is strongly inhibited. As noted by Ziogas and Georgopoulos,[31] the growth of *U. maydis* in the presence of chloramphenicol, a mitochondrial protein synthesis inhibitor, induces the formation of a second cyanide-insensitive system which, unlike the constitutive one, is not inhibited by hydroxamates. If an *ants* mutant is grown in the presence of chloramphenicol, it develops high rates of cyanide-insensitive respiration.[29]

When mitochondrial preparations of an *ants* mutant and a wild-type strain were compared with respect to inhibitor sensitivity, an approximately fivefold increase in carboxin concentration was required to achieve the same amount of inhibition of succinate oxidation in the mutant vs. wild-type mitochondria.[29] This certainly indicates that the *ants* mutation affects some property of Complex II. It seems, therefore, that a site in Complex II, responsible for low resistance to carboxin, is part of the constitutive, hydroxamate-sensitive, alternate electron transport system. It must be noted that it is the genetic information for the synthesis of the constitutive system which is lost as a result of the *ants* mutation and that the cyanide-insensitive respiration of mutant cells grown in the presence of chloramphenicol is mediated solely by the inducible system.[29]

U. maydis strains with higher resistance to carboxin than the *ants* mutants do not differ from the wild-type strain in their sensitivity to antimycin A, cyanide, and azide and were designed as *oxr* because, at the time, a 1,4-oxathiin ring was thought essential for fungitoxicity.[22] Two allelic mutations at the *oxr*-1 locus have been identified and designated by Georgoloulos and Ziogas,[28] as *oxr*-1A and *oxr*-1B having moderate and high resistance to carboxin, respectively. Dosage-response measurements showed an excellent correlation between growth and SDC activity with respect to carboxin sensitivity of wild-type *oxr*-1A and *oxr*-1B strains.[29] Therefore, these mutations are responsible for a modification of Complex II which is different from the one caused by the *ants* mutation, as it does not affect the alternate electron transport system.

As mentioned previously, the SDC of *oxr* mitochondria differs from that of wild-type mitochondria not only in carboxin sensitivity, but also in loss of activity at room temperature.[22,23] Later studies by White et al.[32] showed that, with the use of specific carboxamide structures, a distinction can be made *between oxr*-1A and *oxr*-1B mitochondria. Thus, the *oxr*-1A mutant SDC which is moderately resistant to carboxin is highly sensitive to 4′-phenylcarboxin, which has almost no effect on the wild-type SDC. In contrast, the enzyme complex from *oxr*-1B mutants is highly resistant to both carboxin and 4′-phenylcarboxin.[32] Many specific types of molecular structures (primarily 4′-substituted carboxanilides) have shown a strong selectivity for the mutated *U. maydis* SDCs and have actually reversed the effect of carboxin-selected mutations in the SDCs.[33-35]

The SDC from an *oxr*-1B mutant was distinctly less sensitive to TTF than the wild-type SDC, indicating a common mechanism of action for carboxamides and TTF on the enzyme complex.[28] Also observed was that the oxidation of NADH by the alternate pathway is more sensitive to TTF in the presence of a carboxin concentration sufficient to saturate the SDC site, which suggests that the two inhibitors compete for this site. However, a marked difference was found between the inhibitors with respect to succinate or NADH oxidation via the alternate pathway; while TTF inhibits the constitutive cyanide-insensitive pathway at a site beyond the NADH branch, carboxin does not.[36] A schematic representation of the electron transport system in *U. maydis* as indicated by the use of inhibitors on mitochondria from wild-type and carboxin-selected mutant strains is given in Figure 2.

In *Aspergillus nidulans,* mutations for carboxin resistance were first noted by van Tuyl.[37] A more detailed study was published later by Gunatilleke et al.[24] Mutation at any of three loci reduces carboxin inhibition of succinate dehydrogenase. None of the mutations recognized by selection on carboxin-containing medium affects sensitivity of *A. nidulans* to antimycin A, cyanide, or azide, as does the *ants* mutation in *U. maydis*. The three mutations differ in the degree of resistance to carboxin and, as in the case of *U. maydis,*[33-35] specific carboxamides can be found which may alleviate or reverse the effect of a given mutation.[13,32]

IV. ACTION ON THE SDC IN FUNGI, BACTERIA, AND HIGHER PLANTS

Carboxamides appear to be most active and possibly selective toward Basidiomycete fungi. At one time, in fact, it was thought that the molecular site of action for carboxin might be unique to Basidiomycetes.[18] However, as mentioned previously, some have proven

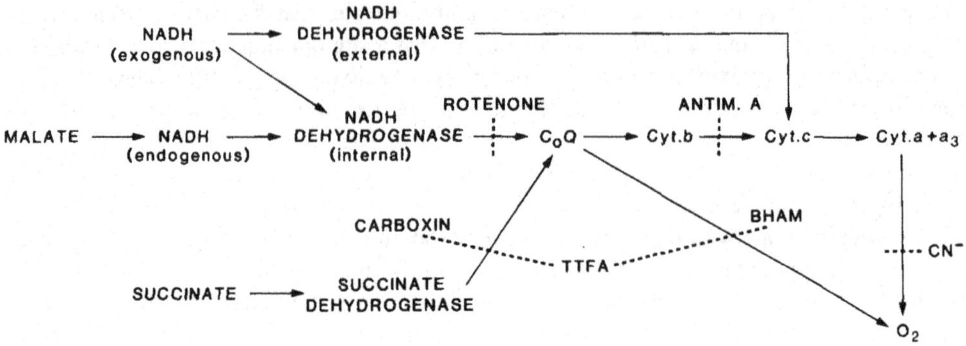

FIGURE 2. Schematic representation of the mitochondrial electron transport chain in *Ustilago maydis*. (From Ziogas, B. N. and Georgopoulos, S. G., *Pestic. Biochem. Physiol.*, 11, 208, 1979.)

TABLE 1
The Effect of Carboxin on Succinate Oxidation by Non-Mammalian Enzyme Preparations

Enzyme source	Type of preparation	Enzyme assay oxidant[a]	K_i or I_{50}[b] value (μM)		Ref.
Ustilago maydis	Mitochondria	O_2	K_i,	0.32	17
	Soluble	PMS-DCIP	No inhibition		23
	Mitochondria	DCIP	K_i,	0.5	18
	Mitochondria	DCIP	I_{50},	0.5	19
Cryptococcus laurentii	Mitochondria	DCIP	I_{50},	0.55	19
Rhizoctonia solani	Mitochondria	DCIP	I_{50},	0.68	40
Aspergillus nidulans	Mitochondria	Cytochrome c	I_{50}	0.50	24
	Mitochondria	DCIP	I_{50},	1.9	13
Fusarium oxysporum	Mitochondria	DCIP	I_{50},	5.9	40
	Mitochondria	DCIP	I_{50},	2.8	42
Gaeumannomyces graminis	Mitochondria	DCIP	I_{50},	1.84	40
Glomerella cingulata	Mitochondria	DCIP	I_{50},	10.0	42
Penicillium chrysogenum	Mitochondria	DCIP	I_{50},	13.0	42
Trichoderma viride	Mitochondria	DCIP	I_{50},	46.0	42
	Submitochondria	DCIP	I_{50},	200.0	43
Trametes versicolor	Submitochondria	DCIP	I_{50},	10.0	43
Saccharomyces cerevisiae	Mitochondria	O_2	K_i,	270.0	17
Micrococcus denitrificans	Membrane	O_2	I_{50},	16.0	44
Paracoccus denitrificans	Membrane	DB-DCIP	K_i,	13.2[c]	45
Phaseolus vulgaris	Mitochondria	DCIP	26% inhibition at 100 μM		20
Solanum tuberosum	Mitochondria	O_2	K_i,	200.0	39

[a] The oxidants are listed under "Abbreviations".
[b] I_{50} (μM): Concentration of compound which inhibits the rate of DCIP, cytochrome c, or O_2 reduction by 50%.
[c] K_i for 3′-methylcarboxin.

effective against non-Basidiomycetes.[9,12,13] Now it seems that cell permeability factors rather than carboxamide affinity for SDCs are a limiting influence on inhibition of fungal growth.[13,32,33]

As shown in Table 1, carboxin can inhibit Complex II activity in subcellular preparations isolated from fungi, bacteria, and higher plants. The purified mammalian SDC is also sensitive.[38] Structural analogues of carboxin and other heterocyclic carboxamides can be found which are highly active against the "carboxin-insensitive" particulate enzyme complexes from certain fungi (wild-type and carboxin-selected mutant strains),[13,32-35] higher plants,[39] and mammalian tissue.[38] This indicates that there are subtle differences, i.e., microheterogeneity in molecular architecture of the enzyme complex, which can drastically affect the inhibitor-receptor site interaction.[40] Such is exemplified, for example, by studies with carboxin-resistant mutants of *U. maydis* where it was observed that certain analogues of carboxin and other carboxanilides were more active against the enzyme complex from the mutants than from the carboxin-sensitive wild-type strain.[32-35] Another element involved in carboxamide inhibition of SDCs seems to be mitochondrial permeability to the inhibitor. For example, the SDC in intact mammalian mitochondria is not highly sensitive to carboxin, yet, when purified, becomes strongly inhibited.[17,38] The same phenomenon may hold for the carboxin-insensitive SDC in yeast mitochondria (see Table 1). Differences in carboxamide sensitivity among various SDCs could also be related partly to the mitochondrial lipid content.[41]

V. STRUCTURAL REQUIREMENTS FOR CARBOXAMIDE INHIBITION OF MITOCHONDRIAL COMPLEX II ACTIVITY

Some of the general conclusions made by White and Thorn[19] regarding the structural requirements of carboxamides for inhibition of SDC activity in mitochondria from *U. maydis* and *C. laurentii* can be modified in light of more recent structure-activity data.[13,32-35,40] In essence, among various types of hetero- and nonheterocyclic carboxamides, the only absolute structural features for high activity are (1) a group such as methyl or iodo at the "*cis*" position (β-carbon of *cis*-croton anilide) or at an adjacent nitrogen relative to the carboxanilido function — substitution at the *cis*-carbon by a hydrogen atom or a large group such as phenyl abolishes activity; (2) the carbonyl oxygen in the carboxamide function is essential for activity — it could not be replaced with sulfur or the carbonyl group with SO_2; (3) the NH group is also essential for activity as shown by results with benzyl-2-methylfur-3-yl ketone, where a methylene group replaces NH — the *intact* carboxamide group appears to be a prerequisite for full activity; (4) while the nitrogen of the carboxanilido group has to be protonated for maximum inhibitory effects, it can be replaced by methoxy, *n*-alkyl, and benzyl and benzoyl groups with some activity loss; and (5) the aromatic phenyl ring is not required for activity since carboxamides with cyclohexyl or long chain N-*n*-alkyl groups are highly active.

Interpretation of the effects of substitution on or for the aniline ring tends to be complex, since inhibitory activity also depends on the type of hetero- or nonhetero ring system attached to the carbonyl group. Considering *only* the carboxin molecule, certain modifications can be made which considerably influence the inhibitory potency with respect to the SDC of the wild-type *U. maydis* strain. Structure activity data for a representative cross-section of such carboxin analogues are shown in Table 2. It can readily be seen that substitution of certain groups into specific carbon positions of the phenyl ring markedly increases inhibitory activity (relative sensitivity level >1.0). High activity was noted with 3′-*n*-hexyloxycarboxin (XIV) and 4′-*n*-hexylcarboxin (XXII) which were 50 times as effective as carboxin (I). In general, alkyl or alkoxy groups at the 4′-position increased the binding affinity but less so than the identical 3′-substitutions. A carboethoxy addition at the 3′-position (XVI) gave about equal inhibition to carboxin, while at the 4′-position the equivalent analogue was over 150-fold less active than (I). Polar 3′-carboxyl (XII) and 4′-hydroxyl (XVIII) groups abolished inhibition as did an electrophilic 4′-nitro (XVII) substituent. Note, however (Table 2), that the 3′-chloro (VII) and 3′-bromo (VIII) analogues were more active than (I). Bulky substituents such as phenyl (III) and benzoyl (V) at the 2′-carbon atom decreased inhibition about fourfold. Replacement of the aromatic ring with long-chain *n*-alkyl groups (>C6) produced active compounds, particularly the N-*n*-tetradecyl analogue (XXXII). The non-

TABLE 2
Structure-Activity Relationships between Selected Oxathiin Carboxamides and the Succinate-2,6-Dichlorophenolindophenol Reductase Activity in Mitochondria from a Wild-Type Strain of *Ustilago maydis*

		U. maydis ATCC 14826	
Compound no.	Structure	I_{50} (μM)[a]	Relative sensitivity[b]
I (carboxin)	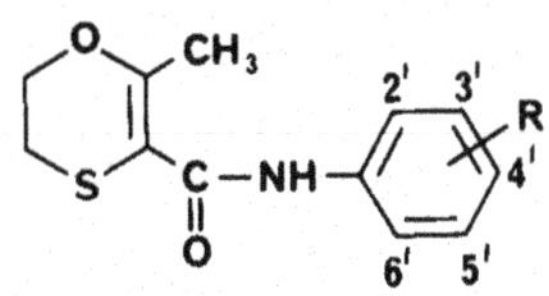	0.36	1.0
	R		
II	2′-NO_2	1.5	0.24
III	2′-C_6H_5	0.55	0.65
IV	2′-$CONH_2$	0.35	1.03
V	2′-COC_6H_5	1.3	0.28
VI	2′-C_2H_5, 6′-C_2H_5	13.0	0.028
VII	3′-Cl	0.18	2.0
VIII	3′-Br	0.073	4.9
IX	3′-NO_2	3.5	0.10
X	3′-CN	5.6	0.064
XI	3′-C≡CH	1.2	0.3
XII	3′-COOH	>500.0	<0.0007
XIII	3′-CH_2OH	2.4	0.15
XIV	3′-*O*-nC_6H_{13}	0.007	51.4
XV	3′-OC_6H_5	0.03	12.0
XVI	3′-$COOC_2H_5$	0.5	0.72
XVII	4′-NO_2	250.0	0.0014
XVIII	4′-OH	>100.0	<0.0036
XIX	4′-CN	110.0	0.0033
XX	4′-iC_3H_7	0.1	3.6
XXI	4′-nC_5H_{11}	0.04	9.0
XXII	4′-nC_6H_{13}	0.007	51.4
XXIII	4′-*O*-nC_6H_{13}	0.045	8.0
XXIV	4′-C_6H_5	4.8	0.075
XXV	4′-$C(C_6H_5)_3$	0.82	0.44
XXVI	4′-COC_6H_5	3.5	0.10
XXVII	4′-$COOC_2H_5$	59.0	0.006
XXVIII	4′-N=NC_6H_5	2.0	0.18
	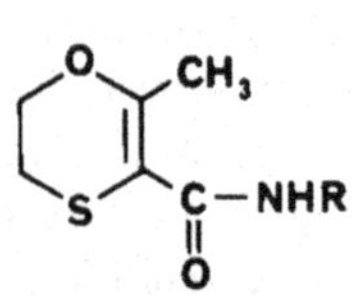		
	R		
XXIX	−C_2H_5	375.0	0.001
XXX	−nC_6H_{13}	2.1	0.17
XXXI	−$nC_{10}H_{21}$	0.22	1.6
XXXII	−$nC_{14}H_{29}$	0.07	5.1
XXXIII	−$(CH_2)_3C_6H_5$	2.8	0.13
XXXIV	−*cyclo*-C_6H_{11}	5.3	0.067

TABLE 2 (continued)
Structure-Activity Relationships between Selected Oxathiin Carboxamides and the Succinate-2,6-Dichlorophenolindophenol Reductase Activity in Mitochondria from a Wild-Type Strain of *Ustilago maydis*

		U. maydis ATCC 14826	
Compound no.	**Structure**	**I_{50} (μM)[a]**	**Relative sensitivity[b]**
XXXV	$-cyclo$-C_8H_{15}	0.82	0.44
XXXVI	$-cyclo$-$C_{12}H_{23}$	6.6	0.055
XXXVII	$-C_{10}H_7$(α-naphthyl)	0.22	1.64

[a] Concentration of oxathiin carboxamide which inhibits the rate of DCIP reduction by 50% as determined from semilog plots of percentage inhibition vs. concentration.

[b] I_{50} of carboxin divided by I_{50} of the analogue.

aromatic cyclo-octyl derivative (XXXV) was about as active as carboxin, while the fused-ring α-naphthyl analogue was two times more active than (I). Replacement of the phenyl group of carboxin with a variety of heterocyclic ring systems gave carboxamides with lower activity.[41]

A structure activity study by ten Haken and Dunn on systemic control of rust diseases by 27 carboxanilides led to the conclusion that a "*cis*-crotonanilide" structure is basic for systemic fungicidal activity.[46] (*Cis*-crotonanilide per se is not active on succinate dehydrogenase.[19]) The primary structure proposed by ten Haken and Dunn[46] is still valid but has now been modified to include active carboxanilides in which a formal carbon-carbon or carbon-nitrogen single bond exists rather than a double bond between carbon atoms. Table 3 shows inhibition data for these types of compounds and SDC activity in wild-type *U. maydis* mitochondria.

Some interesting results were found with carboxanilides derived from 1,4-dihydro-2-methylbenzoic acid. The parent molecule, 1,4-dihydro-2-methylbenzanilide (I), was reasonably active with an I_{50}-value of 20 μM. A 3′-methyl (II) or 4′-*n*-hexyl (III) substituent decreased activity about sevenfold. Compound IV (2-methylthiazole-4-methylene carboxanilide), which lacks a *cis*-methyl group and has a methylene "spacer" between the thiazole ring and carboxanilide group, was suprisingly active. The carbon-nitrogen single bond compound having a methyl group on the N-atom (VI; *N*-methylpyrolle-2-carboxanilide) was active in contrast to the parent anilide (V) with a protonated N-atom. Among some *N*-methylpyrazole carboxanilides (VII to XI), 1,3-dimethylpyrazole-4-carboxanilide (IX) was the best inhibitor, while the closely related isomer 1,4-dimethylpyrazole-3-carboxanilide (VII) where the position of the methyl and carboxanilide substituents are reversed was less active. Compound VII (cf. IX) without the flanking 3-methyl group was inactive. The 1-methylpyrazole-5-carboxanilide analogue (X) was weakly active compared to *N*-methylpyrolle-2-carboxanilide (VI). However, 1,4-dimethylpyrazole-5-carboxanilide (XI) was nine times more active than (X). Inhibition of the *U. maydis* enzyme complex was mirrored in general by a similar inhibition of *R. solani* growth (Table 3).

VI. GENERAL NATURE AND COMPOSITION OF MAMMALIAN SUCCINATE-UBIQUINONE REDUCTASE (COMPLEX II)

Since virtually all of the studies on the mode of action of carboxamides at the molecular level have been accomplished with beef heart Complex II, it seems necessary to present a cursory account of the nature and properties of the mammalian enzyme complex. For more

TABLE 3
Inhibition of Succinate-DCIP Reductase Activity in Mitochondria from *Ustilago maydis* and the Growth of *Rhizoctonia solani* by Heterocyclic Carboxamides Having a Single Bond between the Significant Methyl and Carboxanilido Groups

Compound no. and source	Structure	*U. maydis* ATCC 14826 I_{50} (μM)[a]	*U. maydis* ATCC 14826 Relative sensitivity[b]	*Rhizoctonia solani* EC_{50} (μM)[c]	*Rhizoctonia solani* Relative sensitivity[d]
	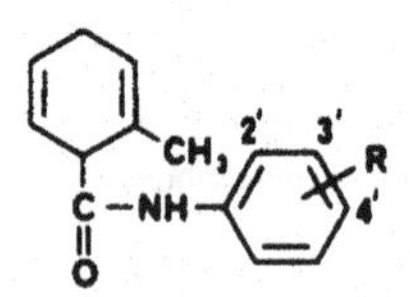				
	R				
I[e]	H	20.0	1.0	1.7	1.0
II[e]	3′-CH_3	135.0	0.15	9.3	0.18
III[e]	4′-nC_6H_{13}	110.0	0.18	>100.0(16)[f]	<0.017
IV[g]	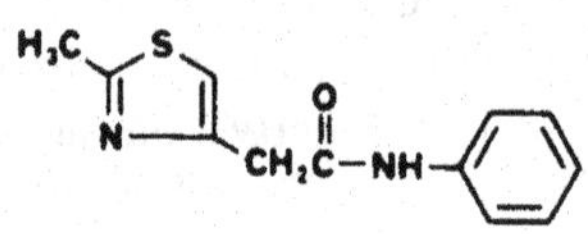	46.0	0.014	—	—
V[h]	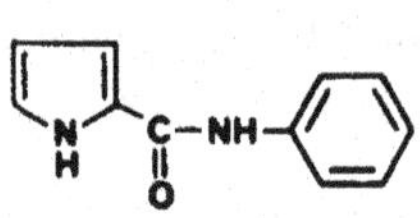	>500.0(38)[f]	1.0	50.0	1.0
VI[h]	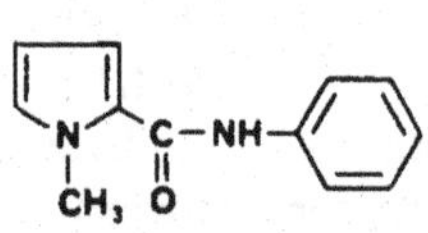	20.0	>25.0	7.0	7.1
VII[i]	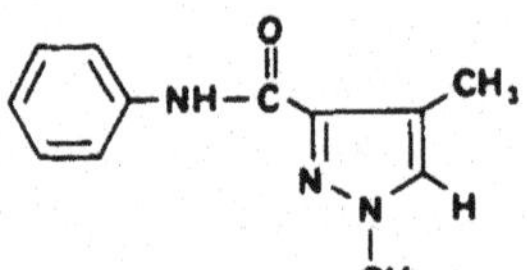	48.0	>10.4	320.0	1.6
VIII[i]	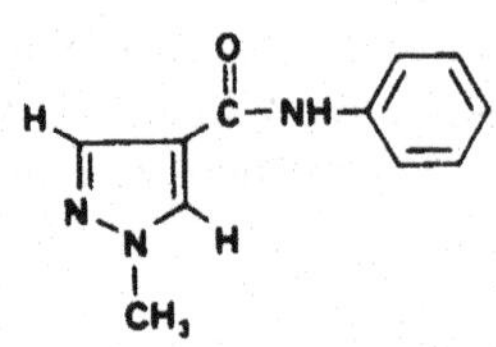	>500.0	1.0	>500.0	1.0

TABLE 3, continued
Inhibition of Succinate-DCIP Reductase Activity in Mitochondria from *Ustilago maydis* and the Growth of *Rhizoctonia solani* by Heterocyclic Carboxamides Having a Single Bond between the Significant Methyl and Carboxanilido Groups

Compound no. and source	Structure	*U. maydis* ATCC 14826		*Rhizoctonia solani*	
		I_{50} (μM)[a]	Relative sensitivity[b]	EC_{50} (μM)[c]	Relative sensitivity[d]
IX[i]	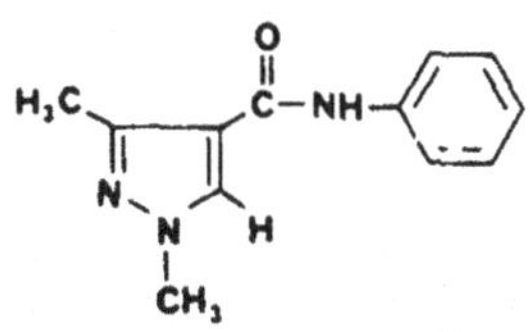	5.7	>88.0	20.0	>25.0
X[i]	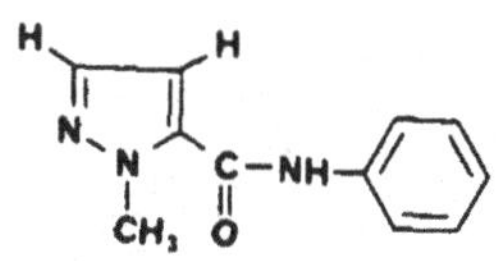	120.0	1.0	200.0	1.0
XI[i]	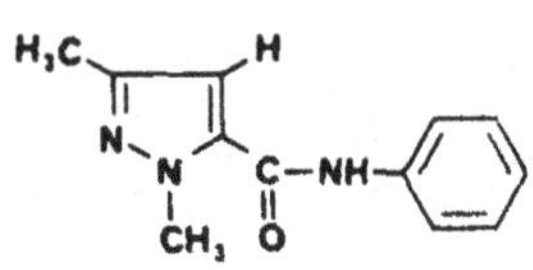	13.0	9.2	40.0	5.0

[a] I_{50} (μM): concentration of compound which inhibits the rate of DCIP reduction by 50% as determined from semi log plots of percentage inhibition vs. conc. of compound.
[b] Relative sensitivity: ratio of the concentration of the parent carboxanilide to the analog concentration giving 50% inhibition of DCIP reduction.
[c] EC_{50} (μM): effective concentration which gives 50% inhibition of radial mycelial growth.
[d] Relative sensitivity: defined by footnote (b) except that inhibition of DCIP reduction is replaced by inhibition of fungal growth.
[e] Synthesized by G. A. White, Research Centre, Agriculture Canada, London, ON, Canada.
[f] Value in parentheses is the percentage inhibition of enzyme activity of growth at the I_{50} (μM) or EC_{50} (μM) value shown in the table.
[g] Uniroyal Chemical, Division of Uniroyal, Inc., Naugatuck, CT.
[h] Synthesized by G. D. Thorn, Research Centre, Agriculture Canada, London, ON, Canada.
[i] Synthesized by J. L. Huppatz, CSIRO, Division of Plant Industry, Canberra City, Australia.

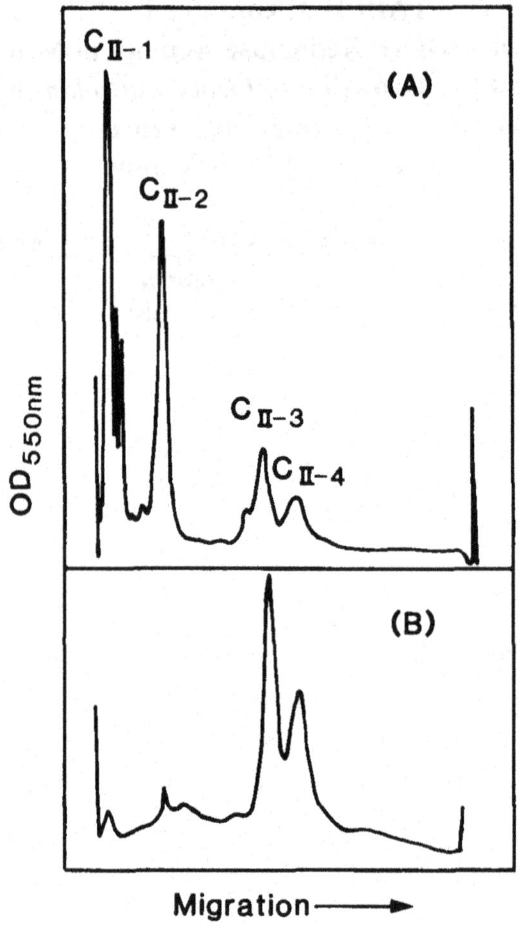

FIGURE 3. Gel scans of (A) Complex II and (B) a preparation of peptides C_{II-3} and C_{II-4} following SDS-polyacrylamide electrophoresis. (From Ackrell, B. A. C., Ramsay, R. R., Kearney, E. B., Singer, T. P., White, G. A., and Thorn, G. D., *Function of Quinones in Energy Conserving Systems,* Trumpower, B. L., Ed., Academic Press, New York, 1982, 319. With permission.)

extensive details, one can refer to a number of review articles on succinate dehydrogenase.[47-49]

Bovine heart Complex II has four polypeptides with apparent molecular masses of 70, 30, 13.5, and 7 kDa.[50] The first two are the C_{II-1} (FP) and C_{II-2} (IP) subunits of succinate dehydrogenase and the other two comprise the binding polypeptides denoted as C_{II-3} and C_{II-4}.[50,51] Figure 3A shows the SDS-polyacrylamide gel scan of a normal preparation of Complex II.[52] A gel pattern of the detergent extract (Figure 3B) shows peptides C_{II-3} and C_{II-4} as being virtually free of contaminating peptides. Present at low and variable concentration is cytochrome b and substantial amounts of phospholipids, mostly lecithin, and no ubiquinone. One mole of soluble SDH contains 1 mol of covalently bound S-α-N(3)-histidyl FAD, 7 to 8 g atoms of Fe, and 7 to 8 mol of acid-labile sulfide (S^{2-}).[51] Polypeptide C_{II-1} has the covalently bound FAD and four equivalents each of Fe and S^{2-}, and peptide C_{II-2}

contains three Fe and three S^{2-} per mole.[51] The 30-kDa subunit is considered to possess a high potential (HiPIP) iron-sulfur center designated as S-3, and the other Fe-S centers are on the 70-kDa subunit of succinate dehydrogenase.[53] The current status regarding the complicated nature of the three iron-sulfur clusters in succinate dehydrogenase has been reviewed by Singer and Johnson.[54] A functional thiol group has also been detected in the enzyme complex.[55] Soluble succinate dehydrogenase can transfer electrons to PMS but not ubiquinone or DCIP. Full succinate-ubiquinone and succinate-DCIP reductase activities are restored when the soluble enzyme is reconstituted with polypeptides C_{II-3} and C_{II-4}.[56]

Complex II preparations also have a b-type cytochrome (b_{560}) distinct from the b-cytochrome of Complex III.[57] Considerable discussion and disagreement have ensued about the role of the b-type cytochrome in succinate ubiquinone reductase preparations ranging from the view that it is an impurity of Complex II to an intrinsic and required component for Q reductase activity. One proposal is that the b-cytochrome is unique to Complex II and contains two polypeptides of 15.5 and 13.5 kDa.[58] It has also been suggested that a 15-kDa polypeptide (Q_p) capable of binding ubiquinone and distinct from the b-cytochrome is responsible for conferring Q-reductase activity to soluble succinate dehydrogenase.[59-61] Another idea is that the most basic unit for restoring Q-reductase activity contains two polypeptides of 13.5 (C_{II-3}) and 7 kDa (C_{II-4}) in association with the 70- and 30-kDa units of the dehydrogenase.[62] The hydrophobic polypeptide C_{II-3} unit, which may carry the b_{560} heme, is essential for Q-reductase activity.[52] Notwithstanding, there is no firm evidence that the b_{560} heme functions as an obligatory electron transfer component in Complex II.

An interesting model of the topographical arrangement of Complex II components in the mitochondrial inner membrane has been presented by Girdlestone et al.[63]

VII. INHIBITION KINETICS

A. KINETICS OF DYE REDUCTION

Kinetic experiments with oxathiin carboxamides showed that I_{50} values found with the succinate-DCIP reductase assay equaled inhibition constants (K_i's) obtained when PMS was used as the primary electron acceptor in the dye assay.[32] Figure 4 compares the inhibition by 3′-*n*-hexyloxy-3-methylthiophene-2-carboxanilide of succinate-PMS-DCIP and succinate-DCIP reductase activities in *U. maydis* mitochondria.[33] Activity is expressed relative to a control rate of 100%. Differences in the left and right ordinate scales are due to the fact that only 50% of the total activity is inhibited in the PMS-DCIP assay, whereas inhibition of DCIP reduction is complete. Therefore, 50% inhibition in the PMS-saturated assay actually means 100% inhibition of the carboxamide-sensitive site. The inhibition curve found with the thiophene carboxamide was identical for both types of assays (Figure 4). A Dixon plot (Figure 5) of the data in Figure 4 was linear, indicating, at least for this carboxamide, simple saturation kinetics. The SDC of *A. nidulans* behaves identically to the SDC of *U. maydis* in that thiophene carboxamides, as does carboxin, interfere with electron transfer to one of two apparent sites of PMS reduction in the succinate-ubiquinone reductase complex.[13] The succinate-PMS activity of beef heart complex II is also inhibited to a maximum of about 50% by carboxins and TTF and to 100% in the succinate-DCIP assay.[38]

The kinetics of carboxamide and TTF inhibition of succinate oxidation in various types of SDC preparations from heart muscle as studied by Mowery et al.[64] The inhibition kinetics depended both on the type of enzyme preparation and the dye assay system used. Thus, the inhibition kinetics of succinate-PMS activity in ETP by TTF, carboxin, and carboxin analogues was different to that observed for "typical Complex II". Some carboxamides, e.g., 4′-phenoxycarboxin, showed uncompetitive inhibition at low concentrations and a slope effect at higher concentration with PMS as the variable substrate in double reciprocal plots.[64]

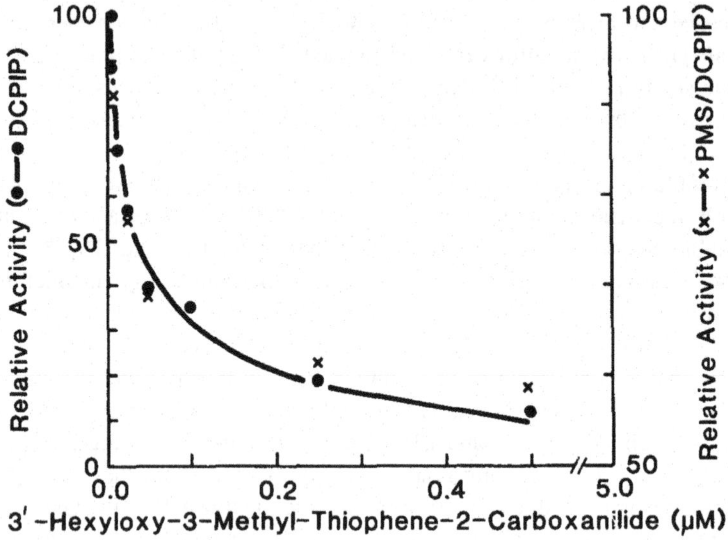

FIGURE 4. Titration of succinate-PMS (X) and succinate-DCIP (●) reductase activities in wild-type *U. maydis* mitochondria by 3′-hexyloxy-3-methylthiophene-2-carboxanilide. (From White, G. A. and Thorn, G. D., *Pestic. Biochem. Physiol.*, 14, 26, 1980. With permission.)

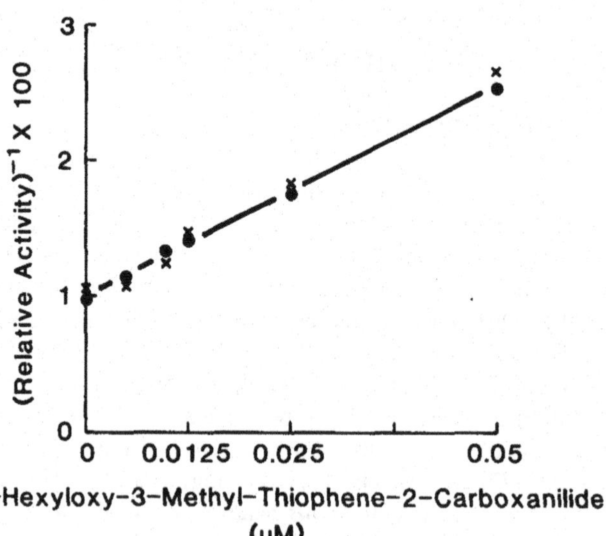

FIGURE 5. Dixon plot of the data in Figure 4.

Electron paramagnetic resonance results from studies of the effects of 3′-methylcarboxin on kinetics of the reduction and reoxidation of Fe-S components of Complex II are shown in Figure 6.[65] The reoxidation of both the ferredoxin-type Fe-S center (center 1) and the superoxidized (S-3; HiPIP) center by DPB is markedly inhibited by 3′-methylcarboxin and TTF with only a small effect, if any, on the reduction rates of these Fe-S centers by succinate. The proposal by Schewe et al.[66] that carboxin inhibits between center 1 and S-3 is not

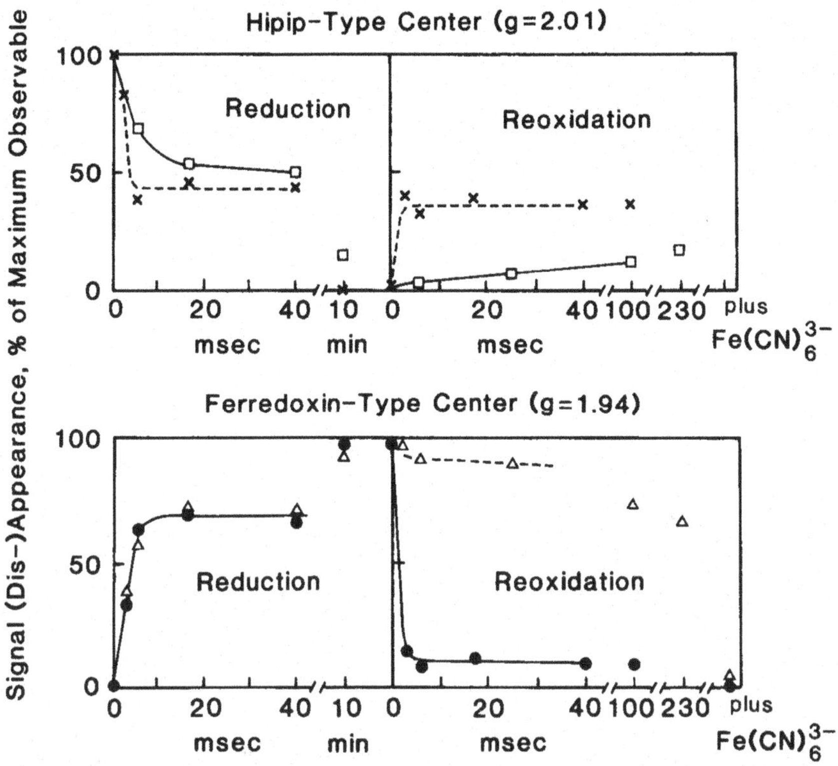

FIGURE 6. Effect of 3′-methylcarboxin on kinetics of the reduction and reoxidation of the ferredoxin-type (S-1) and HiPIP-type (S-3) centers of Complex II. For reduction, oxidized Complex II was rapidly mixed with 100 m*M* succinate, final concentration, at 15°C. In reoxidation experiments, a sample of Complex II, reduced with 1.2 m*M* succinate, was shot at 15°C against 200 μ*M* DPB, final concentration. Symbols: (X) HiPIP control; (□) HiPIP in presence of 25 μ*M* 3′-methylcarboxin; (●) g = 1.94 control; (△) g = 1.94 in presence of 25 μ*M* 3′-methylcarboxin. (From Ackrell, B.A.C., Kearney, E. B., Coles, C. J., Singer, T. P., Beinert, T. P., Wan, Y.-P., and Folkers, K., *Arch. Biochem. Biophys.*, 182, 107, 1977. With permission.)

substantiated by the data in Figure 6. Exposure of Complex II to 0.4 vol % ethanol produced an immediate line broadening of the HiPIP signal and the added presence of 3′-methylcarboxin caused a further broadening. When the signal widening is considered and corrections made, there is no definite evidence for the destruction of the HiPIP center. Even though 3′-methylcarboxin strongly inhibits the oxidation of both Fe-S centers by DPB, the overall effect is on center S-3 and the effect on center S-1 probably reflects an obligatory oxidation route via S-3, and not a direct influence on the ferredoxin-type center. In the experiments shown in Figure 6, ferricyanide acts as a chemical reoxidant of Fe-S centers 1 and 3.

B. KINETICS OF COENZYME Q_2 REDUCTION

The succinate-ubiquinone reductase preparation of Tushurashvili et al.[67] slowly reduces WB or DCIP dyes in the absence of exogenous quinone. Kinetic studies of quinone interaction with the enzyme-specific site demonstrate simple saturation kinetics of Q_2 reduction in the absence of added Triton X-100 detergent (Figure 7). When QH_2 rapidly reduces WB in aqueous solution, the hyperbolic dependence of the reaction rate on Q_2 appears to correspond to quinone saturation of the specific binding site with a K_m (Q_2) value of 0.5 μ*M*. Tushurashvili et al.[67] noted that the reduction of Coenzyme Q_2 by succinate-ubiquinone reductase was nearly completely inhibited by carboxin with a K_i value of 1 μ*M* determined from a Dixon plot (Figure 8). The inhibition was independent of the concentration of Q_2

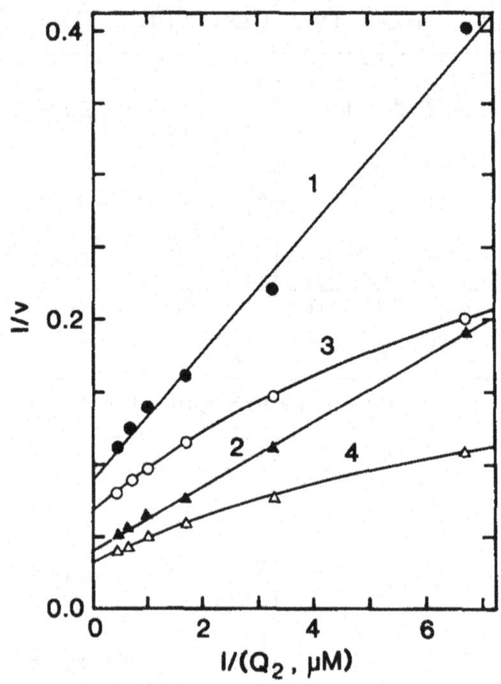

FIGURE 7. The kinetics of Q_2 reduction by succinate-ubiquinone reductase. Q_2-mediated WB reduction rates were determined as a function of Q_2 concentration without added Triton X-100 (lines 1 and 2) or with 0.004% Triton X-100 (curves 3 and 4). For curves 1 and 3, 2 μM carboxin was present in the reaction mix. (From Tushurashvili, P. R., Gavrikova, E. V., Ledenev, A. N., and Vinogradov, A. D., *Biochim. Biophys. Acta,* 809, 145, 1985. With permission.)

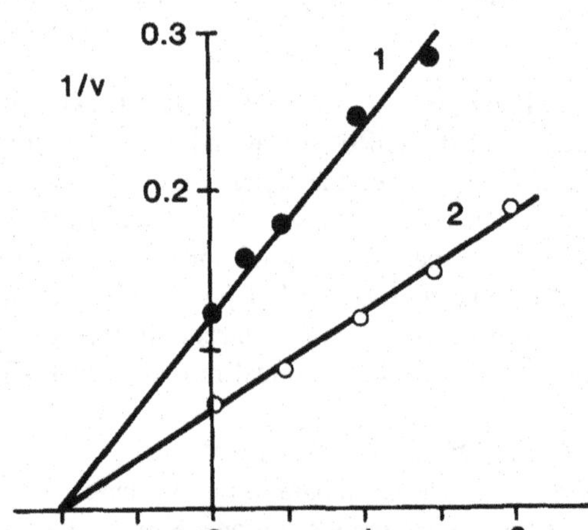

FIGURE 8. Noncompetitive inhibition of succinate-ubiquinone reductase by carboxin. Activities were determined with 0.3 μM (1) and 1.0 μM (2) Coenzyme Q_2. (From Tushurashvilli, P. R., Garvrikova, E. V., Ledenev, A. N., and Vinogradov, A. D., *Biochim. Biophys. Acta,* 809, 145, 1985. With permission.)

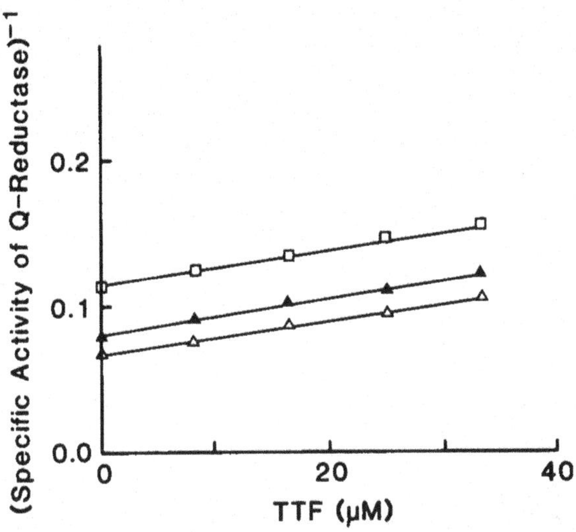

FIGURE 9. Inhibition of succinate-ubiquinone reductase activity of Complex II. Thenoyltrifluoroacetone (TTF) concentration as shown and zero (△), 1.7 (▲), and 4.5 (□) μM amounts of azidocarboxin. (From Ackrell, B. A. C., Ramsay, R. R., Kearney, E. B., Singer, T. P., White, G. A., and Thorn, G. D., *Function of Quinones in Energy Conserving Systems*, Trumpower, B. L., Ed., Academic Press, New York, 1982, 319. With permission.)

showing that carboxin acts kinetically as a simple noncompetitive inhibitor of Q_2 reductase as does TTF. Pure noncompetitive inhibition of the enzyme was found with TTF (K_i, 8 μM). Carboxin also does not compete with succinate or DCIP in the succinate-DCIP reductase assay of Complex II activity.[42]

It has been reported that the apparent K_i for carboxin in reconstituted succinate ubiquinone reductase is nondependent on the amount of a fraction containing QPs and endogenous ubiquinone.[68] As Tushurashvili et al.[67] point out, the latter finding disagrees with the proposal of a competition between ubiquinone (bulk phase or bound semiquinone form) and carboxin or TTF.

VIII. BINDING STUDIES

A. EQUILIBRIUM DIALYSIS

Kinetic evidence that azidocarboxin, as with other carboxanilides, competes with TTF for the binding site involved in the inhibition of succinate oxidation was obtained by assaying the succinate-Coenzyme Q reductase activity of bovine heart Complex II at fixed concentrations of quinone (DPB) and 3′-azidocarboxin (K_i = 2.5 μM at 38°C) and different concentration of TTF.[52] The parallel Dixon plots in Figure 9 indicate that TTF and azidocarboxin are mutually exclusive inhibitors, which is in agreement with the observation that excess TTF displaces [G-^{3}H] TCF from specific but not unspecific binding sites in Complex II.[69]

A method devised for measurement of the binding of [G-^{3}H] TCF involved the presence and absence of a concentration of TTF (200 μM) sufficient to block succinate oxidation without added TCF.[69] The fact that TTF does not appear to significantly interfere with unspecific binding allows quantitation of TCF binding at the site(s) responsible for inhibition of succinate oxidation. Binding curves for [G-^{3}H] TCF to Complex II as obtained by equilibrium dialysis studies are shown in Figure 10. The double reciprocal plot (inset) corresponds

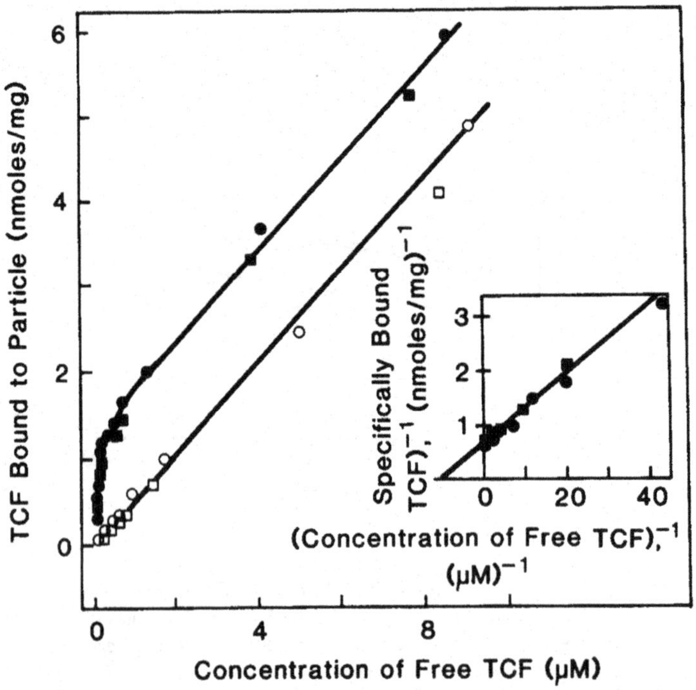

FIGURE 10. Effect of TTF on the binding of [^{3}H]TCF to Complex II determined by equilibrium dialysis and by centrifugation. (■) equilibrium dialysis without TTF; (●) centrifugation without TTF; (□) equilibrium dialysis with TTF; (○) centrifugation with TTF. Inset: a double-reciprocal plot showing the specifically bound component of TCF obtained by subtracting the amount of TCF bound in the presence of TTF from that bound in its absence using equilibrium dialysis (■) or centrifugation (●). (From Coles, C. J., Singer, T. P., White, G. A., and Thorn, G. D., *J. Biol. Chem.*, 253, 5573, 1978.)

to the specific binding site(s) derived from the difference between the two binding curves. The number of binding sites/mole Complex II was estimated by a Scatchard plot (Figure 11) of the data from Figure 10 as 0.5 mol/mol histidyl-FAD. The data obtained by the equilibrium dialysis and centrifugation methods agree well. In contrast to the linear type of binding noted with nonphosphorylating inner membrane preparations (ETP), the binding of TCF to Complex II is biphasic. The Scatchard plot shows that low level inhibitor binding at low concentrations is at the specific site responsible for inhibition, with a K_D of approximately $1.0 \pm 0.1 \times 10^{-7}$ *M* at 4°C, which is in agreement with a $K_i = 1.1 \times 10^{-7}$ *M* for inhibition of the succinate-DCIP reaction by the same enzyme preparation.

Further evidence that TCF binds at the same site as carboxin and structurally related carboxamides can be seen from results in Figure 12 where the effect of unlabeled 4′-hexoxycarboxin (a potent carboxin analogue) on the binding of [G-^{3}H] TCF to Complex II was studied.[69] The carboxin analogue displaced labeled TCF from the specific binding site as does TTF without affecting the unspecific binding. The Scatchard plot of the data was the same as in Figure 10. It thus appears that the inhibition of succinate oxidation in Complex II by TCF, TTF, and 4′-hexoxycarboxin accrues from binding at the same site.

Because TTF can displace carboxanilides from the specific binding site, it permits a means of quantitating the amount of [G-^{3}H] azidocarboxin bound at the site.[52] Equilibrium dialysis (dark) was done at a series of [G-^{3}H] azidocarboxin concentrations with and without a large excess of TTF (Figure 13). In the presence of TTF (closed symbols) only nonspecific binding occurred, and since this was not easily saturated, a linear relation was obtained

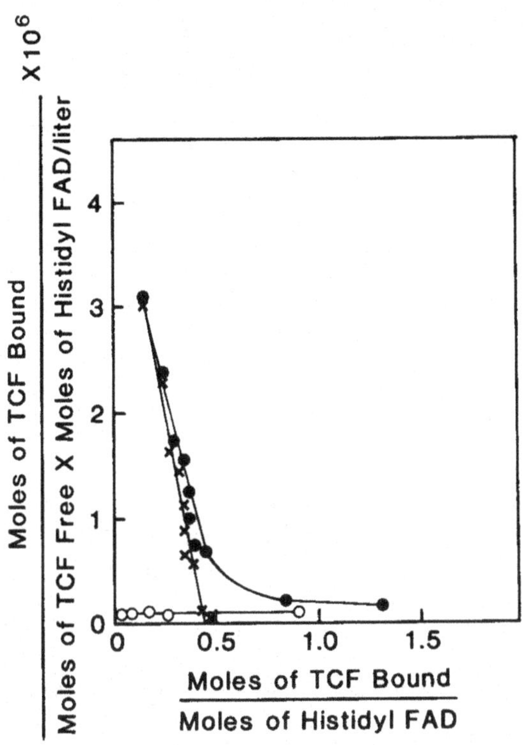

FIGURE 11. Scatchard plots of the data in Figure 10 obtained in the absence (●) or presence (○) of 200 μM TTF by equilibrium dialysis. The specifically bound component (X) is calculated as in Figure 10, consistent with the treatment of Scatchard plots involving two independent binding sites. (From Coles, C. J., Singer, T. P., White, G. A., and Thorn, G. D., *J. Biol. Chem.*, 253, 5573, 1978.)

between free and bound [G-^{3}H] azidocarboxin. In the absence of TTF, both specific and nonspecific binding occurred, producing a biphasic binding curve (Figure 13, open symbols). Specific binding is the difference between the two curves. Figure 13A represents the results of equilibrium dialysis experiments with [G-^{3}H] azidocarboxin and Complex II. A Scatchard plot of the equilibrium dialysis data in Figure 13 gave a dissociation constant (K_D) of 1.2 μM and a binding titer of 0.6 mol of azidocarboxin specifically bound per mole of succinate dehydrogenase as determined by histidylflavin analysis. The stoichiometry is about the same as noted for the binding of [G-^{3}H] TCF to Complex II (0.5 mol/mol histidylflavin).[69]

When Complex II is resolved into its components (succinate dehydrogenase and peptides C_{II-3} and C_{II-4}), the specific binding site for azidocarboxin is nonfunctional (Table 4).[70] Soluble succinate dehydrogenase, which reduces DCIP only with added PMS, is not inhibited by carboxins or TTF and shows no specific binding for azidocarboxin. If the pure enzyme is recombined with the peptides with complete reconstitution of Q-reductase activity, specific binding is restored (Figure 13B) with approximately the same stoichiometry as for unresolved Complex II (Table 4). Dissociation constant (K_D) values calculated from Scatchard plots for the unresolved (~0.5 μM at 4°C) and reconstituted Complex II preparations (1.2 μM at 16°C) agreed satisfactorily with each other when corrected for differences in temperature and with K_i values found from kinetic experiments. The results indicate that the dehydrogenase subunits and peptides C_{II-3} and C_{II-4} form the correct orientation on recombination and have the same physical arrangement as Complex II with respect to the binding site of TTF and carboxanilides.

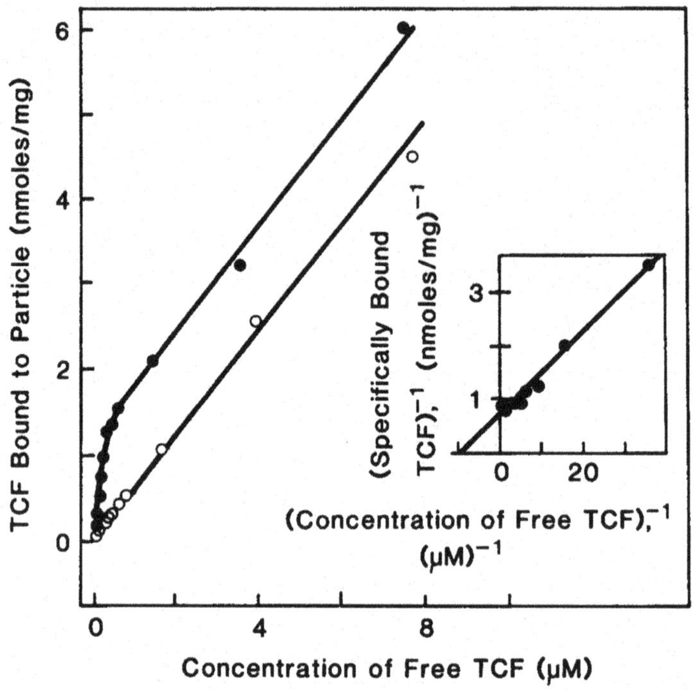

FIGURE 12. Effect of 4′-hexoxycarboxin on the binding of TCF to Complex II. Binding curves are in the absence (●) and presence (○) of 3 μM 4′-hexoxycarboxin. Inset: double reciprocal plot derived from this data showing the specifically bound component of TCF. (From Coles, C. J., Singer, T. P., White, G. A., and Thorn, G. D., *J. Biol. Chem.*, 253, 5573, 1978.)

The results in Table 4 also show that Complex II and the isolated binding peptides, in contrast to the purified dehydrogenase, bind azidocarboxin nonspecifically. The nonspecific binding to unresolved Complex II (12 to 13 mol/mol of flavin at 50 μM inhibitor) was about one fifth as much in the reconstituted system even though the peptide and phospholipid composition of both preparations are similar. A lower value for nonspecific binding also noted in the isolated peptide fraction may be due to an association of the hydrophilic azidocarboxin molecule with deoxycholate detergent used to provide stability to peptides C_{II-3} and C_{II-4}.

Figure 13C represents the data obtained by recombining the enzyme with a preparation of peptide C_{II-4} freed of C_{II-3} by prior treatment with chymotrypsin. The lack of specific binding in this case demonstrated the positive roles of the enzyme and peptide C_{II-3} in the binding of inhibitors at the inhibition site, but permitted no assessment of the contributions of peptide C_{II-4} or phospholipids. A further point is that combination of a chymotrypsin-digested peptide preparation (C_{II-3} free) with the dehydrogenase does not cause the "low K_m" ferricyanide reductase activity to disappear.[52,71] Peptide C_{II-3} appears to mask the ferricyanide reaction suggesting that C_{II-3} is proximal to the HiPIP center S-3 of the enzyme in the membrane. Grivennikova and Vinogradov have shown that carboxin protects against an alkali-induced dissociation of succinate dehydrogenase from submitochondrial particles.[72] Inactivation of succinate-ubiquinone reductase following rapid dilution in an assay without Triton X-100 was due likely to a dissociation of the dehydrogenase from the complex.[67] The emergence of ferricyanide reductase activity induced by rapid dilution of succinate-ubiquinone reductase is sensitive to carboxin in agreement with the alkali-induced results.[67]

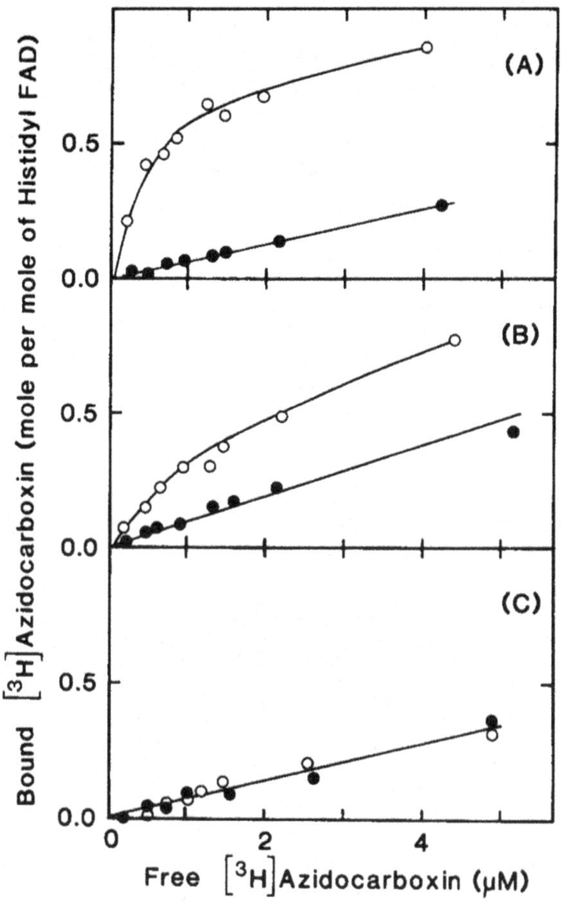

FIGURE 13. Binding of [^{3}H]-azidocarboxin to Complex II (A), mixtures of succinate dehydrogenase plus peptide preparation (B), and succinate dehydrogenase plus chymotrypsin-treated peptide preparation (C). Equilibrium dialysis experiments (dark) with (●) and without (○) 500 μM TTF. (From Ackrell, B. A. C., Ramsay, R. R., Kearney, E. B., Singer, T. P., White, G. A., and Thorn, G. D., *Function of Quinones in Energy Conserving Systems*, Trumpower, B. L., Ed., Academic Press, New York, 1982, 319, With permission.)

B. PHOTOAFFINITY LABELING

The distribution of radioactivity in the components of Complex II is shown in Table 5 as calculated from photoaffinity labeling experiments using [G-^{3}H]-azidocarboxin with and without TTF.[70] About 90% of the Q-reductase activity was irreversibly blocked after UV irradiation of Complex II and azidocarboxin. A very limited amount of specific binding to the 70s and 30s units of the dehydrogenase was found, while larger quantities of inhibitor were bound covalently to peptides C_{II-3} and C_{II-4}. It is evident that both peptides were labeled extensively as well as the nonstaining region of the gel (phospholipids). The amount of phospholipid labeling was low on a molar basis, although a large fraction of azidocarboxin was observed in the phospholipid band due to their high concentration in Complex II relative to peptides C_{II-3} and C_{II-4}.

A separation of specific from nonspecific binding without using TTF as a displacement agent was found possible by measuring the dependence of [G-^{3}H] azidocarboxin into each

TABLE 4
Specific and Nonspecific Binding of Azidocarboxin to Complex II and Its Components

Sample	Specific binding,[a] mol/mol flavin	Nonspecific binding,[b] mol/mol flavin
Complex II	0.6 ± 0.10	12.8
Reconstituted system	0.5 ± 0.12	2.7
Binding peptides[c]	0	1.6
Succinate dehydrogenase	0	0.06

Note: The data come from the equilibrium dialysis experiments described in Figure 1 of Ramsay et al.[70]

[a] Specific binding is calculated from the intercept of the Scatchard plot ± the standard error of the estimate.
[b] In the presence of 50 μM [^{3}H]azidocarboxin.
[c] The amount of binding peptides used was the same as in the reconstituted system. The molar concentration is assumed to be the same.

TABLE 5
Photoaffinity Labeling of Complex II

Component	Azidocarboxin bound[a] (nmol)		
	50 μM Azidocarboxin	50 μM Azidocarboxin + 2 mM TTF	Difference
70s	0.26	0.16	0.10
30s	0.13	0.03	0.10
$C_{II\text{-}3}$ + $C_{II\text{-}4}$	0.91	0.51	0.40
Phospholipids	2.97(0.06)[b]	1.21(0.02)[b]	1.76(0.04)[b]

[a] Per 250 μg protein (≃1 nmol histidyl flavin).
[b] Per mole of phospholipid, assuming 52 mol/mol flavin.[56]

component of Complex II on the concentration of azidocarboxin.[70] If binding was totally nonspecific, incorporation would be linearly dependent on azidocarboxin concentration, but in the case of specific plus nonspecific binding a hyperbolic curve would be superimposed on the line. In Figure 14 it is shown that binding to the 70s and 30s subunits is linearly dependent on azidocarboxin concentration (i.e., unspecific binding). Incorporation into peptides $C_{II\text{-}3}$ and $C_{II\text{-}4}$ and into phospholipids produced a hyperbolic component, characteristic of saturable, specific binding. Phospholipids thus appear to compose part of or to be near the binding site. The dehydrogenase 70s and 30s subunits are not in close proximity to the site for specific labeling although required for the site's existence. The peptide units failing to bind azidocarboxin unless a catalytically reconstitutionable succinate-coenzyme Q reductase unit is present suggests that the combination of the dehydrogenase with the binding peptides somehow changes the conformation of the peptides to restore the binding site for carboxanilides and TTF.

IX. COMMENTS ON THE STOICHIOMETRY OF CARBOXANILIDE BINDING TO COMPLEX II AND MOLECULAR MECHANISMS OF INHIBITION

As mentioned previously, calculations from Scatchard plots gave an equilibrium binding titer of 0.5 mol of TCF and 0.6 mol of azidocarboxin specifically bound per mole of succinate

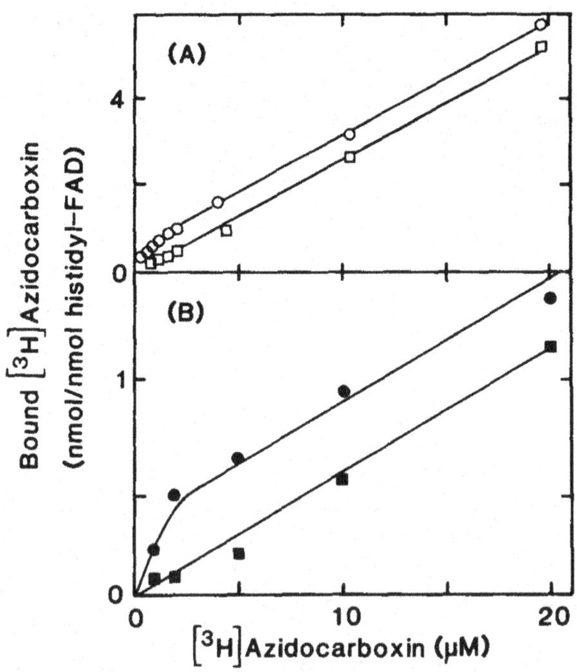

FIGURE 14. Covalent binding of [^{3}H]-azidocarboxin to Complex II components. (■) 70s subunit of succinate dehydrogenase; (●) 30s subunit of succinate dehydrogenase; (○) peptides $C_{II\text{-}3}$ + $C_{II\text{-}4}$; (□) phospholipids. (From Ramsay, R. R., Ackrell, B. A. C., Coles, C. J., Singer, T. P., White, G. A., and Thorn, G. D., *Proc. Natl. Acad. Sci. U.S.A.*, 78, 825, 1981. With permission.)

dehydrogenase (measured as histidylflavin). It was expected that the binding ratio would be 1:1. Following irradiation of Complex II, a total recovery of enzyme activity was obtained, indicating that TTF completely protected the inhibition site from being covalently bound by the photoaffinity carboxanilide. The noted decrease in radioactivity in peptides $C_{II\text{-}3}$ and $C_{II\text{-}4}$ (Table 5) was equivalent to 0.4 mol of azidocarboxin. This amount is unrelated to the ~90% inhibition found after irradiation in the absence of TTF unless total inhibition is obtained by 0.5 mol of specifically bound azidocarboxin per mole of Complex II.

Although the reason for the substoichiometric binding is not understood, a distinct possibility is that 2 mol of succinate dehydrogenase plus other membrane polypeptides and lipids organize the binding site, i.e., a dimeric form of Complex II.[69,73] Thus, 1 mol of azidocarboxin would be bound per mole of dimer, i.e., 1 mol/2 mol of histidylflavin or HiPIP center S-3 (assuming that there is 1 mol of histidylflavin per mole of HiPIP center).

Schewe et al.[74] have proposed a somewhat tenuous model for the binding of carboxin in Complex II. The model is not entirely consistent with current knowledge regarding the action of carboxin and other heterocyclic carboxamides on Complex II. Notwithstanding that nothing is known about the tertiary structure of the iron-sulfur center S-3 in relation to the 30s subunit of succinate dehydrogenase, peptides $C_{II\text{-}3}$ plus $C_{II\text{-}4}$ (b-type cytochrome), ubiquinone, and phospholipids, the stoichiometry of the receptor model of Schewe et al.[74] requires 2 mol of carboxin bound per mole of center S-3 (or per mole of histidyl-FAD), which does not agree with the actual binding data. Also, the coordination binding of carboxin to the iron atoms in center S-3 through the carbonyl oxygen and amide-N atoms as proposed by Schewe et al.[74] might be expected to affect the shape of spin resonance signal for center S-3, but this was not observed either for 3′-methylcarboxin or TTF on the succinate-linked reduction of iron-sulfur centers in Complex II.[65]

All considered, there is no compelling reason to believe that the heterocyclic ring system of carboxin is proximate to the iron-sulfur center S-3. This redox center probably occupies a strongly hydrophobic region which includes part of the 30s subunit, ubiquinone, phospholipids, and the smaller associated membrane polypeptides. The amino acid composition and sequence of the smaller subunits of beef heart Complex II might reveal that these contain hydrophobic segments which could form part of the binding site for carboxamides and TTF.[58] Wood et al.[75] have determined the nucleotide sequence encoding the small subunits of the succinate dehydrogenase of *Escherichia coli.* Two genes, sdhC and sdhD, encode strongly hydrophobic proteins of M_r 14,167 and 12,792, respectively. The dramatic effects on SDC inhibition by certain substituent groups on the phenyl ring of different heterocyclic carboxanilides suggest that the molecule attached to the amide-nitrogen (e.g., phenyl, 3′-*n*hexyloxyphenyl, etc.) locates in a hydrophobic "cleft" between center S-3 and Q and blocks electron transfer. In this case, the ring system attached to the carbonyl group could lie in a less hydrophobic region of the binding site. That the substituent group on the amide-N is important for binding is shown by high inhibition values for the N-*n*-tetradecyl analogue of a carboxin and succinate -DCIP reductase activity in mitochondria from *A. nidulans* and inhibition of Complex II activity in *U. maydis* mitochondria by 3′-*n*-hexyl-2,4,5-trimethylfuran-3-carboxanilide.[13,34] In fact, carboxin substituted with a large lipophilic 4′-trityl group [$C(C_6H_5)_3$] has an I_{50} of 0.82 μM (Table 2). The high degree of specificity and inhibitory activity shown by 4′-phenyl-substituted carboxanilides toward carboxin-resistant *mutant* SDCs of *U. maydis* also shows that the nonheterocyclic "end" of the molecule exerts a large effect on binding and inhibition of electron flow form the HiPIP iron-sulfur center to Coenzyme Q.

Possibly, ubiquinone (Q_p?) is prevented from reacting with center S-3 by inhibitor interference with membrane phospholipid function as hinted at by the high activity of lipophilic carboxamides and the extent of photoaffinity labeling of phospholipids (Table 5). Rather than invoking a complicated mechanism of action for carboxamides, the compounds may simply be "lodged" in Complex II in such a way as to hinder the conformational interaction of succinate dehydrogenase and the associated binding polypeptides which allow reaction with ubiquinone. The mechanism of action of TTF is still undefinitive. Using bovine succinate-cytochrome c reductase complex, Trumpower and Simmons have proposed that TTF inhibits by intercalation between the high potential Fe-S center S-3 of succinate dehydrogenase and the spin-coupled ubiquinone pair ($QH\cdot \rightarrow QH_2$) near this center.[76] This has been questioned by Salerno and Ohnishi who found that TTF could induce destabilization of a radical signal assigned to a pair of specifically bound ubisemiquinone molecules.[77]

X. DIRECTION OF FUTURE RESEARCH STUDIES WITH CARBOXAMIDES

There are abundant possibilities for future research with carboxamides not only at the molecular level, but, from a practical viewpoint, studies directed at the extension of the fungitoxic spectrum to non-Basidiomycetes and control of mutational resistance to commercially used carboxamides. A computer modeling and QSAR approach might be useful in such studies. Apparent selective inhibition or specificity for carboxin-resistant SDCs of *U. maydis* has been shown with a number of analogues of carboxin and other heterocyclic and nonheterocyclic carboxanilides, primarily 4′-substituted derivatives.[33-35] It is interesting that some diverse types of carboxamide structures have been found which give weak inhibition of SDC activity, yet are strong inhibitors of fungal growth.[35] This situation could indicate multiple reaction sites and/or metabolic conversion of the carboxamides to an active metabolite.

Some exciting experiments await to be done at the molecular level in order to understand the nature of inhibition of Complex II by carboxins. For instance, the photoaffinity [^3H] 3′-azidocarboxin analogue could be used to identify the exact binding region when amino acid sequences in the small polypeptide (C_{II-3} + C_{II-4}) subunits of fungal or mammalian SDCs are known. Work is in progress to isolate and sequence the genes which confer resistance to carboxin in *U. maydis.*[78] This might be potentially useful in identifying the carboxin binding region of the polypeptides (wild-type and carboxin-resistant mutant strains) encoded by the genes. However, a mutational alteration in the amino acid sequence of the small subunits of Complex II may conceivably affect their conformation with respect to the larger SD subunits and not be involved in the actual carboxamide binding. Nonetheless, sequencing of photoaffinity-carboxin-labeled polypeptide(s) would be required to verify the genetic coding data. Another research area to be considered involves electron paramagnetic resonance studies of spin-spin interaction between nitroxide-labeled carboxamides and the Fe-S centers and low-spin heme iron in Complex II. Thus, if spin-labeled carboxamide locates near a center like S-3, it may be possible to find the interatomic distance between the inhibitor and the iron-sulfur cluster. Such experiments have been initiated recently in collaboration with M. K. Johnson, University of Georgia, Athens, and B. A. C. Ackrell, VA Medical Center, San Francisco. Finally, paramagnetic resonance could be used to probe effects of carboxamides on ubiquinone function in Complex II as was done with TTF.[76,77]

Abbreviations used: DB, 2,3-dimethoxy-5-methyl-6-decyl-1,4 benzoquinone; DCIP, 2,6-dichlorophenolindophenol; DPB, 2,3-dimethoxy-5-methyl-6-pentyl-1,4 benzoquinone; ETP, nonphosphorylating inner membrane preparation; FP, flavoprotein; Fe-S, iron sulfur; HiPIP, superoxidized high potential iron-sulfur cluster (center S-3); IP, nonheme iron protein; PMS, phenazine methosulfate; Q, Coenzyme Q_{10}; Q_2, Coenzyme Q_2 (2,3-dimethoxy-5-methyl-6-decyl-1,4 benzoquinone); Qp, Q binding protein; SDC, succinate dehydrogenase complex (Complex II, succinate-ubiquinone reductase); TCF, 2,4,5-trimethylfuran-3-carboxanilide; TTF, thenoyltrifluoroacetone; WB, Wurster's blue.

NOTE ADDED IN PROOF

Recently, Keon et al.[79] have succeeded in isolating a gene, Cbx^r, which confers resistance to carboxin in a mutant strain (No. 92) of *U. maydis.*[32] The gene encodes a mutant form of the iron-sulfur (IP) subunit of Complex II. The sequence of the Cbx^r gene showed a high degree of homology to Sdh Fe-S subunit genes from other organisms. However, in the Fe-S cluster S-3, histidine was replaced with a leucine residue and threonine with a serine residue. Subsequent to the publicaton of Keon et al.,[79] sequencing of the wild-type gene has shown that the mutation is a single amino acid change in the third Fe-S cluster.[80] Here, the wild-type gene has a histidine residue rather than leucine. When the wild-type gene was mutagenized *in vitro* at this site to convert histidine back to a leucine, transformation of *U. maydis* with this mutant gene confers resistance.

The significance of these changes in the amino acid composition at this region of IP to the mode of action or resistance to carboxin is not understood at present. It is possible that a modification of the structure of the IP subunit in the mutant strain affects the binding of carboxamides to peptides C_{II-3} and C_{II-4} upon association with polypeptide subunits C_{II-1} (FP) and C_{II-2} (IP) to form a functional Complex II.[56] It should be noted that photoaffinity labeling experiments with azidocarboxin have shown that specific binding is with the membrane peptide components of Complex II and not the FP or IP subunits.[70] Undoubtedly, genes encoding for mutational alteration in the smaller membrane peptides will be isolated from other carboxin-resistant strains of *U. maydis* and the affected amino acid sequence region(s) may involve a Q binding site.[59-61]

REFERENCES

1. **Fargher, R. G., Galloway, L. D., and Probert, M. E.,** The action of certain substances on the growth of mould fungi, *Mem. Shirley Inst.*, 9, 37, 1930.
2. **von Schmeling, B. and Kulka, M.,** Systemic fungicidal activity of 1,4-oxathiin derivatives, *Science*, 152, 659, 1966.
3. **Woodcock, D.,** Structure-activity relationships, in *Systemic Fungicides*, Marsh, R. W., Ed., Longman, London, 1972, chap. 3.
4. **Kuhn, P. J.,** Mode of action of carboxamides, in *Mode of Action of Antifungal Agents*, Trinci, A. P. J. and Ryley, J. F., Eds., Cambridge University Press, Cambridge, England, 1984, chap. 8.
5. **Kulka, M. and von Schmeling, B.,** Carboxin fungicides and related compounds, in *Modern Selective Fungicides*, Lyr, H., Ed., Longman Scientific and Technical, Essex, England, 1987, chap. 9.
6. **Pring, R. J. and Richmond, D. V.,** An ultrastructural study of the effect of oxycarboxin on *Uromyces phaseoli* infected leaves of *Phaseolus vulgaris*, *Physiol. Plant Pathol.*, 8, 155, 1976.
7. **Richmond, D. V. and Pring, R. J.,** Some cytological effects of systemic fungicides on fungi and plants, *Neth. J. Plant Pathol.*, 83(Suppl. 1), 403, 1977.
8. **Hirooka, T., Miyagi, Y., and Araki, F.,** Effect of flutolanil on the infection process of *Rhizoctonia solani*, in *Abstr. Paper, 5th Int. Congr. Plant Pathology*, 1988, 22.
9. **Edgington, L. V. and Barron, G. L.,** Fungitoxic spectrum of oxathiin compounds, *Phytopathology*, 57, 1256, 1967.
10. **Snel, M., von Schmeling, B., and Edgington, L. V.,** Fungitoxicity and structure-activity relationships of some oxathiin and thiazole derivatives, *Phytopathology*, 60, 1164, 1970.
11. **Pommer, E.-H. and Kradel, J.,** Mebenil (BAS 3050F), a new compound with specific action against some basidiomycetes, in *Proc. 5th British Insecticide and Fungicide Conf.*, Vol. 2., British Crop Protection Council, 1969, 563.
12. **Pommer, E.-H.,** The systemic activity of a new fungicide of the furane carbonic acid aniline group (BASF 3191F), in *Proc. 2nd Int. IUPAC Congr. Pesticide Chemicals*, Vol. 5, Tahori, A. S., Ed., Gordon and Breach, New York, 1971, 397.
13. **White, G. A. and Georgopoulos, S. G.,** Thiophene carboxamide fungicides: structure-activity relationships with the succinate dehydrogenase complex from wild-type and carboxin-resistant mutant strains of *Aspergillus nidulans*, *Pestic. Biochem. Physiol.*, 25, 188, 1986.
14. **Mathre, D. E.,** Mode of action of oxathiin systemic fungicides. I. Effect of carboxin and oxycarboxin on the general metabolism of several basidiomycetes, *Phytopathology*, 60, 671, 1970.
15. **Ragsdale, N. N. and Sisler, H. D.,** Metabolic effects related to the fungitoxicity of carboxin, *Phytopathology*, 60, 1422, 1970.
16. **Georgopoulos, S. G. and Sisler, H. D.,** Gene mutation eliminating antimycin A-tolerant electron transport in *Ustilago maydis*, *J. Bacteriol.*, 103, 745, 1970.
17. **Mathre, D. E.,** Mode of action of oxathiin systemic fungicides. III. Effect on mitochondrial activities, *Pestic. Biochem. Physiol.*, 1, 216, 1971.
18. **White, G. A.,** A potent effect of 1,4-oxathiin systemic fungicides on succinate oxidation by a particulate preparation from *Ustilago maydis*, *Biochem. Biophys. Res. Commun.*, 44, 1210, 1971.
19. **White, G. A. and Thorn, G. D.,** Structure-activity relationships of carboxamide fungicides and the succinic dehydrogenase complex of *Cryptococcus laurentii* and *Ustilago maydis*, *Pestic. Biochem. Physiol.*, 5, 380, 1975.
20. **Ulrich, J. T. and Mathre, D. E.,** Mode of action of oxathiin systemic fungicides. V. Effect on the electron transport system of *Ustilago maydis* and *Saccharomyces cerevisiae*, *J. Bacteriol.*, 110, 628, 1972.
21. **Lyr, H., Ritter, G., and Casperson, G.,** Wirkungsmechanismus des Systemischen Fungicids Carboxin, *Z. Allg. Mikrobiol.*, 12, 271, 1972.
22. **Georgopoulos, S. G., Alexandri, E., and Chrysayi, M.,** Genetic evidence for the action of oxathiin and thiazole derivatives on the succinic dehydrogenase system of *Ustilago maydis* mitochondria, *J. Bacteriol.*, 110, 809, 1972.
23. **Georgopoulos, S. G., Chrysayi, M., and White, G. A.,** Carboxin resistance in the haploid, the heterozygous diploid and the plant-parasitic dicaryotic phase of *Ustilago maydis*, *Pestic. Biochem. Physiol.*, 5, 543, 1975.
24. **Gunatilleke, I. A. U. N., Arst, H. N., Jr., and Scazzochio, C.,** Three genes determine the carboxin sensitivity of mitochondrial succinate oxidation in *Aspergillus nidulans*, *Genet. Res. Cambridge*, 26, 297, 1975.
25. **Iida, W.,** On the tolerance of plant pathogenic fungi and bacteria to fungicides in Japan, *Jpn. Pestic. Inf.*, 23, 13, 1975.
26. **Gullino, M. L. and Garibaldi, A.,** Failure of oxycarboxin to control populations of carnation rust *(Uromyces caryophyllinus)* present in northern Italy, *Chem. Control Newsl.*, 9, 24, 1987.

27. **Leroux, P. and Berthier, G.,** Resistance to carboxin and fenfuram in *Ustilago nuda* (Jens.) Rostr., the causal agent of barley loose smut, *Crop Prot.*, 7, 16, 1988.
28. **Georgopoulos, S. G. and Ziogas, B. N.,** A new class of carboxin-resistant mutants of *Ustilago maydis, Neth. J. Plant Pathol.*, 83(Suppl. 1), 235, 1977.
29. **Ziogas, B. N. and Georgopoulos, S. G.,** Mitochondrial electron transport in a carboxin-resistant, antimycin-A-sensitive mutant of *Ustilago maydis, Pestic. Biochem. Physiol.*, 22, 24, 1984.
30. **Henry, M. F. and Nyms, E. J.,** Cyanide-insensitive respiration. An alternative mitochondrial electron transport pathway, *Sub Cell. Biochem.*, 4, 1, 1975.
31. **Ziogas, B. N. and Georgopoulos, S. G.,** Chloramphenicol induction of a second cyanide- and azide-insensitive mitochondrial pathway in *Ustilago maydis, Biochim. Biophys. Acta,* 592, 223, 1980.
32. **White, G. A., Thorn, G. D., and Georgopoulos, S. G.,** Oxathiin carboxamides highly active against carboxin-resistant succinic dehydrogenase complexes from carboxin-selected mutants of *Ustilago maydis* and *Aspergillus nidulans, Pestic. Biochem. Physiol.*, 9, 165, 1978.
33. **White, G. A. and Thorn, G. D.,** Thiopene carboxamide fungicides: structure-activity relationships with the succinate dehydrogenase complex from wild-type and carboxin-resistant mutant strains of *Ustilago maydis, Pestic. Biochem. Physiol.*, 14, 26, 1980.
34. **White, G. A.,** Furan carboxamide fungicides: structure-activity relationships with the succinate dehydrogenase complex in mitochondria from a wild-type strain and a carboxin-resistant mutant strain of *Ustilago maydis, Pestic. Biochem. Physiol.*, 31, 129, 1988.
35. **White, G. A.,** Substituted 2-methylbenzanilides and structurally related carboxamides: inhibition of complex II in mitochondria from a wild-type strain and a carboxin-resistant mutant strain of *Ustilago maydis, Pestic. Biochem. Physiol.*, 34, 255, 1989.
36. **Ziogas, B. N. and Georgopoulos, S. G.,** The effect of carboxin and of thenoyltrifluoroacetone on cyanide-sensitive and cyanide-resistant respiration of *Ustilago maydis* mitochondria, *Pestic. Biochem. Physiol.*, 11, 208, 1979.
37. **van Tuyl, J. M.,** Acquired resistance to carboxin in *Aspergillus nidulans, Neth. J. Plant Pathol.*, 81, 122, 1975.
38. **Mowery, P. C., Ackrell, B. A. C., Singer, T. P., White, G. A., and Thorn, G. D.,** Carboxins: powerful selective inhibitors of succinate oxidation in animal tissues, *Biochem. Biophys. Res. Commun.*, 71, 354, 1976.
39. **Day, D. A., Arron, G. P., and Laties, G. G.,** The effect of carboxins on higher plant mitochondria, *FEBS Lett.*, 85, 99, 1978.
40. **White, G. A., Phillips, J. N., Huppatz, J. L., Witrzens, B., and Grant, S. J.,** Pyrazole carboxanilide fungicides. I. Correlation of mitochondrial electron transport inhibition and antifungal activity, *Pestic. Biochem. Physiol.*, 25, 163, 1986.
41. **White, G. A.,** unpublished results, 1989.
42. **White, G. A.,** unpublished results, 1978.
43. **Müller, W., Schewe, T., Lyr, H., and Zanke, D.,** Wirkungsmechanismus der Atmungshemmung durch die System Fungizide der Carboxingruppe. Wirkung von Oxathiinderivaten und-Analoga auf nicht phosphorylierende submitochondriale Partikeln aus Rinderherz sowie *Trametes versicolor* und *Trichoderma viride, Z. Allg. Mikrobiol.*, 17, 359, 1977.
44. **Tucker, A. N. and Lillich, T. T.,** Effect of the systemic fungicide carboxin on electron transport function in membranes of *Micrococcus denitrificans, Antimicrob. Agents Chemother.*, 6, 572, 1974.
45. **Pennoyer, J. D., Ohnishi, T., and Trumpower, B. L.,** Purification and properties of succinate-ubiquinone oxidoreductase complex from *Paracoccus denitrificans, Biochim. Biophys. Acta,* 935, 195, 1988.
46. **ten Haken, P. and Dunn, C. L.,** Structure-activity relationships in a group of carboxamides systemically active against broad bean rust *(Uromyces fabae)* and wheat rust *(Puccinia recondita)*, in *Proc. 6th British Insecticide and Fungicide Conf.*, Vol. 2, British Crop Protection Council, 1971, 453.
47. **Singer, T. P., Gutman, M., and Massey, V.,** Succinate dehydrogenase, in *Iron-Sulfur Proteins*, Vol. 1, Lovenburg, W., Ed., Academic Press, New York, 1973, 227.
48. **Ohnishi, T.,** Mitochondrial iron-sulfur flavodehydrogenases, in *Membrane Proteins in Energy Transduction*, Capaldi, R. A., Ed., Marcel Dekker, New York, 1979, 1.
49. **Hederstedt, L. and Rutberg, L.,** Succinate dehydrogenase — a comparative review, *Microbiol. Rev.*, 4, 542, 1981.
50. **Capaldi, R. A., Sweetland, J., and Merli, A.,** Polypeptides in the succinate-coenzyme Q reductase segment of the respiratory chain, *Biochemistry*, 16, 5707, 1977.
51. **Davis, K. A. and Hatefi, Y.,** Succinate dehydrogenase. I. Purification, molecular properties, and substructure, *Biochemistry*, 10, 2509, 1971.
52. **Ackrell, B. A. C., Ramsay, R. R., Kearney, E. B., Singer, T. P., White, G. A., and Thorn, G. D.,** Two small peptides from complex II and their role in the reconstitution of Q reductase activity and in the binding of TTF, in *Function of Quinones in Energy Conserving Systems*, Trumpower, B. L., Ed., Academic Press, New York, 1982, 319.

53. **Ohnishi, T. and Salerno, J. C.,** Iron-sulfur clusters in the mitochondrial electron-transport chain, in *Iron-Sulfur Proteins,* Spiro, T. P., Ed., John Wiley & Sons, New York, 1982, 286.
54. **Singer, T. P. and Johnson, M. K.,** The prosthetic groups of succinate dehydrogenase: 30 years from discovery to identification, *FEBS Lett.,* 190, 189, 1985.
55. **Kenney, W. C.,** The reaction of N-ethylmaleimide at the active site of succinate dehydrogenase, *J. Biol. Chem.,* 250, 3089, 1975.
56. **Ackrell, B. A. C., Ball, M. B., and Kearney, E. B.,** Peptides from complex II active in reconstitution of succinate-ubiquinone reductase, *J. Biol. Chem.,* 255, 2761, 1980.
57. **Davis, K. A., Hatefi, Y., Poff, K. L., and Butler, W. L.,** The b-type cytochromes of bovine heart mitochondria: absorption spectra, enzymatic properties, and distribution in the electron transfer complexes, *Biochim. Biophys. Acta,* 325, 341, 1973.
58. **Hatefi, Y. and Galante, Y. M.,** Isolation of cytochrome b_{560} from complex II (succinate-ubiquinone oxidoreductase) and its reconstitution with succinate dehydrogenase, *J. Biol. Chem.,* 255, 5530, 1980.
59. **Yu, L. and Yu, C. A.,** Interaction between succinate dehydrogenase and ubiquinone-binding protein from succinate-ubiquinone reductase, *Biochim. Biophys. Acta,* 593, 24, 1980.
60. **Yu, C. A. and Yu, L.,** Isolation and properties of a mitochondrial protein that converts succinate dehydrogenase into succinate-ubiquinone oxidoreductase, *Biochemistry,* 19, 3579, 1980.
61. **Yu, L. and Yu, C. A.,** Quantitative resolution of succinate-cytochrome c reductase into succinate-ubiquinone and ubiquinol-cytochrome c reductases, *J. Biol. Chem.,* 257, 2016, 1982.
62. **Merli, A., Capaldi, R. A., Ackrell, B. A. C., and Kearney, E. B.,** Arrangement of complex II (succinate-ubiquinone reductase) in the mitochondrial inner membrane, *Biochemistry,* 18, 1393, 1979.
63. **Girdlestone, J., Bisson, R., and Capaldi, R.,** Interaction of succinate-ubiquinone reductase (complex II) with (arylazido) phospholipids, *Biochemistry,* 20, 152, 1981.
64. **Mowery, P. C., Steenkamp, D. J., Ackrell, B. A. C., Singer, T. P., and White, G. A.,** Inhibition of mammalian succinate dehydrogenase by carboxins, *Arch. Biochem. Biophys.,* 178, 495, 1977.
65. **Ackrell, B. A. C., Kearney, E. B., Coles, C. J., Singer, T. P., Beinert, T. P., Wan, Y.-P., and Folkers, K.,** Kinetics of the reoxidation of succinate dehydrogenase, *Arch. Biochem. Biophys.,* 182, 107, 1977.
66. **Schewe, T., Rapoport, S., Böhme, G., and Kunz, W.,** Zum Angriffspunkt des Systemfungizids Carboxin in der Atmungskette, *Acta Biol. Med. Ger.,* 31, 73, 1973.
67. **Tushurashvili, P. R., Gavrikova, E. V., Ledenev, A. N., and Vinogradov, A. D.,** Studies on the succinate dehydrogenating system. Isolation and properties of the mitochondrial succinate-ubiquinone reductase, *Biochim. Biophys. Acta,* 809, 145, 1985.
68. **Vinogradov, A. D., Gavrikov, V. G., and Gavrikova, E. V.,** Studies on the succinate dehydrogenating system. II. Reconstitution of succinate-ubiquinone reductase from the soluble components, *Biochim. Biophys. Acta,* 592, 13, 1980.
69. **Coles, C. J., Singer, T. P., White, G. A., and Thorn, G. D.,** Studies on the binding of carboxin analogs to succinate dehydrogenase, *J. Biol. Chem.,* 253, 5573, 1978.
70. **Ramsay, R. R., Ackrell, B. A. C., Coles, C. J., Singer, T. P., White, G. A., and Thorn, G. D.,** Reaction site of carboxanilides and of thenoyltrifluoroacetone in complex II, *Proc. Natl. Acad. Sci. U.S.A.,* 78, 825, 1981.
71. **Vinogradov, A. D., Gavrikova, E. V., and Goloveshkina, V. G.,** A new ferricyanide reactive site in soluble succinate dehydrogenase, *Biochem. Biophys. Res. Commun.,* 65, 1264, 1975.
72. **Grivennikova, V. G. and Vinogradov, A. D.,** Interaction of mitochondrial succinate:ubiquinone reductase with thenoyltrifluoroacetone and carboxin, *Biokhimiya,* 50(3), 395, 1985.
73. **Coles, C. J., Tisdale, H. D., Kenney, W. C., and Singer, T. P.,** Studies on succinate dehydrogenase. XXI. Quaternary structure of succinate dehydrogenase, *Physiol. Chem. Phys.,* 4, 301, 1972.
74. **Schewe, T., Müller, W., Lyr, H., and Zanke, D.,** Ein molekulares Rezeptormodell für Carboxin, in *Systemfungizide,* 5th Int. Symp. Reinhardsbrunn, Lyr, H. and Polter, C., Eds., Akademie Verlag, Berlin, 1979, 241.
75. **Wood, D., Darlison, M. G., Wilde, R. J., and Guest, J. R.,** Nucleotide sequence encoding the flavoprotein and hydrophobic subunits of the succinate dehydrogenase of *Esherichia coli, Biochem. J.,* 222, 519, 1984.
76. **Trumpower, B. L. and Simmons, Z.,** Diminished inhibition of mitochondrial electron transfer from succinate to cytochrome c by thenoyltrifluoroacetone induced by antimycin, *J. Biol. Chem.,* 254, 4608, 1979.
77. **Salerno, J. C. and Ohnishi, T.,** Studies on the stabilized ubisemiquinone species in the succinate-cytochrome c reductase segment of the intact mitochondrial membrane system, *Biochem. J.,* 192, 769, 1980.

78. **Hargreaves, J.,** personal communication, 1990.
79. **Keon, J. P. R., White, G. A., and Hargreaves, J. A.,** Isolation, characterization and sequence of a gene conferring resistance to the systemic fungicide carboxin from the maize smut pathogen, *Ustilago maydis*, *Curr. Genet.*, 19, 475, 1991.
80. **Hargreaves, J.,** personal communication, 1991.

Chapter 2

TARGET SITES OF HYDROXYPYRIMIDINE FUNGICIDES

D. W. Hollomon

TABLE OF CONTENTS

I. INTRODUCTION

Routine screening of analogues of the insecticide diazonin (Figure 1) during the 1960s at Jealott's Hill Research Station identified a series of 2-amino-4-hydroxypyrimides with both protectant and systemic activity against several important powdery mildews.[1] Further structure-activity studies improved these fungicidal and systemic properties, and led to the introduction of dimethirimol,[2,3] ethirimol,[4] and eventually bupirimate[5] (Figure 1). Dimethirimol has been used largely as a soil drench on cucurbits, while ethirimol has been available, mainly in the U.K., as a seed treatment formulation for barley powdery mildew (*Erysiphe graminis* f. sp. *hordei*) control. Bupirimate, the sulfamate ester of ethirimol, penetrates woody tissue, and is used for mildew control in top fruit and ornamentals. A detailed account of the practical applications of hydroxypyrimide fungicides has recently been published elsewhere,[6] and their chemical, physical, and toxicological properties are given in Table 1.

Mammalian toxicities are low and few harmful effects have been noted, either against nontarget organisms in the ecosystem, or host plants, if used at recommended application rates. Although there may be *in vitro* effects of these hydroxypyrimides on some fungal pathogens,[7] in practical terms use is restricted to control of members of the Erysiphales. Thus, hydroxypyrimides are well suited to integrated control systems. It is their extreme specificity, coupled with the need to understand the mechanisms of resistance, that maintains interest in this fungicide group.

II. TRANSPORT, METABOLISM, AND DEGRADATION

Availability of hydroxypyrimides in soils depends very much on moisture, acidity, and organic matter content, and strong adsorption of ethirimol on acid peat prevents use in these soils.[8,9] Within herbaceous plants, including cereals, hydroxypyrimides move in the transpiration stream, accumulating around leaf margins and even exuding onto leaf surfaces. Hydroxypyrimides are less mobile in woody plants,[10] although bupirimate seems sufficiently volatile to be redistributed as vapor across leaf surface.[5] Lipophilicity may play some role in the accumulation of hydroxypyrimides by haustoria,[11] but accumulation occurs regardless of whether fungicides are protonated or not, and it is not correlated with fungitoxicity.[12] Degradation in both cucumber and barley leaves involves formation of more polar metabolites resulting from *N*-dealkylation, hydroxylation of the 5-*n*-butyl group,[13] and conjugation to form glucosides. The half-life within plants is no more than 4 d,[14] although some of the metabolites are active against powdery mildews. Hydroxypyrimides polymerize on exposure to UV light in concentrated solutions.[15] On leaf surfaces the combination of sunlight, oxygen, and a suitable absorber of radiation such as chlorophyll, should produce singlet oxygen which then photo-oxidizes hydroxypyrimides by attacking the 5,6-pyrimidine double bond to give hydroperoxides.[16,17] None of these degradation products have so far been detected on leaf surfaces in practice, although they do occur under laboratory conditions.

III. MECHANISM OF ACTION

A. WHEN IS GROWTH ARRESTED?

Powdery mildews invade their hosts through the cuticle. A combination of physical force exerted by the appressorium, an essential component of the infection process, and the early release of hydrolytic enzymes such as esterases (cutinase)[18] dissolve the cuticle and allow penetration. Cytoplasmic aggregates accumulate beneath the penetration peg,[19] and in incompatible infections these papillae may stop invasion. It is formation of appressoria which is the most sensitive developmental stage inhibited by hydroxypyrimides. Appressoria nor-

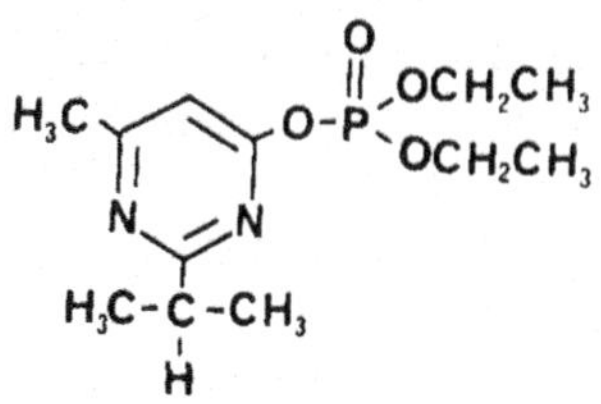

Diazinon

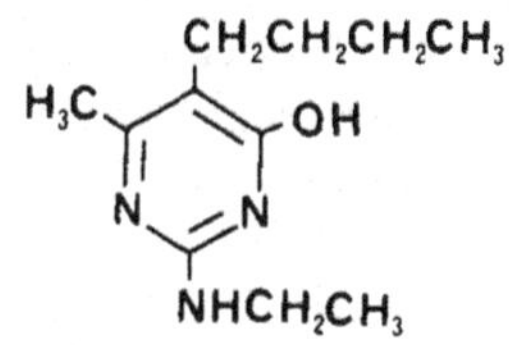

Ethirimol

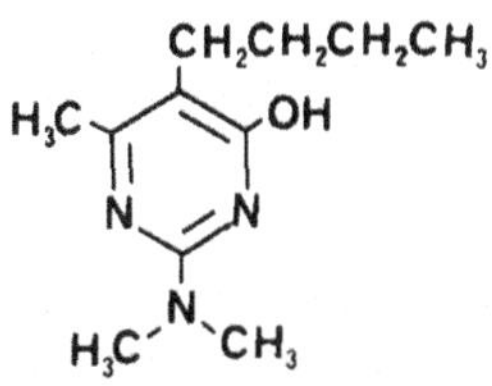

Dimethirimol

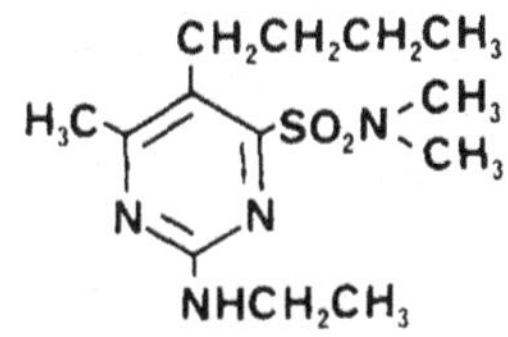

Bupirimate

FIGURE 1. Chemical structures of pyrimidine insecticides and fungicides.

TABLE 1
Some Chemical, Physical, and Toxicological Properties of Hydroxypyrimidine Fungicides

	Ethirimol	Dimethirimol	Bupirimate
Code name	PP 149	PP 675	PP 588
Trademarks	Milgo, Milgo E, Milstem, Milcurb super	Milcurb	Nimrod
Melting point	150—160°C Phase change at 140°C	102°C	50—51°C
Vapor pressure	267 μPa at 25°C	1.46 mPa at 30°C	67 μPa at 20°C
Solubility at room temp	200 mg/l water sparingly soluble in acetone, slightly soluble in diaceton, alcohol, chloroform, trichloroethylene, and aqueous solutions of strong acids and bases	1.2 g/l water 45 g/l acetone 1.2 kg/l chloroform 65 g/l ethanol 360 g/l xylene	22 mg/ml water soluble in most organic solvents except paraffin hydrocarbons
Acute oral LD50 for rats	6340 mg/kg (female rat)	2350 mg/kg	4000 mg/kg (female rat)
No effect level in 2-year feeding tests	200 mg/kg diet	300 mg/kg diet	100 mg/kg diet

Compiled from Ref. 39.

mally form about 16 h after inoculation, but some critical biochemical events occur much earlier, since ethirimol prevented appressoria formation only when applied within the first 8 h after inoculation.[12] Well before appressoria form, primary germ tubes may be involved in the uptake of water and solutes from host cells, events which must precede appressoria formation and penetration.[21] Indeed, almost simultaneously with landing on the leaf surface conidia release esterases, signifying early contact between host and pathogen.[18] Clearly, there is ample scope for fungicides to gain entry into the germinating conidium and block the infection process. Once penetration has occurred and haustoria formed, hydroxypyrimides inhibit, although less effectively, further growth across the leaf surface, and reduce sporulation. The practical significance of these effects on later stages of disease development on fungicide performance is not clear.

B. INHIBITION OF ADENOSINE DEAMINASE

Because powdery mildews are obligate pathogens, conventional approaches to identifying a mode of action through *in vitro* studies were not practicable. Some indication of the likely biochemical mode of action of hydroxypyrimide fungicides arose through attempts to reduce their toxicity with various metabolites (see Hollomon and Schmidt for references). Several powdery mildews were examined by different researchers and, although results varied between systems, a common feature throughout was that adenine and adenosine were always good reversal agents. Some purine analogues, including isopentenyl adenine, kinetin (6-furfuryl adenine), and 6-methyl purine, prevented appressoria formation on barley, and ethirimol-resistant mildew strains were cross-resistant to them.[22] Techniques were developed to allow measurement of incorporation of both adenine and adenosine into nucleic acids, events which were halted in the presence of ethirimol. Synthesis of inosine and adenosine nucleotides was blocked, suggesting that the purine salvage pathway (Figure 2) was inhibited by hydroxypyrimides during primary infection.[23] Thirteen enzymes involved in this pathway were assayed using crude homogenates from barley mildew conidia, and, of those in which activity was detected, only adenosine deaminase (ADAase, EC 3.5.4.4) was inhibited by ethirimol. ADAase catalyzes the largely irreversible deamination of adenosine (and deoxyadenosine) to yield inosine (deoxyinosine) and channels purines derived from the breakdown of nucleic acids from 5′-AMP(5′-dAMP) to 5′-GMP (5′-dGMP). Structure-activity studies involving 18 different ethirimol analogues showed that poor enzyme inhibitors were also poor fungicides (Table 2). For the few exceptions conversion to ethirimol within barley leaves during the bioassay may account for anomalous results.

Inhibition of ADAase by hydroxypyrimides is pH dependent, with a K_i of 2.32×10^{-5} *M* for ethirimol at pH 7.8, and deoxyadenosine as a substrate.[24] Enzymatic deamination of adenosine proceeds by direct water attack on the leaving amino group with a 1,6-hydrated intermediate.[25] Analogues of possible intermediates, such as 2′-deoxycoformycin and 6-hydroxymethyl-1,6-dihydropurine ribonucleoside, are potent ADAase inhibitors, acting probably as transition state analogues which form stable bonds between the 6-OH and a cystein within the enzyme's active site.[26] The cysteine SH serves as a proton donor to N-1 of the substrate, allowing formation of a nucleophile with C-6. This proton also serves to hydrogen bond to the thiolate group from which it originated. Similar interactions may occur with the OH of hydroxypyrimide fungicides, and certainly this function is important for ADAase inhibition by ethirimol.[23] Substitution of the 4-OH in ethirimol with Cl, OCH_3, CH_3, or $OCH_2OCOC_6H_5$ destroyed ADAase activity totally, although fungicidal activity was not always reduced accordingly (Table 2). However, the 4-chloro and 4-benzoyloxymethoxy derivatives may well be converted during the course of bioassay to fungicidal hydroxypyrimidines.[27] Unlike 2′-deoxycoformycin, ethirimol is not a tight binding inhibitor of ADAase; its action is not competitive, and the shape of ethirimol hardly resembles that of adenosine. Correct enzyme binding also depends on the 5-*n*-butyl group which may ensure adequate lipid solubility for passage through membranes. Hydroxylation of this *n*-butyl group not only destroys fungicidal activity, but also ADAase is no longer inhibited in cell-free extracts (Table 3), indicating that this group is probably buried within the enzyme, with the 2-amino group exposed toward the surface.

C. ADAase CHARACTERISTICS

ADAase has been intensively studied in mammalian systems since a deficiency in this enzyme is associated with severe immunodeficiency. Inhibitors of ADAase are also used to

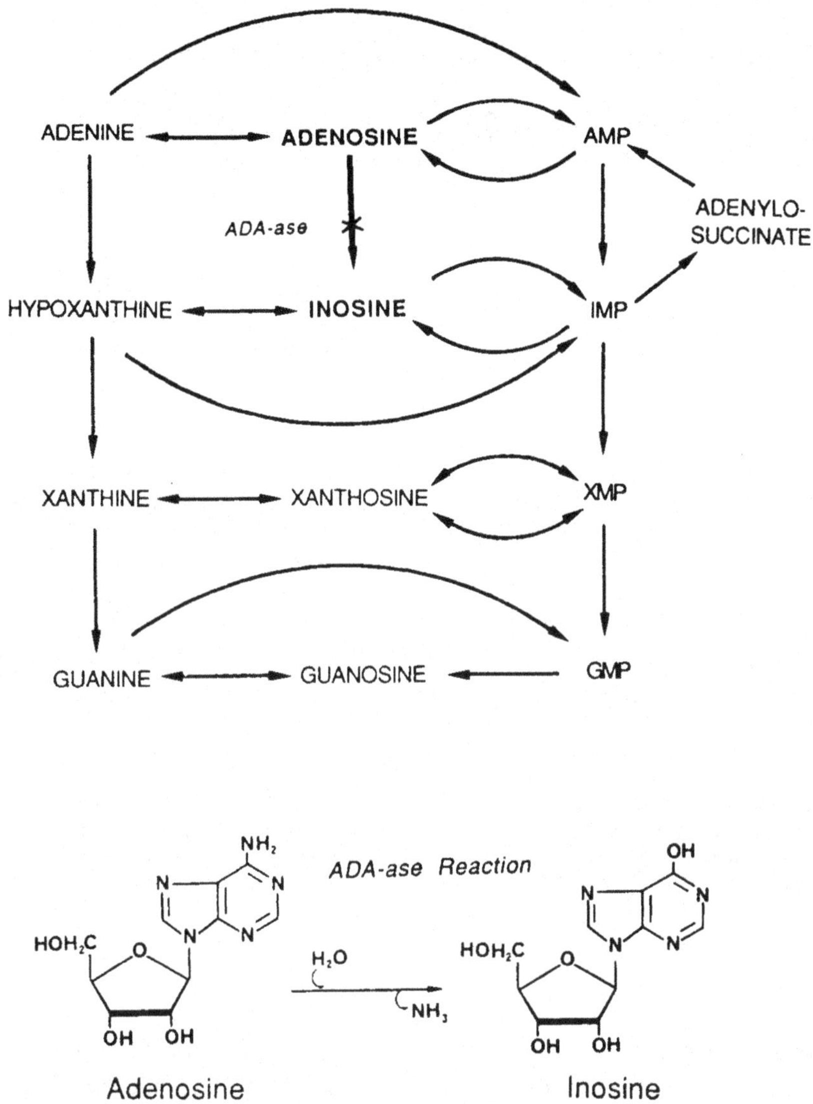

FIGURE 2. Purine salvage pathway. Not all steps are necessarily important in powdery mildews.

prevent deamination of nucleotide analogues used in cancer chemotherapy. In mammals, ADAase has a single catalytic subunit (M_r 36,000 Da), which is associated with a large accessory protein (M_r up to 280,000 Da) whose function is as yet unknown. Full length DNA clones of human ADAase were amongst the first mammalian genes to be sequenced.[28] Purification of microbial ADAases has been less successful, since activity is generally lost once the catalytic subunit is removed from its accessory protein. In both mildew-infected barley and in conidia, ADAase is associated with a single large protein (approximately 300,000 Da), which shows other characters in common with the mammalian enzyme, such as substrate specificity and a requirement for SH groups. Curiously, ADAase from infected plants is always slightly larger than the conidial enzyme (Figure 3).

D. WHAT DOES ADAase DO?

Why inhibition of ADAase should affect mildew growth is not clear. *E. graminis* appears to be a purine auxotroph, and, apart from reutilizing purines already present in conidia,

TABLE 2
Effect of Hydroxypyrimidines on ADAase and Mildew Growth

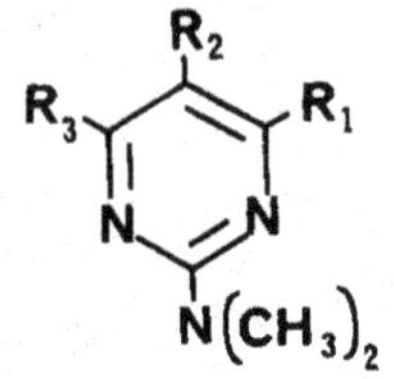

R_1	R_2	R_3	ADAase inhibition	Bioassay[a,b] Appressoria formation	Colony length
OH	nC_3H_9	CH_3 (dimethirimol)	1a	1a	1a
OH	H	CH_3	NA[c]	NA	NA
OH	CH_3	CH_3	NA	NA	NA
OH	$CH(CH_3)_2$	CH_3	NA	NA	NA
OH	C_6H_{13}	CH_3	0.05b	0.048b	0.16a
OH	n-C_3H_9	H	0.06b	0.020d	NA
OH	H	H	NA	NA	NA
OH	n-C_3H_9	C_2H_5	0.83a	0.769a	10.0b
OH	n-C_3H_9	CH_3NH	NA	0.200c	0.55a
OH	n-C_3H_9	CH_3S	NA	NA	10.0b
Cl	n-C_3H_9	CH_3	NA	0.100c	10.0b
OCH_3	n-C_3H_9	CH_3	NA	0.001e	1.6a
$OCH_2OCOC_6H_5$	n-C_3H_9	CH_3	NA	0.800a	10.0b
CH_3	n-C_3H_9	CH_3	NA	0.002c	1.6a

[a] Activities are expressed relative to dimethirimol and shown as reciprocals. Compounds with lower activities than dimethirimol are both poorer fungicides and less active ADAase inhibitors. Values for dimethirimol are k_i for ADA-ase = 5.4×10^{-4} M; ED_{50} appressoria formation = 1.2×10^{-6} M; colony length = 5.4×10^{-4} M.

[b] Values followed by the same letter are not significantly different ($p = 0.01$).

[c] NA: no activity.

Data are primarily from Ref. 23.

disease development requires a continuous supply of purines from the host. Without access to these purines infection is unlikely to succeed. ADAase may play some part in linking the host's purine metabolism with that of the mildew pathogen,[29] enabling purines to be reutilized through the salvage pathway. ADAase is one of several enzymes that control adenosine (deoxyadenosine) levels, which in turn regulate the activity of many enzymes involved in C-1 metabolism. Finally, breakdown of AMP increases the energy charge[30] and stimulates many of the enzymes required to activate metabolism in germinating conidia.

IV. SPECIFICITY OF ACTION

A. TOWARD NONTARGET ORGANISMS

Conversion of the larger 5′-AMP pool to 5′-GMP is an important feature of the purine salvage pathway, although deamination can occur at several different steps. In plants ADAase has never been conclusively detected. Instead, plants deaminate at the nucleotide level, but this AMP aminohydrolase is unaffected by ethirimol. ADAase is present in many fungi and mammals, but it is only the mildew ADAase that is significantly inhibited by ethirimol (Table 4).[24] This unusual form of ADAase accounts for the specificity of ethirimol toward mildew. Even between mildew species and genera there are differences in the sensitivity of ADAase to ethirimol, which correlate with the effectiveness of ethirimol against each mildew (Table 4).

TABLE 3
Effect of 4-Hydroxy-5-*n*-Butyl-6-Methyl Pyrimidines on ADAase and Mildew Growth

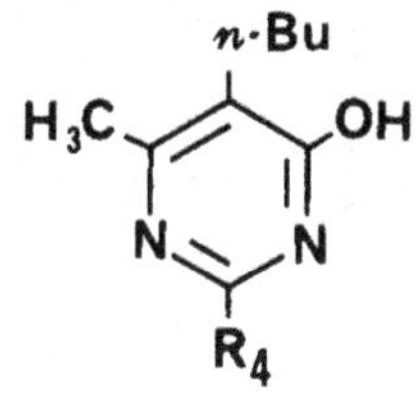

R_4	Partition coefficient[a]	ADAase inhibition	Bioassay[b,c] Appressoria	Bioassay[b,c] Colony length
H	1.38	50.00a	0.006a	0.77ab
NH_2	1.50	2.00b	0.003a	0.38ab
OH	1.26	0.20c	0.002a	NA
SH	1.81	0.05	NA[11]	NA
$NHCH_3$	1.64	1.67c	0.22b	0.15b
$NH(C_2H_5)$ (ethirimol)	2.21	50.00a	1.00c	16.67c
$N(CH_3)_2$ (dimethirimol)	1.93	1.00d	1.00c	1.00a

[a] Octanol/water $\log_{10}$.
[b] Activities are expressed as in Table 2, relative to dimethirimol.
[c] Values followed by the same letter are not significantly different ($p = 0.01$).
[d] NA: no activity.

Data are primarily from Ref. 23.

B. RESISTANCE WITHIN TARGET SPECIES

Resistance to dimethirimol soon followed its introduction for cucumber powdery mildew *(Sphaerotheca fuliginea)* control, and Milcurb was withdrawn from use throughout north-western Europe.[32] Control of mildew on field-grown cucurbits has not encountered similar problems, and hydroxypyrimides are still used widely in these situations. However, a gradual decrease in sensitivity to ethirimol of barley powdery mildew led to performance difficulties in the field some 3 to 4 years after introduction,[33] and seed treatment use was restricted for a while to spring-sown crops in an effort to reduce the frequency of resistant strains. Various techniques were used to monitor hydroxypyrimidine sensitivity and, depending on the assay used, resistant strains were up to 100 times less sensitive than wild types.[34] Cross-resistance occurred between all hydroxypyrimides, although resistance to bupirimate has not yet been a practical problem. Resistance is stable, under genetic control, although the number of genes involved is certainly more than two, and probably many.[31] The biochemical mechanisms involved remain elusive. Any detoxification to nonfungicidal metabolites was not correlated with sensitivity,[40] and neither are there kinetic changes in the target ADAase which might account for resistance.[24]

Resistance to sterol 14-demethylase-inhibiting fungicides (DMIs) also developed in barley powdery mildew,[35] and many of these isolates showed negatively correlated cross-resistance to ethirimol.[36] In experiments aimed at identifying possible mechanisms of resistance in barley powdery mildew to DMIs,[37] we attempted to measure the extent of mildew infection either by incorporation of [^{3}H]-methionine into protein, or [^{3}H]-adenosine into nucleic acid. While this was satisfactory for methionine incorporation, this approach was soon abandoned for adenosine since incorporation varied significantly between strains despite apparently similar infection levels. It is possible that, in some mildew strains, ethirimol

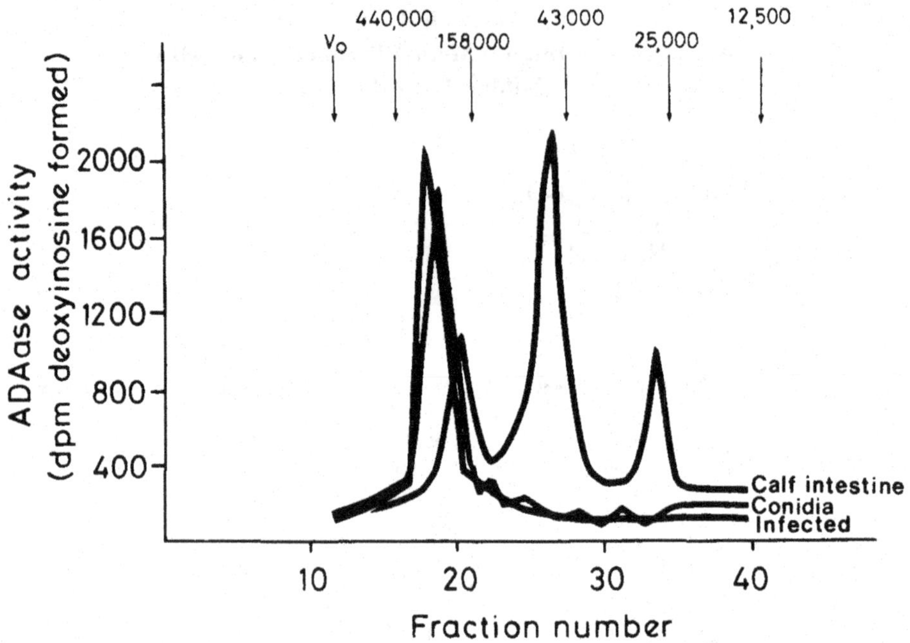

FIGURE 3. Gel filtration of ADAase on Sephacryl S-300. Postmitochondrial supernatants were fractionated with ammonium sulfate (25 to 50%). Precipitates were dissolved in a small volume of 0.05 *M* Tris HCl, pH 7.4 containing 10% glycerol and 10 m*M* 2-mercaptoethanol, and separated on a Sephacryl S-300 column (50 × 1.5 cm). Column was eluted with the same buffer (15 ml/h) and 10-ml fractions collected. ADAase activity was assayed according to the method used in Ref. 23. Calf intestine ADAase was purchased from Sigma, U.K.

resistance reflects more efficient uptake of adenosine from the host, but more detailed studies are needed to prove this.

V. FUTURE DIRECTIONS

Resistance to DMI fungicides in powdery mildews rekindled interest in hydroxypyrimides, and, as a mixture partner with flutriafol, ethirimol has played a significant role in barley mildew control in the U.K. during the last 6 years.[38] The extreme specificity of hydroxypyrimidines toward powdery mildews limits their wider use, but ADAase is present as a potential target site in many other fungi. Detailed knowledge of the target site in other pathogens may identify broader spectrum hydroxypyrimidine analogues. Increased knowledge of ADAase and its regulation may also help counter-resistance. However, microbial ADAases are difficult to purify to homogeneity, and the obligate nature of powdery mildews makes them uninviting candidates for conventional biochemistry.

Recombinant DNA technology offers to overcome many of these difficulties, and is already being applied successfully in our laboratory to isolate and clone from cereal mildews the sterol 14-demethylase gene, the target enzyme for DMI fungicides. Like most genes, ADAase will have conserved regions, and probes obtained from mammalian sources, which are already available,[28] may well be used to identify mildew DNA fragments containing the ADAase gene. Mildew genes may soon be expressed in other fungi, making it possible to identify a target site through gene fusion and deletion experiments. From the DNA sequence and a computer program giving protein structure, it should be possible to generate a three-dimensional picture of the target site in mildew, and compare it with the same region in other ADAases. To develop a broad-spectrum hydroxypyrimidine fungicide in this rational way presents an exciting future challenge to molecular biologists, chemists, and biologists.

TABLE 4
Effect of Ethirimol on ADAase from Different Sources

	ADAase activity[a]		
	No ethirimol	Ethirimol (200 μg ml^{-1})	Inhibition (%)
Powdery mildews			
Erysiphe graminis conidia	48.0	0.8	98
E. pisi conidia and mycelium	6.7	0.4	94
E. polygoni conidia	60.7	38.1	37
Sphaerotheca fuliginea conidia	8.9	4.5	42
Unidentified mildew from *Senecio vulgaris*	20.5	4.3	21
Other fungi			
Penicillium sp. mycelium	76.9	72.2	6
Cladosporium cucumerinum spores and mycelium	11.4	14.8	None
Fusarium sp. mycelium	34.9	33.5	4
Uromyces appendiculatus germlings	4.4	4.4	None
U. fabae germlings	8.5	7.5	12
Puccinia obtegens germlings	4.3	4.7	None
Rhynchosporium secalis spores and mycelium	2.3	2.5	None
Neurospora crassa mycelium	8.6	8.2	9
Saccharomyces cerevisiae	0.8	0.7	12
Streptomyces scabies	7.9	8.5	None
Plants			
Healthy barley	[b]	[b]	[b]
Mildew infected	4.8	2.8	42
Healthy peas	[b]	[b]	[b]
Mildew infected	3.2	0.8	25
Others			
Calf intestine (purified)	54,100	48,300	11
Myzus persicae	[b]	[b]	[b]
Musca domestica	33.2	30.4	9

[a] Picomole substrate converted per mg protein per 10 min at 25°C. Adenosine concentrations in all assays were 3.6×10^{-4} *M*. For other experimental details see Ref. 24.

[b] ADAase activity not detected.

Data from Ref. 24.

REFERENCES

1. **Snell, B. K., Elias, R. S., and Freeman, P. F. H.,** Pyrimidine derivatives and the use thereof as fungicides, *GBP*, 1, 182 and 584, 1966.
2. **Elias, R. S., Shephard, M. C., Snell, B. K., and Stubbs, J.,** 5-*n*-butyl-2-dimethylamino-4-hydroxy-6-methylpyrimidine: a systemic fungicide, *Nature*, 219, 1160, 1968.
3. **Geoghegan, M. J. A.,** Pyrimidine fungicide, *Proc. 5th Br. Insectic. Fungic. Conf.*, p. 333, 1969.
4. **Bebbington, R. M., Brooks, D. H., Geoghegan, M. J., and Snell, B. K.,** Ethirimol: a new systemic fungicide for the control of cereal powdery mildews, *Chem. Ind.*, p. 1512, 1969.
5. **Finney, J. R., Farrell, G. M., and Bent, K. J.,** PP 588-bupirimate: a new fungicide for the specific control of powdery mildews on apples and other crops, *Proc. 8th Br. Instectic. Fungic. Conf.*, p. 667, 1975.
6. **Hollomon, D. W. and Schmidt, H.-H.,** 2-Aminopyrimidine fungicides, in *Modern Selective Fungicides*, Lyr, H., Ed., Longman, Chicago, 1987, 283.
7. **Grindle, M.,** Variations among field isolates of *Botrytis cinerea* in their sensitivity to antifungal compounds, *Pestic. Sci.*, 12, 305, 1981.

8. **Collier, G. F., Graham-Bryce, I. J., Knight, A. G., and Coutts, J.,** Direct observation of the distribution of radiolabelled ethirimol in soil by resin impregnation and autoradiography, *Pestic. Sci.*, 10, 50, 1979.
9. **Graham-Bryce, I. J. and Coutts, J.,** Interactions of pyrimidine fungicides with soil and their influence on uptake by plants, *Proc. 6th Br. Insectic. Conf.*, p. 419, 1971.
10. **Sheppard, M. C.,** Barriers to uptake and translocation in herbaceous and woody plants, *Proc. 7th Br. Insectic. Fungic. Conf.*, p. 841, 1975.
11. **Manners, J. M. and Gay, J.,** Fluxes and accumulation of ethirimol in haustoria of *Erysiphe pisi* and protoplasts of *Pisum sativum, Ann. Appl. Biol.*, 96, 283, 1980.
12. **Hollomon, D. W.,** Antifungal activity of substituted 2-amino-pyrimidines, in *Mode of Action of Antifungal Agents*, Trinci, A. P. J. and Ryley, J. F., Eds., Cambridge University Press, Cambridge, 1984, 185.
13. **Cavell, B. D., Hemmingway, R. J., and Teal, G.,** Some aspects of the metabolism and translocation of the pyrimidine fungicides, *Proc. 6th Br. Insectic. Fungic. Conf.*, p. 431, 1971.
14. **Calderbank, A.,** Metabolism and mode of action of dimethirimol and ethirimol, *Acta Phytopathol. Acad. Sci. Hung. (Budapest)*, 99, 255, 1981.
15. **Wells, C. H., Pollard, S. J., and Sen, D.,** Phytochemistry of some systemic pyrimidine fungicides, *Pestic. Sci.*, 10, 171, 1979.
16. **Dixon, S. R. and Wells, C. H. J.,** Chlorophyll sensitised photodegradation of 2-dimethylamino-5,6-dimethylpyrimidin-4-ol, *Pestic. Sci.*, 21, 155, 1987.
17. **Dixon, S. R. and Wells, C. H. J.,** Products from the sensitised photo-oxidation of 2-dimethylamino-5,6-dimethylpyrimidin-4-ol, *Pestic. Sci.*, 25, 255, 1989.
18. **Nicholson, R. L., Yoshioka, H., Yamaoka, N., and Kunoh, H.,** Preparation of the infection court by *Erysiphe graminis*. Release of esterase enzyme from conidia in response to a contact stimulus, *Exp. Mycol.*, 12, 336, 1988.
19. **Aist, J. R. and Israel, H. W.,** Autofluorescence and ultraviolet-absorbing components in cell walls and papillae of barley coleoptiles and their relationship to disease resistance, *Can. J. Bot.*, 64, 266, 1986.
20. **Kunoh, H., Yamamori, K., and Ishizaki, H.,** Cytological studies of the early stages of powdery mildew in barley and wheat. VIII. Autofluorescence at penetration sites of *Erysiphe graminis hordei* on living barley coleoptiles, *Physiol. Plant Pathol.*, 21, 373, 1982.
21. **Carver, J. L. W. and Bushnell, W. R.,** The probable role of primary germ-tubes in water uptake before infection by *Erysiphe graminis, Physiol. Plant Pathol.*, 23, 229, 1983.
22. **Hollomon, D. W.,** Evidence that ethirimol may interfere with adenine metabolism during primary infection of barley powdery mildew, *Pestic. Biochem. Physiol.*, 10, 181, 1979.
23. **Hollomon, D. W. and Chamberlain, K.,** Hydroxypyrimidine fungicides inhibit adenosine deaminase in barley powdery mildew, *Pestic. Biochem. Physiol.*, 16, 158, 1981.
24. **Hollomon, D. W.,** Specificity of ethirimol in relation to inhibition of the enzyme adenosine deaminase, *Proc. Br. Crop Prot. Conf. Pests Dis.*, p. 251, 1979.
25. **Jones, W., Kurz, L. C., and Wolfenden, R.,** Transition-state stabilization by adenosine deaminase: 1,6 addition of water to purine ribonucleoside, the enzymes affinity for 6-hydroxy-1,6-dihydropurine ribonucleoside, and the effective concentration of substrate water at the active site, *Biochemistry*, 28, 1242, 1989.
26. **Weiss, P. M., Cook, P. F., Hermes, J. D., and Cleland, W. W.,** Evidence from nitrogen-15 and solvent deuterium isotope effects on the chemical mechanism of adenosine deaminase, *Biochemistry*, 26, 7378, 1987.
27. **Woodcock, D.,** Structure-activity relationships, in *Systemic Fungicides*, Marsh, R. W., Ed., Longmans, London, 1972, 34.
28. **Wiginton, D. A., Kaplan, D. J., States, J. C., Akeson, A. L., Perme, C. H., Bilyk, I. J., Vaughan, A. J., Lattier, D. L., and Hutton, J. J.,** Complete sequence and structure of the gene for human adenosine deaminase, *Biochemistry*, 25, 8234, 1986.
29. **Butters, J. A., Burrell, M. M., and Hollomon, D. W.,** Purine metabolism in barley powdery mildew and its hosts, *Physiol. Plant Pathol.*, 27, 65, 1985.
30. **Atkinson, D. E.,** The energy charge of the adenylate pool as a regulatory parameter: interaction with feedback monitors, *Biochemistry*, 7, 4030, 1968.
31. **Hollomon, D. W.,** Genetic control of ethirimol resistance in a natural population of *Erysiphe graminis* f. sp. *hordei, Phytopathology*, 71, 536, 1981.
32. **Bent, K. J., Cole, A. M., Turner, J. A. W., and Woolner, M.,** Resistance of cucumber powdery mildew to dimethirimol, *Proc. 6th Br. Insectic. Fungic. Conf. Pests Dis.*, p. 331, 1977.
33. **Brent, K. J.,** Case study 4: powdery mildews of barley and cucumber, in *Fungicide Resistance in Crop Protection*, Dekker, J. and Georgopoulos, S. G., Eds., Centre for Agricultural Publishing and Documentation, Wageningen, 1982, 219.
34. **Hollomon, D. W.,** Behaviour of a barley powdery mildew strain tolerant to ethirimol, *Proc. 8th Br. Insectic. Fungic. Conf.*, p. 51, 1975.
35. **Fletcher, J. S. and Wolfe, M. S.,** Insensitivity of *Erysiphe graminis* f. sp. *hordei* to triadimenol and other fungicides, *Proc. 1981 Br. Crop Prot. Conf. Pests Dis.*, p. 633, 1981.

36. **Butters, J. A., Clark, J., and Hollomon, D. W.,** Resistance to inhibitors of sterol biosynthesis in barley powdery mildew, *Meded. Fac. Landbouwwet. Rijksuniv. Gent,* 49/2a, 143, 1984.
37. **Hollomon, D. W., Butters, J. A., and Hargreaves, J. A.,** Current status and biochemical basis of resistance to sterol biosynthesis inhibiting fungicides, in *Fundamental and Practical Approaches to Combating Resistance,* Green, M. B., Moberg, W. K., and LeBaron, H., Eds., American Chemical Society, Washington, D.C., 1990, 199.
38. **Northwood, P. J., Paul, J. A., Gibbard, M., and Noon, R. A.,** FF 4050 seed treatment — a new approach to control barley diseases, *Proc. Br. Crop Prot. Conf. Pests Dis.,* p. 47, 1984.
39. **Worthing, C. R. and Walker, S. B.,** *The Pesticide Manual — a World Compendium,* 7th ed., British Crop Protection Council, London, 1983.
40. **Hollomon, D. W.,** unpublished data.

Chapter 3

TARGET SITES OF TUBULIN-BINDING FUNGICIDES

Hideo Ishii

TABLE OF CONTENTS

I. INTRODUCTION

The benzimidazole fungicides, with benomyl and thiophanate-methyl as prominent examples (Figure 1), were developed in the late 1960s and have been used extensively for the control of fungal diseases on a broad variety of important crops. The primary mode of action of all benzimidazoles has been identified as the specific binding to the β-tubulin subunit of fungal tubulin and, consequently, an interference with microtubule assembly. Microtubules are major components of the fungal cytoskeleton and are involved in meiosis and mitosis, both of which are blocked in the presence of benzimidazoles.[1-4]

Shortly after the introduction of benomyl, first cases of resistance were reported. This development continued, and during the last two decades, resistance of plant pathogenic fungi to benzimidazoles has increased throughout the world, making continued control of many diseases with this class of fungicides difficult.[5] In fields where resistant phenotypes became predominant, use of benzimidazoles was discontinued or restricted, and compounds with different modes of action were introduced. Regardless of discontinued use, field monitoring data have demonstrated that benzimidazole resistance was quite persistent. More rational and direct approaches would thus be required to overcome the problem. The mode of action of benzimidazoles has been treated in several comprehensive and recent reviews[1-4] and thus will not be described again in detail. Rather, an attempt will be made to summarize more recent advances made in the understanding of benzimidazole resistance.

II. GENETICS OF BENZIMIDAZOLE RESISTANCE

The genetics of fungicide resistance are often discussed in connection with the development of resistance in the field. Genetics are also regarded as an important factor, when the risk of fungicide resistance is assessed. Control of resistance by a single gene is reflected in a discontinuous frequency distribution of resistant genotypes in field populations. High levels of resistance can be acquired in one-step mutations, and under the selection force of the fungicide, the population can change qualitatively toward the resistant subpopulation, leading to a serious loss of fungicide efficacy.[5] The best known example for this scenario is provided by the benzimidazoles. In case of a polygenic control, the development of resistance is expressed by a gradual shift of the frequency distribution toward resistance, and the sterol biosynthesis inhibitors are frequently given as an example for this quantitative response of populations.[5] However, evidence exists for the presence of more complex situations not reflected by this simple scheme of resistance development.

As listed in Table 1, benzimidazole resistance of plant pathogenic fungi is normally based on a mutation of one major chromosomal gene. These mutations, however, can be qualitatively different. For example, in *Venturia nashicola* the scab fungus of Japanese pear, all three different levels of resistance — high, medium, and low — are caused by allelic mutations in a single chromosomal gene.[17] In other words, the manifestation of different levels of resistance is controlled by one of the multiple alleles at the same locus. With this fungus, high frequencies of resistant genotypes can be selected through both qualitative and quantitative population responses.[20]

As mentioned above, benzimidazoles exert their antifungal activities by binding to β-tubulin.[1-4] The molecular mechanisms of resistance proposed for plant pathogenic fungi are summarized in Table 2. Although several mechanisms have been reported in the past, a decreased binding of the inhibitor to tubulin-like proteins was always involved. In *V. nashicola*, for example, the binding of ^{14}C-carbendazim to tubulin-like proteins was lower in benzimidazole-resistant than in sensitive isolates, suggesting that a decreased affinity of the inhibitor to the target was closely correlated with resistance.[24] Therefore, a mutation might have changed the structure of the benzimidazole binding site at β-tubulin (Figure 2).

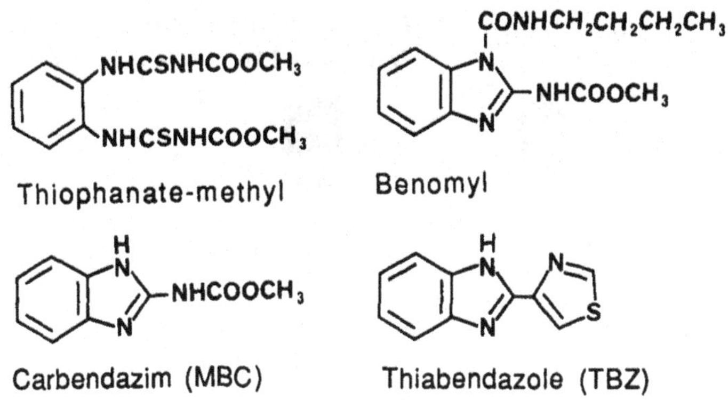

FIGURE 1. Structures of benzimidazole fungicides.

TABLE 1
Genetics of Benzimidazole Resistance in Phytopathogenic Fungi

Pathogen	Number of resistance genes	Ref.
Ceratocystis ulmi	1	6
Fusarium oxysporum	2	7
Mycosphaerella fijiensis	C[a] (?)	8
Penicillium flavus	1	9
P. italicum	2	10
Septoria nodorum	1	11
Ustilago hordei	1	12
U. maydis	1	13
U. violacea	1	14
Venturia inaequalis	1	15, 16
V. nashicola	1	17
V. pirina	1	18

[a] Cytoplasmic gene(s)

Modified from Grindle.[19]

TABLE 2
Molecular Mechanism of Benzimidazole Resistance in Phytopathogenic Fungi

Proposed mechanism	Pathogen	Ref.
Decreased binding affinity to target site	*Botrytis cinerea*	21, 22
	Gibberella fujikuroi	22
	Penicillium brevicompactum	23
	P. corymbiferum	23
	P. expansum	23
	Venturia nashicola	24
Reduced uptake and decreased binding affinity	*Botrytis cinerea*	25
	Fusarium lycopersici	25
	B. cinerea	26
Increased nonspecific and decreased specific binding	*F. oxysporum*	27

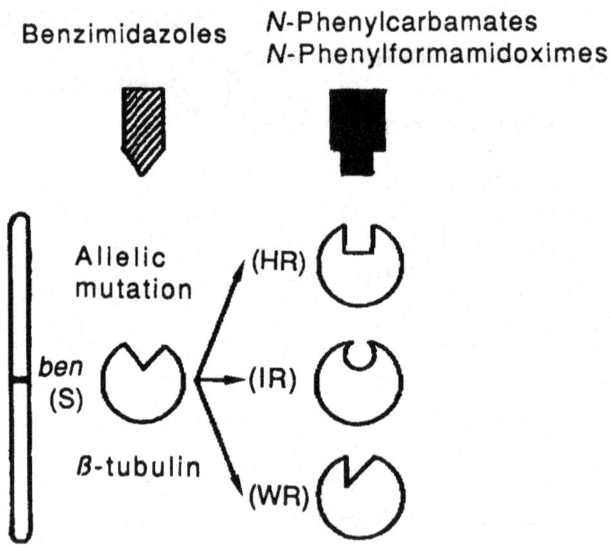

FIGURE 2. Mechanism of resistance of *Venturia nashicola* to benzimidazoles. HR, highly resistant; IR, intermediately resistant; WR, weakly resistant; S, sensitive.

TABLE 3
Sequence Homology between β-Tubulin of *Septoria nodorum* and Other Organisms

Organism	Gene	Nucleic acid level (%)[a]	Amino acid level (%)
Aspergillus nidulans	*ben*A	88.4	95.5
A. nidulans	*tub*C	74.8	86.8
Neurospora crassa	*tub2*	88.3	96.4
Schizosaccharomyces pombe	*nda3*	69.3	81.3
Saccharomyces cerevisiae	*TUB2*	68.4	80.7
Candida albicans	*TUB2*	65.0	84.4
Chicken brain		75.6	85.2

[a] Only coding regions are considered.

From Cooley and Caten.[11]

The cloning of β-tubulin genes has recently been accomplished for several plant pathogenic fungi such as *Erysiphe graminis* f. sp. *hordei*,[28] *Septoria nodorum*,[11] and *V. inaequalis*.[29] The high degree of homology between the β-tubulin sequence of *S. nodorum* and other organisms is summarized in Table 3. Extensive homologies among fungi are apparent at both the nucleic acid and amino acid level. A similar high homology has also been reported for the β-tubulin gene cloned from the barley powdery mildew fungus.[28]

III. STRATEGIES TO COPE WITH BENZIMIDAZOLE RESISTANCE

To cope with the problem of fungicide resistance, strategies such as alternating or mixing unrelated fungicides are usually recommended.[5] However, these strategies would only delay and not avoid the emergence of resistance. The search for chemicals acting on resistant

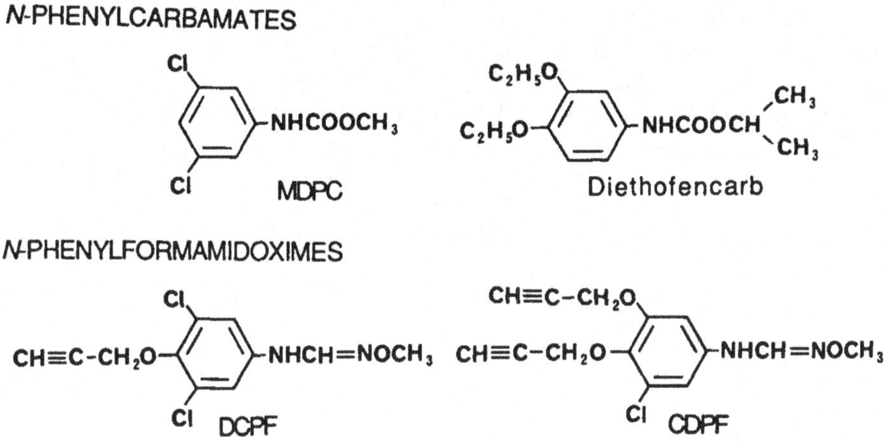

FIGURE 3. Structures of inhibitors with negative cross-resistance to benzimidazoles.

phenotypes more effectively than on sensitive isolates (negative cross-resistance) might be a promising alternative. Leroux and Gredt[30] were the first to report a negative cross-resistance between benzimidazoles and *N*-phenylcarbamate herbicides such as barban and chlorpropham. Some benzimidazole-resistant strains of *Botrytis cinerea* and *Penicillium expansum* were highly sensitive to these *N*-phenylcarbamates. Subsequently, a series of nonherbicidal *N*-phenylcarbamates such as *N*-(3,5-dichlorophenyl) carbamate (MDPC) and diethofencarb, but also *N*-phenylformamidoximes such as *N*-(3,5-dichloro-4-propynyloxyphenyl)-*N'*-methoxyformamidine (DCPF) and *N*-(3-chloro-4,5-dipropynyloxyphenyl)-*N'*-methoxyformamidine (CDPF) (Figure 3), were synthesized and tested in Japan. All of these compounds had a higher antifungal activity against benzimidazole-resistant phenotypes than they had against sensitive wild-type isolates.[31,32]

The vast majority of benzimidazole-resistant isolates of *B. cinerea* are highly resistant and, at the same time, sensitive to *N*-phenylcarbamates and *N*-phenylformamidoximes.[31,32] Consequently, diethofencarb has been introduced in some countries for the control of the benzimidazole-resistant population of the gray mold fungus, in mixture with a benzimidazole.[5] This example has been considered as a first case of successful chemical interference with the phenomenon of fungicide resistance, based on a compound with negative cross-resistance. The concept might not be suitable for pathogens other than *B. cinerea,* and even there, isolates with resistance to both benzimidazoles and *N*-phenylcarbamates have been reported.[5] Other samples have demonstrated that negative cross-resistance of these compounds is not a necessity. For example, with *V. nashicola*, increased sensitivity to *N*-phenylcarbamates and *N*-phenylformamidoximes was restricted to isolates with high levels of benzimidazole resistance; the activity on isolates with intermediate and low levels of resistance was not increased.[17,22,34]

It appears important to examine the question of whether increased sensitivity to *N*-phenylcarbamates and *N*-phenylformamidoximes is determined by a pleiotropism of the same allele coding for benzimidazole resistance. Fujimura et al.[35] have summarized their study on the molecular genetics of increased sensitivity of a benzimidazole-resistant mutant of *Neurospora crassa* to diethofencarb.[35] A highly resistant mutant was obtained from a diethofencarb-resistant wild-type strain, and the mutant was, as expected, highly sensitive to diethofencarb. The β-tubulin gene form this mutant was cloned and sequenced and compared with the respective sequence obtained for the wild-type strain. Only one glutamic acid at position 198 was changed to glycine in tubulin expressed in the benzimidazole-resistant mutant. The authors suggested that this amino acid might be in the binding region for both

TABLE 4
Inheritance of Sensitivities to Carbendazim and MDPC in Random Ascospore Progenies of *Venturia nashicola*

Cross	Number of progenies			
	HR,S[a]	HR,R[b]	IR,R[c]	S,R[d]
HR,S × IR,R	47	10	43	0
HR,S × S,R	47	10	0	43

[a] HR,S = highly resistant to carbendazim and sensitive to MDPC.
[b] HR,R = highly resistant to carbendazim and resistant to MDPC.
[c] IR,R = intermediately resistant to carbendazim and resistant to MDPC.
[d] S,R = sensitive to carbendazim and resistant to MDPC.

From Ishii and Van Raak.[34]

carbendazim and diethofencarb, and that the nature of the amino acid, either glutamic acid or glycine, might be responsible for the differential binding of the two inhibitors.

Because direct biochemical data on the binding of diethofencarb have not yet been reported, this simple binding model remains in a speculative stage. For example, Groves et al.[21] found that binding of another *N*-phenylcarbamate (EPC) to tubulin-like proteins was not obtained in a highly benzimidazole-resistant isolate of *B. cinerea,* regardless of the presence of negative cross-resistance to EPC. For the *N*-phenylformamidoximes, on the other hand, binding of ^{14}C-DCPF to cell-free protein derived from a highly carbendazim-resistant isolate of *V. nashicola* was higher than to protein from a sensitive isolate, or from isolates with intermediate and low levels of benzimidazole resistance, all of which are insensitive to DCPF.[22] Furthermore, sensitivity of this fungus to *N*-phenylcarbamates and *N*-phenylformamidoximes is controlled by a single gene.[34,36] However, when a highly benzimidazole-resistant but MDPC-sensitive isolate was crossed with either an isolate with an intermediate level of resistance or a sensitive wild-type isolate, a portion of the progeny was resistant to both carbendazim and MDPC (Table 4). These double-resistant phenotypes were also detected in pear orchards, although MDPC has never been used. Therefore, more subtle mechanisms might be involved in determining fungal sensitivities to *N*-phenylcarbamates and *N*-phenylformamidoximes.

IV. RHIZOXIN AS A NOVEL MICROTUBULE INHIBITOR

The macrolide compound rhizoxin (Figure 4) was isolated from the culture filtrate of *Rhizopus chinensis,*[37,38] and its inhibitory activity against tubulin polymerization and microtubule formation was characterized with a porcine system.[39] Rhizoxin inhibited mycelial growth of *V. nashicola, B. cinerea* (Table 5), and other fungi, regardless of the presence and the level of benzimidazole resistance.[40,41] Interestingly, even *Phytophthora, Pythium,* and *Alternaria* species, which are naturally insensitive to benzimidazoles, were sensitive to rhizoxin.[42] Treatment of pear leaves with rhizoxin remarkably suppressed black spot caused by *Alternaria alternata* Japanese pear pathotype.

Takahashi et al.[43] concluded that the site of rhizoxin action in *Aspergillus nidulans* was β-tubulin, but it differed from that of other tubulin inhibitors such as benzimidazoles and

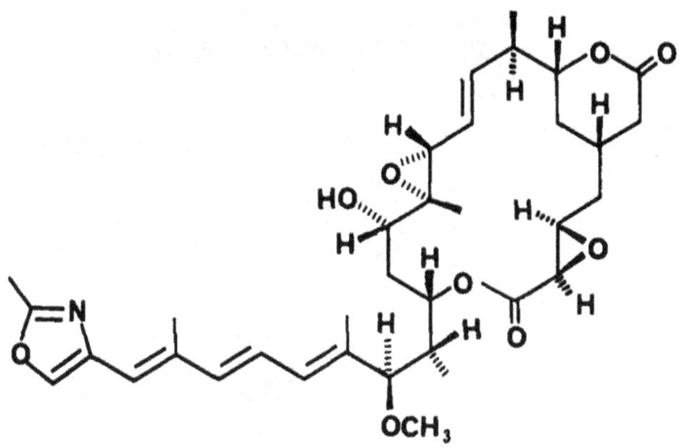

Rhizoxin

FIGURE 4. Structure of rhizoxin.

TABLE 5
Sensitivity of *Botrytis cinerea* Isolates to Carbendazim and Rhizoxin

	EC_{50} (μg/ml) of:	
Isolate	**Carbendazim**	**Rhizoxin**
4519-1 (HR)[a]	>800	<0.01
850447 (MR)[a]	75.4	<0.01
4319-1 (S)[a]	<0.19	<0.01

[a] HR = highly carbendazim-resistant, MR = moderately carbendazim-resistant, S = sensitive.

From Ishii et al.[41]

griseofulvin. Biochemical studies suggested a similar situation for plant pathogenic fungi, indicating a novel target site at β-tubulin suitable to be utilized in the control of fungal diseases (Figure 5). The precise rhizoxin binding site has been identified for *A. nidulans.*[44] The β-tubulin gene from a rhizoxin-resistant mutant was cloned and sequenced, and asparagine at position 100 in a wild-type isolate was changed to isoleucine in the resistant mutant. Furthermore, a DNA fragment which contained the mutated region was excised and substituted for the corresponding region in the wild-type DNA. The transformant was, as expected, resistant. It was concluded that the sensitivity of organisms to rhizoxin was exclusively determined by the 100th amino acid in β-tubulin. The ease by which the rhizoxin-resistant mutant could be generated, and the involvement of a simple point mutation in resistance are very similar to benzimidazole resistance, and the risk of rhizoxin resistance might be similarly high.

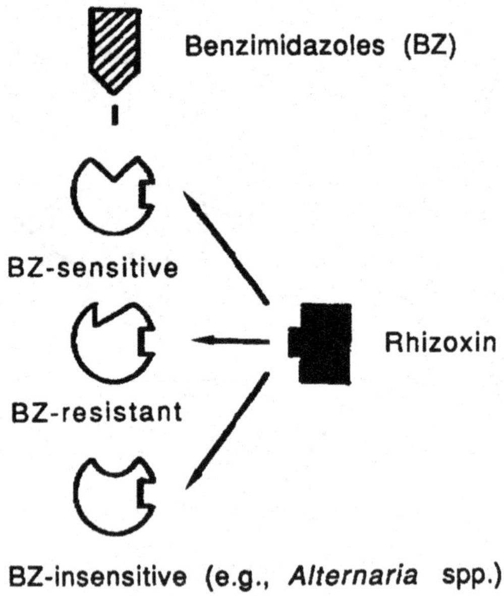

FIGURE 5. Interaction of benzimidazoles (BZ) and rhizoxin with binding sites (β-tubulin) in fungi.

REFERENCES

1. **Corbett, J. R., Wright, K., and Baillie, A. C.,** Compounds interfering with cell division, in *The Biochemical Mode of Action of Pesticides,* 2nd ed., Academic Press, London, 1984, 202.
2. **Davidse, L. C.,** Benzimidazole fungicides: mechanism of action and biological impact, *Annu. Rev. Phytopathol.,* 24, 43, 1986.
3. **Davidse, L. C.,** Biochemical aspects of benzimidazole fungicides — action and resistance, in *Modern Selective Fungicides — Properties, Applications, Mechanisms of Action,* Lyr, H., Ed., Longman, London, 1987, 245.
4. **Davidse, L. C.,** Advances in understanding fungicidal modes of action and resistance, in *Pesticide Science and Biotechnology,* Greenhalgh, R. and Roberts, T. R., Eds., Blackwell Scientific, Oxford, 1987, 169.
5. **Köller, W.,** Fungicide resistance in plant pathogens, in *CRC Handbook of Pest Management in Agriculture,* Vol. 2, 2nd ed., Pimentel, D., Ed., CRC Press, Boca Raton, FL, 1991, 679.
6. **Brasier, C. M. and Gibbs, J. N.,** MBC tolerance in aggressive and non-aggressive isolates of *Ceratocystis ulmi, Ann. Appl. Biol.,* 80, 231, 1975.
7. **Molnár, A., Hornok, L., and Pesti, M.,** The high level of benomyl tolerance in *Fusarium oxysporum* is determined by the synergistic interaction of two genes, *Exp. Mycol.,* 9, 326, 1985.
8. **Stover, R. H.,** Extranuclear inherited tolerance to benomyl in *Mycosphaerella fijiensis* var. *difformis, Trans. Br. Mycol. Soc.,* 68, 122, 1977.
9. **Katan, T., Dunn, M. T., and Papavizas, G. C.,** Genetics of fungicide resistance in *Talaromyces flavus, Can. J. Microbiol.,* 30, 1079, 1984.
10. **Beraha, L. and Garber, E. D.,** A genetic study of resistance to thiabendazole and sodium *o*-phenylphenate in *Penicillium italicum* by the parasexual cycle, *Bot. Gaz.,* 141, 204, 1980.
11. **Cooley, R. N. and Caten, C. E.,** Cloning and characterisation of the β-tubulin gene and determination of benomyl resistance in *Septoria nodorum, Found. Biotech. In. Ferm. Res.,* 6, 207, 1989.
12. **Ben-Yephet, Y., Henis, Y., and Dinoor, A.,** Genetic studies on tolerance of carboxin and benomyl at the asexual phase of *Ustilago hordei, Phytopathology,* 64, 51, 1974.
13. **Van Tuyl, J. M.,** Genetics of fungal resistance to systemic fungicides, *Meded. Landbouwhogesch. Wageningen,* 77-2, 1, 1977.
14. **Garber, E. D., Eng, C., Puscheck, E. E., Weil, M., and Ward, S.,** Genetics of *Ustilago violacea.* XII. Half tetrad analysis and double selection, *Bot. Gaz.,* 143, 524, 1982.

15. **Katan, T., Shabi, E., and Gilpatrick, J. D.**, Genetics of resistance to benomyl in *Venturia inaequalis* isolates from Israel and New York, *Phytopathology*, 73, 600, 1983.
16. **Shabi, E., Katan, T., and Marton, K.**, Inheritance of resistance to benomyl in isolates of *Venturia inaequalis* from Israel, *Plant Pathol.*, 32, 207, 1983.
17. **Ishii, H., Yanase, H., and Dekker, J.**, Resistance of *Venturia nashicola* to benzimidazole fungicides, *Meded. Fac. Landbouwwet. Rijksuniv. Gent*, 49/2a, 163, 1984.
18. **Shabi, E. and Katan, T.**, Genetics, pathogenicity, and stability of carbendazim-resistant isolates of *Venturia pirina*, *Phytopathology*, 69, 267, 1979.
19. **Grindle, M.**, Genetic basis of fungicide resistance, in *Combating Resistance to Xenobiotics*, Ford et al., Eds., Ellis Horwood, Chichester, 1987, 74.
20. **Ishii, H., Udagawa, H., Yanase, H., and Yamaguchi, A.**, Resistance of *Venturia nashicola* to thiophanate-methyl and benomyl: build-up and decline of resistance in the field, *Plant Pathol.*, 34, 363, 1985.
21. **Groves, J. D., Fox, R. T. V., and Baldwin, B. C.**, Modes of action of carbendazim and ethyl *N*-(3,5-dichlorophenyl) carbamate on field isolates of *Botrytis cinerea*, in *Proc. 1988 British Crop Protection Conf. — Pests and Diseases*, BCPC Publications, Croydon, 1988, 100.
22. **Ishii, H. and Takeda, H.**, Differential binding of a *N*-phenylformamidoxime compound in cell-free extracts of benzimidazole-resistant and -sensitive isolates of *Venturia nashicola*, *Botrytis cinerea* and *Gibberella fujikuroi*, *Neth. J. Plant Pathol.*, 95(Suppl. 1), 99, 1989.
23. **Davidse, L. C. and Flach, W.**, Interaction of thiabendazole with fungal tubulin, *Biochim. Biophys. Acta*, 543, 82, 1978.
24. **Ishii, H. and Davidse, L. C.**, Decreased binding of carbendazim to cellular protein from *Venturia nashicola* and its involvement in benzimidazole resistance, in *Proc. 1986 British Crop Protection Conf. — Pests and Diseases*, BCPC Publications, Surry, 1986, 567.
25. **Gessler, C., Sozzi, D., and Kern, H.**, Benzimidazole fungicides: mode of action and problems (in German), *Ber. Schweiz. Bot. Ges.*, 90, 45, 1980.
26. **Tripathi, R. K. and Schlösser, E.**, The mechanism of resistance of *Botrytis cinerea* to methyl benzimidazol-2-yl carbamate (MBC), *Z. Pflkrankh. Pflschutz*, 89, 151, 1982.
27. **Gasztonyi, M., Josepovits, G., Molnár, A., and Hornok, L.**, Biochemical background of resistance to benomyl in genetically different strains of *Fusarium oxysporum*, *Pestic. Biochem. Physiol.*, 29, 17, 1987.
28. **Sherwood, J. E. and Somerville, S. C.**, Sequence of the *Erysiphe graminis* f. sp. *hordei* gene encoding β-tubulin, *Nucleic Acids Res.*, 18, 1052, 1990.
29. **Koenraadt, H., Somerville, S. C., and Jones, A. L.**, Cloning of the β-tubulin gene from benomyl-sensitive and benomyl-resistant field strains of *Venturia inaequalis*, *Pest Resistance Manage. Newsl.*, 2, 40, 1990.
30. **Leroux, P. and Gredt, M.**, Negatively correlated cross resistance between benzimidazole fungicides and *N*-phenylcarbamate herbicides, in *Systemische Fungizide und Antifungale Verbindungen. 6th Int. Symp.*, Vol. 1, Lyr, H. and Polter, C., Eds., Akademie-Verlag, Berlin, 1982, 297.
31. **Kato, T., Suzuki, K., Takahashi, J., and Kamoshita, K.**, Negatively correlated cross-resistance between benzimidazole fungicides and methyl *N*-(3,5-dichlorophenyl) carbamate, *J. Pestic. Sci.*, 9, 489, 1984.
32. **Nakata, A., Sano, S., Hashimoto, S., Hayakawa, K., Nishikawa, H., and Yasuda, Y.**, Negatively correlated cross-resistance to *N*-phenylformamidoximes in benzimidazole-resistant phytopathogenic fungi, *Ann. Phytopathol. Soc. Jpn.*, 53, 659, 1987.
33. **Leroux, P., Gredt, M., Massenot, F., and Kato, T.**, Activity of the phenylcarbamate S 32165 on *Botrytis cinerea*, causative agent of grapevine grey mould (in French), in *Proc. Bordeaux Mixture Centenary Meeting*, Vol. 2, Smith, I. M., Ed., BCPC Publications, Croydon, 1985, 443.
34. **Ishii, H. and Van Raak, M.**, Inheritance of increased sensitivity to *N*-phenylcarbamates in benzimidazole-resistant *Venturia nashicola*, *Phytopathology*, 78, 695, 1988.
35. **Fujimura, M., Oeda, K., Inoue, H., and Kato, T.**, Mechanism of action of *N*-phenylcarbamates in benzimidazole-resistant *Neurospora* strains, in *Managing Resistance to Agrochemicals*, Green, M. B., LeBaron, H. M., and Moberg, W. K., Eds., American Chemical Society, Washington, D.C., 1990, 224.
36. **Ishii, H., Tomikawa, A., Van Raak, M., and Inoue, I.**, Potential limitation of the use of *N*-phenylcarbamates and *N*-phenylformamidoximes as a benzimidazole resistance "breaker", *Plant Pathol.*, submitted.
37. **Sato, Z., Noda, T., Goh, N., Kobayashi, H., and Okuda, S.**, A toxic compound produced by *Rhizopus chinensis* (in Japanese), *Ann. Phytopathol. Soc. Jpn.*, 44, 349, 1978.
38. **Iwasaki, S., Kobayashi, H., Furukawa, J., Namikoshi, M., Okuda, S., Sato, Z., Matsuda, I., and Noda, T.**, Studies on macrocyclic lactone antibiotics. VII. Structure of a phytotoxin "rhizoxin" produced by *Rhizopus chinensis*, *J. Antibiot.*, 35, 354, 1984.
39. **Takahashi, M., Iwasaki, S., Kobayashi, H., Okuda, S., Murai, T., and Sato, Y.**, Rhizoxin binding to tubulin at the maytansine-binding site, *Biochim. Biophys. Acta*, 926, 215, 1987.

40. **Ishii, H., Iwasaki, S., Sato, Z., and Inoue, I.,** Binding of cellular protein from *Venturia nashicola* isolates to carbendazim: its relationship with sensitivity to *N*-phenylcarbamates, *N*-phenylformamidoximes, and rhizoxin, in *Managing Resistance to Agrochemicals,* Green, M. B., LeBaron, H. M., and Moberg, W. K., Eds., American Chemical Society, Washington, D.C., 1990, 237.
41. **Ishii, H., Takeda, H., Iwasaki, S., and Sato, Z.,** Antifungal activity of the microtubule inhibitor rhizoxin against the benzimidazole-resistant plant pathogenic fungi, in *Systemic Fungicides and Antifungal Compounds,* Lyr, H., Ed., in press.
42. **Ishii, H.,** Genetics of fungicide resistance, in *Abstr. 4th Int. Mycol. Congr.,* 1990, 285.
43. **Takahashi, M., Kobayashi, H., and Iwasaki, S.,** Rhizoxin resistant mutants with an altered β-tubulin gene in *Aspergillus nidulans, Mol. Gen. Genet.,* 220, 53, 1989.
44. **Takahashi, M., Matsumoto, S., Iwasaki, S., and Yahara, I.,** Molecular basis for determining the sensitivity of eucaryotes to the antimitotic drug rhizoxin, *Mol. Gen. Genet.,* 222, 169, 1990.

Chapter 4

TARGET SITES OF FUNGICIDES WITH PRIMARY EFFECTS ON LIPID PEROXIDATION

W. Edlich and H. Lyr

TABLE OF CONTENTS

I. EFFECTS OF DICARBOXIMIDE FUNGICIDES ON SENSITIVE FUNGI

While the biochemical principles of various other modern fungicides had been elucidated, the mode of action of dicarboximide fungicides (DCOFs) remain unresolved for a long time. The group of DCOFs contains hydantoines (iprodione), oxazolidindiones (chlozolinate, vinclozolin), and succinimides (metomeclam, procymidone) (Figure 1). A 3,5-dichlorophenyl group is a highly preserved structural element in all DCOF structures. A similar lack of biochemical information was also apparent for a group of compounds named "aromatic hydrocarbon fungicides" (AHFs). This group of active compounds had been classified according to their biological properties rather than chemical structures. Since they could not be fit into any known scheme, Lyr[1] described AHFs as compounds acting in an unconventional manner. Some typical AHF structures are shown in Figure 2; more detailed information can be obtained from recent review articles.[2,3] In contrast to DCOFs, the AHFs mainly represent substituted aromatic compounds with the exception of etridiazole. However, the aromatic part of AHFs is, similar to the DCOFs, responsible for the antifungal activity.

Similarities between AHFs and DCOFs are also apparent with respect to morphological effects caused by representatives from both groups in sensitive fungi. Moreover, resistant fungal populations are generally cross-resistant to both DCOFs and AHFs.[4-7] Although exceptions have been described, the similarities between the two groups of fungicides were discussed as an indication for a common mode of action.[8] DCOFs have only minor effects on a number of cellular processes such as cell division,[9] biosynthesis of RNA, DNA, and proteins,[10,11] cell wall synthesis,[12] and lipid metabolism.[13,14] On the other hand, morphological studies have shown that treated fungal hyphae abruptly collapse or burst. These drastic effects indicated a complex mode of action sequence with impact on several metabolic events and cell compartments. Pronounced ultrastructural changes, in particular in mitochondria and the endoplasmic reticulum, have been reported for both DCOFs and AHFs.[3,15] Initially, mitochondria appeared swollen dependent on the fungicide concentration, and then cristae became dissolved and vacuolized.

DCOFs and AHFs are similar with regard to their biological activities. The intrinsic activity against a number of phytopathogenic fungi correlates significantly when members from both groups are compared. For example, the correlation coefficient for antifungal activities of vinclozolin and chloroneb was 0.91, and 0.88 for vinclozolin and PCNB. In contrast, the correlation coefficient was significantly lower when vinclozolin and triarimol were compared.[1] Not surprisingly, the physiological and biochemical effects are also very similar.[3,15] Both classes affect main metabolic pathways only to a small degree, but several intracellular structures (mitochondria, endoplasmic reticulum, nucleus) in target cells are altered.

Similar morphological effects have been reported to occur in mitochondria of heart muscle cells under oxidative stress conditions,[16] for example, after treatment with adriamycin or paraquat.[17,18] During the authors' investigations on the mode of action of DCOFs and AHFs, they observed that the content of intracellular lipid peroxides was in good correlation with the fungicide concentration when *Botrytis cinerea* and *Mucor mucedo* were treated with representatives from each group.[15] In these studies the contents of lipid peroxides were determined via quantitative measurements of malon dialdehyde, which is generated during the disintegration of cyclic 5-ring peroxides.[19] They concluded that lipid peroxidation is involved in the mode of action of DCOFs and AHFs, and that lipid peroxidation is initiated in particular in the inner mitochondrial membrane, which is rich in unsaturated fatty acids. The stimulation of lipid peroxidation is not a general effect of fungicide action and is specific to DCOFs and AHFs. Other antifungal inhibitors such as aldimorph, tridemorph, dichlofluanid, and metalaxyl had no effects on lipid peroxidations at concentrations sufficient to inhibit growth.

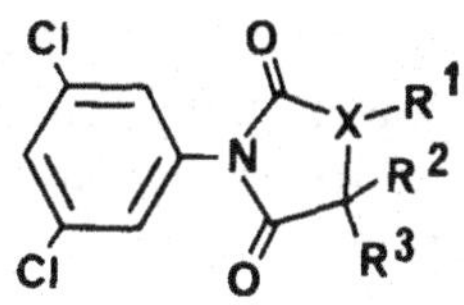

Common name	X	R1	R2	R3
Dimethachlor	CH	H	H	H
Metomeclam	CH	H	CH_2OCH_3	H
Procymidone	C	CH_3	$-CH_2-$*	CH_3
Dichlozoline	O	-	CH_3	CH_3
Myclozolin	O	-	CH_2OCH_3	CH_3
Chlozolinate	O	-	$CO-OC_2H_5$	CH_3
Vinclozolin	O	-	$CH=CH_2$	CH_3
Iprodione	N	$CONHCH(CH_3)_2$	H	H

* 3,4-cyclopropane

FIGURE 1. Structures of dicarboxímide fungicides.

FIGURE 2. Structures of aromatic hydrocarbon fungicides.

II. UNCONTROLLED LIPID PEROXIDATION: A PATHOLOGICAL EVENT

In general, the oxidation of polyunsaturated fatty acids is initiated by the abstration of a hydrogen atom. Thereby, a free radical is generated, which can be converted to a peroxy radical upon reaction with molecular oxygen within the cell. If continued, the process becomes a cascade reaction via permanent dismutations of hydroperoxides. A "catastrophic cycle" is thus initiated by a single initiation reaction, and whole segments of membranes can become structurally and functionally damaged. Moreover, the reaction products of fatty acid peroxidation originating from the disintegration of free peroxy radicals, primarily alkanales, alkenales, or aliphatic ketones,[20] are cytotoxic compounds. In particular, *trans*-4-hydroxy-2-octenal and *trans*-4-hydroxy-2 pentenal are known to be of very high toxici-

TABLE 1
Increase of Malon Dialdehyde Levels in Mycelium of *Mucor mucedo* Caused by Fungicides

Compound	Conc (mg/l)	Relative fresh masse (%)	Relative content of malon dialdehyde (%)
Untreated control	—	100	100
Aldimorph	100	97	80
Biphenyl	10	32	270[a]
Chloroneb	2	55	210[a]
Dicloran	1	54	250[a]
Etridiazole	10	54	220[a]
Iprodione	5	48	245[a]
Metalaxyl	10	92	98
Myclozolin	5	55	275[a]
PCNB	10	27	436[a]
Procymidone	5	63	215[a]
Tridemorph	100	88	127
Vinclozolin	5	46	336[a]
Zineb	50	78	85

[a] Significantly different from an untreated control.

ty.[21-23] Both peroxidation products persist in cells longer than the active oxygen specimens or most free radicals. Consequently, they can diffuse from their site of origin to other sites of action and, thus, have to be regarded extremely destructive.[20]

A prerequisite for the peroxidation of polyunsaturated fatty acids in living cells is either the presence of an active enzyme (lipoxygenase in mammals and higher plants), or a more general oxidative stress situation caused by free radicals or by reduced oxygen specimens. Superoxide anions, the main intermediates in oxygen activation, are generated by autoxidation of hydroquinones, leukoflavins, catecholamines, thiols, ferredoxins, or hemoproteins. An alternative route of active oxygen generation is a single electron transfer during the catalytic action of enzymes such as xanthine oxidase, several flavoprotein dehydrogenases, as well as some oxidases and hydroxylases.[24] Superoxide production appears to comprise the obligatory and initial event in the creation of oxidative stress conditions in living cells.[25] Superoxide radicals originating from superoxide are unstable in aqueous solutions and are spontaneously converted to hydroxyl radicals. As a complicating factor, active oxygen specimens are interconvertable, either catalyzed by enzymes (superoxide dismutase) or alternative catalysts such as complexed ferrous ions. Precise analytical identification is thus hampered and difficult. The generation of free organic radicals proceeds most likely in an identical way through hydrogen abstraction, either during the contact of organic molecules with active oxygen or catalyzed by electron-tranferring enzymes such as NADPH-dependent cytochrome c reductase.

Living cells are affected by oxidative damage whenever the aggressive oxidative agents are not inactivated, or when uncontrolled oxygen activation is not prevented. Radicals play an important role in the entire field of stress physiology and comprise primary events in aging (as it relates to energy-rich irradiation of living cells), in the metabolism of xenobiotics, and some other pathological events. Free radicals attack oxidatively labile compounds such as polyunsaturated fatty acids, but they also affect cell constituents with low chemical reactivity (lignin, cell wall components, proteins, nucleic acids, etc.). Since radicals have short half-lives and thus cannot diffuse over long distances, damages to biological membranes are restricted to small regions surrounding the sites of radical origin.[26] Some membrane-bound enzymes are inactivated directly (oxidation of thiol groups) or indirectly through conformational changes caused by damages within membrane domains surrounding the enzyme. Membrane damage can lead to the release of lytic enzymes, as reported for proteases by Halliwell and Gutteridge.[27] Highly reactive oxygen specimens are also known to cause single chain breaks and other damages in DNA molecules,[28] as well as chromosome alterations.[29]

III. INDUCTION OF ACTIVE OXYGEN GENERATION BY FUNGICIDES

The induction of radical or active oxygen formation represents a toxic principle that might be well suited to be utilized as the mode of action of antimicrobial agents. In nature, such mechanisms play a role in the defense against microorganisms by mammals (e.g., protection against inflammations by blood neutrophils[30,31]) and plants (e.g., hypersensitive reactions of host plants[32,33]). Some herbicides (diquat, paraquat, phenylethers, etc.) act as cytotoxicants via the induction of radical reactions after enzymatic conversion to free radicals by flavin enzymes.[18,34] Several anticancer drugs (adriamycin, etc.) are activated in a similar way.[17,35] In contrast, other compounds are not directly activated to radicals, but they stimulate the activation of oxygen. Overproduction of reduced oxygen specimens can lead to a breakdown of the oxidative protection system, and these highly reactive agents will attack a variety of cell components. The utilization of such reactions for selective damage of microorganisms is attractive since the the toxic mode of action is predominant only in the interaction between the target organism and the active compound. Because of the absence of suitable targets or for other reasons, these active compounds can be less toxic to nontarget organisms, and a high degree of selectivity might be achieved. Recent results have shown that DCOFs and AHFs mediate an increase of active oxygen specimens, which is responsible for the peroxidation of phospholipids described above.

The various species of fungi have a variety of different enzymes active in oxygen reduction. Wood-decomposing fungi in particular produce superoxide, supporting the extracellular biodegradation of lignin and other cell wall components. For example, in *Phanerochaete chrysosporium* a fatty acid CoA-oxidase generates superoxide.[36] In *B. cinerea,* the major part of superoxide released by the infection hyphae in order to mazerate the infected host tissue is derived from sugar oxidases (glucose-1-oxidase, xylose oxidase, galactose oxidase), and some sugar oxidases are significantly stimulated by DCOFs. The induction of lipid peroxidation by DCOFs and AHFs is most likely not caused via a direct conversion of these fungicides to free organic radicals. Rather, these compounds seem to mediate oxidative damage of several cell compartments by a stimulation of oxygen activation. The DCOF-treated cells counteract this event by a significant increase in catalase activity (Figure 3). Apparently, the increased levels of superoxide are similar to those found by Natvig[38] in studies with other fungal organisms.

The concentration of activated molecules that induce lipid oxidation is normally lower than that required for the inhibition of mycelial growth. As long as the capacity of the antioxidative defense system warrants adequate protection, it will be impossible to induce cytotoxic effects by increasing the concentration of lipid peroxides until a certain threshold level is reached. For example, phospholipids containing peroxidized fatty acids can be repaired by the action of phospholipase A_2. In case of a too rapid buildup of oxidative stress, however, the oxygen-inactivating enzymes will be overrun and oxidative thresholds will be surpassed. Undisturbed propagation of lipid hydroperoxides takes place under these conditions. Consequently, structural disintegration of membranes, loss of enzyme activities, release of membrane-bound enzymes, and membrane fluidity changes will occur. Most of the morphological effects of DCOFs described in the literature can be explained by these oxidative events.

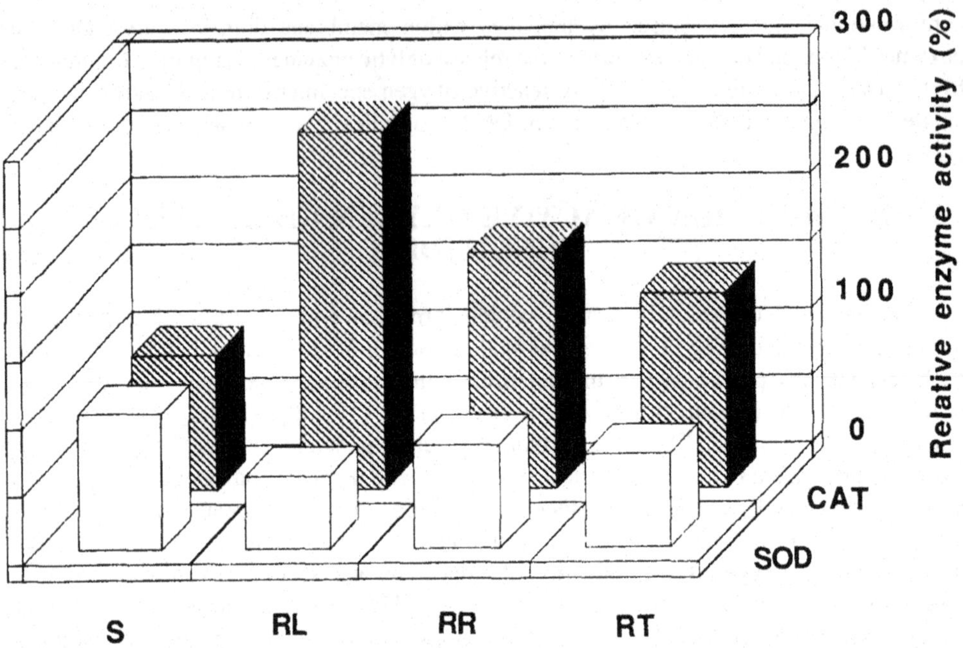

FIGURE 3. Relative activities of catalase (CAT) and superoxide dismutase (SOD) in homogenates derived from sensitive and DCOF-resistant strains of *Botrytis cinerea*. S, sensitive wild-type strain; RL, highly resistant laboratory mutant; RR, medium resistant field isolate*; RT, rarely found resistant field isolate*. *Isolates obtained from Dr. Lorenz, BASF, Limburgerhof, FRG.

Strong evidence for the accuracy of the above hypothesis is derived from the fact that antioxidants decrease or even nullify the oxidative stress imposed on DCOF-treated fungal cells. The antioxidant most widely used is α-tocopherol, a very effective antagonist of lipid peroxidation. Due to its hydrophobicity and chemical structure α-tocopherol is readily incorporated into membranes (e.g., in mammals), where it acts as a protective agent. Each molecule of α-tocopherol is able to prevent the oxidative attack of 200 fatty acid molecules.[39] The presence of α-tocopherol in fungi has not been demonstrated yet.[40] However, Hughes and Tove[41] reported the presence of several compounds similar to α-tocopherol. The fact that the inhibition of spore germination and hyphal growth by DCOFs as well as the morphological alterations are minimized or nullified by α-tocopherol (Table 2) represents strong evidence for the decisive participation of oxidative reactions in the fungicidal mode of action of these compounds. In addition to α-tocopherol, other antioxidants such as β-carotene, propylgallate, phenylthiourea, and tris (hydroxymethyl)-amino methane had similar DCOF-antagonizing effects (Table 2).

IV. INTERACTION OF DICARBOXIMIDE FUNGICIDES AND RELATED COMPOUNDS WITH FLAVIN ENZYMES

DCOF fungicides have no significant effect on major metabolic pathways in fungi, but they clearly affect NADPH-dependent cytochrome c reductases. This enzyme is membrane-bound in *B. cinerea* and *M. mucedo*.[42] All of the AHFs and DCOFs tested (chloroneb, etridiazole, PCNB, iprodione, myclozolin, procymidone, vinclozolin) inhibit the enzymatic cytochrome c reduction in cell-free systems derived from both fungi, with K_i values <10 mM in most cases (Figure 4). Originally, NADPH-dependent cytochrome c reductases were defined as cytochrome P-450-reducing enzymes. More recent studies, however, revealed that they have important additional roles,[43] for example, in the metabolism of xenobiotics,[44]

TABLE 2
Effect of Antioxidants Action of Dicarboximide Fungicides in Cultures of *Botrytis cinerea*

Antioxidant	Concentration	Growth inhibition (mg/l) (%)	Theoretical value for growth inhibition (%)
Untreated control	—	0	—
Without compound	—	32	32
Ascorbic acid	500	38[a]	68
β-Carotin	10	40[a]	76
2,5-Dimethyl furane	250	69	78
Dimethyl sulfoxime	1000	13	19
Ethanol	1000	9	16
Propylgallate	500	55[a]	76
Histidin	250	41	42
Luminol	500	59[a]	77
N,N-Dimethyl-*p*-phenylendiamine	500	55[a]	86
p-Phenylendiamine	500	36[a]	69
1-Phenyl-2-thiourea	250	39[a]	73
α-Tocopherol acetate	250	15[a]	35
	500	2[a]	33
Trihydroxyaminoethane	500	41[a]	72

Note: Growth was measured as diameter of colonies on agar plates; the medium contained 0.5 mg/l vinclozolin. Calculation of the theoretical value of growth inhibition was done according to Colby.[80]

[a] Significantly different from theoretical value.

reduction of organic nitro compounds,[45,46] heme degradation,[47,48] and ubiquinone biosynthesis.[49] It is feasible to suggest that other functions still remain unknown, and thus it is not surprising that the physiological role of the enzyme in *B. cinerea* has not yet been elucidated. Similar uncertainties exist with regard to the number of DCOF target sites. For example, the inhibition of cytochrome c reductase by DCOFs leads to a simultaneous stimulation of NADPH oxidation.[50] Since the respective studies have been performed with only partially purified extracts derived from *B. cinerea,* it remains unknown whether the inhibitors interact with only one or several enzymes. Regardless of this open question, however, evidence exists that the inhibition of the NADPH-oxidizing enzymes contributes to the fungicidal mode of action. For example, the DCOF-induced NADPH oxidation described above is reduced in the presence of piperonyl butoxide and SKF 525A, both known inhibitors of mammalian cytochrome P-450 monooxygenases. This *in vitro* effect is paralleled by an antagonistic effect of these compounds on mycelial growth inhibition by DCOFs,[51] indicating a decisive role of this phenomenon in the fungicidal action.

Although the inhibition of cytochrome c reductase in *B. cinerea* is unambiguous, studies with commercial sources of flavin-containing enzymes revealed that both DCOFs and AHFs can also affect other members of this enzyme class (Figure 5). These effects are enzyme-specific. Whereas vinclozolin stimulated the activities of ascorbate oxidase and glucose-1-oxidase, it strongly inhibited both cytochrome c reductase and D-amino acid oxidase, with minimum inhibitory concentrations in the range of 10^{-5} *M*. It is possible that the interaction of the inhibitors with different flavin enzymes could have additional effects on fungal metabolism and, thus, could account for fungicidal side effects.

It is known from studies with mammalian systems that NADPH-dependent cytochrome c reductase plays an important role in the generation of active oxygen specimens.[52,53] The

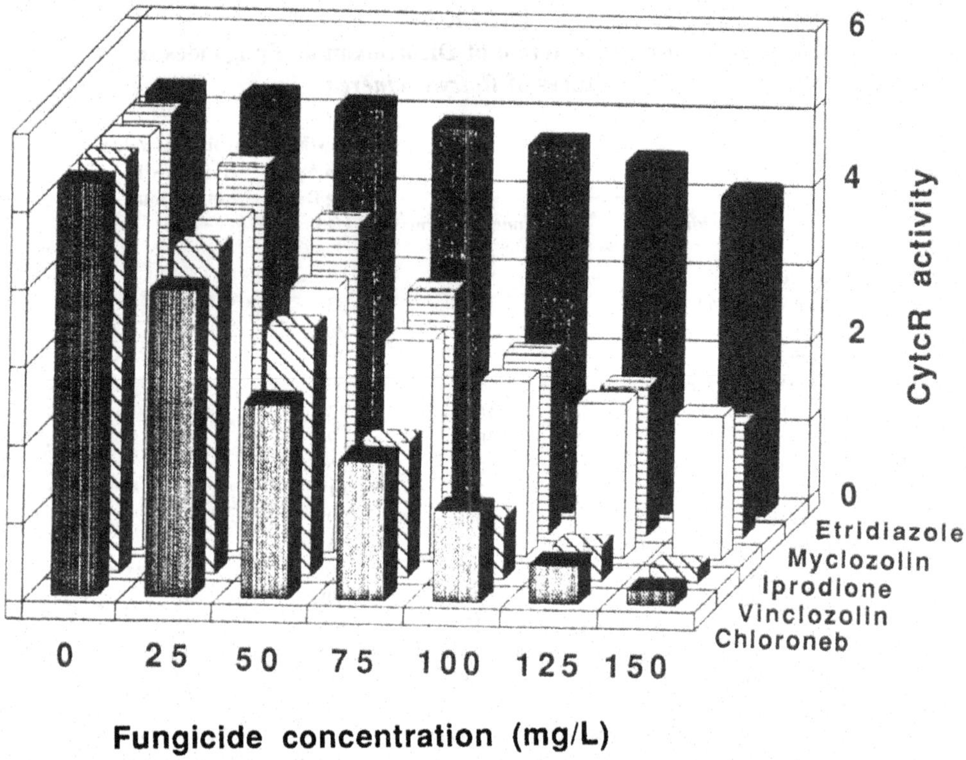

FIGURE 4. Inhibition of NAD(P)H-dependent cytochrome c reductase (CytcR) in homogenates from *Botrytis cinerea*.

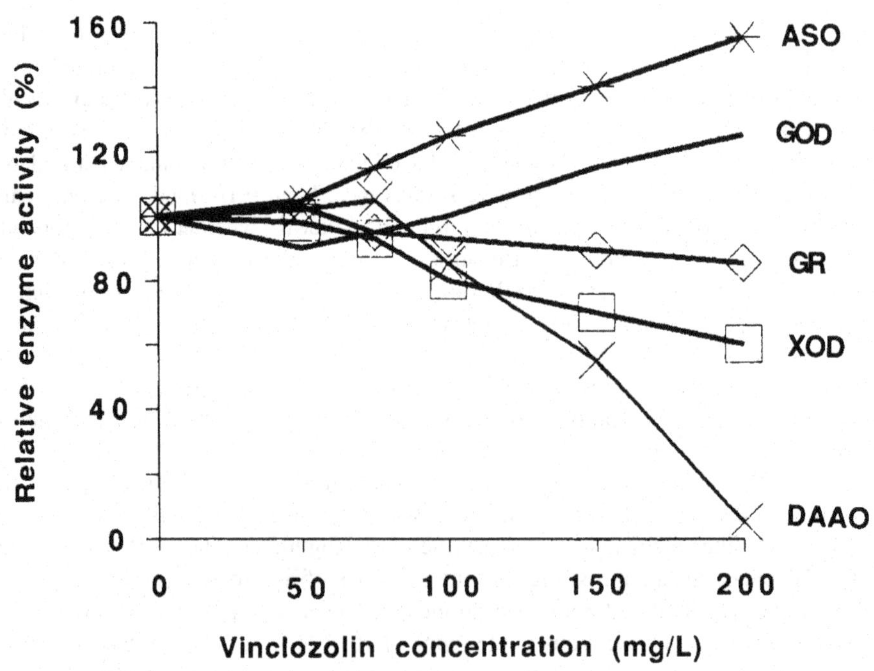

FIGURE 5. Effect of vinclozoline on the relative activities of some flavin-containing enzymes (commercial sources). XOD, xantine oxidase; GR, oxidized glutathione reductase; (GOD, glucose-1-oxidase; ASO, ascorbate oxidase; DAAO, D-amino acid oxidase.

reductase produces under aerobic conditions hydrogen peroxide in cooperation with a quinone and, in an one-step electron transfer, hydroxyl radicals that dismutate to superoxide.[54] Reactions of the FENTON-type are possibly involved in the radical formation via the reduction of complexed ferrous ions.[55] In addition, free radicals of NAD+ and NADP+ can be generated.[56] The efficacy of oxygen reduction is determined by the availability of alternative electron acceptors. Substrate analogues that are only slowly or not at all converted by flavin enzymes have been shown to increase the rate of oxygen activation under certain conditions.[52] These substrate analogues probably cause conformational changes of the enzymes that hinder the optimal orientation of NAD(P)H required for flavin reduction. Such a model would explain why DCOFs and AHFs significantly stimulate the enzymatic oxidation of NAD(P)H. Since the binding regions of some flavin enzymes appear to be similar, it is possible that several enzymes interact with the same inhibitors. Alternatively, the binding regions could not be identical, and the inhibitors would interact primarily with one enzyme. In addition, the affinity of the affected enzyme to its native electron acceptor might be changed rendering oxygen to become the favored substrate.[57]

The 3,5-dichlorophenyl feature of DCOFs appears mandatory for an efficient interaction of the inhibitor with the flavine enzyme. Takayama and Fujinami[58] reported that the antifungal activities of 1-(3,5-dichlorophenyl)-2,5-pyrolidindiones and 3-(3,5-dichlorophenyl)-2,4-oxazolidindiones against *B. cinerea* are critically influenced by only small modifications of the imido ring substitution. These findings mirror the empirical experience with all DCOFs in commercial use. Compounds of the AHF group such as tolclofos methyl and dicloran also contain a 3,5-dihalogenated aromate, although some such as chloroneb depart from the structural rule. We found a relatively strong binding of [R-^{14}C]-vinclozoline to glucose-1-oxidase. The labeled inhibitor could be partly displaced with unlabeled chloroneb (as well as iprodione) when applied in 100-fold access. This result supports the hypothesis that both compounds interact with the same or a very similar inhibitor binding site and explains the striking similarities with regard to the mode of action, the biological efficacy, and the fungicidal spectrum, in spite of the structural dissimilarities.

We also have found that some other 3,5-dihalogenated phenyl compounds were active as fungicides (Table 3). The halogenation pattern rather than the substitution in 1-position had a strong impact on the fungicidal activity against *B. cinerea*. All active compounds tested induced morphological changes in treated mycelia similar to DCOFs, with the exception of the thiocyanic acid phenyl esters. In this case, a number of metabolic pathways were affected in addition to a stimulation of lipid peroxidation. A laboratory strain of *B. cinerea* with high levels of vinclozolin resistance was cross-resistant to all of the 3,5-dihalogenated aromatic compounds tested. The resistance factors, however, were variable, with the thiocyanic and cyanic acid phenyl esters almost equally active with both the sensitive and the resistant strain. In summary, all of the findings support the importance of the 3,5-dihalogenated phenyl ring for the effective interaction of the inhibitors with its target.

V. REASONS FOR SELECTIVE INHIBITOR ACTION

Although oxidative stress conditions mediated by AHFs and DCOFs lead to the pathological effects observed in susceptible fungal cells, these inhibitors are not generally toxic to microorganisms, higher plants, and mammals (Table 4). To the contrary, both AHFs and DCOFs are rather selective fungicides. The reason for this high degree of selectivity remains largely unknown. Several explanations are possible:

1. The presence and distribution of target sites are specific for certain fungal organisms.
2. The binding affinity of the inhibitor to the target site is different in different fungi.
3. The intracellular inhibitor concentrations are different, depending on the specifics of uptake, metabolism, and efflux patterns.

TABLE 3
Antifungal Activity of Some Substituted Aromates Against *Botrytis cinerea*

R_1	R_2	R_3	R_4	R_5	R_6	ED_{50} (mg/l) S	ED_{50} (mg/l) R
SCN	H	Br	NH_2	Br	H	0.1	2
CN	H	Br	NH_2	Br	H	0.2	30
NO_2	H	Br	NH_2	Br	H	2	70
CH_3	H	Br	NH_2	Br	H	0.9	45
CO_2H	H	Br	NH_2	Br	H	30	70
CO_2H	H	Cl	NH_2	Cl	H	10	100
SCN	H	Br	H	Br	H	0.1	0.8
CN	H	Br	H	Br	H	0.4	35
NO_2	H	Br	H	Br	H	0.8	20
CH_3	H	Br	H	Br	H	50	25
CO_2	H	Br	H	Br	H	5	50
NO_2	H	Cl	H	Cl	H	4	55
NH_2	H	Br	H	Br	H	2.5	50
NH_2	H	Cl	H	Cl	H	7	50
SCN	H	Cl	H	Cl	H	0.4	1
CN	H	Cl	H	Cl	H	3.5	80

S — dicarboximide-sensitive wild strain
R — highly DCOF-resistant laboratory strain
Antifungal activity is expressed as ED_{50} value of growth inhibition in agar plate tests.
All compounds were synthesized by Dr. H. J. Brückner, Institute for Plant Protection Research, Kleinmachnow.

4. The capacity of the defense systems to counteract oxidative stress and subsequent events is different.

A comparison of vinclozolin effects on a DCOF-sensitive *(B. cinerea)* and an insensitive fungal species *(Phytophthora infestans)* revealed that a strong stimulation of lipid peroxidation was only observed with *B. cinerea*. The respective inhibitor effect was very small in the case of *P.infestans*. In contrast, etridiazole and chloroneb increased the levels of lipid peroxides in both organisms, accompanied by the inhibition of fungal growth. Although a low level of unsaturated fatty acids in membranes might serve as an explanation for the insensitivity of some fungi (e.g., *Fusarium oxysporum*), this phenomenon is not generally the cause for AHF and DCOF selectivity. For example, Dorobek and Lyr[59] reported that the levels of polyunsaturated fatty acids in mitochondrial membranes of *P. infestans* were even higher than that in the DCOF-sensitive *B. cinerea*. Although not all potential causes of selective inhibitor action have been investigated, the presence of respective target sites and variable inhibitor binding affinities appears to be most important. The final clarification of these phenomena will comprise the subject of future studies.

VI. PROTECTION AGAINST OXIDATIVE STRESS AND MECHANISM OF RESISTANCE TO DICARBOXIMIDES

The effective protection against active oxygen specimens, which are continuously generated during normal metabolism, is a matter of survival for all aerobic organisms. These oxidative agents are highly reactive and thus represent a permanent danger for the precisely interconnected and tightly regulated metabolic reactions. Any uncontrolled accumulation of

TABLE 4
Comparison of Antifungal Activities and Mammalian Toxicity of Some AHFs and DCOFs

	Chlo.	Dicl.	Etri.	Tolc.	Ipro.	Vinc.
			Antifungal Activity			
ED_{50} (mg/l)						
Botrytis cinerea	5	1	20	1	0.5	1
Fusarium oxysporum	500	100	70	100	25	30
Mucor mucedo	3	0.6	8	20	8	10
Rhizoctonia solani	15	100	70	0.1	15	20
Sclerotinia sclerot.	5	2	10	1.5	0.5	1
			Toxicity			
Acute oral LD_{50} (mg/kg)	11,000	4,040	1,100	n.d.[a]	3,500	10,000
Acute dermal LD_{50} (mg/kg)	5,000	5,000	Irritations	n.d.	1,000	Slow irritations
Fish toxicity LC_{50} (mg/l)	4,200	n.d.	7.5	2.1	6.7	26
Bird toxicity LD_{50} (mg/kg)	5,000	n.d.	n.d.	5,000	n.d.	n.d.

[a] No data available.

Abbreviations: Chlo. — Chloroneb, Dicl. — Dicloran, Etri. — Etridiazole, Tolc. — Tolclophos methyl, Ipro. — Iprodione, Vinc. — Vinclozolin

active oxygen can lead to irreversible damages, and all aerobic organisms have evolved a complex oxygen disactivation system containing both enzymatic and chemical components (Figure 6). Most of the respective enzymes are induced under oxidative stress conditions. Thus, a certain level of stress tolerance can be achieved, which is, in particular, important for facultative aerobic organisms.[38,60] Catalase, superoxide dismutase, and several peroxidases represent the central enzymatic elements of the protective system. Lipid peroxidation is suppressed by the combined action of catalase and superoxide dismutase.[61]

Under physiological conditions the basal levels of these enzymes are sufficient to protect polyunsaturated acids from peroxidative destruction.[61-63] However, the enzymes may be synthesized or activated whenever higher levels of activities are required.[64-67] This type of regulatory mechanism implies, however, that the protective machinery can be overrun under sudden oxidative stress situations, and even the protective enzymes themselves can be oxidatively destroyed. Normally, catalase acts as protective enzyme by scavenging hydroxyl radicals and by degrading lipid hydroperoxides, besides its function in hydrogen peroxide cleavage.[68,69] Superoxide dismutase detoxifies mainly the highly reactive superoxide radicals. Both glutathione and glutathione-dependent peroxidases are probably integrative parts of the fungal protective system. It is known that oxidative stress leads to an increased formation of mixed disulfides composed of proteins and glutathione. Apparently, the protection of thiol groups from attack by reduced oxygen is of crucial importance.[70-72]

The presence of all protective components described above has been proven for *B. cinerea* and *Mucor* spp.[50,72] Since the DCOFs and AHFs mode of action is based on the generation of active oxygen specimens, it can be speculated that changes of the protective system toward higher efficacy would lead to resistance. Indeed, alterations of catalase, phenol peroxidase, and superoxide dismutase could be observed for several DCOF-resistant strains of *B. cinerea*. In particular, the significantly increased levels of catalase in a highly

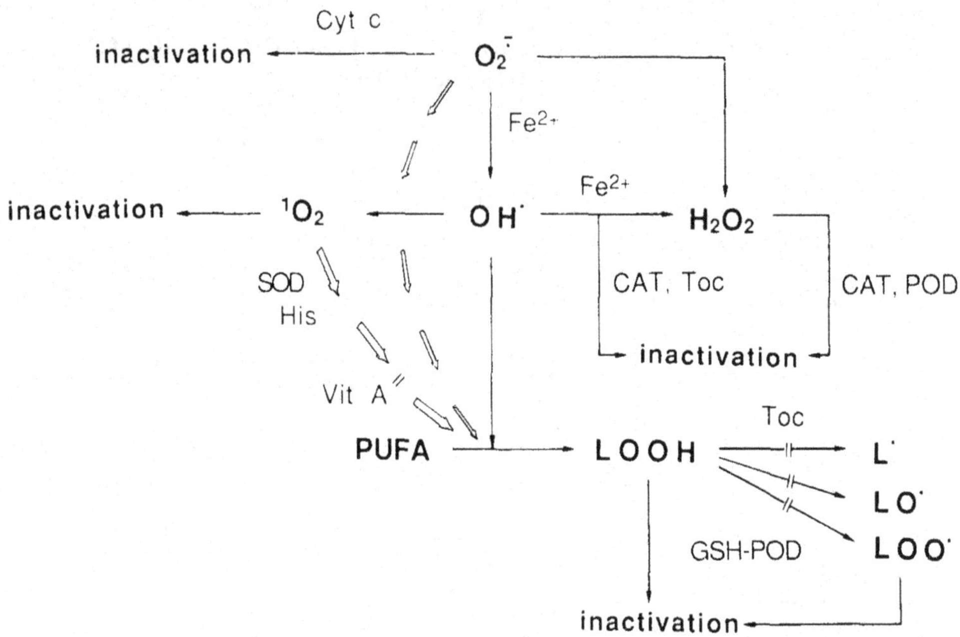

FIGURE 6. Components of the protective system against active oxygen. Cytc, cytochrome c; SOD, superoxide dismutase; HIS, histidine; CAT, catalase; TOC, α-tocopherole; POD, peroxidase; PUFA, polyunsaturated fatty acids; LOOH, lipid hydroperoxide; GSH-POD, oxidized glutathione reductase.

resistant isolate could account for resistance. However, a conclusive correlation between levels of such enzymes and the degree of resistance has not been found in a study employing 23 DCOF-resistant strains of *B. cinerea* (field isolates and laboratory mutants).[73] In some isolates, cytochrome c reductase and NADPH oxidase activities were decreased when compared to the sensitive wild-type strain. However, again there was no correlation between these factors and the degree of resistance. A loss of the target site can be excluded as a cause for resistance since cytochrome c reductase is an essential enzyme for *B. cinerea.* Most likely, a complex of several factors is responsible for the expression of resistance to this class of fungicides. Alterations of protective enzymes and altered affinities of the inhibitors to their targets appear to play a role, in addition to other factors such as the content and localization of polyunsaturated fatty acids, the content of radical scavengers and antioxidants, and the efficacy of repair mechanisms. Thus far, most of these potential causes for resistance have not been the subject of detailed investigations. However, considering the diversity of the system that protects the fungal cell from oxidative stress, and considering the great genetic variability of fungal species like *B. cinerea,* it appears feasible to assume that a variety of factors contribute to the mechanisms of resistance found in field populations. Based on our current experience, both the total number and the efficacy of these resistance factors can vary from strain to strain. The development of practical resistance has been reported for some of the diseases controlled with DCOFs,[74] and a clearer understanding of the molecular events responsible for resistance is urgently needed.

VII. LIPID PEROXIDATION AS A SECONDARY EFFECT OF OTHER FUNGICIDES

Recent studies have demonstrated that lipid peroxidation also occurs under the action of fungicides unrelated to DCOFs and AHFs. In *Mucor mucedo* and *Ustilago maydis,* sterol demethylation inhibitors (see Chapter 7) caused a concentration-dependent increase of lipid peroxides, although the doses required were higher than those reported for the inhibition of sterol demethylation.[75] However, the *M. mucedo*, the level of peroxides correlated well with the inhibition of growth based on mycelium dry weight.[15,76] This mode of action was confirmed for several DMIs such as prochloraz and other fungi such as *Pseudocercosporella herpotrichoides*.[77] Chloroneb and DCOF-resistant strains of *M. mucedo* were not cross-resistant to the sterol demethylation inhibitors, indicating that the primary mechanism of growth inhibition is different from the DCOF mode of action. However, scavengers such as α-tocopherol and β-carotene reduced the growth inhibition caused by sterol demethylation inhibitors to a certain extent and decreased the content of lipid peroxides to a normal level. This indirect evidence indicates that lipid peroxidation is at least contributing to the growth inhibition caused by demethylation inhibitors.[77]

Additional fungicides for which lipid peroxidation has been reported as part of their modes of action are draxolon *(M. mucedo)*[78] and dimethoporph *(Phytophthora megasperma)*.[79] The latter effect has been confirmed for *P.infestans*.[82] Again, no cross-resistance of these compounds to DCOFs and AHFs could be observed, and the relative importance of lipid peroxidation as part of their modes of action remains to be elucidated.

In summary, lipid peroxidation has been observed in the interaction of fungi with several groups of unrelated fungicides. However, the primary events that initiate these reactions appear to be quite different from each other and are thus far poorly understood. Presumably, several modes of action can be operative at fungicide concentrations higher than those used for practical control, and in contrast to AHFs and DCOFs lipid peroxidation appears to comprise a secondary effect not related to the primary mode of action. Nevertheless, these secondary effects can contribute to the fungicidal action.

REFERENCES

1. **Lyr, H.,** Mechanism of action of fungicides, *Plant Dis.*, 1, 239, 1977.
2. **Lyr, H.,** Aromatic hydrocarbon fungicides, in *Modern Selective Fungicides — Properties, Application and Mechanism of Action*, Lyr, H., Ed., Gustav Fischer Verlag, Jena, 1987, 63.
3. **Lyr, H.,** Mechanism of action of aromatic hydrocarbon fungicides, in *Modern Selective Fungicides — Properties, Application and Mechanism of Action*, Lyr, H., Ed., Gustav Fischer Verlag, Jena, 1987, 75.
4. **Lyr, H. and Bischoff, G.,** Zum Problem der Resistenzbildung von Pilzen gegenüber modernen Fungiziden, *Biol. Rundsch.*, 18, 365, 1980.
5. **Leroux, P. and Besselat, B.,** Pourriture grise: la resistance aux fungicides de *Botrytis cinerea, Phytoma Def. Cult.*, 359, 25, 1984.
6. **Gullino, M. L., Romano, M. L., and Garibaldi, A.,** Characterization of dicarboximide-resistant strains of *Botrytis cinerea* Pers. naturally occurring in Italy, *Meded. Landbouwwet. Rijksuniv. Gent*, 47, 781, 1982.
7. **Leroux, P. and Fritz, R.,** Antifungal activities of dicarboximides and aromatic hydrocarbons and resistance to these fungicides, in *Mode of Action of Antifungal Agents*, Trici, A. P. J. and Ryley, J. F., Eds., Cambridge University Press, Cambridge, 1984, 207.
8. **Beever, R. E. and Byrde, R. J. W.,** Resistance to dicarboximide fungicides, in *Fungicide Resistance in Crop Protection*, Dekker, J. and Georgopoulos, S. G., Eds., Centre for Agricultural Publishing and Documentation, Wageningen, 1982, 101.
9. **Georgopoulos, S. G., Sarris, M., and Ziogas, B.,** Mitotic instability in *Aspergillus nidulans* caused by the fungicides iprodione, procymidone and vinclozolin, *Pestic. Sci.*, 10, 389, 1979.
10. **Fritz, R., Leroux, P., and Gredt, M.,** Mechanisms de l'action fongitoxique de la procymidone de la vinclozolin et du dicloran sur *Botrytis cinerea, Phytopathol. Z.*, 90, 152, 1977.
11. **Pappas, A. C. and Fisher, D. J.,** A comparison of the mechanism of action of vinclozolin, procymidone, iprodione and prochloraz against *Botrytis cinerea, Pestic. Sci.*, 10, 239, 1979.
12. **Albert, G.,** Sphaeroblastenbildung bei *Botrytis cinerea*, hervorgerufen durch Vinclozolin, *Z. Pflanzenkr. Pflanzenschutz*, 88, 337, 1981.
13. **Buchenauer, H.,** Preliminary studies on the mode of action of vinclozolin, *Meded. Fac. Landbouwwet. Rijksuniv. Gent*, 41, 1509, 1976.

14. **Hisada, Y. and Kawase, Y.,** Morphological studies on antifungal action of N-(3′,5′-dichlorophenyl)-1,2-dimethyl-cyclopropane-1,2-dicarboximide on *Botrytis cinerea, Ann. Phytopathol. Soc. Jpn.*, 43, 151, 1977.
15. **Edlich, W. and Lyr, H.,** Mechanism of action of dicarboximide fungicides, in *Modern Selective Fungicides — Properties, Application and Mechanism of Action,* Lyr, H., Ed., Gustav Fischer Verlag, Jena, 1987, 107.
16. **Flamigni, F., Guanieri, C., Toni, R., and Caldera, C. M.,** Effect of oxygen radicals on heart mitochondrial function in α-tocopherol deficient rabbits, *Int. J. Vitam. Nutr. Res.*, 52, 402, 1982.
17. **Mimnaugh, E. G., Trush, M. A., Ginsburg, E., Hirokata, Y., and Gram, T. E.,** The effects of adriamycin in vitro and in vivo on hepatic microsomal drug metabolizing enzymes, *Toxicol. Appl. Pharmacol.*, 61, 313, 1981.
18. **Ueda, S., Hayashi, T., and Namiki, M.,** Effect of ascorbic acid on lipid autoxidation in a model food system, *Agric. Biol. Chem.*, 50, 1, 1986.
19. **Slater, T. F.,** Overview of methods used for detecting lipid peroxidation, *Methods Enzymol.*, 105, 283, 1984.
20. **Poli, G., Dianzani, M. U., Cheeseman, K. H., Slater, T. F., Lang, J., and Esterbauer, H.,** Separation and characterization of the aldehydic produce of lipid peroxidation stimulated by carbon tetrachloride or ADP iron in isolated rat hepatocytes and rat liver microsomal suspensions, *Biochem. J.*, 227, 629, 1985.
21. **Pryor, N. A.,** *Free Radicals in Biology,* Academic Press, New York, 1977.
22. **Dianzani, M. U.,** Biochemical effectors of saturated and unsaturated aldehydes, in *Free Radicals, Lipid Peroxidation and Cancer,* McBrian, D. C. H. and Slater, T. F., Eds., Academic Press, New York, 1982.
23. **Esterbauer, H.,** Aldehydic products of lipid peroxidation, in *Free Radicals, Lipid Peroxidation and Cancer,* McBrian, D. C. H. and Slater, T. F., Eds., Academic Press, London, 1982.
24. **Fridovich, I.,** Superoxide dismutases, *Annu. Rev. Biochem.*, 44, 147, 1975.
25. **Holtzman, J. L.,** Role of reactive oxygen and metabolite binding in drug toxicity, *Life Sci.*, 30, 1, 1982.
26. **Werner, P.,** Zum Wirkungsmechanismus des systemischen Fungizids Chloroneb und zu den möglichen Ursachen erzeugter Resistenz gegenüber Mucor mucedo (L.) Fres., Dissertation, Martin Luther University, Halle Wittenberg, 1979.
27. **Halliwell, B. and Gutteridge, J. M. C.,** Role of iron on oxygen radical reactions, *Methods Enzymol.*, 105, 47, 1984.
28. **Mello Filho, A. C., Hoffmann, M. E., and Meneghini, R.,** Cell killing and DNA damage by hydrogen peroxide are mediated by intracellular iron, *Biochem. J.*, 218, 273, 1984.
29. **Hassan, H. M. and Moody, C. S.,** Determination of the mutagenicity of oxygen free radicals using microbial systems, *Methods Enzymol.*, 105, 254, 1984.
30. **Green, T. R. and Wu, D. E.,** Purification and resolution of NADH diaphorase activity from NADPH diaphorase-linked/oxygen oxidoreductase activity from human neutrophils, *Biochim. Biophys. Acta,* 831, 74, 1985.
31. **Griffin, B. W. and Ashley, P. L.,** Evidence for a radical mechanism of halogenation of monochlorodimedone catalyzed by chloroperoxidase, *Arch. Biochem. Biophys.*, 233, 188, 1984.
32. **Doke, N.,** Involvement of superoxide anion generation in the hypersensitive response of potato tuber tissues to infection with an incompatible race of *Phytophthora infestans* and to the hyphal wall components, *Physiol. Plant Pathol.*, 23, 345, 1983.
33. **Doke, N.,** Generation of superoxide anion by potato tuber protoplasts during the hypersensitive responses to hyphal wall components of *Phytophthora infestans* and specific inhibition of the reaction by suppressors of hypersensitivity, *Physiol. Plant Pathol.*, 23, 359, 1983.
34. **Klimek, J., Schaap, A. P., and Kimura, R.,** Effect of paraquat on cytochrome P-450-dependent lipid peroxidation in bovine adrenal cortex mitochondria, *Biochim. Biophys. Acta,* 752, 127, 1983.
35. **Bhyan, D. K., Bhyan, P. K., and Bhyan, K. C.,** Adriamycin-induced cataract: involvement of a free radical mechanism, in *Oxygen Radicals in Chemistry and Biology,* Bors, W., Saran, M., and Tait, D., Eds., Walter de Gruyter, Berlin, 1984.
36. **Green, R. V. and Gould, J. M.,** Substrate-induced hydrogen peroxide production in mycelia from lignin-degrading fungus *Phanerochaete chrysosporium, Biochem. Biophys. Res. Commun.*, 117, 275, 1983.
37. **Edlich, W., Lorenz, G., Lyr, H., Nega, E., and Pommer, E.-H.,** New aspects on the infection mechanism of *Botrytis cinerea* Pers., *Neth. J. Plant Pathol.*, 95(Suppl. 1), 53, 1989.
38. **Natvig, D. O.,** Cooperative biochemistry of oxygen toxicity in lactic acid-forming aquatic fungi, *Arch. Microbiol.*, 132, 107, 1982.
39. **Fukuzawa, K., Takase, S., and Tsukatani, H.,** The effect of concentration on the antioxidant effectiveness of α-tocopherole in lipid peroxidation induced by superoxide free radicals, *Arch. Biochem. Biophys.*, 240, 117, 1985.
40. **Taketomi, H., Soda, K., and Katsui, G.,** Results of screening test in tocopherols in microbial realm, *Vitamins,* 57, 133, 1983.
41. **Hughes, P. E. and Tove, S. B.,** Occurrence of α-tocopherolquinone and α-tocopherolquinol in microorganisms, *J. Bacteriol.*, 15, 1397, 1982.

42. **Edlich, W. and Lyr, H.,** Effectors of various flavin containing enzymes initiating a lipid peroxidation in fungi, *Hoppe Seyler 2. Physiol. Chem.*, 367, 257, 1986.
43. **Winston, G. W. and Cederbaum, A. I.,** NADPH-dependent production of oxy radicals by purified components of the rat liver mixed function oxidase system, *J. Biol. Chem.*, 258, 1514, 1983.
44. **Miwa, G. T., West, S. B., Walsh, J. S., and Lu, A. Y. H.,** Drug residue formation from ronidazole, a 5-nitroimidazole, *Chem. Biol. Interact.*, 41, 297, 1982.
45. **Harada, N. and Omura, T.,** Participation of cytochrome P-450 in the reduction of nitro compounds by rat liver microsomes, *J. Biochem.*, 87, 1539, 1980.
46. **West. S. B., Wislocki, P. G., Wolf, F. J., and Lu, A. Y. N.,** Drug residue formation from ronidazole, a 5-nitroimidazole, *Chem. Biol. Interact.*, 41, 281, 1982.
47. **Komiyama, T., Sanada, M. T., Kobashi, K., and Yoshimoto, A.,** Enhanced production of ethylene from methional by iron chelates and heme containing proteins in the system consisting of quinone compounds and NADPH-cytochrome P-450 reductase, *Biochem. Pharmacol.*, 34, 977, 1985.
48. **Yoshinaga, T.,** A comparative study of heme degradation by NADPH-cytochrome c reductase alone and by the complex heme oxygenase system, *J. Biol. Chem.*, 257, 7794, 1982.
49. **Terriere, C., Giordano, G., Haddock, B., and Azoulay, E.,** Hydroxylation of 2-octaphenylphenol in *Escherichia coli* K-12, *Biochem. Biophys. Res. Commun.*, 111, 830, 1983.
50. **Edlich, W., Lorenz, G., Pommer, E.-H., and Lyr, H.,** Studies on mode of action of vinclozolin and other dicarboximide fungicides, in preparation.
51. **Sisler, H.,** Dicarboximide fungicides: mechanism of action and resistance, in *Fungicide Resistance in North America*, Delp, C. J., Ed., APS Press, St. Paul, 1988, 52.
52. **Aust, S. D., Roerig, D. L., and Pederson, T. C.,** Evidence for superoxide generation by NADPH-cytochrome c reductase of rat liver microsomes, *Biochem. Biophys. Res. Commun.*, 47, 1133, 1972.
53. **Boesterling, B. and Trudell, J. R.,** Spin trap evidence for production of superoxide radical anion by purified NADPH-cytochrome P-450 reductase, *Biochem. Biophys. Res. Commun.* 93, 569, 1984.
54. **Komiyama, T., Kickuchi, T., and Sugiura, V.,** Generation of hydroxyl radical by anticancer quinone drugs, carbazilquinone, mitomycin c, aclacinomycin A and adriamycin, in the presence of NADPH-cytochrome P-450 reductase, *Biochem. Pharmacol.*, 31, 3651, 1982.
55. **Krikun, G. and Cederbaum, A. I.,** Evaluation of microsomal pathway of oxidation of alcohols and hydroxyl radical scavenging agents with carbon monoxide and cobalt protoporphyrin. IX, *Biochem. Pharmacol.*, 34, 2929, 1985.
56. **Avigliano, L., Carelli, V., Casini, A., Finazzi-Agro, A., and Liberatore, F.,** Oxidation of NAD dimers by horseradish peroxidases, *Biochem. J.*, 226, 391, 1985.
57. **Massey, V., Palmer, G., and Ballou, D.,** On the reaction of reduced flavins and flavoproteins with molecular oxygen, in *Flavins and Flavoproteins*, Slater, E. C., Ed., BBA Library 3, 1966.
58. **Takayama, C. and Fujinami, A.,** Quantitative structure-activity relationships of antifungal N-phenylsuccimides and N-phenyl-1,2-dimethylcyclopropane dicarboximides, *Pestic. Biochem. Physiol.*, 12, 163, 1979.
59. **Dorobek, F. and Lyr, H.,** Determination of unsaturated fatty acids in mitochondria from various fungal species with lipoxygenase, *Tagungsber. Akad. Landwirtschaftswiss. DDR*, 253, 345, 1987.
60. **Friedovich, I.,** Overview: biological sources of superoxide, *Methods Enzymol.*, 105, 1984.
61. **Girotti, A. W. and Thomas, J. P.,** Superoxide and hydrogen peroxide-dependent lipid peroxidation in intact and triton-dispersed erythrocyte membranes, *Biochem. Biophys. Res. Commun.*, 118, 474, 1984.
62. **Van Sluys, M.-A., Oliveira, R. C., Felzenszwalb, I., and Menck, C. F. M.,** Ascorbate copper induced DNA lesions and repair in *Escherichia coli* K12, *Carcinogen*, 7, 197, 1986.
63. **Thomas, C. E., Morehouse, L. A., and Aust, S. D.,** Ferritin and superoxide-dependent lipid peroxidation, *J. Biol. Chem.*, 260, 3275, 1985.
64. **Kulakova, S. M. and Gogotov, I. N.,** Wirkung von Sauerstoff und Substraten auf Wachstum und die Aktivität von Superoxiddismutase und Katalasen in Mikroorgansimen, *Mikrobiologiva*, 51, 21, 1982.
65. **Meir, E. and Yagil, E.,** Further characterization of the two catalases in *Escherichia coli*, *Curr. Microbiol.*, 12, 315, 1985.
66. **Allgod, G. S. and Perry, J. J.,** Paraquat toxicity as effect of hydrogen peroxide on thermophilic bacteria, *J. Free Radicals Biol. Med.*, 1, 233, 1985.
67. **Nations, C., Allen, R. G., Farmer, K. J., Toy, P. L., and Sohal, R. S.,** Superoxide dismutase activity during the plasmodial life cycle of *Fusarium polycephalum*, *Experimentia*, 42, 64, 1986.
68. **Edwards, J. C. and Quinn, P. J.,** Decomposing potassium peroxychromate produces hydroxyl radical that can peroxidize the unsaturated fatty acids of phospholipids dispersions, *J. Lipid Res.*, 23, 994, 1982.
69. **Gutteridge, J. M. C., Beard, A. P. C., and Quinlan, G. J.,** Superoxide dependent lipid peroxidation: problems with the use of catalase as specific probe for fenton-derived hydroxyl radicals, *Biochem. Biophys. Res. Commun.*, 117, 901, 1983.
70. **Flohe, L. and Günzler, W. A.,** Glutathione: metabolism and function, in *Kroc Foundation Series*, Vol. 6, Arias, I. M. and Jacoby, W. B., Eds., Raven Press, New York, 1976.

71. **Beloqui, O. and Cederbaum, A. I.,** Prevention of microsomal production of hydroxyl radicals, but not lipid peroxidation by the glutathione-glutathione peroxidase system, *Biochem. Pharmacol.*, 35, 2663, 1986.
72. **Aisaka, K., Uwajima, T., and Terada, O.,** Purification and properties of glutathion peroxidase (EC 1.11.1.9) from *Mucor hiemalis*, *Agric. Biol. Chem.*, 47, 1107, 1983.
73. **Edlich, W., Lyr, H., and Nega, E.,** Some aspects for the biochemical reasons of dicarboximide resistance in *Botrytis cinerea*, in preparation.
74. **Pommer, E.-H. and Lorenz, G.,** Dicarboximide fungicides, in *Modern Selective Fungicides — Properties, Application and Mechanism of Action*, Lyr, H., Ed., Gustav Fischer Verlag, Jena, 1987, 91.
75. **Lyr, H. and Edlich, W.,** Lipid peroxidation, a consequence of the mechanism of action of aromatic hydrocarbon and dicarboximide fungicides and a side effect in DMI fungicides, *Brighton Crop Prot. Conf.*, 1, 879, 1986.
76. **Lyr, H.,** Lipid peroxidation: a side effect of sterol demethylation inhibitor fungicides in *Mucor mucedo* (L.) Fres. and *Ustilago maydis*, *Pestic Biochem. Physiol.*, 32, 197, 1988.
77. **Lyr, H. and Müller, H. M.,** Induction of lipid peroxidation, a secondary mechanism of action in DMI and other fungicides?, *Tagungsber. Akad. Landwirtschaftsmiss. DDR*, in press.
78. **Lyr, H. and Casperson, G.,** Anomalous cell wall synthesis in *Mucor mucedo* (L.) Fres. induced by some fungicides and other compunds related to the problem of dimorphismus, *Z. Allg. Mikrobiol.*, 22, 245, 1982.
79. **Kuhn, P. J.,** Studies on the antifungal activity in vitro of dimethomorph, a novel fungicide active against downy mildews and Phytophthora, *Tagungsber. Akad. Landwirtschaftswiss. DDR*, in press.
80. **Colby, S. R.,** Calculating synergistic and antagonistic responses of herbicide combinations, *Weeds*, 15, 20, 1967.
81. **Perkow, W.,** *Wirksubstanzen der Pflanzenschutz- und Schädlingsbekämpfungsmittel*, Verlag Paul Parey, Berlin, 1985.
82. **Edlich, W. and Lyr, H.,** unpublished results.

Chapter 5

TARGET SITES OF FUNGICIDES TO CONTROL OOMYCETES

J. M. Griffith, A. J. Davis, and B. R. Grant

TABLE OF CONTENTS

I. WHAT ARE THE OOMYCETES?

A. INTRODUCTION

The taxon Oomycota (Oomycetes) contains some very important pathogens of plants and animals. Perhaps the most infamous of these belong to the family Pythiaceae, and particularly to the genera *Pythium* and *Phytophthora,* many of which are extremely destructive pathogens of plants. The economic importance of the Pythiaceae, together with the ease with which they may be cultured, has resulted in their being studied more intensively than other Oomycetes. However, the Pythiaceae have characteristics which are not shared by other Oomycete families, e.g., the absence of a functional biosynthetic pathway for sterols. Therefore, it cannot be assumed that the Pythiaceae are typical members of the Oomycetes and that conclusions based on studies of the family are valid for the entire group.

When considering the topic for this chapter, together with some of the problems which have occurred in the provision of lasting chemical control of this group of microorganisms, the authors felt it essential to devote a significant space to reviewing recent developments in the unraveling of the evolutionary history of the Oomycetes. It is now clear that although these organisms are traditionally referred to as fungi, their evolutionary history and therefore their genomic composition, as well as their biochemistry, set them apart from most other organisms in this kingdom. Although there may be merit in retaining the Oomycetes within a general grouping of microorganisms which possess a mycelial stage,[1] failure to appreciate their true evolutionary relationships can only retard progress toward discovering aspects of their biochemistry and biology, which will assist in their control.

The Oomycetes are "fungi" in the sense that they are heterotrophic, eukaryotic organisms which obtain their food by absorption, that they possess a mycelial stage, and that they produce spores by asexual and sexual means.[2-6] Until very recently they were placed in the kindgom Fungi, as a class in the division Mastigomycotina. Other members of this class were the Chitridiomycetes, the Hyphochytriomycetes, and the Plasmodiophoromycetes.[5-10] All of these organisms produce flagellated asexual spores, termed zoospores (mastigo = a whip), and at some stage in their life cycle most require water, hence the colloquial names "zoosporic fungi" or "water molds". Each class within the Mastigomycetes was separated on the basis of its types of flagella.[11] The Oomycetes have a tinsel and a whiplash flagellum of unequal length. Some taxonomists have also included the Myxomycetes (slime molds) and the Zygomycetes within the Mastigomycetes, and, as such, this group of organisms has also been known as the Phycomycetes or "lower fungi".[8-11] The summary of some of the various classification systems and terminology used to 1978 is given by Whittaker and Margulis.[12]

B. MODERN CLASSIFICATION SCHEMES

Classification schemes for many groups of organisms are currently the subject of drastic revisions. Advances in techniques for studying the ultrastructure of cells, in molecular biology and computer technology have enabled comparisons to be made between organisms at a level that could not have been envisaged a few years ago.[13-26] This is particularly true for those organisms for which there is a poor fossil record, such as the fungi.[27-36] The schemes in use at the moment are particularly confusing to the nonspecialist reader, since different research groups not only suggest that there should be different numbers of kingdoms and phyla within these kingdoms, but use different names for these taxa.[37-41] It seems likely that a unified classification scheme, embodying all living organisms, is still some years away.

Recent evidence, which will be presented in this review, has convinced most biologists that the Oomycetes should be placed in the kingdom Protista (also termed the Protoctista).[42-45] This is despite the fact that the Oomycetes have developed a fungal-like nutrition and mycelial stage in their life cycle. There is, however, no consensus of opinion on the position of the Oomycetes within this kingdom. Corliss[39] has continued to use the term "Mastigomycetes" as a "superphylum" (division) and has placed the Oomycetes in the phylum Oomycota, as have Margulis and Schwartz,[40] although they have abandoned the concept of a division. In contrast, Cavalier-Smith[46,47] has suggested that the Oomycetes be separated from other protozoa into a new subkingdom, the Chromophyta (later renamed the Chromista), of which they would form a subphylum, the Pseudofungi.

C. EVIDENCE THAT THE OOMYCETES ARE NOT "TRUE FUNGI"

1. Morphology

The precise relationship of the Oomycetes to other fungi and to the heterokont algae, which they resemble in many respects, has been debated for over a century.[11] The Oomycetes appear to be quite distinct both from the higher fungi and from the other Mastigomycetes.[11,29,34,35,42,48] The life cycle of the Oomycetes is primarily diploid, while that of the higher fungi is haploid.[22,49-51] Oomycete sexual reproduction is oogamous and is unique in its lack of a motile stage.[11,51] Their mitochondrial christae are tubular, while those of the higher fungi are flattened.[11,33,37] Certain structures of the zoospore (flagellar apparatus, the presence of K-bodies,[31] and the flagella rootlet morphology) show similarities to heterokont algae (the Chrysophaceae, the Phaeophyceae, and the Xanthophyceae).[13,15,18,29,52]

2. Cell Walls

The mycelia of Oomycetes are "coenocytic", that is, they are multinucleate and lack cross-walls.[40,43] Their cell walls contain hydroxyproline[53-57] and are predominantly composed of β-1,3- and β-1,6-glucans and cellulose rather than chitin.[58-62] The presence of cellulose has been considered to indicate algal ancestry,[11,59-61,63,64] but the discovery of Oomycetes with both cellulose and chitin,[65-67] and the discovery of cellulose in certain ascomycetes and chitin in certain algae, have led Barr[11] to suggest that cell wall structure may be a much more variable characteristic than previously thought.

3. Metabolic Pathways

The main storage product in Oomycetes is not glycogen, but mycolaminarin, a soluble β-1,3-glucan which is similar to the laminarins found in several groups of algae.[64] Acyclic polyols (sugar alcohols) are widely distributed throughout higher fungi, yet they are absent from the Oomycetes.[68-70] The Oomycetes do not contain the granules of long chain polyphosphate ("metachromatic granules") found in a wide range of higher fungi.[71] Short chain polyphosphates of up to eight phosphate groups in length have, however, been detected recently in the Oomycete, *Phytophthora palmivora,*[72] and also in other members of the genus.[397]

In addition, the Oomycetes possess several biosynthetic pathways and enzymes with properties which differ from their counterparts in the higher fungi.[42,60,63,73-75] They synthesize lysine by the α,ε-diaminopimelic acid pathway, whereas higher fungi use the α-aminoadipic route. The Oomycetes are able to synthesize D(−)lactic acid. The enzymes for tryptophan biosynthesis, NADP-linked isocitrate dehydrogenases, malate dehydrogenases, and NAD-linked glutamic dehydrogenases, have properties different from those of higher fungi. The Oomycetes also possess a cyanide-insensitive superoxide dismutase, which is typical of prokaryotes and lower eukaryotic algae and not of the higher fungi, which possess a cyanide-sensitive superoxide dismutase.[76]

Oomycetes also synthesize a number of peculiar lipids which are not found in higher fungi, such as γ-linolenic acids, and unusual long chain polyunsaturated fatty acids[63,77-81] and plasmalogens.[82] Phosphonolipids have also been reported in species of *Phytophthora* and *Pythium*.[82-86] These unusual phospholipids are characterized by a C–P bond in place of C–O–P bond. They have not been identified to date in higher fungi, but are commonly found in protozoa, mollusks, and coelenterates.[87-89] It should be noted, however, that Niere et al.[72] were unable to detect phosphonolipids in *Phytophthora* spp. by ^{31}P-NMR. This anomaly has not yet been explained. Recently, Lhomme et al.[90] have identified an unusual inositol sphingophospholipid from *Phytophthora capsici*; the major fatty acid in this lipid (4-hydroxy-2-docosenoic acid) has not been reported in higher fungi.

4. Sterol Metabolism

The sterol metabolism of the Oomycetes also differs dramatically from that of the higher fungi.[51,91-95] Several laboratories[93,95-97] have shown that the Oomycetes may be divided into two groups: those which are able to synthesize sterols *de novo* from mevalonate (the Saprolegniales, the Leptomitales, and the Lagenidiales) and those which are not (the Pythiaceae and some other individual species belonging to other families, e.g., *Lagenidium giganteum*). The most common sterols found in the sterol-synthesizing Oomycetes have been 24-methylene cholesterol and fucosterol. A number of other sterols, including ergosterol, have also been found in varying amounts.[91-96] The anomalous presence of ergosterol in some strains of Oomycetes is now thought to result from contamination from other fungal spores or from media containing ergosterol.[92,93] Oomycetes have the ability to scavenge exogenous sterols; a novel lipoprotein which binds exogenous sterols has been isolated from these organisms.[98] The anomalous presence of ergosterol in the organism described as "*Zoophagus insidians*"[99] has recently been resolved by its excision from the Oomycetes.[100,101]

Sterol biosynthesis has been considered to be a useful phylogenic marker: photosynthetic organisms produce cycloartenol as their first polycyclic intermediate, whereas nonphotosynthetic organisms produce lanosterol.[102-104] The higher fungi have retained the lanosterol pathway. Elucidation of the pathway for sterol biosynthesis in sterol-synthesizing Oomycetes was thus of considerable interest. Work by Bu'Lock and Osagie[105] and Warner and colleagues[106-108] suggested that the Oomycetes metabolized cycloartenol in preference to lanosterol, although cycloartenol synthesis itself was not shown. These workers suggested that sterol biosynthesis most likely occurred via the cycloartenol pathway. Subsequently, Nes et al.[109] and Berg et al.[110,111] showed that sterol biosynthesis in the Oomycetes occurred by the lanosterol pathway. They found that although Oomycetes were capable of metabolizing cycloartenol, there was no evidence that they could synthesize it. Based on these observations, they suggested that the Oomycetes were derived from a nonphotosynthetic, lanosterol-synthesizing prokaryote,[109] rather than a photosynthetic algal (eukaryote) ancestor. This suggestion, however, is at variance with the phylogenic trees which have subsequently been derived using rRNA sequence analysis (discussed below).[112] Further elucidation of the sterol biosynthetic pathway has shown that the substrate for the C-24 transalkylation is zymosterol, rather than lanosterol.[95]

In the 1960s, a number of laboratories reported that the addition of sterols to basal media stimulated the vegetative growth of *Phytophthora* spp. and *Pythium* spp. and were required for reproduction.[51,91-95,102] A variety of triterpenoids, which have sterol-like properties, also stimulated growth, but specific sterols were required to induce oospore production.[92,113-115] The effect of sterols (and triterpenoids) on vegetative growth is thought to be the result of membrane stabilization and an increase in the rate of hyphal growth.[51,95,115-117] The requirement for added sterols for sexual reproduction is believed to result from their role as precursors for the synthesis of steroid hormones, analogous to the oogoniols and antheridiol in *Achyla*.[118-120] Nes and colleagues have shown that the metabolism of sterols involved in sexual reproduction in *Phytophthora cactorum* differs significantly from that in *Achyla* and *Saprolegnia*.[121] They have suggested that the Oomycetes are of polyphyletic ancestry and may have evolved by convergent, rather than parallel, evolution. Once again, however, this suggestion has not been borne out by recent molecular biology studies, which indicated that the Oomycetes are a monophyletic group.[112]

The Pythiaceae synthesize squalene, but lack squalene epoxidase.[109,122,123] However, they still possess remnants of the sterol biosynthetic pathway which enable them to carry out certain transformations when supplied with exogenous sterols.[92,94,95] The ability of the Pythiaceae to grow in the absence of added sterols is unusual among eukaryotes,[102,103,124] and is believed to be one of the reasons why these organisms have been so difficult to control with many of the fungicides in common use. Species of *Tetrahymena* and some *Mycoplasma* are also able to grow in the absence of sterols. In both of these genera, as with the Pythiaceae, the sterol pathway is not totally absent, and compounds formed from sterol precursors, such as tetrahymanol and carotenol,[102,103] appear to replace the sterols in their structural role in membranes. Certain triterpenoids may even play a similar role in bacteria.[125,126] Recently, long chain fatty alcohols (phytophthorols) have been identified in sterol-requiring *Phytophthora* spp., which may fulfill a similar role.[127-128]

It has been generally accepted that the Pythiaceae require sterol supplementation for sexual reproduction.[51,91-95] However, recent work by Ko[129,130] has suggested that lecithins, in particular those containing dioleoyl-1-α-phosphatidylcholine, can replace sterols in growth media. A similar sterol-sparing effect of phospholipids has been observed in sterol auxotrophs of *Tetrahymena* and *Mycoplasma*.[131,132] Subsequently, Kerwin and Duddles[86] found that trace amounts of sterols in inocula act synergistically with the phospholipid to induce oosporogenesis. Sterols, at levels as low as 1 ng/l, added together with the appropriate phospholipid, significantly increased oosporogenesis. Other evidence suggested that the phospholipid itself was hydrolyzed in the growth media and only the fatty acid moiety was taken up by the organism. Phospholipids in the absence of sterols were not able to induce oospore formation. Kerwin and Duddles therefore concluded that oospore induction showed an absolute requirement for sterols, albeit at very low levels. They postulated that at these levels, sterols acted as "sparkers" with a critical domain role, similar to that for sterols proposed in other fungi.[133] The dramatic enhancement of growth, following the addition of low levels of sterols after a period of sterol-less growth, has been observed by others.[95] Addition of unsaturated fatty acids and their esters has also been shown to enhance growth rates and oospore formation.[134]

5. Molecular Biology

Work in the 1970s on guanosine-cytosine ratios and rRNA sedimentation coefficients indicated that the Oomycetes probably had different phylogenic origins from the higher fungi.[27,42,48,135] However, it was not until the 1980s that nucleic acid sequencing and computer technology were sufficiently advanced to provide a basis for suggesting the evolutionary origins of the Oomycetes. Several phylogenic trees based on macromolecular sequences have been published. Ribosomal RNAs are particularly useful for this purpose as they are found in all organisms and their function is constant. They also possess a number of highly conserved and partially conserved sequences which allow phylogenic relationships between different organisms to be compared. They also have the advantage that they can be sequenced directly. Different methods have been used to analyze the data and the interested reader is referred to reviews by Woese,[136,138] Olsen,[24,137] and Krishnan et al.[139] for a discussion. Phylogenic trees, based on sequences derived from Oomycetes, have been published.[19,20,26,112,139-141] These have been based on 5S, 5.8S, and on the larger 16S rRNA sequences. In general, the phylogenic trees support the conclusions drawn from the biochemical and morphological evidence presented above, and indicate that the Oomycetes are a monophyletic group, more akin to the heterokont (chromophyte) algae (brown algae, golden-brown algae, and diatoms) than to the higher fungi. Hori and Osawa[19] and Krishnan et al.,[139] using 5S rRNA sequences, however, found no close association between the Oomycete *Saprolegnia* and the chromophytes, but the latter workers did find such a relationship based on the 16S rRNA sequence data. This observation supports the suggestion that, in some cases, the larger sequences of 16S rRNA may give more meaningful results.[25] Methods of treatment of the data are also important.[26]

Delineation of both genera and species within the Oomycetes has also proved difficult. The structures of sporangia and oospores, which are traditionally used for taxonomic purposes, are very similar in many species, and, in some cases, oospores have not been found. There is also often a high degree of variability within a species in morphology, cultural and physiological characteristics, pathogenicity, and resistance to antibiotics. Some Oomycetes are also obligate parasites, e.g., the Peronosporaceae (see below). The diploid state of the mycelial stage is believed to be responsible for the immense variation in phenotype observed. This is particularly true of the important plant pathogens, *Phytophthora* and *Pythium*.[49,50,142-149] Within a species, the identification of subgroups for epidemiological purposes, and the terminology used to describe them, are likely to be of increasing importance in the future.[150]

Information, derived from studies of mitochondrial DNA (mtDNA) sequences, has allowed some progress in clarifying the phylogenic relationships between species and genera within the Oomycetes, particularly the Pythiaceae. These studies, however, can only be applied to those organisms which may be readily cultured in the laboratory. The relatively small size of mtDNA has made it an attractive candidate for such studies.[33] The mtDNAs of Oomycetes[151-156] and those of the closely related Hyphochytriomycetes[157] were found to contain a large inverted repeat sequence. McNabb and Klassen[156] found that the organization of the Oomycete mitochondrial genome differed from any other known system and most closely resembled chloroplast DNA. They suggested that the presence of an inverted repeat is an ancestral character of the Oomycetes, and that its absence from *Phytophthora* spp.[158,159] and certain *Apodachlya*[154,156,159] is the result of loss during evolution.

Restriction fragment length polymorphism studies and mapping of DNA molecules have also been used to determine the relationships between species and genera.[160-165] It seems likely that many of the relationships which have been assigned previously will need to be revised.[162,163] Likewise, other new techniques, such as the development of species-specific monoclonal antibodies,[166-170] DNA probes,[171-173] enzyme-linked adsorbent assays (ELISAs),[174,175] and gel electrophoresis of proteins and DNA[176-178] may also help in the assignment of the individual isolate to a particular species.

Sequences for gene coding for certain proteins from Oomycetes also reflect the extent of the evolutionary separation between the Oomycetes and the higher fungi. That for a heat-shock protein from *Bremia lactucae,* Hsp70d, was as similar to the heat-shock protein of yeast as it was to a plant protein (72% and 74%, respectively).[179] The sequence analysis of the *Phytophthora megasperma* gene for actin revealed that this protein differed considerably from all other actin proteins which have been sequenced to date.[180]

II. WHAT MEMBERS OF THE OOMYCETES ARE PATHOGENIC AND WHAT DISEASES DO THEY CAUSE?

There are more than 800 described species of Oomycetes, divided into over 70 different genera.[8-10,43,181,182] They are mainly saprophytic and are found living in fresh water, salt water, and in the soil. Some members, however, are parasites of plants, animals, and microorganisms.

Currently, there are four orders recognized within the class of Oomycetes: the Saprolegniales, the Leptomitales, the Lagenidiales, and the Peronosporales. The Thraustochytiales are now considered not to be members of the class and have been transferred to the Laybrinthulales,[14,29,30] although even this classification may not be satisfactory.[20,35] Recent information is currently challenging the assignment of the four remaining orders. In the past few years, Dick and co-workers have described two species of Oomycetes, isolated in Australia from diseased Kikuyu grass[183] and sugar cane,[184] which do not fit into any of the four orders and which, they believe, will require a major reappraisal of the orders. Since the identity of the different orders within the Oomycetes is in such a state of flux, the authors will continue to use the four traditional names.

A. THE PERONOSPORALES

The order Peronosporales consists of three well-defined families: the Pythiaceae (consisting of seven genera, including *Phytophthora* and *Pythium*);[181,182,185-188] the Peronosporaceae (commonly called the downy mildews and consisting of some ten genera including *Bremia, Peronospora, Plasmopara, Pseudoperonospora,* and *Sclerospora*); and the Albugonaceae (consisting of one genus — the *Albugo* spp. or white rusts).[181,182,189-193] The Peronosporales are mainly plant pathogens, attacking the dicotyledons, although some species do attack monocotyledons.[181,183,194-196] Some species, particularly members of the Peronosporaceae, tend to be obligate parasites of plants.[193] They cause diseases of seedlings and mature plants, attacking leaves, buds, fruits, pods, stems, crowns, and roots. Members of this order are responsible for the destruction of many valuable plant crops throughout the world.[181,197-201] New host plants[194,195,202] and new species[183,203-205] continue to be discovered. Some species attack only one or a very limited number of hosts, whereas others have a very wide range.[206,207] For instance, *Phytophthora vignea* causes stem and root rot of cowpea only, whereas *Phytophthora cinnamomi* is one of the most ubiquitous and destructive plant pathogens known, causing diseases of almost 1000 different species of plants throughout the world.[143,207-209] Some of the Peronosporales are very destructive pathogens and under favorable conditions have been known to give rise to pandemics, resulting in great economic losses. The most notorious member of the order is *Phytophthora infestans,* the cause of late blight of potatoes, which was responsible for the Irish potato famine last century.[210-214] Other pandemics have occurred.[211,212] Toward the end of the last century, downy mildew, caused by *Plasmopara viticola,* spread rapidly through the vineyards of Europe until it was halted by Millardet's discovery of Bordeaux Mixture.[210,212,215] In Europe, pandemics of downy mildew of hops (caused by *Pseudoperonospora humuli*) occurred from 1924 to 1926.[210] There have also been pandemics of blue mold of tobacco (caused by *Peronospora tabacina*), both in Europe in the 1960s[210] and in Canada in 1979.[41] In this century in Australia, *P. cinnamomi* has devastated much of the native flora, particularly the jarrah forests of Western Australia.[209,216]

One member of the Peronosporales, *Pythium insidiosum* (syn. *P. destruens*), causes disease in mammals.[217-219] This disease, known by many local names, is now termed "pythiosis". It is prevalent in tropical and subtropical regions, and has been recorded in horses, dogs, cattle, and recently humans where it is associated with thalassemia;[220] one case of corneal pythiosis has been described.[221] There is some evidence that cats and sheep may also be hosts. The disease presents as large, ulcerated cutaneous and subcutaneous granulomas. In dogs, gastrointestinal lesions may also occur. Secondary lesions involving the lymph nodes and lungs have also been recorded. To date, there is no effective treatment. Surgery plus amphotericin B[218] and immunotherapy have been tried, the latter with some possible benefit. Since amphotericin B binds to membranes which contain sterols,[222] the use of amphotericin B as chemotherapy seems to be of doubtful benefit against the Pythiaceae, which are capable of vegetative growth in the absence of sterols. Its use probably arose from a misunderstanding of the nature of these organisms, and provides an example of how using the term "fungi", without qualification, to describe the Oomycetes can result in the wrong treatment being given.

The Peronosporales are sensitive to many antibiotics which are effective against bacteria.[223] Yet, because of their ability to grow vegetatively in the absence of sterols, they are insensitive to agents which act by interfering with sterol metabolism and which are effective against other fungi (see below). Some Peronosporales are particularly sensitive to streptomycin and this antibiotic has been used to control downy mildew of hops (caused by *Pseudoperonospora humuli*).[182,198] The mode of action is not yet known. It has been suggested that the streptomycin competes for Ca^{2+} ions on the mycelial surface.[224] These authors are not aware of any studies which indicate whether any of these antibiotics are effective in the treatment of pythiosis.

B. THE SAPROLEGNIALES

Most members of the Saprolegniales are saprophytic, but some are parasites of other microorganisms (including other Oomycetes) and algae, higher plants, fungi, and animals.[10,43,182,184,225-227] Diseases of economic importance occur both in the agricultural and in the fishing industries. Certain *Achyla* sp. are pathogens of rice and *Aphanomyces* sp. cause diseases of peas, beans, sugar beets, and radish.[181,228] Species of *Saprolegnia, Achyla,* and *Aphanomyces* are major parasites of fish and their eggs, and are a problem particularly of aquaria and fish hatcheries. In some cases, some of these organisms may not be true primary pathogens, but gain entry to the fish via trauma to the outer skin.[218,229,230] The disease, ulcerative mycosis, which is a major disease of esturine fish in the U.S., is believed to be caused by salt-tolerant species of *Aphanomyces.*[231,232] Other *Saprolegnia* species cause diseases in eggs and larval forms of marine crustaceae.[233,234] *Aphanomyces astaci* has devastated the European freshwater crayfish industry.[235-238] It was probably introduced to Europe from the New World, as the European crayfish is highly susceptible to this organism, while the North American species is resistant. The Japanese and Australian freshwater crayfish are also highly susceptible to *A. astaci,* but there appear to be no records of outbreaks of disease caused by this organism to date in these countries.[239,240] The development of intensive aquaculture of both marine and freshwater species, coupled with breeding programs selecting for rapid growth rates, but ignoring the natural resistance against these organisms in wild-type populations, can be expected to result in future epidemics caused by these organisms requiring the design of new types of fungicide. Some members of the Saprolegniales (*Leptolegnia chapmanii*) and the Lagenidiales (*Lagenidium giganteum*), which parasitize mosquito larvae, are currently being investigated as a means of controlling the spread of malaria.[241-243]

C. THE LEPTOMITALES

The members of the order Leptomitales are known only as saprophytes on plant debris.[10,43,182,244] Although they play a role in recycling plant debris, they are not of economic importance. *Leptomitus lacteus,* an obligate fermentor, is common in sewage-polluted waters.[6,43]

D. THE LAGENIDIALES

Members of the order Lagenidiales are parasitic on other microorganisms (including other zoosporic fungi), higher plants (including pollen grains), and animals (especially arthropods, rotifers, and nematodes).[10,43,182,233,245] *Lagena radicola,* an obligate parasite, infects many mono- and dicotyledons.[181] It is believed to be widely distributed throughout Europe and northern America, and is now recognized as a primary invader of wheat and barley roots.[246] It has been often neglected as it is frequently associated with *Pythium* spp., which it closely resembles. There is also some evidence that it may be a vector for soil-borne viruses.[246] A new marine species of *Petersenia, P. palmariae,* which was isolated on Vancouver Island, Canada, has been recognized as a parasite of the red alga *Palmaria mollis.*[247,248] There is concern that the Oomycete may be introduced into the east coast of Canada where the seaweed is harvested commercially.

III. WHY ARE THE OOMYCETES DIFFICULT TO CONTROL?

The parasitic species within the Oomycetes are particularly successful and have proved difficult to control. As stated in the introduction to this chapter, most studies have been carried out with members of the Pythiaceae because of their continuing economic importance and the relative ease with which they may be cultivated in the laboratory (particularly *Phytophthora*). At present, there is insufficient information to know to what extent observations made on the Pythiaceae can be extrapolated to other members of the Oomycetes.

A. PERSISTENCE

The Oomycetes have developed an extremely varied range of tactics to ensure survival under various environmental conditions.[185,216] Mycelia (possibly protected within plant debris), zoospores, sporangia, chlamydospores, and particularly oospores are all capable of surviving outside the host, though for different periods.[213,216,249-251] *Phytophthora* spp. are able to survive well in soils, particularly water-logged soils which do not favor plant growth and which reduce other competitors.

One of the most characteristic features of the Oomycetes is their ability, with few exceptions,[198] to produce zoospores. The release of zoospores is dependent on the presence of water.[254-259] The ability to produce zoospores is believed to be one of the main reasons why the Peronosporales, in particular, are such successful parasites of plants.[257-259] Zoospores are produced in sporangia (*Aphanomyces* are an exception and form zoospores in the thallus).[252] Some sporangia, particularly those of species which infect the leaves of plants rather than the roots, are deciduous and are capable of being dispersed by wind (e.g., those of *P. infestans, Phytophthora erythroseptica, Peronospora tabacina*). Depending on the environmental conditions, these may either germinate directly or release zoospores.[249,253] The explosive nature of outbreaks of disease, under favorable weather conditions or saturated soils, suggests that these do not result from single infections, but rather are the result of a buildup of secondary inocula, either of sporangia or zoospores.[249] Some species of Saprolegniales produce two types of zoospore, which are produced in succession; some species are capable of repeated cycles of emergence and encystment, so increasing their chances of survival.[260,261]

Most studies of zoospores have been carried out on species of *Phytophthora*. Zoospores of *Phytophthora* are attracted to the plants by chemotaxis.[258,259] Following recognition of specific receptors on the plant surface,[260-267] the zoospores encyst, form a cell wall,[64] produce a germ tube, and initiate infection of the plant. A similar process of attraction and encystment has been observed in *A. astaci* toward its host, the crayfish,[268] and *Pythium insidiosum* to animal hairs.[269] There is some evidence to suggest that zoospores, which have reached a potential site of infection, may be able to attract other zoospores to the area (adelphotaxis) and so increase their infection potential.[270] Under conditions of nutrient deprivation, the cysts, rather than producing a germ tube, may form a microsporangium, from which a single zoospore emerges. Chlamydospores, likewise, can undergo alternative forms of germination, depending on nutrient conditions, so increasing the chances of survival.[216]

While zoospores are capable of swimming only a few centimeters, most are believed to be spread further by water currents or splashes of rain.[249,258,259] There have been several reports of zoospores remaining motile for several days.[258,259] In the soil, the ability of a plant pathogen to move actively for even only a short distance may be sufficient to initiate new infection, and thus provide considerable advantage over other competitors.[249]

B. ABSENCE OF TARGET SITES FOR MANY CONVENTIONAL FUNGICIDES

The peculiar biochemistry of the Oomycetes has meant that they have been able to evade the action of many of the antifungal agents currently in use. Because most Oomycetes lack chitin (see above), they are insensitive to antifungal agents targeted against chitin synthetase, namely, the polyoxins and the nikkomycins, which act by mimicking its substrate UDP-GlcNac.[271] In response to fungal attack, plants themselves synthesize enzymes (chitinases and β-1,3-glucanases) which are targeted at fungal cell walls.[272,275] Since it appears that the activity of both enzymes may be required for fungal inhibition to occur, Oomycetes, whose cell walls lack chitin, are able to evade this defense response.

The Oomycetes are also resistant to the benzimidazoles (benomyl and thiophanate-methyl).[276,277] In the plant, these compounds are converted to carbendazim, which binds to fungal tubulin; this indicates that the structure of Oomycete tubulin is different from that of other fungi (compare with actin, cited above).[180] Another derivative, nocodazole, acts in the same manner, but has a wider range of activity affecting both animal and fungal cells, and also displays activity against Oomycetes.[276] The sterol-auxotrophic Oomycetes (largely members of the Pythiaceae, described above) are able to grow vegetatively in the absence of sterols. They are therefore able to evade the action of the polyene antibiotics (nystatin and amphotericin B), which act by binding to sterols in membranes.[222] This means that these Oomycetes, many of which are plant pathogens, are also resistant to the action of saponin, a natural detergent found in many plants.[278-280] Likewise, the morpholine fungicides (tridemorph, dodemorph, fenpriomorph, and others), and the triazole fungicides and the pyrimidines, all of which interfere with ergosterol biosynthesis, though at different sites, are ineffective against the Pythiaceae.[281-286] The Pythiaceae also lack squalene epoxidase (see above).[121,122] This enzyme is the site of action of the thiocarbamates (tolnaftate and tolciclate) and the allylamines (naftifine and terbinafine), compounds which have been developed successfully for treatment of a number of clinical conditions.[287] As far as the authors of this review are aware, there have been no reports that the Oomycetes are sensitive to compounds which are believed to act by inhibiting phospholipid synthesis (edifenphos, iprobenfos, validacin),[288] nor to the carboxins which interfere with energy metabolism of fungal mitochondria,[289,290] nor to melanin biosynthesis inhibitors, which are involved in appressorium function.[291-294]

C. VARIABILITY

The diploid state of the coenocytic mycelium and the existence of somatic recombination are believed to make the Oomycetes extremely versatile pathogens, able to adapt rapidly to changes in the environment.[149,192,198,295] These properties enable the Oomycetes to accumulate mutations within their genome, in a way not possible in a haploid organism, increasing the possibility of overcoming resistance genes introduced into the genome of a host.

The Oomycetes have proved particularly difficult organisms to study genetically.[149,192,193,295] This has hampered our knowledge of the interactions between the host and the pathogen at the molecular level. Some progress has been made recently in this area,[296-298] particularly with lettuce downy mildew.[179,193] Analysis of rDNA from *Pythium ultimum* has revealed subrepeats which may be of importance in the control of transcription and hence pathogenicity.[299] There is some suggestion that enhanced virulence of *Phytophthora infestans* may be associated with double-stranded RNA.[300]

IV. FUNGICIDES ACTIVE AGAINST THE OOMYCETES

As described in the preceding sections, the peculiar biochemistry of the Oomycetes, the result of the fact that they have a different evolutionary history to the higher fungi, has rendered them insensitive to most of the broad-spectrum fungicides currently available. Prior to the discovery of selective anti-Oomycete agents, disease control was achieved with non-selective biocides such as methyl bromide, a soil sterilant, or protectant foliar fungicides such as mancozeb and metiram.[198,199] It was not until the late 1970s that systemic and selective fungicides were developed specifically for use against the pathogenic Oomycetes.

The subject of selective, systemic fungicides active against the Oomycetes has been reviewed recently by several authors.[198-201,301] Other reviews have concentrated on the mode of action of specific classes of compounds currently in use, particularly the phenylamides,[302] the phosphonates,[303,304] and synergistic mixtures of fungicides.[305,306] Two other recognized classes of compounds in use against the Oomycetes are the carbamates and the cyano-oximes.

The first systemic anti-Oomycete fungicide produced was chloroneb. It proved moderately fungistatic to *Pythium* diseases and still enjoys limited use. Chloroneb was soon followed by the compounds etridiazole and hymexazol. All three are still in use, but will only be discussed briefly here. Chloroneb and hymexazol also provide protection against fungi other than the Oomycetes.

A. HYMEXAZOL

Hymexazol was introduced by the Sankyo company[307] and was first used for the control of *Fusarium* disease in cucumber plants.[308] It is also used for the control of soil-borne diseases caused by *Pythium, Rhizoctonia,* and *Fusarium* spp. on rice seedlings, cucumber, tomato, and beets.[309] Hymexazol is rapidly translocated and metabolized to its *N*- and *O*-glucoside forms. The *O*-glucoside form is the active metabolite. Investigations into its mode of action suggest that it interferes with DNA synthesis.[309]

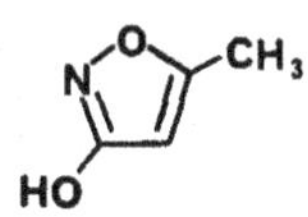

Hymexazol

Hymexazol has been used as an experimental tool to isolate *Phytophthora* spp.[310] from soil and plant samples as it was believed to selectively inhibit *Pythium* and *Mortierella* species. Ho[311] reported that some species of *Phytophthora* were also inhibited by hymexazol and more recently the sensitivities of various Oomycetes to hymexazol have been compared.[312] A relationship between hymexazol sensitivity and the taxonomic grouping with the Oomycetes was shown.

B. ETRIDIAZOLE

Etridiazole is recommended for the control of *Phytophthora* and *Pythium* spp. of cotton, ornamentals, and fruit and vegetable crops. It is used mainly to control fungi in the soil[301] and is considered to be only slightly systemic. In mode of action studies[313] it was demonstrated that etridiazole inhibited the electron transport system between cytochrome b and c. It was also shown that *Pythium* spp. developed tolerance to the fungicide. Inhibition by etridiazole can be circumvented by a ubiquinone shunt bypassing the cytochromes,[314] although endogenous levels of ubiquinone are probably too low for this alternate pathway to reduce the usefulness of the fungicide. Despite the ease with which tolerance to etridiazole developed in the laboratory, no field resistance problems have arisen in a decade of use.[301]

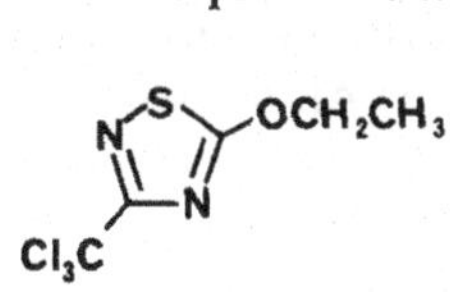

Etridiazole

C. PHENYLAMIDES

The phenylamide fungicides include the acylalanines, butyrolactones, and one representative of the oxazolidinones. These compounds act almost exclusively against the Peronosporales.[302] Metalaxyl is probably the best known acylalanine and most widely studied of the phenylamides. It has a moderately wide spectrum of activity, providing control against species of *Phytophthora, Pythium, Pseudoperenospora, Plasmopara,* and *Peronospora*.

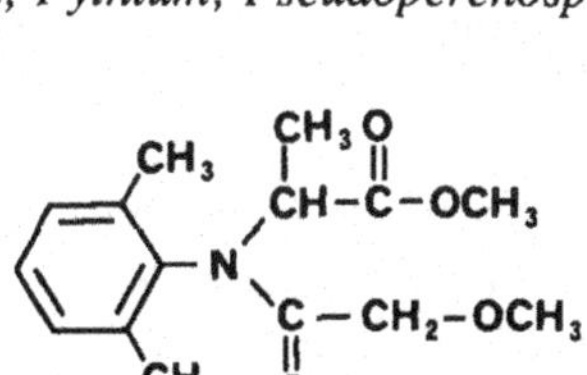

Metalaxyl

Metalaxyl is translocated readily in the xylem and shows good protective activity. It is also effective as a curative, but is not recommended for use in this mode. Resistance to phenylamides has developed with relative ease. Within a year of its release for use on potatoes in the Netherlands, resistance to metalaxyl was reported in *P. infestans* populations[315] and resistance of *Pseudoperonospora cubensis* to metalaxyl evolved within a similar time frame in Israel.[316] The rapid development of widespread resistance to metalaxyl led to a concentrated research effort into its mode of action. It is now considered to act by interfering with the RNA polymerase-I-template complex of sensitive fungi, inhibiting ribosomal RNA synthesis. Resistance is apparently the result of a muation event which leads to a change at the binding site.[317]

In metalaxyl-resistant isolates, the phenylamide binding site has a low affinity for metalaxyl. Currently, workers are trying to identify the phenylamide receptor, and the nature of the binding between it and the phenylamides.[318] This information may allow the design of a phenylamide that will not exert such a strong selection pressure on a single component of the pathogen-fungicide interaction. The mechanism of uptake of metalaxyl and its translocation within the plant have not been studied to the same extent and little information appears to be available on the subject.

Cross-resistance to metalaxyl has been demonstrated for other phenylamides.[319] Benalaxyl, cyprofuram, and oxadixyl were all less effective against a metalaxyl-resistant strain of *Phytophthora megasperma* f. sp. *medicaginis in vitro* than a susceptible isolate.[302] The phenomenon of cross-resistance has led to the conclusion that all of the phenylamides have a similar mode of action. However, the degree of cross-resistance between the different phenylamides is not necessarily correlated and could be explained by alternate structures of the phenylamide binding site.[319] A second site of action has been postulated for two other phenylamides, cyprofuram[320] and benalaxyl,[317] although there appears to be no practical benefit to be derived from this property.[322]

The ease with which pathogens develop resistance to the phenylamides places severe restrictions on their use. In the Netherlands a strategy has been developed whereby potato growers may use metalaxyl mixtures no more than twice in one season, and only in critical situations.[322] More generally, the Fungicide Resistance Action Group (FRAC) has advocated the use of only prepacked mixtures for foliar use, together with a limited number of applications per season, and no curative use and no soil treatments for control of airborne pathogens.[323] For the control of soilborne diseases, seed treatment is also an effective way of directing these fungicides to their target.[324] All of these strategies are designed to reduce selection pressure on the pathogen and check resistance buildup within the pathogen population. Whether they will prove effective, in the long term, against a group of organisms as flexible as the Oomycetes remains to be seen.

D. CARBAMATES

The carbamates play a relatively minor role in Oomycete control. The carbamic acid esters, propamocarb and prothiocarb, were developed for control of *Pythium* and *Phytophthora* species which cause root diseases of ornamentals and vegetables. Schwinn and Urech[199] rate the carbamates as having weak, slow acropetal translocation. Due to their poor translocation, the carbamates are usually applied directly to the target organ, the roots, as a soil drench or root dip, and their use has been limited to the control of disease in labor-intensive, high value crops. The lifetime of the carbamates as chemical controls may be extended as a result of the development of resistance against phenylamides in the Oomycetes. Mixtures of propamocarb and mancozeb are being used on potato crops in Israel where metalaxyl resistance is evident.[325] Although propamocarb was found to have limited curative efficacy in the laboratory, its synergism with mancozeb resulted in effective control of late blight.[325]

$(CH_3)_2N\text{-}(CH_2)_3\text{-}NH\text{-}CO\text{-}O\text{-}CH_2\text{-}C_2H_5$ Propamocarb

$(CH_3)_2N\text{-}(CH_2)_3\text{-}NH\text{-}CO\text{-}S\text{-}C_2H_5$ Prothiocarb

The mode of action of the carbamates is yet to be elucidated. Mode of action studies have been conducted using propamocarb,[326] as prothiocarb is metabolized to ethyl mercaptan after soil application.[301] Ethyl mercaptan itself may be fungitoxic. The carbamates are believed to affect the biosynthesis of a membrane component. Burden et al.[326] have demonstrated an alteration in fatty acid composition in *Pythium* and *Phytophthora* species treated with propamocarb. Previously propamocarb had been shown to induce growth inhibition and membrane leakiness. It might be inferred that the alteration in fatty acid composition is the result of altered acyl groups in the membrane phospholipids, which in turn alters the fluidity of the membranes.

E. CYANO-OXIMES

Cymoxanil, {2-cyano-*N*-[(ethylamino)carbonyl]-2-(methoxyamino)} acetimide, is the only representative of the cyano-oximes that is used as an Oomycete fungicide. It has proved effective against *Phytophthora infestans* attacks on potato and tomato and *Plasmopara viticola* of grape vines[327] and has protectant as well as curative activity. Cymoxanil is translocated acropetally,[327,328] but protectant activity is short-lived (4 to 6 d) as cymoxanil is metabolized quickly in plants.[301] The protection afforded by cymoxanil may be extended by using it in combination with a second fungicide such as mancozeb. Mixtures of cymoxanil with mancozeb, or mancozeb and oxadixyl, significantly extend the period over which 90% efficacy of control is maintained, compared with the use of mancozeb or cymoxanil alone.[329] When used as a part of a three-component mixture with oxadixyl and mancozeb, in the glasshouse, 97% control of *P. infestans* on tomato was achieved 14 days after fungicide application. The next best result obtained was 25% disease control after 9 days, using a cymoxanil/oxadixyl mix.[328] It was also demonstrated that synergism existed in the cymoxanil/oxadixyl/mancozeb combination, whether applied as a single formulation or in its component parts. When applied as three separate components, the synergistic ratio for control of phenylamide-resistant isolates was greater than that observed for phenylamide-sensitive ones. The synergism has been attributed to the interaction between cymoxanil and mancozeb. The mechanism by which the synergism is effected is unknown, as is the mode of action of cymoxanil.

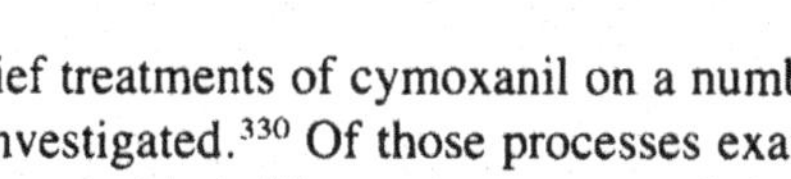

Cymoxanil

The effect of brief treatments of cymoxanil on a number of biochemical pathways in *P. infestans* has been investigated.[330] Of those processes examined, nucleic acid synthesis was found to be the most sensitive. The authors reported that mycelial growth and germ tube formation were also inhibited. The failure of *Peronospora viticola* zoospores to germinate in the presence of the fungicide has also been reported. Where germ tubes were produced after treatment with cymoxanil, they were malformed.[327]

F. PHOSPHONATES

In 1977 aluminum tris-*O*-ethyl phosphonate (fosetyl-Al), trade name "Aliette", was released by Rhone-Poulenc for control of the Oomycetes. It was promoted for use against *Plasmopara viticola* and against *Phytophthora* species responsible for a range of diseases in tropical and temperate crops.[331] Although it was reported at this time that fosetyl Al was quickly degraded to phosphonate (phosphonic acid) in the plant, the initial research into the activity of fosetyl Al appeared to largely ignore this information. The observation that protection of plants continued after fosetyl Al could no longer be found in planta contributed to the controversy which existed over the mode of action of these compounds.[304,332,333] However, it is now accepted that the phosphonate produced by the metabolism of the phosphonate ester provided the protection and it has been shown that potassium phosphonate is as effective as fosetyl Al as a fungicide.[334,335]

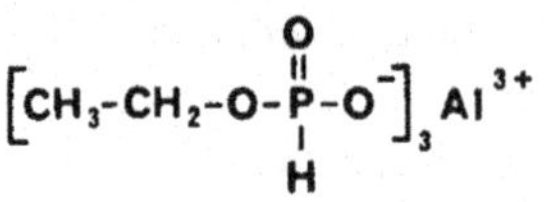

Fosetyl-Al

Phosphonates are the salts or esters of phosphonic acid, which exists in a tautomeric equilibrium with phosphorous acid. The terms phosphonate and phosphite have been used synonymously in recent literature. Organophosphorous compounds containing a carbon-phosphorus bond are also referred to as phosphonates. These compounds have quite different properties from the salts or esters of phosphonic acid. The ambiguity caused by this terminology is still causing confusion and the reader is referred to Guest and Grant[304] for a full discussion. In the context of this chapter, the term phosphonate refers only to the salts of phosphonic acid.

Phosphonates, like most of the fungicides active against the Oomycetes, are highly water soluble. However, unlike most others they can be translocated in both xylem and phloem.[331,336,337] Due to these properties, workers have been able to directly inject aqueous solutions of potassium phosphonate into tree trunks to provide disease control, circumventing the need for foliar absorption which is generally slow unless wetting agents are used. Control of *Phytophthora cinnamomi* in avocado trees[333,338] has been achieved in this way. Australia, and those countries covered by Australia's patent filing system, are the only countries in which potassium phosphonate itself is registered as a fungicide. It has been used effectively to treat root and heart rots of a number of subtropical crops and has been investigated for control of root rots of temperate crops.[339] Progress in the use of phosphonic acid directly for the control of a wide variety of plant diseases has been reported in a recent symposium (*Australasian Plant Pathology*, 19(4), 1990).

Since phosphonate is transported in the phloem, application of the fungicide can be timed to coincide with the target organ of the plant becoming a metabolic sink. Considerable work remains to be done before there is sufficient information to allow this methodology to be used most effectively.

Phosphonates are generally considered to be effective against members of the Peronosporales, and particularly against the genus *Phytophthora,* although there is sufficient evidence available to indicate that they probably have a far wider spectrum of activity.[304] Even within the genus *Phytophthora*, species differ greatly in their sensitivity to phosphonate *in vitro*.[303,334,340,341] The species of *Phytophthora* have been grouped according to their sensitivity to phosphonate.[341] Those having the highest sensitivity include: *P. cinnamomi, P. citrophthora, P. parasitica,* and *P. palmivora*; those with medium sensitivity: *P. cactorum, P. cryptogea,* and *P. vignae;* and those with low sensitivity: *P. megasperma, P. infestans,* and *P. capsici*. The sensitivity of *Pythium* to phosphonate has not been investigated in the same detail as has *Phytophthora,* but generally the genus appears less sensitive.[342]

It has recently been established that the concentration of phosphate in the culture media determines the extent to which phosphonate is assimilated by fungi grown in culture.[343-345] The Oomycetes possess both a high and low affinity uptake system for phosphate[343-345] and phosphonate competes for these transporter sites. Although millimolar concentrations of phosphonate are required to inhibit phosphate uptake, only micromolar amounts of phosphate are needed to inhibit phosphonate transport.[344] Phosphonate uptake in plants has also been reported to be mediated by a carrier system,[346] but studies on leaf disks suggest that uptake may not be dependent on phosphate concentration.[347] In fungi, the degree of phosphate/phosphonate antagonism appears to differ between species and even between races within a species.[398] The concentration of phosphate present in many types of agar used for solidifying media is quite sufficient to alter the extent of phosphonate toxicity observed *in vitro*, and this has doubtlessly contributed to some of the variation observed in these tests. Studies of the effect which variation in phosphate concentration has on phosphonate protection in plants are urgently needed.

As indicated earlier, the mode of action of phosphonate has generated considerable debate. Originally, there were two opposing schools of thought. One believed that phosphonate acted directly on the fungus, while the other proposed that the action of phosphonate was indirect, stimulating the defense responses within the plant which in turn inhibited the pathogen invasion. The proponents of a direct mode of action accepted that changes in the host defense response occurred in the presence of phosphonate, but did not consider these to be essential to phosphonate activity.[201,348,349] Those who championed the indirect mode of action admitted the possibility of a direct inhibition of fungal growth by phosphonate *in vitro*, but argued that this was of secondary importance, since in many situations phosphonate concentrations in plant tissues did not reach the levels to provide the protection observed.[350-353] In the early presentations of the hypothesis for an indirect mode of action, it was proposed that the phosphonate altered the plant metabolism rather than that of the pathogen,[354,355] but this has now been shown to be unlikely. Fungal mutants tolerant to phosphonate were produced *in vitro*,[342,356-358] and they have been shown to be less sensitive to phosphonate in plants.[356,357]

It now appears that the mode of action of phosphonate is complex. The site of phosphonate action almost certainly lies within the pathogen. However, the precise targets, and the way in which exposure of the pathogen to phosphonate at sublethal concentrations results in an alteration in the host defense response are still unknown. However, a considerable body of evidence has accumulated in recent years which suggests how the effects might be achieved.

Exposure of pathogenic fungi to phosphonate has impact on their morphology, both *in vitro* and within the plant,[359-361] and on their intermediary metabolism *in vitro*.[72,363-366] These changes include alteration in polyphosphate and triacyglycerol composition.[72,365,366] Changes were observed at levels of phosphonate that did not inhibit growth in culture. Other changes indicating alteration of amino acid metabolism and protein composition were also observed.[366] Therefore phosphonate can cause changes in metabolism in at least some members of the Oomycetes at concentrations insufficient to inhibit growth.

At higher concentrations sufficient to inhibit growth in culture, exposure to phosphonate for 25 h resulted in reductions in the nucleotide pool and increased activity of the pentose phosphate pathway.[363] In shorter-term experiments of up to 3 h, a decreased nucleotide pool consistent with an inhibition of adenylate synthesis was observed for *P. palmivora*, when mycelia had been starved of phosphate to induce rapid uptake of phosphonate.[367] It was suggested that the enzyme 5′-phosphoribosyl-pyrophosphate synthetase might be one of the targets of phosphonate action. However, in these shorter-term experiments, neither changes in respiration rate nor in the activity of the pentose phosphate pathway were observed.[368] There is no evidence to suggest that phosphonate inhibits the phosphorylation of ADP, either at substrate phosphorylation sites or at the mitochondrial ATPase.

There is considerable evidence to support the hypothesis that phosphonate treatment results in a more active plant defense reaction following challenge by a pathogen. Moreover, with one exception,[348] experimental results have shown that inhibition of the host defense reaction by the application of exogenous inhibitors results in a reduced protection by phosphonate.[354,355,361,365,370-375] It has also been shown that phosphonate was unable to control infection in killed tissue.[340] If it is accepted that the target sites for phosphonate are located within the pathogen, an explanation of the mechanism of the increased activity of the dynamic defense response within the host must be provided. This explanation will require a more detailed understanding of the interactions between host and parasite than is currently available. A testable hypothesis is that the elicitor activity of the pathogen is increased. Increases in the production or activity of elicitors from *Phytophthora* spp. have been observed in culture filtrates where mycelia were exposed to phosphonate at concentrations insufficient to inhibit growth.[376,377] The compounds which acted as elicitors in these experiments were not isolated, so it is not possible to distinguish whether there was actually an increase in compounds acting as elicitors or a decrease in compounds acting as suppressors of elicitor activity in the plant tissue. Either effect would produce the results observed, altering the interaction between pathogen and host, changing a compatible (disease) response to an incompatible (no disease) response.

There is evidence that molecules which could have a role as suppressors exist on the surface of Oomycetes. It has been shown that *Phytophthora* spp. produce high molecular weight molecules on their surfaces. These are water soluble and appear to be polysaccharides or glycoproteins.[364,378,379] It has also been shown that these same surface molecules, prepared from *P. infestans,* alter the response produced in wounded potato leaves. This ability is specific to the surface molecules from *P. infestans,* the late blight pathogen of potato.[380] Surface molecules from *P. cinnamomi,* which is not a pathogen of potato, failed to produce the effect. When phosphonate was added to cultures at levels which are noninhibitory to mycelial growth, the production of these surface molecules was greatly reduced.[364,379] Further research is required to determine whether the capacity of phosphonate to inhibit production of these surface molecules in *P. infestans* occurs in other species which are controlled by phosphonate application.

The evidence to date suggests that phosphonate activity is best explained by a combination of both direct and indirect modes of action. The contribution of each will depend on the sensitivity of the pathogen to direct inhibition by phosphonate in the specific environment under consideration and on the strength of the host plant's defense response in the tissue under attack. Two quite different genomes are therefore involved in the expression of sensitivity to phosphonate, the pathogen's and the host's, each offering unique opportunities for manipulation.

The interaction of phosphonate with the products of two separate genomes would also provide an explanation for the slow rate at which resistance appears to develop against this fungicide. There have been no confirmed reports of field resistance to phosphonate and a claim[381] that cross-resistance between metalaxyl and Aliette had occurred in *P. infestans* and *Pseudoperonospora cubensis* has not been substantiated in other experiments.[319,356,382-384] The available evidence suggests that there is no cross-resistance between metalaxyl and the phosphonates.[384] Although mutants with reduced sensitivity to phosphonate have been produced, the phosphonate concentration required to inhibit mycelial growth in these mutants still falls within the range of sensitivities observed in natural isolates.[304]

Although phosphonates have been in use for more than a decade, their full potential has not been fully explored. The spectrum of their activity is not known, and little serious attempt has been made to explore how activity might be modified by subtle alteration of chemical structure. Perhaps the most exciting conclusion to be drawn from the limited work carried out to date is that other chemicals may exist which act in a similar manner, and which may induce protection against fungi which belong to quite different groups. In this respect, two recent reports[385,386] showing that metalaxyl can, under certain circumstances, stimulate host defenses, and that these host defenses contribute to disease control, are important.

G. FUNGICIDES UNDER DEVELOPMENT

1. Andoprim

Although much research is directed toward developing more effective analogues of existing chemical control agents, (e.g., metalaxyl[387] and cymoxanil,[388]) there have been some developments of novel fungicides active against the Oomycetes. One of these is andoprim, 2-*p*-methoxyanilino-4,6-dimethyl pyrimidine, which has been developed by VEB Fahlberg-List Magdeburg (GDR). Grunwaldt et al.[389] reported that andoprim has a highly selective activity against *Phytophthora infestans* on potato, while *Pythium* species are much less sensitive. Lyr and co-workers found that andoprim affects the uptake of leucine and the nucleosides uridine and thymidine.[399] The decreased uptake of precursors is reflected in decreased synthesis of proteins and nucleic acids. Lyr's data suggest that andoprim affects active transport processes across the cell membrane. Passive uptake processes are not impaired.

2. Dimethomorph

Dimethomorph is a cinnamic acid derivative, (E,Z) 4-[3-(4-chlorophenyl)-3-(3,4-dimethoxyphenyl)-acryloyl]-morpholine (CME 151), which was developed by Shell Forschung GmbH[390] and has been evaluated for use against *Plasmopora viticola* and *P. infestans* on tomato and potato. It is ineffective against the genus *Pythium*. Its curative and protective activity against *P. viticola in vitro* was as good as that of metalaxyl.[391] It also significantly inhibited sporulation. The mode of action of dimethomorph is unknown. It does not display cross-resistance to phenylamides.

3. Pyridazines

The haloalkyl pyridazines are also being investigated for their activity on the Oomycetes. Hackler et al.[392] have demonstrated that of a series of these compounds tested, 3,6-dichloro-4-(2-chloro-1,1-dimethylethyl)pyridazine was the most active on *Pseudoperonospora cubensis*.

4. ICIA0001

(*RS*)-4-chloro-*N*-[cyano(ethoxy)methyl]benzamide is a benzamide compound. In both *in vitro* and glasshouse trials it has been shown to have a longer lasting curative activity than cymoxanil on *Plasmopara viticola.*[393] The compound also displays a protectant activity against *P. viticola,* although this activity has short persistance (biological half-life 3.4 days). The workers developing this compound are advocating its use in conjunction with multisite protectants for the control of *P. viticola.*[394] It has no cross-resistance with the phenylamides. No other report has appeared which suggests its mode of action.

V. THE FUTURE

There seems little doubt that members of the Oomycetes will continue to present problems in plant production for as far as can be seen into the future. There has been no recent breakthrough in the development of a new group of fungicides with the ability to control their members effectively. The phosphonates appear to offer some hope in this regard, but for the present, ignorance about their real site of action, which is clearly not going to be a simple matter to determine, has limited progress in this field. There is no evidence that this line of research is being actively pursued, and only three or four laboratories have published substantial information in the area during the last 5 years.

We have suggested in this chapter that one of the problems which has retarded progress in the development of new chemicals targeted against the Oomycetes is the failure to appreciate the fact that they are simply not fungi in the same sense that the Ascomycetes and Basidiomycetes are fungi. This fundamental difference explains why many extremely effective chemotherapeutic agents that control pathogens belonging to the higher fungi are ineffective against these organisms.

However, it is well known that problems viewed from a different perspective open opportunities. The production of zoospores is an essential feature of dispersion and infection of the Oomycetes. This process is only now being understood at the molecular level and if progress continues, may present several opportunities for specific intervention.[395,396] As mentioned in the first part of the chapter, the actin protein of the Oomycetes appears to be quite different from any described in other organisms to date. The development of agents which bind to the unique sequences in this protein offers a further opportunity for specific control.

The real role of sterols in the biology of some members of the Oomycetes demands further attention. It is difficult to envisage that the membrane systems of these organisms, which contain such an extremely wide variety of membrane-bound structures, can function in the total absence of any compound playing the role of sterols in the normal eukaryotic cell. If there is such a compound it needs to be examined carefully in order to determine its potential as the target site for a fungicidal "magic bullet".

The metabolism of soluble polyphosphates found in some Oomycetes also appears to warrant further attention. It has only recently been recognized that these compounds are present in some of these organisms and account for the major part of the phosphorus within the organism.[72] Recent work from the authors' laboratory[400] suggests that either polyphosphates or pyrophosphates play a much more important role in the energy metabolism of the Oomycetes than they do in the higher fungi, where they appear to function largely as storage reserves for phosphorus. When the details of these metabolic pathways and their respective enzymes are known, they may present unique targets for new fungicides.

If Oomycetes are really best regarded as protozoa, it may also be possible to use compounds that have already proven successful against Oomycetes in plants in the control of protozoan-induced diseases of humans, some of which are notoriously difficult to control. Again, the fact that the Oomycetes have until now been regarded as plant pathogenic fungi and therefore fall within the province of plant pathology, may well have meant that protozoologists have not considered testing compounds like the phenylamides against protozoa.

ACKNOWLEDGMENTS

Work in the authors' laboratory described in this article has been supported by grants from the Rural Industries Research and Development Corporation, The Reserve Bank, Albright & Wilson Ltd. and Albright & Wilson (Aust) Ltd., Edgell-Birds Eye, and the Australian Research Grants Commission. Information on work not published at the time this manuscript was prepared was kindly provided by Professor H. Lyr and Professor F. Schwinn.

REFERENCES

1. **Bartnicki-Garcia, S.,** What are fungi? in 4th Int. Congr. of Plant Pathology, Abstract, Regensburg, Germany, 1990, 8.
2. **Hickman, C. J.,** Fungal structure and organization, in *The Fungi: An Advanced Treatise,* Vol. 1, Ainsworth, G. C. and Sussman, A. S., Eds., Academic Press, New York, 1965, chap. 2.
3. **Talbot, P. H. B.,** *Principles of Fungal Taxonomy,* Macmillan, London, 1971, chap. 1.
4. **Alexopoulos, C. J. and Mims, C. W.,** *Introductory Mycology,* 3rd ed., John Wiley & Sons, London, 1979, chap. 1.
5. **Barr, D. J. S.,** Eumycota, in *Synopsis and Classification of Living Organisms,* Vol. 1, Parker, S. P., Ed., McGraw-Hill, New York, 1982, 178.

6. **Deacon, J. W.,** Introduction to modern mycology, in *Basic Microbiology*, Vol. 7, 2nd ed., Wilkinson, J. F., Ed., Blackwell Scientific, Oxford, 1984, chap. 1.
7. **Talbot, P. H.,** *Principles of Fungal Taxonomy*, Macmillan, London, 1971, chap. 6.
8. **Sparrow, F. K.,** Mastigomycotina (zoosporic fungi), in *The Fungi: An Advanced Treatise*, Vol. 4B, Ainsworth, G. C., Sparrow, F. K., and Sussman, A. S., Eds., Academic Press, New York, 1973, chap. 4.
9. **Alexopoulos, C. J. and Mims, C. W.,** *Introductory Mycology*, 3rd ed., John Wiley & Sons, London, 1979, chap. 8.
10. **Dick, M. W.,** Oomycetes, in *Synopsis and Classification of Living Organisms*, Vol. 1, Parker, S. P., Ed., McGraw-Hill, New York, 1982, 179.
11. **Barr, D. J. S.,** *Zoosporic Plant Pathogens: A Modern Perspective*, Buczacki, S. T., Ed., Academic Press, London, 1983, chap. 2.
12. **Whittaker, R. H. and Margulis, L.,** Protist classification and the kingdoms of organisms, *BioSystems*, 10, 3, 1978.
13. **Cavalier-Smith, T.,** The evolutionary origin and phylogeny of microtubules, mitotic spindles and eukaryote flagella, *BioSystems*, 10, 93, 1978.
14. **Taylor, F. J. R.,** Problems in the development of an explicit hypothetical phylogeny of the lower eukaryotes, *BioSystems*, 10, 67, 1978.
15. **Hibberd, D. J.,** The structure and phylogenetic significance of the flagellar transition region in the chlorophyll c-containing algae, *BioSystems*, 11, 243, 1979.
16. **Heath, I. B.,** Variant mitoses in lower eukaryotes: indicators of the evolution of mitoses?, *Int. Rev. Cytol.*, 64, 1, 1980.
17. **Heath, I. B.,** An investigation of protistan phylogeny using a numerical taxonomy (cluster) analysis of mitotic systems, *BioSystems*, 14, 261, 1981.
18. **McQuade, A. B.,** Origins of the nucleate organisms. II, *BioSystems*, 16, 39, 1983.
19. **Hori, H. and Osawa, S.,** Evolutionary change in 5S rRNA secondary structure and a phylogenic tree of 352 5S rRNA species, *BioSystems*, 19, 163, 1986.
20. **Wolters, J. and Erdmann, V. A.,** Cladistic analysis of 5S rRNA and 16S rRNA secondary and primary structure — the evolution of eukaryotes and their relation to archibacteria, *J. Mol. Evol.*, 24, 152, 1986.
21. **Barrett, J. A.,** Molecular variation and evolution, in *Evolutionary Biology of the Fungi*, Rayner, A. D. M., Brasier, C. M., and Moore, D., Eds., British Mycology Society Symp. Series No. 12, Cambridge University Press, Cambridge, 1987, chap. 6.
22. **Dick, M. W.,** Sexual reproduction: nuclear cycles and life histories with particular reference to lower eukaryotes, *Biol. J. Linn. Soc.*, 30, 181, 1987.
23. **Erdmann, V. A., Wolters, J., Pieler, T., Digweed, M., Specht, T., and Ulbrich, N.,** Evolution of organisms and organelles as studied by comparative computer and biochemical analyses of ribosomal 5S RNA structure, *Ann. N.Y. Acad. Sci.*, 503, 103, 1987.
24. **Olsen, G. J.,** Earliest phylogenetic branchings: comparing rRNA evolutionary trees inferred with various techniques, *Cold Spring Harbor Symp. Quant. Biol.*, 52, 825, 1987.
25. **Sogin, M. L. and Grunderson, J. H.,** Structural diversity of eukaryotic small subunit ribosomal RNAs, *Ann. N.Y. Acad. Sci.*, 503, 125, 1987.
26. **Wolters, J. and Erdmann, V. A.,** Cladistic analysis of ribosomal RNAs — the phylogeny of eukaryotes with respect to the endosymbiotic theory, *BioSystems*, 21, 209, 1988.
27. **Lovett, J. S. and Haselby, J. A.,** Molecular weights of the ribosomal ribonucleic acid of fungi, *Arch. Microbiol.*, 80, 191, 1971.
28. **Heath, I. B.,** The possible significance of variation in the mitotic systems of the aquatic fungi (Phycomycetes), *BioSystems*, 7, 351, 1975.
29. **Barr, D. J. S.,** The phylogenetic and taxonomic implications of flagella rootlet morphology among zoosporic fungi, *BioSystems*, 14, 359, 1981.
30. **Barr, D. J. S. and Allan, P. M. E.,** A comparison of the flagellar apparatus in *Phytophthora*, *Saprolegnia*, *Thraustochytrium* and *Rhizidiomyces*, *Can. J. Bot.*, 63, 138, 1984.
31. **Powell, M. J., Lehnen, L. P., and Bortnick, R. N.,** Microbody-like organelles as taxonomic markers among Oomycetes, *BioSystems*, 18, 321, 1985.
32. **Walker, W. F.,** 5S and 5.8S RNA sequences and protist phylogenetics, *BioSystems*, 18, 269, 1985.
33. **Taylor, J. W.,** Fungal evolutionary biology and mitochrondrial DNA, *Exp. Mycol.*, 10, 259, 1986.
34. **Burnett, J. H.,** Aspects of the macro- and microevolution of the fungi, in *Evolutionary Biology of the Fungi*, Rayner, A. D. M., Brasier, C. M., and Moore, D., Eds., British Mycology Society Symp. Series No. 12, Cambridge University Press, Cambridge, 1987, chap. 1.
35. **Beakes, G. W.,** Oomycete phylogeny: ultrastructural perspectives, in *Evolutionary Biology of the Fungi*, Rayner, A. D. M., Brasier, C. M., and Moore, D., Eds., British Mycology Society Symp. Series No. 12, Cambridge University Press, Cambridge, 1987, chap. 27.
36. **Tehler, A.,** A cladistic outline of the Eumycota, *Cladistics*, 4, 227, 1988.

37. **Cavalier-Smith, T.,** Eukaryote kingdoms: seven or nine? *BioSystems,* 14, 461, 1981.
38. **Jeffrey, C.,** Kingdoms, codes and classification, *Kew Bull.*, 37, 403, 1982.
39. **Corliss, J. O.,** The kingdom Protista and its 45 phyla, *BioSystems,* 17, 87, 1984.
40. **Margulis, L. and Schwartz, K. V.,** *Five Kingdoms: An Illustrated Guide to the Phyla of Life on Earth,* 2nd ed., Freeman, New York, 1988, 142.
41. **Kendrick, B.,** *The Fifth Kingdom,* Mycologue Publications, Ontario, 1985, chap. 1.
42. **Ragan, M. A. and Chapman, D. J.,** *A Biochemical Phylogeny of the Protists,* Academic Press, New York, 1978, chap. 12.
43. **von Arx, J. A.,** *The Genera of Fungi Sporulating in Pure Culture,* 3rd ed., J. Cramer, Hirschberg, 1981, 33.
44. **Barnes, R. S. K.,** Ed., *A Synoptic Classification of Living Organisms,* Blackwell Scientific, Oxford, 1984, 3.
45. **Sleigh, M. A.,** *Protozoa and Other Protists,* Hodder and Stroughton, London, 1989, chap. 5.
46. **Cavalier-Smith, T.,** The kingdom Chromista: origin and systematics, in *Progress in Phycological Research,* Vol. 4, Round, F. E. and Chapman, D. J., Eds., Biopress, Bristol, 1986, chap. 4.
47. **Cavalier-Smith, T.,** The origin of fungi and pseudofungi, in *Evolutionary Biology of the Fungi,* British Mycology Society Symp. Series No. 12, Rayner, A. D. M., Brasier, C. M., and Moore, D., Eds., Cambridge University Press, Cambridge, 1987, chap. 23.
48. **Gleason, F.,** The physiology of the lower freshwater fungi, in *Recent Advances in Aquatic Mycology,* Jones, E. B. G., Ed., John Wiley & Sons, New York, 1976, chap. 21.
49. **Dick, M. W.,** Morphology and taxonomy of the Oomycetes, with special reference to Saprolegniaceae, Leptomitaceae and Pythiaceae. II. Cytogenetic systems, *New Phytol.*, 71, 1151, 1972.
50. **Shaw, D. S.,** The cytogenetics and genetics of *Phytophthora,* in *Phytophthora: Its Biology, Taxonomy, Ecology and Pathology,* Erwin, D. C., Bartnicki-Garcia, S., and Tsao, P. H., Eds., American Phytopathology Society, St. Paul, 1983, chap. 6.
51. **Elliott, C.,** Physiology of sexual reproduction in *Phytophthora,* in *Phytophthora: Its Biology, Taxonomy, Ecology and Pathology,* Erwin, D. C., Bartnicki-Garcia, S., and Tsao, P. H., Eds., American Phytopathology Society, St. Paul, 1983, chap. 5.
52. **Bortnick, R. N., Powell, M. J., and Bangert, T. N.,** Zoospore fine structure of the parasite *Olpidiopsis saprolegniae* variety *saprolegniae* (Oomycetes, Langenidiales), *Mycologia,* 77, 861, 1985.
53. **Crook, E. M. and Johnston, I. R.,** The qualitative analysis of the cell walls of selected fungi, *Biochem. J.*, 83, 325, 1962.
54. **Bartnicki-Garcia, S.,** Chemistry of hyphal walls of *Phytophthora, J. Gen. Microbiol.*, 42, 57, 1966.
55. **Novaes-Ledieu, M., Jiménez-Martinez, A., and Villanueva, J. R.,** Chemical composition of hyphal wall of Phycomycetes, *J. Gen. Microbiol.*, 47, 237, 1967.
56. **Vaziri-Tehrani, B. and Dick, M. W.,** Amino acid composition of Oomycete cell walls, *Trans. Br. Mycol. Soc.*, 74, 225, 1980.
57. **Vaziri-Tehrani, B. and Dick, M. W.,** Taxonomic significance of variations between ratios of cell wall amino acids in fungi, with special reference to the Oomycetes, *Trans. Br. Mycol. Soc.*, 74, 231, 1980.
58. **Vaziri-Tehrani, B. and Dick, M. W.,** Neutral and aminosugars from the cell walls of Oomycetes, *Biochem. Syst. Ecol.*, 8, 105, 1980.
59. **Bartnicki-Garcia, S.,** Cell wall chemistry, morphogenesis, and taxonomy of fungi, *Annu. Rev. Microbiol.*, 22, 87, 1968.
60. **Bartnicki-Garcia, S.,** Cell wall composition and other biochemical markers in fungal phylogeny, in *Phytochemical Phylogeny,* Harborne, J. B., Ed., Academic Press, London, 1970, chap. 5.
61. **Bartnicki-Garcia, S.,** The cell wall: a crucial structure in fungal evolution, in *Evolutionary Biology of the Fungi,* Rayner, A. D. M., Brasier, C. M., and Moore, D., Eds., British Mycology Society Symp. Ser. No. 12, Cambridge University Press, Cambridge, 1987, chap. 26.
62. **Fèvre, M., Girard, V., and Nodet, P.,** Cellulose and β-glucan synthesis in *Saprolegnia,* in *Biochemistry of Cell Walls and Membranes in Fungi,* Kuhn, P. J., Trinci, A. P. J., Jung, M. J., Goosey, M. W., and Copping, L. G., Eds., Springer-Verlag, Heidelberg, 1990, chap. 7.
63. **Lé John, H. B.,** Biochemical parameters of fungal phylogenetics, *Evol. Biol.*, 7, 79, 1974.
64. **Bartnicki-Garcia, S. and Wang, M. C.,** Biochemical aspects of morphogenesis in *Phytophthora,* in *Phytophthora: Its Biology, Taxonomy, Ecology and Pathology,* Erwin, D. C., Bartnicki-Garcia, S., and Tsao, P. H., Eds., American Phytopathology Society, St. Paul, 1983, chap. 9.
65. **Dietrich, S. M.,** Carbohydrates from the hyphal wall of some Oomycetes, *Biochim. Biophys. Acta,* 313, 95, 1973.
66. **Lin, C. C., Sicher, R. C., Jr., and Aronson, J. M.,** Hyphal wall chemistry in *Apodachylya, Arch. Microbiol.*, 76, 108, 1976.
67. **Aronson, J. M. and Lin, C. C.,** Hyphal wall chemistry of *Leptomitus lacteus, Mycologia,* 70, 363, 1978.
68. **Pfyffer, G. B. and Rast, D. M.,** The polyol pattern of some fungi not hitherto investigated for sugar alcohols, *Exp. Mycol.*, 4, 160, 1980.

69. **Pfyffer, G. E., Pfyffer, B. U., and Rast, D. M.,** The polyol pattern, chemotaxonomy, and phylogeny of the fungi, *Sydowia Ann. Mycol.*, 39, 160, 1986.
70. **Pfyffer, G. E., Boraschi-Gaia, C., Weber, B., Hoesch, L., Orpin, C. G., and Rast, D. M.,** A further report on the occurrence of acyclic sugar alcohols in fungi, *Mycol. Res.*, 94, 219, 1990.
71. **Chilvers, G. A., Lapeyrie, F. F., and Douglass, P. A.,** A contrast between Oomycetes and other taxa of mycelial fungi in regard to metachromatic granule formation, *New Phytol.*, 99, 203, 1985.
72. **Niere, J. O., Griffith, J. M., and Grant, B. R.,** ^{31}P NMR studies on the effect of phosphite on *Phytophthora palmivora*, *J. Gen. Microbiol.*, 136, 147, 1990.
73. **Vogel, H. J., Thompson, J. S., and Shockman, G. D.,** Characteristic metabolic patterns of prokaryotes and eukaryotes, in *Organization and Control in Prokaryotes and Eukaryotes*, Charles, H. P. and Knight, B. C. J. G., Eds., 20th Symp. Soc. Gen. Microbiol., Cambridge University Press, Cambridge, 1970, 107.
74. **Lé John, H. B.,** Enzyme regulation, lysine pathways and cell wall structures as indicators of major lines of evolution in fungi, *Nature*, 231, 164, 1971.
75. **Hendrix, J. W.,** Physiology and biochemistry of growth and reproduction in *Pythium*, *Proc. Am. Phytopathol. Soc.*, 1, 207, 1974.
76. **Natvig, D. O. and Gleason, F. H.,** Oxygen uptake by obligately-fermentative aquatic fungi: absence of a cyanide-sensitive component, *Arch. Microbiol.*, 134, 5, 1983.
77. **Shaw, R.,** The occurrence of γ-linolenic acid in fungi, *Biochim. Biophys. Acta*, 98, 230, 1965.
78. **Erwin, J. A.,** Comparative biochemistry of fatty acids in eukarotic microorganisms, in *Lipids and Biomembranes of Eukaryotic Microorganisms*, Erwin, J. A., Ed., Academic Press, New York, 1973, chap. 2.
79. **Wassef, M. K.,** Fungal lipids, *Adv. Lipid Res.*, 15, 159, 1977.
80. **Lösel, D. M.,** Fungal lipids, in *Microbial Lipids*, Vol. 1, Routledge, C. and Wilkinson, S. G., Eds., Academic Press, San Diego, 1988, chap. 10.
81. **Weete, J. D., Fuller, M. S., Huang, M. Q., and Gandhi, S.,** Fatty acids and sterols of selected Hyphochytriomycetes and Chytridiomycetes, *Exp. Mycol.*, 13, 183, 1989.
82. **Wassef, M. K. and Hendrix, J. W.,** Ceramide aminoethylphosphonate in the fungus *Pythium prolatum*, *Biochim. Biophys. Acta*, 486, 172, 1977.
83. **Hendrix, J. W. and Rouser, G.,** Polar lipids of *Phytophthora parasitica* var. *nicotianae* in comparison with those of other selected fungi, *Mycologia*, 68, 354, 1976.
84. **Creamer, J. R. and Bostock, R. M.,** Characterization and biological activity of *Phytophthora infestans* phospholipids in the hypersensitive response of the potato tuber, *Physiol. Mol. Plant Pathol.*, 28, 215, 1986.
85. **Moreau, R.,** The involvement of membrane-degrading enzymes during infection of potato leaves by *Phytophthora infestans*, in *Ecology and Metabolism of Plant Lipids*, Fuller, G. and Nes, W. D., Eds., ACS Symp. Series 325, American Chemical Society, Washington, D.C., 1987, chap. 21.
86. **Kerwin, J. L. and Duddles, N. D.,** Reassessment of the role of phospholipids in sexual reproduction by sterol-auxotrophic fungi, *J. Bacteriol.*, 171, 3831, 1989.
87. **Rosenberg, H.,** Phosphonolipids, in *Form and Function of Phospholipids*, BBA Library Vol. 3, Ansell, G. B., Dawson, R. M. C., and Hawthorne, J. N., Eds., Elsevier, New York, 1973, chap. 12.
88. **Hori, T. and Nozawa, Y.,** Phosphonolipids, in *Phospholipids*, New Comprehensive Biochem. Vol. 4, Hawthorne, J. N. and Ansell, G. B., Eds., Elsevier, New York, 1982, chap. 3.
89. **Moschidis, M. C.,** Phosphonolipids, *Prog. Lipid Res.*, 23, 223, 1985.
90. **Lhomme, O., Bruneteau, M., Costello, C. E., Mas, P., Molot, P.-M., Dell, P. R., Tiller, P. R., and Michel, G.,** Structural investigations and biological activity of inositol sphingophospholipids from *Phytophthora capsici*, *Eur. J. Biochem.*, 191, 203, 1990.
91. **Hendrix, J. W.,** Sterols in growth and reproduction of fungi, *Annu. Rev. Phytopathol.*, 8, 111, 1970.
92. **Nes, W. D.,** Uniformity vs. diversity in the structure, biosynthesis, and functions of sterols, in *Isopentenoids in Plants: Biochemistry and Function*, Nes, W. D., Fuller, G., and Tsai, L.-S., Eds., Marcel Dekker, New York, 1984, 325.
93. **Weete, J. D.,** Structure and function of sterols in fungi, *Adv. Lipid Res.*, 23, 115, 1989.
94. **Elliott, C. G.,** Sterols in fungi: their functions in growth and reproduction, *Adv. Microb. Physiol.*, 15, 121, 1977.
95. **Nes, W. D.,** Biosynthesis and requirement for sterols in the growth and reproduction of Oomycetes, in *Ecology and Metabolism of Plant Lipids*, Fuller, G. and Nes, W. D., Eds., ACS Symp. Series 325, American Chemical Society, Washington, D.C., 1987, chap. 19.
96. **Domnas, A. J., Srebo, J. P., and Hicks, B. F.,** Sterol requirement for zoospore formation in the mosquito parasitizing fungus *Lagenidium giganteum*, *Mycologia*, 69, 875, 1977.
97. **Kerwin, J. L. and Washino, R. K.,** Sterol induction of sexual reproduction in *Lagenidium giganteum*, *Exp. Mycol.*, 7, 109, 1984.
98. **Warner, S. A., Eierman, D. F., Domnas, A. J., and Dean, D. D.,** A novel lipoprotein from Oomycete fungi, *Exp. Mycol.*, 10, 315, 1986.

99. **Warner, S. A., Sovocool, G. W., and Domnas, A. J.,** Ergosterol, the unusual dominant sterol of the pythiaceous fungus *Zoophagus insidians, Phytochemistry,* 21, 2135, 1982.
100. **Coffel, R., Hudspeth, M. E. S., and Meganathan, R.,** Biochemical evidence for the exclusion of *Zoophagus insidians* from the Oomycetes, *Mycologia,* 82, 326, 1990.
101. **Dick, M. W.,** The systematic position of *Zoophagus insidians, Mycol. Res.,* 94, 347, 1990.
102. **Nes, W. R.,** Role of sterols in membranes, *Lipids,* 9, 596, 1974.
103. **Bloch, K. E.,** Sterol structure and membrane function, *CRC Crit. Rev. Biochem.,* 14, 47, 1983.
104. **Nes, W. R.,** The biochemistry of plant sterols, *Adv. Lipid Res.,* 15, 233, 1977.
105. **Bu'Lock, J. D. and Osagie, A. U.,** Sterol biosynthesis via cycloartenol in *Saprolegnia, Phytochemistry,* 15, 1249, 1976.
106. **Warner, S. A. and Domnas, A. J.,** Evidence for cycloartenol-based sterol synthetic pathway in *Lagenidium* spp., *Exp. Mycol.,* 5, 184, 1981.
107. **Warner, S. A., Eierman, D. F., Sovocool, G. W., and Domnas, A. J.,** Cycloartenol-derived sterol biosynthesis in the Peronosporales, *Proc. Natl. Acad. Sci. U.S.A.,* 79, 3769, 1982.
108. **Warner, S. A., Sovocool, G. W., and Domnas, A. J.,** Sterols of selected species of Oomycetes and Hypochytridiomycetes, *Mycologia,* 75, 285, 1983.
109. **Nes, W. D., Le, L. R., Berg, G. W., and Kerwin, J. L.,** A comparison of cycloartenol and lanosterol biosynthesis and metabolism by the Oomycetes, *Experientia,* 42, 556, 1986.
110. **Berg, L. R., Patterson, G. W., and Lusby, W. R.,** Effects of triarimol and tridemorph on sterol biosynthesis in *Saprolegnia ferax, Lipids,* 18, 448, 1983.
111. **Berg, L. R. and Patterson, G. W.,** Phylogenic implications of sterol biosynthesis in the Oomycetes, *Exp. Mycol.,* 10, 175, 1986.
112. **Förster, H., Coffey, M. D., Elwood, H., and Sogin, M. L.,** Sequence analysis of the small subunit ribosomal RNAs of three zoosporic fungi and implications for fungal evolution, *Mycologia,* 82, 306, 1990.
113. **Nes, W. D., Patterson, G. W., and Bean, G. A.,** Effect of steric and nuclear changes in steroids and triterpenoids on sexual reproduction in *Phytophthora cactorum, Plant Physiol.,* 66, 1008, 1980.
114. **Elliott, C. G. and Knights, B. A.,** Uptake and interconversions of cholesterol and cholesteryl esters by *Phytophthora cactorum, Lipids,* 16, 1, 1981.
115. **Nes, W. D. and Heftmann, E.,** A comparison of triterpenoids with steroids as membrane components, *J. Nat. Prod.,* 44, 377, 1981.
116. **Nes, W. D.,** The effect of steroids and their solubilizing agents on mycelial growth of *Phytophthora cactorum, Lipids,* 14, 458, 1979.
117. **Ko, W. H.,** Hormonal heterothallism and homothallism in *Phytophthora, Annu. Rev. Phytopathol.,* 26, 57, 1988.
118. **Lösel, D. M.,** Functions of lipids: specialized roles in fungi and algae, in *Microbial Lipids,* Vol. 2, Ratledge, C. and Wilkinson, S. G., Eds., Academic Press, San Diego, 1989, chap. 18.
119. **Horton, J. S. and Horgen, P. A.,** Molecular cloning of cDNAs regulated during steroid-induced sexual differentiation in the aquatic fungus *Achlya, Exp. Mycol.,* 13, 263, 1989.
120. **Nes, W. D. and Stafford, A. F.,** Evidence for metabolic and functional discrimination of sterols by *Phytophthora cactorum, Proc. Natl. Acad. Sci. U.S.A.,* 80, 3227, 1983.
121. **Nes, W. D. and Le, P. H.,** Regulation of sterol biosynthesis in *Saprolegnia ferax* by 25-azacholesterol, *Pestic. Biochem. Physiol.,* 30, 87, 1988.
122. **Wood, S. G. and Gottlieb, D.,** Evidence from cell-free systems for differences in the sterol biosynthetic pathway of *Rhizoctonia solani* and *Phytophthora cinnamomi, Biochem. J.,* 170, 355, 1978.
123. **Gottlieb, D., Knaus, R. J., and Wood, S. G.,** Differences in the sterol synthesizing pathways of sterol-producing and non-sterol-producing fungi, *Phytopathology,* 68, 1168, 1978.
124. **Vanden Bossche, H.,** Importance and role of sterols in fungal membranes, in *Biochemistry of Cell Walls and Membranes in Fungi,* Kuhn, P. J., Trinci, A. P. J., Jung, M. J., Goosey, M. W., and Copping, L. G., Eds., Springer-Verlag, Heidelberg, 1990, chap. 10.
125. **Taylor, R. F.,** Bacterial triterpenoids, *Microb. Rev.,* 48, 181, 1984.
126. **Poralla, K. and Kannenberg, E.,** Hopanoids: sterol equivalents in bacteria, in *Ecology and Metabolism of Plant Lipids,* Fuller, G. and Nes, W. D., Eds., ACS Symp. Series 325, American Chemical Society, Washington, D.C., 1987, chap. 15.
127. **Stafford, A. E. and Nes, W. D.,** Biosynthesis of long-chain fatty alcohols in *Phytophthora cactorum, Exp. Mycol.,* 7, 374, 1983.
128. **Nes, W. D.,** Phytophthorols — novel lipids produced by *Phytophthora cactorum, Lipids,* 23, 9, 1988.
129. **Ko, W. H.,** Stimulation of sexual reproduction of *Phytophthora cactorum* by phospholipids, *J. Gen. Microbiol.,* 131, 2591, 1985.
130. **Ko, W. H.,** Sexual reproduction of *Pythium aphanidermatum:* stimulation by phospholipids, *Phytopathology,* 76, 1159, 1986.
131. **Guyer, W. and Bloch, K.,** The sterol requirement of *Tetrahymena setosa* HZ-1, *J. Protozool.,* 30, 625, 1983.

132. **Dahl, J. and Dahl, C.,** Coordinate regulation of unsaturated phospholipid, RNA, and protein synthesis in *Mycoplasma capricolum* by cholesterol, *Proc. Natl. Acad. Sci. U.S.A.*, 80, 692, 1983.
133. **Parks, L., Rodriguez, R. J., and Low, C.,** An essential fungal growth factor derived from ergosterol: a new end product of sterol biosynthesis in fungi?, *Lipids*, 21, 89, 1986.
134. **Kerwin, J. L. and Washino, R. K.,** Regulation of oosporogenesis by *Lagenidium giganteum:* promotion of sexual reproduction by unsaturated fatty acids and sterol availability, *Can. J. Microbiol.*, 32, 294, 1986.
135. **Storck, R. and Alexopoulos, C. J.,** Deoxyribonucleic acid of fungi, *Bacteriol. Rev.*, 34, 126, 1970.
136. **Pace, N. R., Olsen, G. J., and Woese, C. R.,** Ribosomal RNA phylogeny and the primary lines of evolutionary descent, *Cell*, 45, 325, 1986.
137. **Olsen, G. J.,** Phylogenic analysis using ribosomal RNA, in *Methods in Enzymology*, Vol. 164, Noller, H. F. and Moldive, K., Eds., Academic Press, San Diego, 1988, chap. 53.
138. **Woese, C. R.,** Bacterial evolution, *Microbiol. Rev.*, 51, 221, 1987.
139. **Krishnan, S., Barnabas, S., and Barnabas, J.,** Interrelationships among major protistan groups based on a parsimony network of 5S rRNA sequences, *BioSystems*, 24, 135, 1990.
140. **Grunderson, J. H., Elwood, H., Ingold, A., Kindle, K., and Sogin, M. L.,** Phylogenetic relationships between chloroplasts, Chrysophytes, and Oomycetes, *Proc. Natl. Acad. Sci. U.S.A.*, 84, 5823, 1987.
141. **Battacharya, D. and Druehl, L. D.,** Phylogenetic comparison of the small-subunit ribosomal DNA sequence of *Costaria costata* (Phaeophyta) with those of other algae, vascular plants and Oomycetes, *J. Phycol.*, 24, 539, 1988.
142. **Caten, C. E. and Jinks, J. L.,** Spontaneous variability of single isolates of *Phytophthora infestans.* I. Cultural variation, *Can. J. Bot.*, 46, 329, 1968.
143. **Zentmeyer, G. A.,** The world of *Phytophthora*, in *Phytophthora: Its Biology, Taxonomy, Ecology and Pathology*, Erwin, D. C., Bartnicki-Garcia, S., and Tsao, P. H., Eds., American Phytopathology Society, St. Paul, 1983, chap. 1.
144. **Brasier, C. M.,** Problems and prospects in *Phytophthora* research, in *Phytophthora: Its Biology, Taxonomy, Ecology and Pathology*, Erwin, D. C., Bartnicki-Garcia, S., and Tsao, P. H., Eds., American Phytopathology Society, St. Paul, 1983, chap. 27.
145. **Erwin, D. C.,** Variability within and among species of *Phytophthora*, in *Phytophthora: Its Biology, Taxonomy, Ecology and Pathology*, Erwin, D. C., Bartnicki-Garcia, S., and Tsao, P. H., Eds., American Phytopathology Society, St. Paul, 1983, chap. 11.
146. **Waterhouse, G. M., Newhook, F. J., and Stamps, D. J.,** Present criteria classification of *Phytophthora*, in *Phytophthora: Its Biology, Taxonomy, Ecology and Pathology*, Erwin, D. C., Bartnicki-Garcia, S., and Tsao, P. H., Eds., American Phytopathology Society, St. Paul, 1983, chap. 10.
147. **Hansen, E. M.,** Speciation in *Phytophthora:* evidence from the *Phytophthora megasperma* complex, in *Evolutionary Biology of the Fungi*, Rayner, A. D. M., Brasier, C. M., and Moore, D., Eds., British Mycology Society Symp. Series No. 12, Cambridge University Press, Cambridge, 1987, chap. 22.
148. **Klassen, G. R., McNabb, S. A., and Dick, M. W.,** Comparison of physical maps of ribosomal DNA repeating units in *Pythium, Phytophthora* and *Apodachyla*, *J. Gen. Microbiol.*, 133, 2953, 1987.
149. **Shaw, D. S.,** The *Phytophthora* species, *Adv. Plant Pathol.*, 6, 27, 1988.
150. **Brasier, C. M. and Rayner, A. D. M.,** Wither terminology below the species level in the fungi?, in *Evolutionary Biology of the Fungi*, Rayner, A. D. M., Brasier, C. M., and Moore, D., Eds., British Mycology Society Symp. Series No. 12, Cambridge University Press, Cambridge, 1987, chap. 25.
151. **Boyd, D. A., Hobman, T. C., Grueke, S. A., and Klassen, G. R.,** Evolutionary stability of mitochondrial DNA organization in *Achyla*, *Can. J. Biochem. Cell Biol.*, 62, 571, 1984.
152. **Hudspeth, M. E. S., Shumard, D. S., Bradford, C. J. R., and Grossman, L. I.,** Organization of *Achlya* mtDNA: a population with two orientations and a large inverted repeat containing the rRNA genes, *Proc. Natl. Acad. Sci. U.S.A.*, 80, 142, 1983.
153. **Shumard, D. S., Grossman, L. I., and Hudspeth, M. E. S.,** *Achlya* mitochondrial DNA: gene localization and analysis of inverted repeats, *Mol. Gen. Genet.*, 202, 16, 1986.
154. **Förster, H., Kinscherf, T. G., Leong, S. A., and Maxwell, D. P.,** Molecular analysis of the mitochondrial genome of *Phytophthora*, *Curr. Genet.*, 12, 215, 1987.
155. **McNabb, S. A., Boyd, D. A., Belkhiri, A., Dick, M. W., and Klassen, G. R.,** An inverted repeat comprises more than three-quarters of the mitochondrial genome in two species of *Pythium*, *Curr. Genet.*, 12, 205, 1987.
156. **McNabb, S. A. and Klassen, G. R.,** Uniformity of mitochondrial DNA complexity in Oomycetes and the evolution of the inverted repeat, *Exp. Mycol.*, 12, 233, 1988.
157. **McNabb, S. A., Eros, R. W., and Klassen, G. R.,** Presence and absence of large inverted repeats in the mitochondrial DNA of Hyphochytriomycetes, *Can. J. Bot.*, 66, 2377, 1987.
158. **Shumard-Hudspeth, D. S. and Hudspeth, M. E. S.,** Genic rearrangements in *Phytophthora* mitochondrial DNA, *Curr. Genet.*, 17, 413, 1990.
159. **Klimczak, L. J. and Prell, H. H.,** Isolation and characterization of mitochondrial DNA of the Oomyceteous fungus *Phytophthora infestans*, *Curr. Genet.*, 8, 323, 1984.

160. **Förster, H., Kinschief, T. G., Leong, S. A., and Maxwell, D. G.,** Restriction fragment length polymorphisms of the mitochondrial DNA of *Phytophthora megasperma* isolated from soybean, alfalfa, and fruit trees, *Can. J. Bot.*, 67, 529, 1987.
161. **Belkhiri, A. and Dick, M. W.,** Comparative studies of the DNA of *Pythium* species and some possibly related taxa, *J. Gen. Microbiol.*, 134, 2673, 1988.
162. **Förster, H., Kinscherf, T. G., Leong, S. A., and Maxwell, D. P.,** Estimation of relatedness between *Phytophthora* species by analysis of mitochondrial DNA, *Mycologia*, 80, 466, 1988.
163. **Förster, H., Oudemans, P., and Coffey, M. D.,** Mitochondrial and nuclear DNA diversity within six species of *Phytophthora, Exp. Mycol.*, 14, 18, 1990.
164. **Hulbert, S. W. and Michelmore, R. W.,** DNA restriction fragment length polymorphism and somatic variation in the lettuce downy mildew fungus, *Bremia lactucae, Mol. Plant Microbe Interact.*, 1, 17, 1988.
165. **Kaufman, L. and Standard, P. G.,** Specific and rapid identification of medically important fungi by exoantigen detection, *Annu. Rev. Microbiol.*, 41, 209, 1987.
166. **Hardham, A., Suzaki, E., and Perkin, J. L.,** Monoclonal antibodies to isolate-, species-, and genus-specific components on the surface of zoospores and cysts of the fungus *Phytophthora cinnamomi, Can. J. Bot.*, 64, 311, 1986.
167. **Wycoff, K. L., Jellison, J., and Ayers, A. R.,** Monoclonal antibodies to glycoprotein antigens of a fungal plant pathogen, *Phytophthora megasperma* f. sp. *glycinea, Plant Physiol.*, 85, 508, 1987.
168. **Estrada-Garcis, M. T., Green, J. R., Booth, J. M., White, J. G., and Callow, J. A.,** Monoclonal antibodies to cell surface components of zoospores and cysts of the fungus *Phythium aphanidermatum* reveal species-specific antigens, *Exp. Mycol.*, 13, 348, 1989.
169. **Hardham, A. and Suzaki, E.,** Glycoconjugates on the surface of the pathogenic fungus *Phytophthora cinnamomi* studied using fluorescence and electron microscopy and flow cytometry, *Can. J. Microbiol.*, 36, 183, 1990.
170. **Wycoff, K. L. and Ayers, A. R.,** Monoclonal antibodies to surface and extracellular antigens of a fungal plant pathogen, *Phytophthora megasperma* f. sp. *glycinea*, recognize specific carbohydrate epitopes, *Physiol. Mol. Plant Pathol.*, 37, 55, 1990.
171. **Goodwin, P. H., Kirkpatrick, B. C., and Duniway, J. M.,** Cloned DNA probes for identification of *Phytophthora parasitica, Phytopathology*, 79, 716, 1989.
172. **Goodwin, P. H., English, J. T., Neher, D. A., Duniway, J. M., and Kirkpatrick, B. C.,** Detection of *Phytophthora parasitica* from soil and host tissue with a species-specific DNA probe, *Phytopathology*, 80, 277, 1989.
173. **Goodwin, P. H., Kirkpatrick, B. C., and Duniway, J. M.,** Identification of *Phytophthora citrophthora* with cloned DNA probes, *Appl. Environ. Microbiol.*, 56, 669, 1990.
174. **Werres, S. and Casper, R.,** Nachweis von *Phytophthora fragariae* Hickman in Wurzeln der Erdberkulturesorte 'Tenira' mit Hilfe des enzyme-linked immunosorbent assay (ELISA), *J. Phytopathol.*, 118, 367, 1987.
175. **Harrison, J. G., Barker, H., Lowe, R., and Rees, E. A.,** Estimation of amounts of *Phytophthora infestans* mycelium in leaf tissue by enzyme-linked immunosorbent assay, *Plant Pathol.*, 39, 274, 1990.
176. **Hansen, E. M., Brasier, C. M., Shaw, D. S., and Hamm, P. B.,** The taxonomic structure of *Phytophthora megasperma:* evidence for emerging biological species groups, *Trans. Br. Mycol Soc.*, 87, 557, 1987.
177. **Masago, H., Yoshikawa, M., Matsumoto, T., Katsumoto, M., Osata, H., and Kishibe, Y.,** Differentiation and identification of *Phytophthora* species by two-dimensional electrophoresis, *Ann. Phytopathol. Soc. Jpn.*, 55, 336, 1989.
178. **Howlett, B. J.,** An electrophoretic karyotype for *Phytophthora megasperma, Exp. Mycol.*, 13, 199, 1989.
179. **Judelson, H. S. and Michelmore, R. W.,** Structure and expression of a gene encoding heat-shock protein Hsp70 from the Oomycete fungus *Bremia lactucae, Gene*, 79, 207, 1989.
180. **Dudler, R.,** The single-copy actin gene of *Phytophthora megasperma* encodes a protein considerably diverged from any other known actin, *Plant Mol. Biol.*, 14, 415, 1990.
181. **Barr, D. J. S.,** The genera of phytopathogenic zoosporic fungi, in *Zoosporic Plant Pathogens: A Modern Perspective*, Buczacki, S. T., Ed., Academic Press, London, 1983, Appendix III.
182. **Fuller, M. S.,** Representative zoosporic fungi: introduction to organismal section, in *Zoosporic Fungi in Teaching and Research*, Fuller, M. S. and Jaworski, A., Eds., Southeastern Publishing Corp., Athens, GA, 1987, 3.
183. **Dick, M. W., Wong, P. T. W., and Clark, G.,** The identity of the Oomycete causing "Kikuyu Yellows", with a reclassification of the downy mildews, *Bot. J. Linn. Soc.*, 89, 171, 1984.
184. **Dick, M. W., Croft, B. J., Magarey, R. C., De Cock, A. W. A. M., and Clark, G.,** A new genus of the Verrucalvaceae (Oomycetes), *Bot. J. Linn. Soc.*, 99, 97, 1989.
185. **Hendrix, F. F. and Campbell, W. A.,** Pythiums as plant pathogens, *Annu. Rev. Phytopathol.*, 11, 77, 1973.
186. **Waterhouse, G. M.,** Peronosporales, in *The Fungi: An Advanced Treatise*, Vol. 4B, Ainsworth, G. C., Sparrow, F. K., and Sussman, A. S., Eds., Academic Press, New York, 1973, chap. 10.

187. **Hendrix, F. F. and Campbell, W. D.,** Some pythiaceous fungi — new roles for old organisms, in *Zoosporic Plant Pathogens: A Modern Perspective,* Buczacki, S. T., Ed., Academic Press, London, 1983, chap. 4.
188. **Stössel, P.,** Pathogenesis and host-parasite specificity of *Phytophthora,* in *Experimental and Conceptual Plant Pathology,* Vol. 3, Singh, R. S., Singh, U. S., Hess, W. M., and Weber, D. J., Eds., Gorden and Breach, Montreux, 1988, chap. 8.
189. **Frederiksen, R. A. and Renfro, B. L.,** Global status of maize downy mildew, *Annu. Rev. Phytopathol.,* 15, 249, 1977.
190. **Spencer, D. M., Ed.,** *The Downy Mildews,* Academic Press, New York, 1981.
191. **Williams, R. J.,** Downy mildews of tropical cereals, *Adv. Plant Pathol.,* 2, 1, 1984.
192. **Lucas, J. A. and Sherriff, C.,** Pathogenesis and host specificity in downy mildew fungus, in *Experimental and Conceptual Plant Pathology,* Vol. 3, Singh, R. S., Singh, U. S., Hess, W. M., and Weber, D. J., Eds., Gordon and Breach, Montreux, 1988, chap. 10.
193. **Michelmore, R. W., Ilott, T., Hulbert, S. H., and Farrara, B.,** The downy mildews, *Adv. Plant Pathol.,* 6, 53, 1988.
194. **Ploetz, R. C. and Mitchell, D. J.,** Root rot of bamboo palm caused by *Phytophthora arecae, Plant Dis.,* 73, 266, 1989.
195. **Larsson, M. and Gerhardson, B.,** Isolates of *Phytophthora cryptogea* pathogenic to wheat and some other crop plants, *J. Phytopathol.,* 129, 303, 1990.
196. **Sanders, P. L. and Coffey, M. D.,** Laboratory-induced resistance to Fosetyl-Al in a metalaxyl-resistant field isolate of *Pythium aphanidermatum, Plant Dis.,* 74, 690, 1990.
197. **Schwinn, F. J.,** New developments in chemical control of *Phytophthora,* in *Phytophthora: Its Biology, Taxonomy, Ecology and Pathology,* Erwin, D. C., Bartnicki-Garcia, S., and Tsao, P. H. Eds., American Phytopathology Society, St. Paul, 1983, chap. 25.
198. **Bruin, G. C. A. and Edgington, L. V.,** The chemical control of diseases caused by zoosporic fungi, in *Zoosporic Plant Pathogens: A Modern Perspective,* Buczacki, S. T., Ed., Academic Press, London, 1983, chap. 6.
199. **Schwinn, F. J. and Urech, P. A.,** Progress in the chemical control of diseases caused by the Oomycetes, in *Fungicide Chemistry: Advances and Practical Applications,* Green, M. B. and Spilker, D. A., Eds., ACS Symp. Series No. 304, American Chemical Society, Washington, D.C., 1986, chap. 5.
200. **Schwinn, F. J. and Staub, T.,** Phenylamides and other fungicides against Oomycetes, in *Modern Selective Fungicides: Properties, Applications and Mechanisms of Action,* Lyr, H., Ed., Longman Scientific, Harlow, 1987, chap. 17.
201. **Cohen, Y. and Coffey, M. D.,** Systemic fungicides and the control of Oomycetes, *Annu. Rev. Phytopathol.,* 24, 311, 1986.
202. **Hall, G.,** Unusual or interesting records of plant pathogenic Oomycetes, *Plant Pathol.,* 38, 604, 1989.
203. **Pegg, K. G. and Alcorn, J. L.,** *Phytophthora operculata* sp. nov., a new marine fungus, *Mycotaxon,* 26, 99, 1982.
204. **Dick, M. W.,** *Phytophthora undulata* comb. nov., *Mycotaxon,* 35, 449, 1989.
205. **Taylor, P.A. and Greenhalgh, F. C.,** Significance, causes and control of root rots of subterranean clover, in *Temperate Pastures: Their Production, Use and Management,* Wheeler, J. L., Pearson, C. J., and Robards, G. E., Eds., Commonwealth Scientific Research Organization, 1987, 249.
206. **Hickman, C. J.,** *Phytophthora* — plant destroyer, *Trans. Br. Mycol. Soc.* 41, 1, 1958.
207. **Ribeiro, O. K.,** *A Source Book of the Genus Phytophthora,* J. Cramer, Vaduz, 1978.
208. **Zentmeyer, G. A.,** *Phytophthora cinnamomi and the Diseases It Causes,* Monogr. No. 10, American Phytopathology Society, St. Paul, 1980.
209. **Weste, G. and Marks, G. C.,** The biology of *Phytophthora cinnamomi* in Australian forests, *Annu. Rev. Phytopathol.,* 25, 207, 1987.
210. **Klinkowski, M.,** Catastrophic plant diseases, *Annu. Rev. Phytopathol.,* 8, 37, 1970.
211. **Gregory, P. H.,** Some major epidemics caused by *Phytophthora,* in *Phytophthora: Its Biology, Taxonomy, Ecology and Pathology,* Erwin, D. C., Bartnicki-Garcia, S., and Tsao, P. H., Eds., American Phytopathology, Society, St. Paul, 1983, chap. 20.
212. **Large, E. C.,** *The Advance of the Fungi,* Johnathan Cape, London, 1940, chap. 34.
213. **Niederhauser, J. S. and Cobb, W. C.,** The late blight of potatoes, *Sci. Am.,* 250, 100, 1959.
214. **Woodham-Smith, C.,** *The Great Hunger: Ireland 1845—9,* Hamish-Hamilton, London, 1962.
215. **McCallan, S. E. A.,** History of fungicides, in *Fungicides: An Advanced Treatise,* Vol. 1, Torgeson, D. E., Ed., Academic Press, New York, 1967, chap. 1.
216. **Weste, G.,** Population dynamics and survival of *Phytophthora,* in *Phytophthora: Its Biology, Taxonomy, Ecology and Pathology,* Erwin, D. C., Bartnicki-Garcia, S., and Tsao, P. H., Eds., American Phytopathology Society, St. Paul, 1983, chap. 18.
217. **Rippon, J. W.,** *Medical Mycology: The Pathogenic Fungi and the Pathogenic Actinomycetes,* 3rd ed., W. B. Saunders, Philadelphia, 1988, chap. 26.

218. **Smith, J. M. B.,** *Opportunistic Mycoses of Man and Other Animals,* CAB International, Wallingford, 1989, chap. 5.
219. **Connole, M. D.,** Review of animal mycoses in Australia, *Mycopathologia,* 111, 133, 1990.
220. **Sathapatayavongs, B., Leelachaikul, P., Prachaktam, R., Atichartakarn, V., Sriphojanart, S., Trairatvorakul, P., Jirasiritham, S., Nontasut, S., Euvilaichit, C., and Flegel, T.,** Human pythiosis associated with thalassemia hemoglobinopathy syndrome, *J. Infect. Dis.,* 159, 274, 1989.
221. **Tanphaichitra, D.,** AIDS and tropical diseases: melioidosis, pythiosis and penicilliosis, *Arch. AIDS Res.,* 4, 77, 1990.
222. **Kerridge, D. and Whelen, W. L.,** The polyene macrolide antiboitics and 5-fluorocytosine: molecular actions and interactions, in *Mode of Action of Antifungal Agents,* Trinci, A. P. J. and Ryley, J. F., Eds., British Mycology Society Symp., Cambridge University Press, Cambridge, 1984, chap. 16.
223. **Tsao, P. H.,** Factors affecting isolation and quantitation of *Phytophthora* from soil, in *Phytophthora: Its Biology, Taxonomy, Ecology and Pathology,* Erwin, D. C., Bartnicki-Garcia, S., and Tsao, P. H., Eds., American Phytopathology Society, St. Paul, 1983, chap. 17.
224. **Griffin, M. J. and Coley-Smith, J. R.,** Uptake of streptomycin by sporangia of *Pseudoperonospora humili* and the inhibition of uptake by divalent metal cations, *Trans. Br. Mycol. Soc.,* 65, 265, 1975.
225. **Seymour, R. L.,** The genus *Saprolegnia, Nova Hedwigia,* 19, 1, 1970.
226. **Dick, M. W.,** Saprolegniales, in *The Fungi: An Advanced Treatise,* Vol. 4B, Ainsworth, G. C., Sparrow, F. K., and Sussman, A. S., Eds., Academic Press, New York, 1973, chap. 7.
227. **Seymour, R. L.,** *Leptolegnia chapmanii,* an Oomycete pathogen of mosquito larvae, *Mycologia,* 76, 670, 1984.
228. **Allen, R. N., Letham, D. B., Akehurst, A. A., and Say, M. M.,** *Aphanomyces* root rot of bean at Vall, New South Wales, *Aust. Plant Pathol.,* 16, 82, 1987.
229. **Wilson, J.,** Immune response in fish to fungal disease, in *Recent Advances in Aquatic Mycology,* Jones, E. B. G., Ed., John Wiley & Sons, New York, 1976, chap. 22.
230. **Haitai, K., Willoughby, L. G., and Beakes, G. W.,** Some characteristics of *Saprolegnia* obtained from fish hatcheries in Japan, *Mycol. Res.,* 93, 182, 1990.
231. **Dykstra, M. J., Noga, E. J., Levine, J. F., and Moye, D. W.,** Characterization of the *Aphanomyces* species involved with ulcerative mycosis (UM) in menhaden, *Mycologia,* 78, 664, 1986.
232. **Hearth, J. H. and Padgett, D. E.,** Salinity tolerance of an *Aphanomyces* isolate (Oomycetes) and its possible relationship to ulcerative mycoses (UM) of Atlantic menhaden, *Mycologia,* 82, 364, 1990.
233. **Alderman, D. J.,** Fungal diseases of marine animals, in *Recent Advances in Aquatic Mycology,* Jones, E. B. G., Ed., John Wiley & Sons, New York, 1976, chap. 9.
234. **Ludwig-Köhn, H., Jahnke, K.-D., and Bahnweg, G.,** Sterol composition and sterol metabolism of *Haliphthoros milfordensis* and *Atkinsiella dubia,* fungal parasites of marine crustaceans, *Biochim. Biophys. Acta,* 713, 463, 1982.
235. **Unestam, T.,** On the adaption of *Aphanomyces astaci* as a parasite, *Physiol. Plant.,* 22, 221, 1969.
236. **Marren, P.,** The lethal harvest of crayfish plague, *New Sci.,* 109, 46, 1986.
237. **Alderman, D. J. and Polglase, J. L.,** Pathogens, parasites and commensals, in *Freshwater Crayfish: Biology, Management and Exploitation,* Holdich, D. M. and Lowery, R. S., Eds., Croon Helm, London, 1988, chap. 7.
238. **Alderman, D. J. and Polglase, J. L.,** A comparative investigation of the effects of fungicides on *Saprolegnia parasitica* and *Aphanomyces astaci, Trans. Br. Mycol. Soc.,* 83, 313, 1984.
239. **Unestam, T.,** Defence reactions in and susceptibility of Australian and New Guinean freshwater crayfish to European-crayfish-plague fungus, *Aust. J. Exp. Biol.,* 53, 349, 1975.
240. **Sokol, A.,** The Australian yabby, in *Freshwater Crayfish: Biology, Management and Exploitation,* Holdich, D. M. and Lowery, R. S., Eds., Croon Helm, London, 1988, chap. 15.
241. **Martin, R. W. and Miller, C. E.,** Ultrastructure of zoosporogenesis in the endoparasite *Olpidiopsis varians, Mycologia,* 78, 230, 1986.
242. **Lord, J. C. and Fukada, T.,** An ultrastructural study of the invasion of *Culex quinquefasciatus* larvae by *Leptolegnia chapmanii* (Oomycetes: Saprolegniales), *Mycopathologia,* 104, 67, 1988.
243. **Kerwin, J. L., Dritz, D. A., and Washino, R. K.,** Confirmation of the safety of *Lagenidium giganteum* (Oomycetes: Lagenidiales) to mammals, *J. Econ. Entomol.,* 83, 374, 1990.
244. **Dick, M. W.,** Leptomitales, in *The Fungi: An Advanced Treatise,* Vol. 4B, Ainsworth, G. C., Sparrow, F. K. and Sussman, A. S., Eds., Academic Press, New York, 1973, chap. 8.
245. **Sparrow, F. K.,** Lagenidiales, in *The Fungi: An Advanced Treatise,* Vol. 4B, Ainsworth, G. C., Sparrow, F. K., and Sussman, A. S., Eds., Academic Press, New York, 1973, chap. 9.
246. **Barr, D. J. S. and Désaulniers, N. L.,** The life cycle of *Lagena radicicola,* an Oomyceteous parasite of wheat roots, *Can. J. Bot.,* 68, 813, 1990.
247. **van der Meer, J. P. and Pueschel, C. M.,** *Petersenia palmariae* n. sp. (Oomycetes): a pathogenic parasite of the red alga *Palmaria mollis* (Rhodophyceae), *Can. J. Bot.,* 63, 404, 1984.
248. **Pueschel, C. M. and van der Meer, J. P.,** Ultrastructure of the fungus *Petersenia palmariae* (Oomycetes) parasite on the alga *Palmari mollis* (Rhodophyceae), *Can. J. Bot.,* 63, 409, 1984.

249. **Duniway, J. M.,** Role of physical factors in the development of *Phytophthora* diseases, in *Phytophthora: Its Biology, Taxonomy, Ecology and Pathology,* Erwin, D. C., Bartnicki-Garcia, S., and Tsao, P. H., Eds., American Phytopathology Society, St. Paul, 1983, chap. 14.
250. **Lutz, A. and Menge, J.,** Breaking of winter dormancy of *Phytophthora parasitica* propagules using heat shock, *Mycologia,* 78, 148, 1986.
251. **Dick, M. W.,** Coevolution of the Heterokont fungi (with emphasis on the downy mildews and their angiosperm hosts), in *Coevolution of Fungi with Plants and Animals,* Pirozynski, K. A. and Hawksworth, D. L., Eds., Academic Press, London, 1988, chap. 2.
252. **Olsen, L. W., Cerenius, L., Lange, L., and Söderhall, K.,** The primary and secondary spore cyst of *Aphanomyces* (Oomycetes), *Nord. J. Bot.,* 4, 681, 1984.
253. **Aylor, D. E.,** The role of intermittent wind in the dispersal of fungal pathogens, *Annu. Rev. Phytopathol.,* 28, 73, 1990.
254. **Gisi, U., Hemmes, D. E., and Zentmeyer, G. A.,** Origin and significance of the discharge vesicle in *Phytophthora, Exp. Mycol.,* 3, 321, 1979.
255. **Gisi, U. and Zentmeyer, G. A.,** Mechanism of zoospore release in *Phytophthora* and *Pythium, Exp. Mycol.,* 4, 362, 1980.
256. **Ribeiro, O. K.,** Physiology of asexual sporulation and spore germination in *Phytophthora,* in *Phytophthora: Its Biology, Taxonomy, Ecology and Pathology,* Erwin, D. C., Bartnicki-Garcia, S., and Tsao, P. H., Eds., American Phytopathology Society, St. Paul, 1983, chap. 4.
257. **Lange, L. and Olsen, L. W.,** The fungal zoospore — its structure and biological significance, in *Zoosporic Plant Pathogens: A Modern Perspective,* Buczacki, S. T., Ed., Academic Press, London, 1983, chap. 1.
258. **Carlile, M. J.,** Motility, taxis, and tropism in *Phytophthora,* in *Phytophthora: Its Biology, Taxonomy, Ecology and Pathology,* Erwin, D. C., Bartnicki-Garcia, S., and Tsao, P. H., Eds., American Phytopathology Society, St. Paul, 1983, chap. 7.
259. **Carlile, M. J.,** The zoospore and its problems, in *Water, Fungi and Plants,* Ayres, P. G. and Boddy, L., Eds., British Mycology Symp. Series No. 11, Cambridge University Press, Cambridge, 1986, chap. 6.
260. **Cerenius, L. and Söderhall, K.,** Repeated zoospore emergence from isolated spore cysts of *Aphanomyces astaci, Exp. Mycol.,* 8, 370, 1984.
261. **Cerenius, L. and Söderhall, K.,** Repeated zoospore emergence as a possible adaption to parasitism in *Aphanomyces, Exp. Mycol.,* 9, 259, 1985.
262. **Hinch, J. M. and Clarke, A. E.,** Adhesion of fungal zoospores to root surfaces is mediated by carbohydrate determinants of the root slime, *Physiol. Plant Pathol.,* 16, 303, 1980.
263. **Irving, H. R. and Grant, B. R.,** The effects of pectin and plant root surface carbohydrates on encystment and development of *Phytophthora cinnamomi* zoospores, *J. Gen. Microbiol.,* 130, 1015, 1984.
264. **Grant, B. R., Irving, H. R., and Radda, M.,** The effect of pectin and related compounds on encystment and germination of *Phytophthora palmivora* zoospores, *J. Gen. Microbiol.,* 131, 669, 1985.
265. **Longman, D. and Callow, J. A.,** Specific saccharide residues are involved in the recognition of plant root surfaces by zoospores of *Pythium aphanidermatum, Physiol. Mol. Plant Pathol.,* 30, 139, 1987.
266. **Hohl, H. R. and Balsiger, S.,** Surface glycosyl receptors of *Phytophthora megasperma* f. sp. *glycinea* and its soybean host, *Bot. Helv.,* 98, 271, 1988.
267. **Aveling, T. A. S. and Rijkenberg, F. H. J.,** Behaviour of *Phytophthora cinnamomi* zoospores on roots of four avocado cultivars, *J. Phytopathol.,* 125, 157, 1989.
268. **Cerenius, L. and Söderhall, K.,** Chemotaxis in *Aphanomyces astaci,* an arthropod parasitic fungus, *J. Invert. Pathol.,* 43, 278, 1984.
269. **Miller, R. I.,** Investigations into the biology of three 'phycomycotic' agents pathogenic for horses in Australia, *Mycopathology,* 81, 23, 1983.
270. **Thomas, D. D. and Peterson, A. P.,** Chemotactic autoaggregation in the water mould *Achlya, J. Gen. Microbiol.,* 136, 847, 1990.
271. **Gooday, G. W.,** Inhibition of chitin metabolism, in *Biochemistry of Cell Walls and Membranes in Fungi,* Kuhn, P. J., Trinci, A. P. J., Jung, M. I., Goosey, M. W., and Copping, L. G., Eds., Springer-Verlag, Heidelberg, 1990, chap 5.
272. **Schlumbaum, A., Mauch, F., Vögeli, U., and Boller, T.,** Plant chitinases are potent inhibitors of fungal growth, *Nature,* 324, 365, 1986.
273. **Kombrink, E., Schröder, M., and Hahlbrock, K.,** Several "pathogenesis-related" proteins in potato are 1,3-β-glucanases and chitinases, *Proc. Natl. Acad. Sci. U.S.A.,* 85, 782, 1988.
274. **Mauch, F., Mauch-Mani, B., and Boller, T.,** Antifungal hydrolases in pea tissue. II. Inhibition of fungal growth by combinations of chitinase and β-1,3-glucanase, *Plant Physiol.,* 88, 936, 1988.
275. **Bol, J. F., Linthorst, H. J. M., and Cornelissen, B. J. C.,** Plant pathogenesis-related proteins induced by virus infection, *Annu. Rev. Phytopathol.,* 28, 113, 1990.
276. **Davidse, L. C.,** Benzimidazole fungicides: mechanism of action and biological impact, *Annu. Rev. Phytopathol.,* 24, 43, 1986.

277. **Davidse, L. C.,** Biochemical aspects of benzimidazole fungicides — action and resistance, in *Modern Selective Fungicides: Properties, Applications and Mechanisms of Action,* Lyr, H., Ed., Longman Scientific, Harlow, 1987, chap. 16.
278. **Schlösser, E.,** Sterol dependent membrane lytic action of saponins, *Phytopathol. Z.,* 74, 91, 1972.
279. **Roddick, J. D.,** Antifungal activity of plant steroids, in *Ecology and Metabolism of Plant Lipids,* Fuller, G. and Nes, W. D., Eds., ACS Symp. Series No. 325, American Chemical Society, Washington, D.C., 1987, chap. 18.
280. **Christian, D. A. and Hardwinger, L. A.,** Pea saponins in the pea-*Fusarium solani* interaction, *Exp. Mycol.,* 13, 419, 1989.
281. **Berg, D.,** Biochemical mode of action of fungicides, in *Fungicide Chemistry: Advances and Practical Applications,* Green, M. B. and Spilker, D. A., Eds., ACS Symp. Series No. 304, American Chemical Society, Washington, D.C., 1986, chap. 2.
282. **Kerkenaar, A.,** Mechanism of action of morpholine fungicides, in *Modern Selective Fungicides: Properties, Applications and Mechanisms of Action,* Lyr, H., Ed., Longman Scientific, Harlow, 1987, chap. 12.
283. **Buchenauer, H.,** Mechanism of action of triazolyl fungicides and related compounds, in *Modern Selective Fungicides: Properties, Applications and Mechanisms of Action,* Lyr, H., Ed., Longman Scientific, Harlow, 1987, chap. 14.
284. **Scheinpflug, H. and Kuck, K. H.,** Sterol biosynthesis-inhibiting piperazine, pyridine, pyrimidine and azole fungicides, in *Modern Selective Fungicides: Properties, Applications and Mechanisms of Action,* Lyr, H., Ed., Longman Scientific, Harlow, 1987, chap. 13.
285. **Kelly, S. L., Kenna, S., Bligh, H. B. J., Watson, P. F., Stansfield, I., Ellis, S. W., and Kelly, D. E.,** Lanosterol to ergosterol — enzymology, inhibition and genetics, in *Biochemistry of Cell Walls and Membranes in Fungi,* Kuhn, P. J., Trinci, A. P. J., Jung, M. J., Goosey, M. W., and Copping, L. G., Eds., Springer-Verlag, Heidelberg, 1990, chap. 15.
286. **Buchenauer, H.,** Comparative studies on the antifungal activity of triadimefon, triadimenol, fenarimol, naurimol, imazalil and fluotrimazole *in vitro, Z. Pflanzenkr. Pflanzenschutz,* 86, 341, 1979.
287. **Ryder, N. S.,** Squalene epoxidase — enzymology and inhibition, in *Biochemistry of Cell Walls and Membranes in Fungi,* Kuhn, P. J., Trinci, A. P. J., Jung, M. J., Goosey, M. W., and Copping, L. G., Eds., Springer-Verlag, Heidelberg, 1990, chap. 13.
288. **Robson, G. D., Wiebe, M., Kuhn, P. J., and Trinci, A. P. J.,** Inhibitors of phospholipid biosynthesis, in *Biochemistry of Cell Walls and Membranes in Fungi,* Kuhn, P. J., Trinci, A. P. J., Jung, M. J., Goosey, M. W., and Copping, L. G., Eds., Springer-Verlag, Heidelberg, 1990, chap. 17.
289. **Kulka, M. and von Schemling, B.,** Carboxin fungicides and related compounds, in *Modern Selective Fungicides: Properties, Applications and Mechanisms of Action,* Lyr, H., Ed., Longman Scientific, Harlow, 1987, chap. 9.
290. **Schewe, T. and Lyr, H.,** Mechanism of action of carboxin fungicides and related compounds, in *Modern Selective Fungicides: Properties, Applications and Mechanisms of Action,* Lyr, H., Ed., Longman Scientific, Harlow, 1987, chap. 10.
291. **Bell, A. A. and Wheeler, M. H.,** Biosynthesis and functions of fungal melanins, *Annu. Rev. Phytopathol.,* 24, 411, 1986.
292. **Sisler, H. D.,** Control of fungal diseases by compounds acting as antipenetrants, *Crop Prot.,* 5, 306, 1986.
293. **Sisler, H. D. and Ragsdale, N. N.,** Disease control by nonfungitoxic compounds, in *Modern Selective Fungicides: Properties, Applications and Mechanisms of Action,* Lyr, H., Ed., Longman Scientific, Harlow, 1987, chap. 23.
294. **Howard, R. J. and Ferrari, M. A.,** Role of melanin in appressorium function, *Exp. Mycol.,* 13, 403, 1989.
295. **Shaw, D. S.,** The Peronosporales — a fungal geneticist's nightmare, in *Zoosporic Plant Pathogens: A Modern Perspective,* Buczacki, S. T., Ed., Academic Press, London, 1983, chap. 3.
296. **Thompson, W., Mao, O., Kashemsanta, O., and Tyler, B. M.,** Molecular genetics of *Phytophthora megasperma* f. sp. *glycinea, J. Cell. Biochem.,* 13 E, 33, 1989.
297. **Layton, A. C. and Kuhn, D. N.,** Heterokaryote formation by protoplast fusion of drug-resistant mutants in *Phytophthora megasperma* f. sp. *glycinea, Exp. Mycol.,* 12, 180, 1988.
298. **Spielman, L. J., Sweigard, J. A., Shattock, R. C., and Fry, W. E.,** The genetics of *Phytophthora infestans:* segregation of allozyme markers in F2 and backcross progeny and the inheritance of virulence against potato resistance genes *R2* and *R4* in F1 progeny, *Exp. Mycol.,* 14, 57, 1990.
299. **Klassen, G. R. and Buchko, J.,** Subrepeat structure of the intergenic region in the ribosomal DNA of the oomyceteous fungus *Pythium ultimum, Curr. Genet.,* 17, 125, 1990.
300. **Nuss, D. L.,** Significance of dsRNA genetic elements in plant pathogenic fungi, *Annu. Rev. Phytopathol.,* 28, 37, 1990.
301. **Davidse, L. C. and de Waard, M. A.,** Systemic fungicides, *Adv. Plant Pathol.,* 2, 191, 1984.
302. **Davidse, L. C.,** Biochemical aspects of phenylamide fungicides — action and resistance, in *Modern Selective Fungicides,* Lyr, H., Ed., Longman Scientific, Essex, 1987, 275.

303. **Coffey, M. D. and Ouimette, D. G.,** Phosphonates: antifungal compounds against Oomycetes, in *Nitrogen, Phosphorus and Sulphur Utilization by Fungi,* Boddy, L., Marchant, R., and Read, D. J., Eds., Cambridge University Press, Cambridge, 1989, 107.
304. **Guest, D. I. and Grant, B. R.,** The complex action of phosphonates as antifungal agents, *Biol. Rev.,* 66, 159, 1991.
305. **de Waard, M. A.,** Synergism and antagonism in fungicides, *Modern Selective Fungicides,* Lyr, H., Ed., Longman Scientific, Essex, 1987, 355.
306. **Cohen, Y. and Levy, Y.,** Joint action of fungicides in mixtures: theory and practice, *Phytoparasitica,* 18, 159, 1990.
307. The British Crop Protection Council, *The Pesticide Manual — a World Compendium,* Worthing, C. R. and Walker, S. B., Eds., 1987, 314.
308. **Kamimura, S., Nishikawa, M., Saeki, H., and Takahi, Y.,** Absorption and metabolism of 3-hydroxy-5-methylisoxazole in plants and the biological activities of its metabolites, *Phytopathology,* 64, 1273, 1974.
309. **Kamimura, S., Nishikawa, M., and Takahi, Y.,** Mode of action of soil fungicide hymexazol, 3-hydroxy-5-methylisoxazole, on *Fusarium oxysporum* f. *cucumerinum, Ann. Phytopathol. Soc. Jpn.,* 42, 242, 1976.
310. **Tsao, P. H. and Guy, S. O.,** Inhibition of *Mortierella* and *Pythium* in a *Phytophthora*-isolation medium containing hymexazol, *Phytopathology,* 67, 796, 1977.
311. **Ho, H. H.,** Effect of hymexazol on growth and reproduction of low temperature *Phytophthora* species, *Mycopathologia,* 98, 17, 1987.
312. **Kato, S., Coe, R., New, L., and Dick, M. W.,** Sensitivities of various Oomycetes to hymexazol and metalaxyl, *J. Gen. Microbiol.,* 136, 2127, 1990.
313. **Halos, P. M. and Huisman, O. C.,** Inhibition of respiration in *Pythium* species by ethazol, *Phytopathology,* 66, 158, 1976.
314. **Halos, P. M. and Huisman, O. C.,** Mechanism of tolerance of *Pythium* species to ethazol, *Phytopathology,* 66, 152, 1976.
315. **Davidse, L. C., Looyen, D., Turkensteen, L. J., and Van der Wal, D.,** Occurrence of metalaxyl-resistant strains of *Phytophthora infestans* in Dutch potato fields, *Neth. J. Plant Pathol.,* 87, 65, 1981.
316. **Samoucha, Y. and Cohen, Y.,** Occurrence of metalaxyl-resistant isolates of *Pseudoperonospora cubensis* in Israel: a 5-year survey, *Bulletin OEPP/EPPO,* 15, 419, 1985.
317. **Davidse, L. C. and van den Berg-Velthius, G. C. M.,** Biochemical and molecular aspects of the phenylamide fungicide-receptor interaction in plant pathogenic *Phytophthora* spp., in *Signal Molecules in Plants and Plant-Microbe Interactions,* Lugtenberg, B. J. J., Ed., Springer-Verlag, Berlin, 1989, 261.
318. **Davidse, L. C.,** Biochemical basis of resistance to phenylamide fungicides, in *Managing Resistance to Agrochemicals: From Fundamental Research to Practical Strategies,* Green, M. G., Le Baron, H. M., and Moberg, W. K., Eds., ACS Symp. Ser. 421, American Chemical Society, Washington, D.C., 1990, 215.
319. **Diriwachter, G., Sozzi, D., Ney, C., and Staub, T.,** Cross-resistance in *Phytophthora infestans* and *Plasmopara viticola* against different phenylamides and unrelated fungicides, *Crop Prot.,* 6, 250, 1987.
320. **Fisher, D. J. and Hayes, A. L.,** A comparison of the biochemical and physiological effects of the systemic fungicide cyprofuram with those of the related compounds metalaxyl and metachlor, *Crop Prot.,* 4, 501, 1985.
321. **Davidse, L. C., Gerritsma, O. C. M., Ideler, J., Pie, K., and van den Berg-Velthuis, G. C. M.,** Antifungal modes of action of metalaxyl, cyprofuram, benalaxyl and oxadixyl in phenylamide-sensitive and phenylamide-resistant strains of *Phytophthora megasperma* f. sp. *medicaginis* and *Phytophthora infestans, Crop Prot.,* 7, 347, 1988.
322. **Davidse, L. C., Henken, J., van Dalen, A., Jespers, A. B. K., and Mantel, B. C.,** Nine years of practical experience with phenylamide resistance in *Phytophthora infestans* in the Netherlands, *Neth. J. Plant Pathol.,* 95(Suppl. 1), 197, 1989.
323. **Wade, M. and Delp, C. J.,** Aims and activities of industry's fungicide resistance action committee (FRAC), *Bull. OEPP/EPPO,* 15, 577, 1985.
324. **Smith, J. M. and Margot, P.,** The use of metalaxyl as a seed or soil treatment, in *Application to Seeds and Soil,* BCPC Monogr. No. 39, 1987, 41.
325. **Samoucha, Y. and Cohen, Y.,** Toxicity of propamocarb to the late blight fungus on potato, *Phytoparasitica,* 18, 27, 1990.
326. **Burden, R. S., Carter, G. A., James, C. S., Clark, T., and Holloway, P. J.,** Selective effects of propamocarb and prothiocarb on the fatty acid composition of some Oomycetes, *Proc. Int. Crop Prot. Conf. Pests Dis.,* p. 403, 1988.
327. **Klopping, H. L. and Delp, C. J.,** 2-Cyano-N-[(ethylamino)carbonyl]-2-(methoxyimino)acetamide, a new fungicide, *J. Agric. Food Chem.,* 28, 467, 1980.
328. **Samoucha, Y. and Gisi, U.,** Systemicity and persistence of cymoxanil in mixture with oxadixyl and mancozeb against *Phytophthora infestans* and *Plasmopara viticola, Crop Prot.,* 6, 393, 1987.
329. **Samoucha, Y., Levy, R. S., and Cohen, Y.,** Efficacy over time of cymoxanil mixtures on controlling late blight in potatoes incited by a phenylamide-resistant isolate of *Phytophthora infestans, Crop. Prot.,* 7, 210, 1988.

330. **Ziogas, B. N. and Davidse, L. C.,** Studies on the mechanism of action of cymoxanil in *Phytophthora infestans, Pestic. Biochem. Physiol.*, 29, 89, 1987.
331. **Williams, D. J., Beach, B. G. W., Horriere, D., and Marechal, G.,** LS 74-783, a new systemic fungicide with activity against Phycomycete diseases, *Proc. Br. Crop Prot. Conf. Pests Dis.*, p. 565, 1977.
332. **Ouimette, D. G. and Coffey, M. D.,** Phosphonate levels in avocado seedlings and soil following treatment with fosetyl-Al or potassium phosphonate, *Plant Dis.*, 73, 212, 1989.
333. **Whiley, A. W., Pegg, K. G., Saranah, J. B., and Forsberg, L. I.,** The control of *Phytophthora* root rot of avocado with fungicides and the effects of this disease on water relations, yield and ring neck, *Aust. J. Exp. Agric.*, 26, 249, 1986.
334. **Fenn, M. E. and Coffey, M. D.,** Studies on the *in vitro* and *in vivo* antifungal activity of fosetyl-Al and phosphorous acid, *Phytopathology*, 74, 606, 1984.
335. **Pegg, K. G., Whiley, A. W., Saranah, J. B., and Glass, R. J.,** Control of *Phytophthora* root rot of avocado with phosphorous acid, *Australas. Plant Pathol.*, 14, 25, 1985.
336. **Groussol, J., Delrot, S., Caruhel, P., and Bonneman, J.-L.,** Design of an improved exudation method for phloem sap collection and its use for the study of phloem mobility of pesticides, *Physiol. Veg.*, 24, 123, 1986.
337. **Luttringer, M. and de Cormis, L.,** Absorption, degradation et transport du phosethyl-Al et de son metabolite chez la tomate, *Agronomie*, 5, 423, 1985.
338. **Darvas, J. M., Toerien, J. C., and Milne, D. L.,** Control of avocado root rot by trunk injection with phosetyl-Al, *Plant Dis.*, 68, 691, 1984.
339. **de Boer, R. F., Greenhalgh, F. C., Pegg, K. G., Mayers, P. E., Lim, T. M., and Flett, S.,** Phosphorous acid treatments control *Phytophthora* diseases in Australia, *Bull. OEPP/EPPO*, 20, 193, 1990.
340. **Bompeix, G.,** Fungicides and host-parasite interactions: the case of phosphonates, *C. R. Acad. Agric. Fr.*, 75, 183, 1989.
341. **Bompeix, G. and Saindrenan, P.,** In vitro antifungal activity of fosetyl-Al and phosphorous acid on *Phytophthora* species, *Fruits*, 39, 777, 1984.
342. **Sanders, P. L., Coffey, M. D., Greer, G. D., and Soika, M. D.,** Laboratory-induced resistance to fosetyl-Al in a metalaxyl-resistant field isolate of *Pythium aphanidermatum, Plant Dis.*, 74, 690, 1990.
343. **Barchietto, T., Saindrenan, P., and Bompeix, G.,** Characterisation of phosphonate uptake in two *Phytophthora* species and its inhibition by phosphate, *Arch. Microbiol.*, 151, 54, 1989.
344. **Griffith, J. M., Akins, L. A., and Grant, B. R.,** Properties of the phosphite and phosphate uptake transport systems in *Phytophthora palmivora, Arch. Microbiol.*, 152, 430, 1989.
345. **Griffith, J. M., Smillie, R. H., Niere, J. O., and Grant, B. R.,** Effects of phosphate on the toxicity of phosphite in *Phytophthora palmivora, Arch. Microbiol.*, 152, 425, 1989.
346. **Leconte, F., Bonnemain, J. L., Delrot, S., Bompeix, G., and Besson, J.,** Properties and mechanism of phosphite uptake (ambimobile antifungal compound) in leaf tissues, in Int. Con. on Phloem Transport and Assimilate Compartmentation, Cognac, France, August 19 to 24, 1990.
347. **Ouimette, D. G. and Coffey, M. D.,** Symplastic entry and phloem translocation of phosphonate, *Pestic. Biochem. Physiol.*, 38, 18, 1990.
348. **Fenn, M. E. and Coffey, M. D.,** Further evidence for the direct mode of action of fosetyl-Al and phosphorous acid, *Phytopathology*, 75, 1064, 1985.
349. **Fenn, M. E. and Coffey, M. D.,** Quantification of phosphonate and ethyl phosphonate in tobacco and tomato tissues and significance for mode of action of two phosphonate fungicides, *Phytopathology*, 79, 76, 1989.
350. **Saindrenan, P., Barchietto, T., Avelino, J., and Bompeix, G.,** Effects of phosphite on phytoalexin accumulation in leaves of cowpea infected with *Phytophthora cryptogea, Physiol. Mol. Plant Pathol.*, 32, 425, 1988.
351. **Smillie, R. H., Grant, B. R., and Guest, D. I.,** The mode of action of phosphite: evidence for both direct and indirect modes of action on three *Phytophthora* spp. in plants, *Phytopathology*, 79, 921, 1989.
352. **d'Arcy-Lameta, A., Bompeix, G., Reymond, V., and Soulie, M.-C.,** Phosphonate uptake and distribution in healthy and infected leaves of tomato and cowpea, *Plant Physiol. Biochem.*, 27, 919, 1989.
353. **Saindrenan, P. and Bompeix, G.,** Role des phytoalexines dans la response de *Vigna ungiculata* traite par le phosethyl-Al, a l'infection par *Phytophthora cryptogea, C. R. Seances Acad. Sci. Paris*, Ser. III 10, 411, 1986.
354. **Guest, D. I.,** Modification of defence responses in tobacco and capsicum following treatment with fosetyl-Al [aluminum tris(o-ethyl phosphonate)], *Physiol. Plant Pathol.*, 25, 125, 1984.
355. **Guest, D. I., Upton, J. C. R., and Rowan, K. S. R.,** Fosetyl-Al alters the respiratory response in *Phytophthora nicotianae* var. *nicotianae*-infected tobacco, *Physiol Mol. Plant Pathol.*, 34, 257, 1989.
356. **Bower, L. A. and Coffey, M. D.,** Development of laboratory tolerance to phosphorous acid, fosetyl-Al and metalaxyl in *Phytophthora capsici, Can. J. Plant Pathol.*, 7, 1, 1985.
357. **Dolan, T. E. and Coffey, M. D.,** Correlative in vitro and in vivo behaviour of mutant strains of *Phytophthora palmivora* expressing different resistances to phosphorous acid and fosetyl-Na, *Phytopathology*, 78, 974, 1988.

358. **Neema, C., Bompeix, G., and Mourichon, X.,** Induction of *Phytophthora citrophthora* mutants resistant to fosetyl-Al and phosphorous acid and study of their pathogenicity towards citrus, *Fruits,* 43, 499, 1988.
359. **Abu-Jawdah, Y.,** Cell swellings of *Colletotrichum lindemuthianum* (Sacc. & Magn.) Bri. & Cav. induced by treatment with fosetyl-Al, *Phytoparasitica,* 39, 3, 1983.
360. **Dercks, W. and Buchenauer, H.,** Investigations on the influence of aluminum ethyl phosphite on the plant phenolic metabolism in the pathogen-host interactions *Phytophthora fragariae*-strawberry and *Bremia lactucae*-lettuce, *J. Phytopathol.,* 115, 37, 1986.
361. **Guest, D. I.,** Evidence from light microscopy of living tissues that fosetyl-Al modifies the defence response in tobacco seedlings following inoculation with *Phytophthora nicotianae* var. *nicotianae, Physiol. Mol. Plant Pathol.,* 29, 151, 1986.
362. **Nemestothy, G. N. and Guest, D. I.,** Phytoalexin accumulation, phenylalanine ammonia lyase activity and ethylene biosynthesis in fosetyl-Al treated resistant and susceptible tobacco cultivars treated with *Phytophthora nicotianae* var. *nicotianae, Physiol. Mol. Plant Pathol.,* 37, 207, 1990.
363. **Barchietto, T., Saindrenan, P., and Bompeix, G.,** Effects of phosphonate on phosphate metabolism in *Phytophthora citrophthora,* in *7th Int. Congr. of Pesticide Chemistry,* VCH-Verlag, Weinheim, 1990.
364. **Smillie, R. H., Dunstan, R. H., Grant, B. R., Griffith, J. M., Iser, J., and Niere, J. O,** The mode of action of the antifungal agent phosphite, *Bull. OEPP/EPPO,* 20, 185, 1990.
365. **Soulie, M-C., Troton, D., Malfatti, P., Bompeix, G., and Lavalmartin, D.,** Postinfectional changes of lipids and photosynthesis in *Lycopersicon esculentum* susceptible to *Phytophthora capsici, Plant Sci.,* 61, 169, 1989.
366. **Dunstan, R. H., Smillie, R. H., and Grant, B. R.,** The effects of sub-toxic levels of phosphonate on the metabolism and potential virulence factors of *Phytophthora palmivora, Physiol. Mol. Plant Pathol.,* 36, 205, 1990.
367. **Griffith, J. M., Smillie, R. H., and Grant, B. R.,** Alterations in nucleotide and pyrophosphate levels in *Phytophthora palmivora* following exposure to the antifungal agent potassium phosphonate (phosphite), *J. Gen. Microbiol.,* 136, 1285, 1990.
368. **Griffith, J. M., Smillie, R. H., and Grant, B. R.,** Sites of action of the fungicide phosphonate in the fungus *Phytophthora palmivora,* in *Proc. 4th Int. Mycological Congr.,* Abstract, Regensburg, FRG, 1990, 187.
369. **Bashan, B., Levy, Y., and Cohen, Y.,** Variation in sensitivity of *Phytophthora infestans* to fosetyl-Al, *Plant Pathol.,* 39, 134, 1989.
370. **Bompeix, G., Fettouche, F., and Saindrenan, P.,** Mode d'action du phosethyl-Al, *Phytiatr. Phytopharm.,* 30, 257, 1981.
371. **Durand, M. C. and Salle, G.,** Effect du tris-o-ethyl phosphonate d'aluminium sur le couple *L. esculatum-P. capsici,* etude cytologique et cytochemique, *Agronomie,* 1, 723, 1981.
372. **Raynal, G., Ravise, A., and Bompeix, G.,** Action du tris-o-ethylphosphonate d'aluminium (phosethyl d'aluminium) sur la pathogenie de *Plasmopara viticola* et sur la stimulation des reactions de defense de la vigne, *Ann. Phytopathol.,* 12, 163, 1980.
373. **Trique, B., Ravise, A., and Bompeix, G.,** Modulation of infections induced by *Phytophthora* spp. in tomato plants, *Agronomie,* 1, 823, 1981.
374. **Afek, U. and Sztejnberg, A.,** Effects of fosetyl-Al and phosphorous acid on scoparone, a phytoalexin associated with resistance of *Citrus* to *Phytophthora citrophthora, Phytopathology,* 79, 736, 1989.
375. **Khan, A. J. and Ravise, A.,** Stimulation of defence reactions in citrus by fosetyl-Al and fungal elicitors against *Phytophthora* spp., *Br. Crop Prot. Counc. Monogr.,* 31, 281, 1985.
376. **Bompeix, G., Saindrenan, P., and Barchietto, T.,** The release of stress metabolites by Oomycetes and Ascomycetes in the presence of phosphonates, in 5th Int. Congr. on Plant Pathology, Kyoto, Japan, 1988, 245.
377. **Saindrenan, P., Barchietto, T., and Bompeix, G.,** Effects of phosphonate on the elicitor activity in culture filtrates of *Phytophthora cryptogea* on *Vignea, Plant Sci.,* 67, 245, 1990.
378. **Dunstan, R. H. and Grant, B. R.,** The detection of phosphonate-sensitive macromolecules associated with the mycelial surfaces of *Phytophthora palmivora, Aust. N.Z. Soc. Cell Biol. Proc. 8th Annu. Meet.,* 27, 1989.
379. **Dunstan, R. H. and Grant, B. R.,** The presence of phosphonate-sensitive macromolecules on the surface of *Phytophthora* spp., in *Proc. 4th Int. Mycological Congr.,* Abstract, Regensburg, F.R.G., 1990, 181.
380. **Dunstan, R. H., Iser, J., and Grant, B. R.,** Suppression of the wound response in potato leaves by surface components from the mycelium of *Phytophthora infestans,* in *Proc. 4th Int. Mycological Congr.,* Abstract, Regensburg, F.R.G., 1990, 180.
381. **Cohen, Y. and Samoucha, Y.,** Cross resistance to four systemic fungicides in metalaxyl resistant strains of *Phytophthora infestans* and *Pseudoperenospora cubensis, Plant Dis.,* 68, 137, 1984.
382. **Bompeix, G., Clerjeau, M., Lafon, R., and Malfatti, P.,** Downy mildew: review of the effectiveness of fosetyl-Al against anilide resistant strains, *Phytoma Defense Cult.,* 361, 15, 1984.

383. **Clerjeau, M., Moreau, C., Piganeau, B., Bompeix, G., and Malfatti, P.,** Effectiveness of fosetyl-Al against strains of *Plasmopara viticola* and *Phytophthora infestans* that have developed resistance to anilide fungicides, *Br. Crop Prot. Conf. Pests Dis.*, p. 497, 1984.
384. **Abdellaoui-Manne, S., Seng, J.-M., Saindrenan, P., and Bompeix, G.,** Fosetyl-Al is effective against mutants of *Phytophthora capsici* resistant to metalaxyl, *Cryptogamie*, 9, 47, 1988.
385. **Ward, E. W. B.,** Suppression of metalaxyl activity by glyphosphate. Evidence that host mechanisms contribute to metalaxyl inhibition of *Phytophthora megasperma* in soybean, *Physiol. Plant Pathol.*, 25, 381, 1984.
386. **Cahill, D. M. and Ward, E. W. B.,** Effects of metalaxyl on elicitor activity, stimulation of glyceollin production and growth of sensitive and tolerant isolates of *P. megasperma*, *Physiol. Mol. Plant Pathol.*, 35, 97, 1989.
387. **Margot, P., Eckhardt, W., and Dahmen, H.,** CGA 80000: a new phenylamide against soil-borne *Peronosporales*, *Proc. Int. Crop Prot. Conf. Pests Dis.*, p. 527, 1988.
388. **Cheetham, R. C., Crowley, P. J., and Evans, S. E.,** Synthesis and activity of some new cyano-oxime fungicides, in *7th Int. Congr. of Pesticide Chemistry*, Abstract, Hamburg, 1990, 55.
389. **Grunwaldt, G., Lyr, H., Wollgiehn, R., and Klepel, M.,** Andoprim — a novel fungicide with an unconventional mode of action and special selectivity for *Phytophthora* spp., in *7th Int. Congr. of Pesticide Chemistry*, Abstract, Hamburg, 1990, 277.
390. **Albert, G., Curtze, J., and Drandarevski, S. A.,** Dimethomorph (CME 151), a novel curative fungicide, *Proc. Br. Crop Prot. Conf. Pests Dis.*, p. 17, 1988.
391. **Wicks, T. and Hall, B.,** Efficacy of dimethomorph (CME151) against downy mildew of grapevines, *Plant Dis.*, 74, 114, 1990.
392. **Hackler, R. E., Arnold, W. R., Dow, W. C., Johnson, G. W., and Kaster, S. V.,** Structure-activity relationships in (haloalkyl)pyridazines: a new class of systemic fungicides, *J. Agric. Food Chem.*, 38, 508, 1990.
393. **Heaney, S. P., Atger, J. C., and Roques, J. F.,** ICIA0001: a novel fungicide for use against diseases on vines, *Proc. Br. Crop Prot. Conf. Pests Dis.*, p. 245, 1988.
394. **Heaney, S. P., Shephard, M. C., Crowley, P. J., and Shearing, S. J.,** ICIA0001: a novel benzamide fungicide, *Proc. Br. Crop Prot. Conf. Pests Dis.*, p. 551, 1988.
395. **Irving, H. R.,** The Differentiation of *Phytophthora* Zoospores, Ph.D. thesis, University of Melbourne, 1984.
396. **Zhang, Q.,** Zoospore Differentiation and Potential Second Messenger Systems in the Fungus *Phytophthora palmivora*, Ph.D. thesis, University of Melbourne, 1990.
397. **Niere, J. O.,** personal communication, 1990.
398. **Griffith, J. M., Davis, A. J., Snow, A., and Grant, B. R.,** unpublished.
399. **Lyr, H.,** personal communication.
400. **Griffith, J. M.,** unpublished.

Chapter 6

TARGET SITES OF MELANIN BIOSYNTHESIS INHIBITORS

Isamu Yamaguchi and Yasuyuki Kubo

TABLE OF CONTENTS

I. INTRODUCTION

Rice blast disease caused by *Pyricularia oryzae* (perfect sexual state, *Magnaporthe grisea*) is the most serious and damaging of all the plant diseases in rice production, particularly in such temperate and humid climates like Japan. Therefore, the greatest precautions have long been taken in order to control the disease, and many excellent antiblast chemicals have been developed for this particular disease, e.g., blasticidin S, kasugamycin, iprofenphos, edifenphos, isoprothiolane, and probenazole.[1] These conventional chemicals except probenazole can act on the mycelial stage of *P. oryzae* mostly by exhibiting fungicidal effects on the pathogen.

Recent studies proved that nonfungicidal chemicals effectively control rice blast disease and one of the most potent group of such agents belongs to inhibitors of melanin biosynthesis. This group includes compounds such as tricyclazole[2] [5-methyl-1,2,4-triazolo-(3,4-*b*)-benzothiazole], pyroquilon[3] [1,2,5,6-tetrahydropyrrolo-(3,2,1-*i, j*)quinoline-4-one], fthalide[4] (4,5,6,7-tetrachlorophthalide), pp389[5] [4,5-dihydro-4-methyltetrazolo-(1,5-α)quinazolin-5-one], and chlobenthiazone[6] [4-chloro-3-methyl-benzothiazol-2(3H) one]. Tricyclazole, pyroquilon, and fthalide are currently used for practical control of rice blast disease. These chemicals are essentially not fungicidal or only weakly toxic to mycelial growth or spore germination of *P. oryzae*, but possess high protectant activity against the blast disease.[7-9] These inhibitors exhibit high selectivity for some Ascomycetes and imperfect fungi, and under experimental conditions they have been also shown to effectively control *Colletotrichum lagenarium*,[10] and *C. lindemuthianum*.[11] In this chapter a brief summary of the biochemistry and physiology of the target sites of melanin biosynthesis inhibitors (MBIs), an overview of their unique mode of action, and perspectives of related future research are described.

II. MECHANISM OF ACTION OF MELANIN BIOSYNTHESIS INHIBITORS — *PYRICULARIA ORYZAE*

A. FUNGAL APPRESSORIA AND MELANINS

The aerial plant epidermis is generally covered by a cuticle with a layer of epicuticlar wax at the outermost surface. The cuticle membrane is composed mainly of cutin polymer embedded in waxes and is attached to the epidermal cell wall by a layer of pectinaceous material.[12] Typically, the epidermal cell wall of rice is also protected by a silica layer with silicon deposits right beneath the cuticle contributing to the hardness of the plant epidermis.[13] The association of the cuticle and silica layer probably constitutes the main obstacle to the penetration of pathogens into the plant tissue.[14]

An appressorium is the swollen tip of a germ tube or hypha that facilitates attachment and penetration into the host.[15] For pathogenic fungi such as *Pyricularia* and *Colletotrichum* spp., formation of appressoria is obligatory for plant infection. These appressoria are well-defined, dark, thick-walled structures delimited by septa from the germ tubes. With *P. oryzae*, invasion of rice plant is accomplished by an infection peg that is formed at the base of an appressorium developing at the tip of the germ tube. This appears to indicate a critical function of appressoria in the penetration of *P. oryzae* into epidermal cells and, thus, through the rigid surface of rice plants. Studies on the mode of action of MBIs shed light on the biochemical functions of appressoria and revealed the importance of melanin biosynthesis for pathogenicity.

With the appressoria of *P. oryzae*, three developmental stages have been distinguished, i.e., initiation, maturation, and infection peg formation followed by penetration. Once the germ tube apex touches uneven host surfaces with ridges, the germ tube ceases elongating and begins to swell. All the content of a spore cell is vigorously transferred to the swollen tip of the germ tube 5 to 7 h after germination at 25°C. This initiation process is depressed by inhibitors of m-RNA and protein synthesis but not by DNA synthesis inhibitors. Subsequently, the swollen apex is delimited by a septum and it begins to darken and excretes mucilaginous substances to securely fasten the pathogen to the plant.[16] The appressoria formed are spherical or ovoid with diameters ranging from 5 to 12 μm,[17] and pigmentation usually accompanies wall thickening. During early stages of melanization, the walls are grayish black but become dark black at maturity (8 to 13 h). DNA synthesis proceeds in appressoria at this stage and their nuclei appear to divide. At the stage of appressorium maturity, the parent spore cell often becomes empty and sometimes collapses. Then penetration pegs emerge from the underside of appressoria where appressoria are in close contact with the plant surface and where cell walls are thin and mostly not melanized. The infection pegs (diameter, *circa* 0.1 μm) are visible at around 16 h[17] and penetration through the cuticle layer into the epidermal cells of rice plant appears complete after 48 h. Protoplasm of appressoria moves into the infection peg, and appressoria become empty in most cases at this stage.

Appressoria of *P. oryzae* excrete such hydrolytic enzymes as cutinase, pectinase, and cellulase which may cleave ester bonds of the cutin and digest pectin and cellulose in the surface layer and epidermal cell wall of rice plant. In fact, an electron-translucent area is clearly observed around the invading infection pegs by electron microscopy.[13] Thus it is conceivable that these enzymes play an important role in the penetration process of the pathogen; however, it would be difficult to ascribe the penetration ability solely to enzymic action since mycelia of *P. oryzae* can also produce such enzymes, although the penetration from hyphal tips into the plant is rarely observed. In addition, the cuticular wax and silica layer of the plant epidermis may serve as a potent barrier to penetration and shields the ester linkage of cutin and other polymers from enzymatic attack. Therefore, mechanical force exerted by appressoria is necessary for successful penetration of the fungus.[17,17a] It seems likely that vertical pressure of the appressoria on the epidermis may break through the cuticular wax and silica layer, and weaken the surface rigidity of the plant to bring enzymes into easy contact with cuticular and cell wall polymers, leading to penetration of the infection pegs into plant tissue.

Melanin denotes a black pigment of biological origin, but generally melanins are envisioned as pigments consisting of polymeric phenolic compounds.[18] Thus, certain brown, red, yellow, green, or blue pigments with similar chemical structures are occasionally referred to as melanins.[19] Melanins are chemically inert and insoluble in water, aqueous acids, and in most organic solvents. In fact, preparation of melanins is usually carried out by removing or destroying all other biological substances present in cells or tissues. They are found in animals and plants as well as in microbes, and are classified into three types: eumelanins, phaeomelanins, and allomelanins.[20] The eumelanins are black melanins mostly found in animal kingdom, e.g., the pigments of human black hair, the malignant melanoma, and the squid ink. They are derived from tyrosine via 3,4-dihydroxyphenylalanine (DOPA). The pigments in red hair or of bird feathers are commonly called phaeomelanins which are formed by copolymerization of DOPA with cystein in the biosynthetic pathway. The biosyntheses of these indolic melanins are extensively studied and have been shown to be catalyzed by tyrosinase as the key enzyme.[19] The allomelanins, usually black, occur mostly in the plant kingdom, e.g., in the seeds of sunflower or watermelon and in certain fungi, and are formed from diphenol such as catechol.[21] Another phenolic substance, 1,8-dihydroxynaphthalene, has been also recognized as a natural precursor for the allomelanins, particularly in fungal melanins which occur in cell walls and as extracellular polymers formed enzymatically and/or autooxidatively in the medium around cell walls.[22]

The function of melanins has been described in a number of ways: they may serve as a camouflage from predators, a barrier against adverse effects of UV rays in sunlight, and

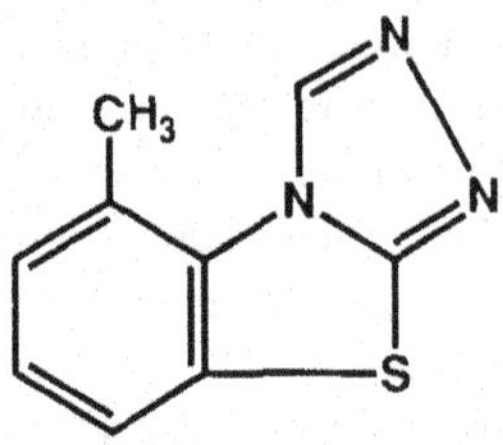

Tricyclazole

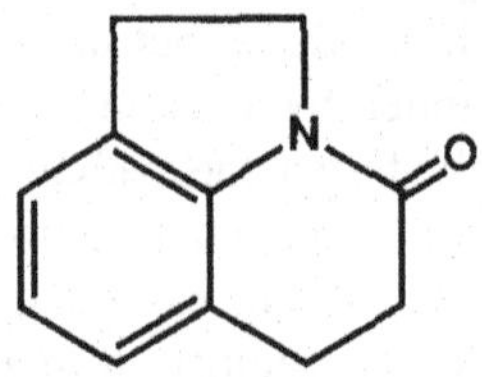

Pyroquilon

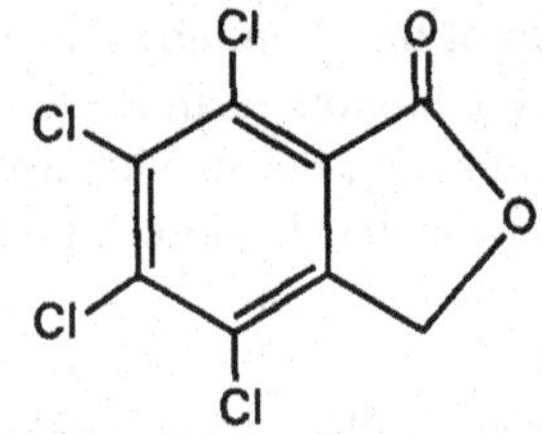

Fthalide

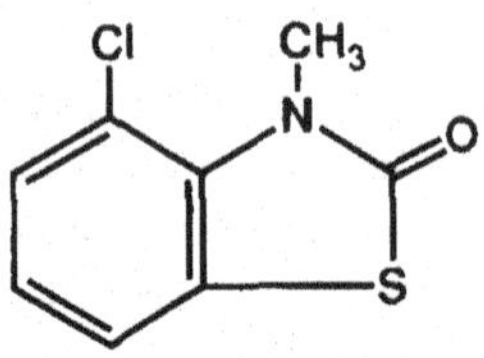

Chlobenthiazone

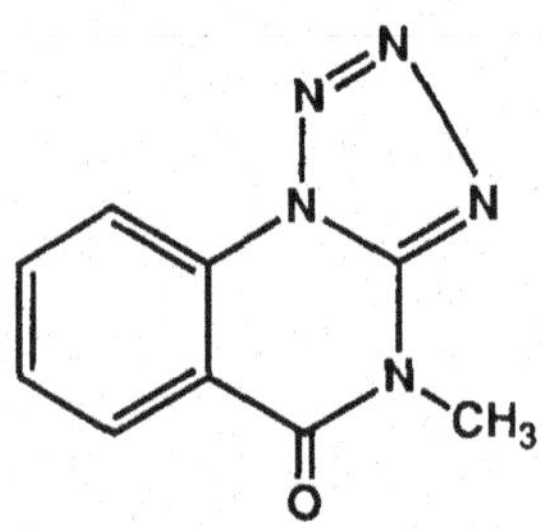

PP 389

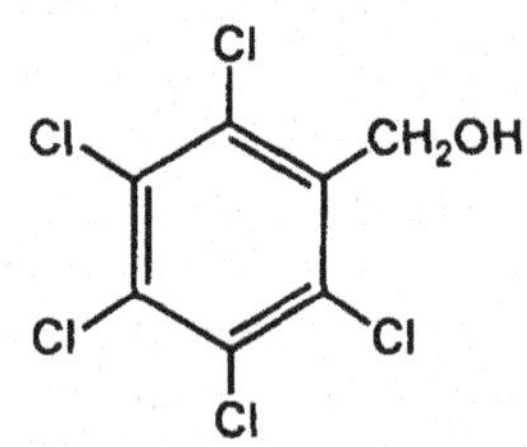

PCBA

FIGURE 1. Chemical structure of antipenetrant fungicides that inhibit fungal melanin biosynthesis.

a potent protectant against enzymatic digestion by microbial attack. In addition, recent studies clarified the importance of fungal melanins for virulence of certain plant pathogens.[22] As described earlier, the appressorial cell wall of *P. oryzae* is vigorously melanized during maturation within a fairly short period, and MBIs can interfere with penetration by the fungus through inhibition of melanin biosynthesis in appressoria as discussed below in detail.

B. MODE OF ACTION

The infection cycle of *P. oryzae* is generally described as follows. The conidial spores which are disseminated from other diseased plants land on and attach to the plant surface. They germinate in an infection drop at high humidity and appropriate temperature (12 to 28°C) conditions. Then melanized appressoria are formed at the apex of germ tubes in order to facilitate penetration as described earlier. The fine threadlike penetration pegs (diameter, ca. 0.1 μm) that succeed to penetrate into the cell lumen, usually of motor cells, become invasion hypha with normal diameters (ca. 2.5 μm). The hyphae invade adjacent cells by overcoming the plant resistance, possibly by excreting phytotoxic substances such as pyricurol, tenuazonic acid, and pyriculariol.[23-25] Finally, mature mycelia in plants cause typical blast lesions from which numerous conidia are formed for second infections.

Chemicals such as tricyclazole, pyroquilon, tetrachlophthalide (fthalide), pp389, chlobenthiazone, and pentachlorobenzyl alcohol (PCBA) specifically interfere with this penetration process in the infection cycle of *P. oryzae* and are actually nonfungitoxic (Figure 1).[17,26-29] Among them, tricyclazole, pyroquilon, and fthalide are of major significance for the control of blast disease; the first two chemicals are systemic compounds which are effective in either submerged or foliar applications, while fthalide is effective only when used as a foliar spray. These chemicals exhibit a marked protective efficacy but have no curative effects on the disease. They act only as potent antipenetrants but not as erradicants against *P. oryzae,* as indicated by the lack of control if the epidermal layer was punctured or when the application of the chemicals was delayed.[17,30] These properties are consistent

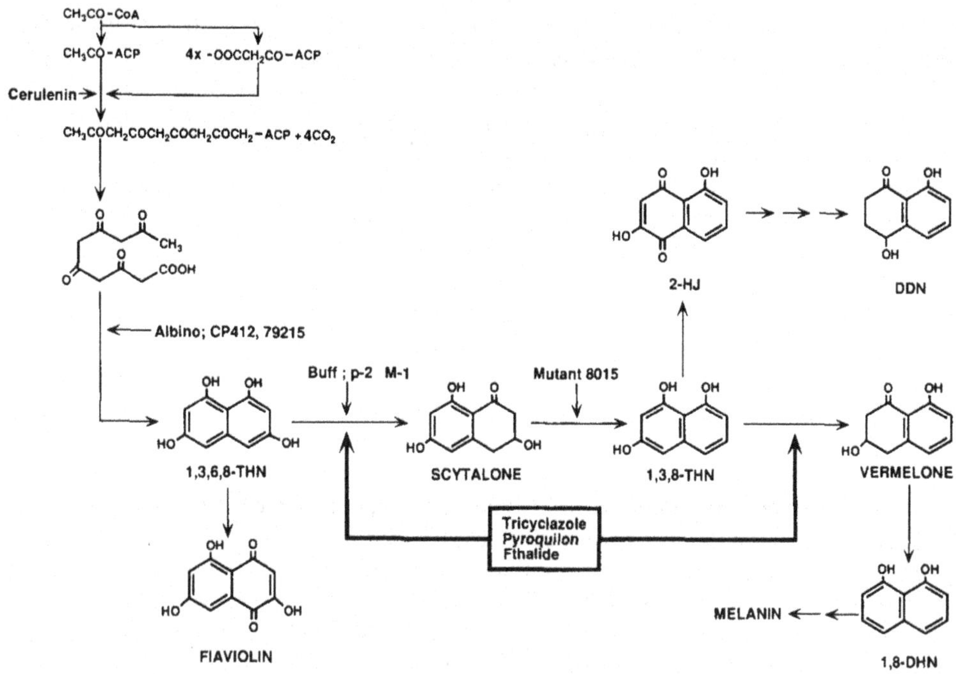

FIGURE 2. Pathway of melanin biosynthesis in *Pyricularia oryzae* and *Colletotrichum* sp., with inhibition sites of MBIs and cerulenin and deficiencies of color mutants. (Revised from Kubo, Y., Katoh, M., Furusawa, I., and Shishiyama, J., Inhibition of melanin biosynthesis by cerulenin in appressoria of *Colletotrichum lagenarium, Exp. Mycol.*, 10, 305, 1986. With permission.)

with a mode of action that specifically involves the appressorial penetration process. The antipenetrant activity was also proven in tests using rice sheaths or cellophane membranes,[26,27] *Bryophyllum* epidermis or Formvar plastic,[17] and onion epidermis.[31,32] The fact that their antipenetrant activity in artificial membrane systems parallels their effects on rice plants strongly suggests that the target site(s) of those chemicals resides inside the pathogen but not the host plant.

When *P. oryzae* is cultured on a nutrient agar, its mycelia generally exhibit black color. The mycelium as well as the culture broth is also markedly melanized in submerged cultures at a later stage of growth in liquid media such as yeast extract-glucose, particularly if the medium is augmented with phosphate and $CaCl_2$.[27] The melanization was recognized to be specifically inhibited by the aforementioned chemicals with antipenetrant activity,[7,8,27,33-35] thus named melanin biosynthesis inhibitors (MBIs). Among the MBIs, tricyclazole, pyroquilon, chlobenthiazone, and pp389 are known to cause a significant change in pigmentation in *P. oryzae,* yielding a reddish to brown color of both the mycelia and the culture media, at concentrations that are not inhibitory to conidial germination or the mycelial growth of the pathogen. In contrast, the effect of fthalide and PCBA appears to be rather weak,[27] leading to brown-gray mycelia at nonfungicidal concentrations.

The experimental results clearly indicated that MBIs inhibited the melanization of mycelia of *P. oryzae* by blocking the polyketide pathway of fungal allomelanin biosynthesis (Figure 2), which resulted in an abnormal accumulation of scytalone, a melanin intermediate, and of additional metabolies of the branched pathway such as 2-hydroxyjuglone (2-HJ; brown pigment) and flaviolin (reddish pigment).[7,27,33] This pathway was first discovered in *Verticillium dahliae* by using melanin-deficient mutants,[36] and was also found to be common in *P. oryzae* and some other imperfect and ascomycetous fungi.[18] The synthesis of scytalone from 1,3,6,8-tetrahydroxynaphthalene (1,3,6,8-THN) via polyketide biosynthesis has been confirmed using ^{13}C- and ^{2}H-labeled acetate in feeding experiments.[37,38] The inhibition sites of MBIs in the pathway, as deduced from the data on metabolite accumulation, were between 1,3,6,8-THN and scytalone and between 1,3,8-trihydroxynaphthalene (1,3,8-THN) and vermelone (Figure 2).[7,27,33,39,40] Since 2-HJ accumulates in fungal cultures to higher levels than flaviolin at low concentrations of MBIs, it has been proposed that the conversion of 1,3,8-THN to vermelone is more sensitive than that of 1,3,6,8-THN to scytalone.

Furthermore, the antipenetrant activity of MBIs was shown to be accompanied by the marked inhibition of appressorial melanization of *P. oryzae* (Figure 3).[17,27] The accumulation of metabolites was also evident during the appressorial maturation in the presence of MBIs (Table 1).[9] This finding comprised the first evidence for an accumulation of metabolites in appressoria and their surroundings, and thus was critical to correlating inhibition of melanin biosynthesis with the antipenetrant activity of MBIs.

Applying electron microscopy, Woloshuk et al.[17] revealed that an electron-dense melanin layer in walls of appressoria of wild strains was absent either in those treated with tricyclazole, or in walls of appressoria of "buff mutants", which are genetically deficient in melanin biosynthesis and nonpathogenic. Their ultrastructural data suggested that melanin deposition at the basal edge of the appressorial walls may strengthen the underwall of appressoria and direct turgor forces by focusing the mechanical pressure toward the epidermal surfaces of the host plant. In addition, while melanized walls of appressoria were impermeable to sucrose and various other solutes, unmelanized ones were shown to be freely permeable, implying that inhibition of melanin biosynthesis caused interference with the generation of osmotic pressure to support the penetration ability of appressoria. From these results, Sisler and coworkers proposed as a mechanism of MBI action the failure of treated appressoria to accomplish epidermal penetration, due to the lack of rigidity and undisturbed architecture of unmelanized appressorial walls.[17]

Their suggestion was supported by the fact that the addition of vermelone or 1,8-dihydroxynophthalene (1,8-DHN), both later intermediates subsequent to the MBI inhibition sites in melanin biosynthesis (Figure 2), rescued appressorial melanization of *P. oryzae* in the presence of MBIs and restored penetration. However, in contrast to the significant recovery of melanization, the restoration of appressorial penetration by these intermediates was not complete and only partial.[9,31,41] Therefore, there may be additional explanations for the antipenetrant action of MBIs, such as an interference with the production of enzymes to digest epidermal surfaces of host plants or the production of some pentaketide metabolites required for the formation of infection pegs. More importantly, since a remarkable accumulation of scytalone, flaviolin, and 2-HJ was observed during the maturation of appressoria in the presence of MBIs,[9] the precursors may cause adverse effects on appressorial function, in particular if they are cytotoxic like many phenols and quinones.[42] In fact, 2-HJ significantly interfered with appressorial penetration into the host plant,[27] as shown in Table 2, presumably by its inhibitory effects on the biochemical functions of membrane systems in *P. oryzae*.[43] Thus, abnormal accumulation of toxic shunt metabolites such as 2-HJ in melanin biosynthesis may result in the self-destruction of penetration activity by appressoria of *P. oryzae*. However, this hypothesis does not necessarily comply with the facts that albino mutants such as CP412 or the wild-type strain treated with cerulenin, an inhibitor of earlier steps in melanin biosynthesis, cannot penetrate the host epidermis,[31] regardless of 2-HJ accumulation. To the contrary, penetration ability of wild-type appressoria treated with MBIs was efficiently restored by 1,8-DHN only when cerulenin was also present (Table 3),[31] and added scytalone even increased the toxicity of tricyclazole toward appressorial penetration of *P. oryzae*.[32] Thus, it is conceivable that accumulation of toxic metabolites may contribute to the antipenetrant activity of MBIs under practical conditions when high concentrations of chemicals must be present for disease control. It is the authors' view that the antipenetrant action of MBIs in *P. oryzae* is ultimately based on the inhibition of appressorial melanin biosynthesis, and that both the resulting lack of wall rigidity and the accumulation of inhibitory metabolites

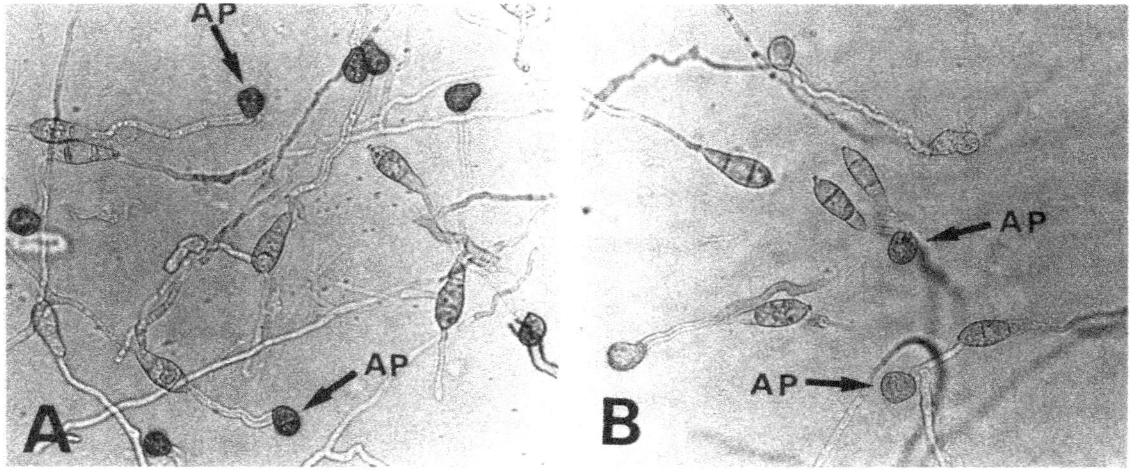

FIGURE 3. Appressorial pigmentation of *Pyricularia oryzae* on cellulose membranes. (A) No treatment; (B) treated with pyroquilon (10 μg/ml). AP, appressoria. (From Yamaguchi, I., Sekido, S., and Misato, T., The effect of nonfungicidal anti-blast chemicals on the melanin biosynthesis and infection by *Pyricularia oryzae*, *J. Pestic. Sci.*, 7, 527, 1982. With permission.)

TABLE 1
Effect of Chemicals on Accumulation of Metabolites during Appressorial Maturation

Chemicals	μg/ml	nmols/10^7 appressoria Scytalone	2-Hydroxyjuglone	Flaviolin
Control		0.85	0.15	0.28
Tricyclazole	1	2.80	1.50	0.40
	10	1.70	2.75	1.70
Pyroquilon	1	1.54	0.69	0.69
	10	2.46	4.20	3.04
Tetrachlorophthalide	50	5.55	0.44	0.61
PCBA	50	1.30	0.61	0.83

Note: Appressoria of *P. oryzae* were formed on cellophane film (26°C, 20 h). Figures indicate the total amounts of each metabolite extracted from the whole incubation system.

From Yamaguchi, I., Sekido, S., and Misato, T., Inhibition of appressorial melanization in *Pyricularia oryzae* by nonfungicidal anti-blast chemicals, *J. Pestic. Sci.*, 8, 230, 1983. With permission.

TABLE 2
Effect of Scytalone and 2-Hydroxyjuglone on Each Developmental Stage of *P. oryzae*

Metabolites	m*M*	% Inhibition: Spore germination	Appressorium formation	Appressorial penetration[a]	Mycelial growth	Spore formation
Scytalone	0.01	0	6	23	2	0
	0.1	0	15	43	11	0
	1.0	2	20	94	89	0
2-Hydroxyjuglone	0.01	14	2	35	8	0
	0.1	17	18	91	87	0
	1.0	100	—	100	98	42

[a] Appressorial penetration was evaluated by the improved cellophane film method; compounds were applied after initial formation of appressoria.

From Yamaguchi, I., Sekido, S., and Misato, T., The effect of nonfungicidal anti-blast chemicals on the melanin biosynthesis and infection by *Pyricularia oryzae*, *J. Pestic. Sci.*, 7, 527, 1982. With permission.

in appressoria prevent fungal penetration, possibly by both mechanisms acting in a synergistic manner.

In addition, Inoue et al. suggested that the antipenetrant action of chlobenthiazone and tricyclazole might be ascribed to inhibition of appressorial adhesion of *P. oryzae* to the leaf surface.[44] They found that appressoria developing on treated barley leaves were easily detached from the plant surface simply by shaking in water. Okuno et al. found that the spores of *P. oryzae* formed in the presence of tricyclazole had a reduced capability to penetrate nitrocellulose membranes,[45] which implies that such spores would fail to establish secondary infections. However, these two modes of action can also be regarded as the result of melanin biosynthesis inhibition, since the accumulation of toxic metabolites apparently causes the deterioration of membrane functions in *P. oryzae*.[43] A different mechanism of tricyclazole action was proposed by Shiba and Nagata.[46] The authors suggested that antifungal substances were produced in rice leaves treated with the chemicals. Although there might

TABLE 3
Restoration of Penetration Ability by 1,8-DHN in Fthalide- or Tricyclazole-Treated Appressoria of *P. oryzae* P-2 in the Presence or Absence of Cerulenin

Cerulenin (μg/ml)	1,8-DHN[a] (g/ml)	Penetration ratio and percentage[b]: Fthalide (μg/ml) 0	Fthalide 0.1	Fthalide 1.0	Fthalide 10	Tricyclazole (μg/ml) 0.1	Tricyclazole 1.0
0	0	686/937 (73)	8/935 (1)	0/955 (0)	0/629 (0)	1/966 (0)	0/620 (0)
0	1.25	433/988 (44)	377/989 (38)	7/978 (1)	0/624 (0)	323/953 (34)	15/633 (2)
1.25	0	2/930 (0)	2/962 (0)	0/950 (0)	0/624 (0)	0/930 (0)	0/630 (0)
1.25	1.25	537/994 (54)	273/955 (29)	259/932 (28)	134/635 (21)	338/984 (34)	96/624 (15)

[a] 1,8-DHN was added to spore suspension 4 h after inoculation.
[b] Penetration ratio is number of appressorial penetrations/number of appressoria observed 40 h after inoculation. Penetration percentage is penetration ratio × 100.

Data from Chida and Silser.[31]

be such an integrated effect in the practical control of the disease *in vivo,* the primary target site of MBIs is definitely located in the pathogen, as shown by the fact that MBIs act in artificial membrane systems just as well as on host plants.

Then, what is the actual target site of MBIs in the melanin biosynthesis pathway in *P. oryzae*? As described earlier, the sensitive steps to MBIs are the conversion of 1,3,6,8-THN to scytalone and 1,3,8-THN to vermelone (Figure 2). The latter step is believed to be more sensitive to MBIs than the former, since 2-HJ accumulates in fungal cultures more significantly than flaviolin at low inhibitor concentrations.[33,39] Both steps are reductive reactions which were, in a cell-free homogenate from *Verticillium dahliae,* catalyzed by reductase(s) requiring NADPH as a cofactor.[47] The reductase activity in extracts from *V. dahliae* was substantially inhibited by MBIs, with a stronger inhibitory effect on the reduction of 1,3,6,8-THN than on 1,3,8-THN. Although such a clearly defined sensitivity difference was not established for *P. oryzae,*[29] Wheeler and Greenblatt suggested that the inhibitors were also more effective in preventing the reduction of 1,3,6,8-THN rather than 1,3,8-THN in extracts of this fungus.[48] This different sensitivity to MBIs *in vitro* might be explained if an alternate pathway to 2-HJ via flaviolin, which was indicated in *Wangeiella dermatitidis,*[49] also exists in *P. oryzae*. Even if the step from 1,3,6,8-THN to scytolane is the most sensitive reaction, the most critical one for total inhibition of melanin biosynthesis appears to be the step from 1,3,8-THN to vermelone. Furthermore, it has not been determined yet whether these two reduction reactions are catalyzed by the same or different enzymes. The findings that both reactions were blocked by MBIs and enzyme levels were lower in extracts derived from buff mutants suggested that they might be catalyzed by the same enzyme.[29,48] However, the authors' recent work using crude enzyme preparation from *P. oryzae* proved that the reaction from 1,3,8-THN to vermelone was significantly inhibited by MBIs in a competitive manner, while 1,3,6,8-THN conversion was not even at high concentrations.[94] This may imply that the two reactions are catalyzed by different enzymes, or the same enzyme with different binding sites for the two substrates. Final answers will require detailed studies with purified reductases and, in addition, with cloned genes coding for the responsible enzymes.

Structural requirements for melanin biosynthesis inhibitors were studied by Inoue et al., who proposed that a benzo-bicyclic ring system, a nitrogen atom at a position alpha to the benzene ring, and small substitution groups at specific positions are important for potent activity of the compounds.[50] From a different point of view, new melanin inhibitors were designed on the assumption that the enzymatic inhibition by known MBIs occurs competitively between the inhibitors and 1,3,8-THN.[51] The synthesized phthalazine derivatives showed high inhibitory activity on the fungal melanization as well as on rice blast. These results suggest that the reductase converting 1,3,8-THN to vermelone may be the actual target site of MBIs.

The melanin biosynthesis pathway from 1,3,6,8-THN to 1,8-DHN involves two alternating reduction and dehydration steps as shown in Figure 2. The dehydration reactions were shown to be insensitive to MBIs in cell-free homogenates from *V. dahliae* and *P. oryzae*.[29,47,48] Since the reactions did not require pyridine nucleotide cofactors and exhibited a temperature sensitivity clearly different from the reductive reactions, they are apparently catalyzed by different enzymes with a specific binding affinity to similar substrates. The dehydratase was purified from a black yeast *Phaeococcomyces* sp.[52] and a rice pathogen *Cochliobolus miyabeanus*.[53] Although purification of this enzyme has not been accomplished for *P. oryzae* and no inhibitors of the dehydratase were reported thus far, such inhibitors would be promising as alternative MBIs and blast control agents.

III. MECHANISM OF ACTION OF MELANIN BIOSYNTHESIS INHIBITORS — *COLLETOTRICHUM* SPECIES

Recent studies on appressorial penetration by *Colletotrichum* species indicated that melanization of appressoria is an essential factor for successful penetration of the host by appressoria. While appressorial melanization has been assumed to contribute to an increased endurance under adverse environmental conditions,[15] the use of several melanin-deficient mutants of *C. lagenarium,* genetically blocked at different steps in the melanin biosynthetic pathway, clarified the essential role of melanin in appressorial penetration.[54-61]

Studies with *C. lagenarium* using melanin-deficient mutants and melanin biosynthesis inhibitors led to a clear evaluation of the effects of the chemicals on melanin biosynthesis and penetration by appressoria, and a mechanism of MBI action could be proposed for this fungus.[59]

A. PENETRATION ABILITY OF MELANIN-DEFICIENT MUTANTS

Colletotrichum spores germinate and form well-developed melanized appressoria. *C. lagenarium* strain 104-T (laboratory strain of Kyoto University) forms abundant spores that differentiate synchronously to form appressoria. The infection process is, very similar to *P. oryzae,* divided into spore germination, appressorium development and melanization, infection peg formation, and the establishment of penetration hyphae. Cellulose membranes simulating the plant cell walls have been used as a substrate that allowed observations of the infection process.[26,62-64] Spores of *C. lagenarium* germinate and form melanized appressoria on the cellulose membranes. Infection hyphae develop from the appressoria and penetrate the membranes.

The importance of appressorial melanization in penetration was first noted for *C. lagenarium* when albino mutants were used.[56] Pigmented appressoria of the wild-type strain 104-T effectively formed penetration hyphae and penetrated cellulose membranes, while albino mutants formed colorless appressoria that germinated laterally to form secondary appressoria. The mutant appressoria rarely penetrated the membranes. Furthermore, it has been shown that the albino mutants are nonpathogenic on their host cucumber leaves.[10,56]

The role of appressorial melanization in penetration was clarified by elucidating the biosynthetic pathway of melanin. A metabolite with melanin precursor activity was isolated from a melanin-deficient mutant 8015 and identified as scytalone.[57] Since scytalone is known as a melanin precursor derived from pentaketide synthesis in *Verticillium dahlia*,[36,39,65-67] *Thielaviopsis basicola*,[68] and *P. oryzae*,[33] the pathway of *C. lagenarium* melanin was considered to be identical or similar[59] (Figure 2). Mutant 8015 is considered to be defective in the enzyme involved in the conversion of scytalone to 1,3,8-trihydroxynaphthalene (1,3,8-THN). Therefore, scytalone accumulates in the medium. Since albino mutants did not accumulate detectable levels of melanin intermediates, the deficient site is thought to be at or before pentaketide cyclization.

Scytalone also restored appressorial melanization of albino mutants, indistinguishable from that of the parent strain. The restored appressoria of albino mutants effectively penetrated both cellulose membranes and plant cell walls. In addition, they did not germinate laterally as did the nonmelanized appressoria.[57] Thus it becomes clear that appressorial melanization is essential for penetration by appressoria of *C. lagenarium*.

The ultrastructural localization of electron-dense material in appressoria of *Colletotrichum* species has been suggested to be representative of melanin.[69-73] The location of melanin in relation to its role in penetration by appressoria was investigated in *C. lagenarium*[55] and *C. lindemuthanium*.[11] In these fungi, the location of melanin was verified by comparing the ultrastructure of appressoria of wild-type strains with those of a melanin-deficient albino mutant or wild-type strains treated with MBIs. For *C. lagenarium*, the site of melanization was determined from observations of melanized appressoria formed by albino mutants in the presence of scytalone. In case of the *C. lagenarium* wild-type strain 104-T, there was a smooth, thin, electron-dense layer just outside of the plasmalemma, whereas the layer was absent in the appressorial wall of an albino mutant. A similar melanin layer was also observed in appressoria of *C. lindemuthianum*.[11] The melanin layer extends around the appressorium except within the immediate region of the pore.[69,71,72,74]

From the above observation, a role for melanin in appressorial penetration could be proposed. During hyphal growth cell wall-degrading enzymes appear to be involved in directing the incipient sites at hyphal branchings,[75-78] and it is known that melanin is resistant to these enzymes.[79-83] It seems possible that lateral germination (branching) of appressoria of *C. lagenarium* is prevented by melanin because it is resistant to lytic enzymes involved in wall softening. Vertical germination (penetration) thus ensues through the pore which is the only nonmelanized region of the appressorium. In the case of *C. lindemuthianum*, melanin is thought to provide the physical rigidity necessary to focus the turgor pressure of the protoplasm onto a small area of the host cuticle to facilitate penetration.[11]

B. MODE OF ACTION

Treatment of wild-type *C. lagenarium* appressoria with tricyclazole inhibited melanization and resulted in colorless appressoria. Such appressoria germinated laterally to form secondary appressoria, similar to those of albino mutants, and exhibited poor penetration efficiency.[10]

The action mechanism of melanin-inhibiting fungicides on *C. lagenarium* was investigated to elucidate the inhibition site within the melanin biosynthetic pathway. A comparison was made with the metabolites accumulating in tricyclazole-treated cultures of the parent strain 104-T, the mutant 8015 defective in conversion of scytalone to 1,3,8-THN, and the albino mutant 79215. A shunt metabolite of melanin biosynthesis was isolated from culture media of the parent strain treated with tricyclazole, and was identified as 3,4-dihydro-4,8-dihydroxy-1(2*H*)naphthalenone (DDN). DDN has been shown to be the last shunt product derived from 1,3,8-THN through inhibition of its reduction to vermelone.[33,39] Since the mutant 8015 lacks the enzyme involved in conversion of scytalone to 1,3,8-THN,[57] the metabolite DDN was not found in the respective culture medium. These results indicated clearly that tricyclazole inhibited the conversion of 1,3,8-THN to vermelone in melanin biosynthesis of *C. lagenarium*[59] (Figure 2).

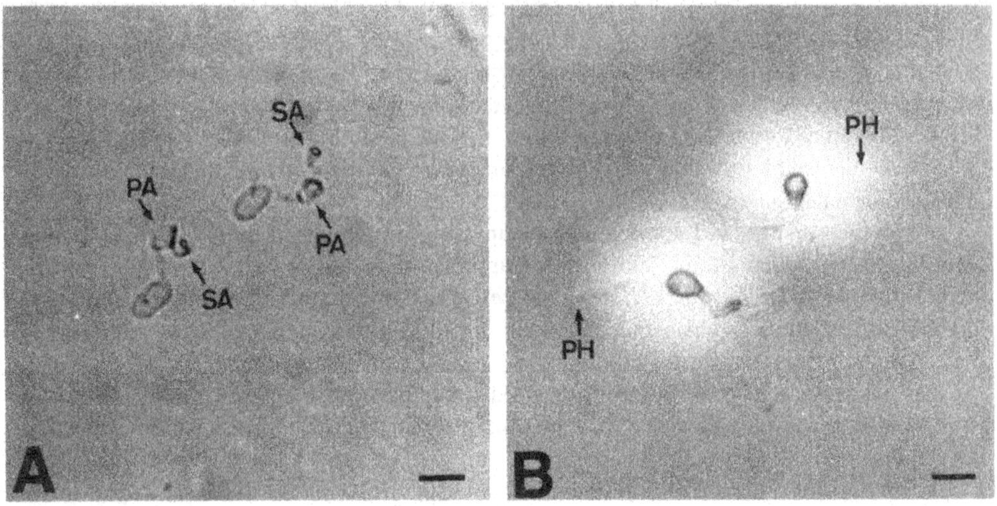

FIGURE 4. Appressorial pigmentation and penetration on cellulose membranes by penetration hyphae of vermelone-treated appressoria of the parent strain 104-T of *Colletotrichum lagenarium* in the presence of pp389. (A) No treatment; (B) treatment with vermelone. Bar markers represent 10 μm. PA, primary appressoria; SA, secondary appressoria; PH, penetration hyphae. (From Kubo, Y., Suzuki, K., Furusawa, I., and Yamamoto, M., Melanin biosynthesis as a prerequisite for penetration by appressoria of *Colletotrichum lagenarium*.: site of inhibition by melanin-inhibiting fungicides and their action on appressoria, *Pestic. Biochem. Physiol.*, 23, 47, 1985. With permission.)

Melanin-inhibiting fungicides, including tricyclazole, pp389, and pyroquilon, inhibited appressorial pigmentation and host penetration by *C. lagenarium*. The application of various melanin precursors to cells in which melanin synthesis was inhibited has elucidated the site of inhibition. Vermelone, a melanin precursor, restored appressorial pigmentation and subsequent penetration of the parent strain 104-T in the presence of tricyclazole, pp389, or pyroquilon at concentrations that normally inhibit appressorial pigmentation and penetration (Figure 4 and Table 4).[59] To the contrary, scytalone restored neither appressorial pigmentation nor penetration. These results indicated again that tricyclazole inhibited conversion of 1,3,8-THN to vermelone, and that lack of melanin in the appressoria is the major reason for penetration failures of the MBI-treated fungus. Colorless appressoria of *P. oryzae* that develop under the action of MBIs or form melanin-deficient mutants cannot penetrate host surfaces, but application of 1,8-dihydroxynaphthalene (1,8-DHN) to the appressoria partially restored the penetration ability.[9,17,27,84] The restoration efficiency with *P. oryzae* was occasionally low compared with that of *C. lagenarium*. Yamaguchi and co-workers reported that melanin biosynthesis intermediates such as 2-hydroxyjuglone (2-HJ) or scytalone accumulated in or around the appressoria due to the inhibition of the melanin biosynthesis and that the metabolites were cytotoxic for appressorial penetration.[9,27,43] It is evident, however, that in the case of *C. lagenarium* melanization of appressoria is of primary importance, since albino mutants do not accumulate cytotoxic intermediates, and are also nonpathogenic.[56,57] In contrast to *C. lagenarium*, it may be necessary to consider the cytotoxic effect of the intermediates in addition to the inhibition of appressorial melanization in *P. oryzae*.

Further evidence that melanization is essential for penetration was provided by the experiments using an antibiotic, cerulenin, which inhibits synthesis of compounds via the polyketide pathway by inhibiting the malonyl-ACP/acyl-ACP condensation reaction as well as that of fatty acid synthesis.[85] Since melanin of *C. lagenarium* is synthesized from the polyketide pathway, the effect of cerulenin on appressorial melanization was investigated.[86] Cerulenin inhibited appressorial melanization effectively, and the colorless appressoria formed on cellulose membranes germinated laterally similar to those of the albino mutants. Such

TABLE 4
Effect of Melanin Precursors on Appressiorial Pigmentation and Penetration by Appressoria in the Presence of Melanin-Inhibiting Chemicals in *Colletotrichum lagenarium*[a]

	Concentration (μM)			Percentage of appressoria with		
Chemicals	Scytalone	Vermelone	Appressorial pigmentation[b]	Penetration hyphae	Haloes	Lateral germination
Tricyclazole	0	0	−	0.9	1.5	82.5
(100 μM)	100	0	−	0.0	0.0	99.2
	500	0	−	0.0	0.0	98.0
	1000	0	±	2.9	3.6	79.0
	0	1	−	3.1	2.1	66.7
	0	5	±	20.3	19.0	31.6
	0	10	+	48.3	40.4	14.2
pp389	0	0	−	0.0	0.0	96.2
(500 μM)	100	0	−	0.5	0.5	97.5
	500	0	−	0.7	5.6	91.8
	1000	0	±	5.1	4.5	89.8
	0	1	−	0.6	0.6	96.8
	0	5	±	26.7	24.8	47.6
	0	10	+	48.4	36.3	8.2
Pyroquilon	0	0	−	0.0	0.0	96.8
(500 μM)	100	0	−	0.0	0.0	99.5
	500	0	−	0.0	0.0	98.0
	1000	0	±	0.0	0.0	96.5
	0	1	−	20.5	18.2	78.2
	0	5	±	29.4	25.2	63.8
	0	10	+	50.5	46.2	8.1
None	0	0	+	75.7	74.0	9.7

[a] Spores were incubated on nitrocellulose membranes for 72 h at 24°C.
[b] Appressoria were darkly pigmented (+), light brown (±), or colorless (−).

From Kubo, Y., Suzuki, K., Furasawa, I., and Yamamoto, M., Melanin biosynthesis as a prerequisite for penetration by appressoria of *Colletotrichum lagenarium:* site of inhibition by melanin-inhibiting fungicides and their action on appressoria, *Pestic. Biochem. Physiol.*, 23, 47, 1985. With permission.

cerulenin-treated appressoria rarely penetrated the membranes. When scytalone, a melanin precursor subsequent to the inhibition site of cerulenin, was treated in the presence of cerulenin, appressorial melanization was restored along with the ability of the appressoria to penetrate cellulose membranes.[86] Thus it is evident that cerulenin inhibited pentaketide and melanin biosynthesis in *C. lagenarium.* Since the inhibition of melanin biosynthesis at early steps of the pathway caused penetration failures without the accumulation of cytotoxic polyketide melanin intermediates, further proof for the essential role of appressorial melanization in penetration is evident.

IV. CONCLUSION

Although many functions of fungal melanins remain to be elucidated, significant progress has been made over the past decade, in particular with regard to their role in fungal pathogenicity. This progress was accomplished with improved biochemical and ultrastructural techniques, and apparently triggered by the development of specific inhibitors of fungal melanin biosynthesis. Nonfungicidal chemicals such as tricyclazole, pyroquilon, and fhthalide block the polyketide pathway, resulting in prevention of melanization in the appressorial walls of *Pyricularia oryzae* and *Colletotrichum* species. Nonmelanized appressorial walls of these pathogens do not appear to have sufficient rigidity and/or proper architecture necessary for the penetration of plant epidermal walls. Moreover, such polyketide metabolites as 2-HJ which accumulate as a result of melanin biosynthesis inhibition may interfere with the appressorial penetration processes in *P. oryzae*.

While tricyclazole is known to inhibit melanin biosynthesis in a number of Ascomyconita and Fungi imperfecti,[17] only an extremely narrow spectrum of plant diseases is controlled. Under practical conditions, control is actually limited to rice blast disease. This is explained by the fact that specific steps in penetration requiring melanin synthesis are critical for the ability of *P. oryzae* to invade rice plants with their rigid epidermis. The specific and selective activity of MBIs may offer advantages over conventional fungi-toxic chemicals which act on biochemical processes common to many nontarget organisms, since it would ordinarily constitute less environmental hazard.

Advances also have been made in the understanding of the enzymology of melanin biosynthesis. The presence of enzymes catalyzing two consecutive reduction and dehydration reactions required for the conversion of 1,3,6,8-THN to 1,8-DHN was demonstrated in cell-free homogenates of several fungi.[29,47-49,52,53]

MBIs specifically inhibit the reductase activity required for the conversion of 1,3,6,8-THN to scytalone and 1,3,8-THN to vermelone, leading to the accumulation of flaviolin and 2-HJ, respectively. Future attempts to purify the enzymes may facilitate the search for new inhibitors. In addition, genetic studies in the blast fungus *Magnaporthe grisea* will trigger progress in the characterization of virulence factor of the fungus.[87-89] In particular the recent development of a genetic transformation system[90] may reveal the complex determinants that control melanin biosynthesis and other pathogenicity factors in the fungus.

Effects of MBIs on other microorganism have been reported; tricyclazole induced melanin shunt products in *Leptosphaeria maculans*[91] and in *Alternaria solani*,[92] and it also inhibited the accumulation of a phytotoxin altersolanol A by *A. solani*[92] and the release of aflatoxin by *Aspergillus* sp.[93] The latter effects of tricyclazole may imply that MBIs have activities other than inhibition of melanin biosynthesis, and they call for further exploration.

Most MBIs have been developed empirically on the basis of conventional screening systems. However, elucidation of fungal metabolisms that are essential for the expression of pathogenicity should provide promising and unique targets for the development of new disease-control agents. Dehydratase(s) of scytalone or vermelone in melanin biosynthesis seems to be one of those potential targets, since the enzyme(s) is unique in 1,8-DHN melanin synthesis and is apparently involved in pathogenicity of *Pyricularia* and *Colletotrichum*. The terminal oxidative enzymes that produce the final melanin polymer from 1,8-DHN or related metabolites would be another MBI target site.

Progress in research on fungal pathogenicity will lead to a clear understanding of the mode of action of fungicides. Such precise understanding established by molecular (genetic), enzymatic, and immunocytologic studies will facilitate new and better means of plant disease control. So far, relatively little is even known about the key enzymes in melanin biosynthesis, and continued research is definitely needed.

REFERENCES

1. **Yamaguchi, I.,** Fungicides for control of rice blast disease, *J. Pestic. Sci.*, 7, 307, 1982.
2. **Froyd, J. D., Paget, C. J., Guse, L. R., Dreikorn, B. A., and Pafford, J. L.,** Tricyclazole: a new systemic fungicide for control of *Piricularia oryzae* on rice, *Phytopathology*, 66, 1135, 1976.
3. **Schwinn, F., Nakamura, M., and Handschin, G.,** CGA 49104, a new systemic fungicide against rice blast, *9th Int. Congr. Plant Protect. Abstr.*, p. 479, 1979.

4. **Nambu, K.,** Rabcide, a newly developed fungicide for controlling rice blast, *Jpn. Pestic. Inf.*, 10, 73, 1972.
5. **Shephard, M. C. and Ramaswamy, S. R.,** Control of rice blast with the systemic fungicide, *3rd Int. Congr. Plant Pathol. Abstr.*, p. 358, 1978.
6. **Inoue, S., Uematsu, T., and Kato, T.,** Mechanism of rice blast control of chlobenthiazone (S-1901), *3rd Int. Mycol. Congr. Abstr.*, p. 481, 1983.
7. **Tokousbalides, M. C. and Sisler, H. D.,** Effect of tricyclazole on growth and secondary metabolism in *Pyricularia oryzae, Pestic. Biochem. Physiol.*, 8, 26, 1978.
8. **Woloshuk, C. P., Wolkow, P. M., and Sisler, H. D.,** The effect of three fungicides, specific for the control of blast disease, on the growth and melanin biosynthesis by *Pyricularia oryzae* Cav., *Pestic. Sci.*, 12, 86, 1981.
9. **Yamaguchi, I., Sekido, S., and Misato, T.,** Inhibition of appressorial melanization in *Pyricularia oryzae* by nonfungicidal anti-blast chemicals, *J. Pestic. Sci.*, 8, 229, 1983.
10. **Kubo, Y., Suzuki, K., Furusawa, I., and Yamamoto, M.,** Effect of tricyclazole on appressorial pigmentation and penetration from apprssoria of *Colletotrichum lagenarium, Phytopathology*, 72, 1198, 1982.
11. **Wolkow, P. M., Sisler, H. D. F., and Vigil, E. L.,** Effect of inhibitors of melanin biosynthesis on structure and function of appressoria of *Colletotrichum lindemuthianum, Physiol. Plant Pathol.*, 23, 55, 1983.
12. **Kolattukudy, P. E.,** Enzymatic penetration of the plant cuticle by fungal pathogens, *Annu. Rev. Phytopathol.*, 23, 223, 1985.
13. **Hashioka, Y.,** Electronmicroscopical studies on rice blast infection, *Ann. Phytopathol. Soc. Jpn.*, 42, 237, 1976.
14. **Suzuki, N.,** Nature of resistance to blast, in *The Rice Blast Disease*, Proc. Int. Symp. IRRI, The Johns Hopkins Univ. Press, Baltimore, 1963, 277.
15. **Emmett, R. W. and Parbery, D. G.,** Appressoria, *Annu. Rev. Phytopathol.*, 13, 147, 1975.
16. **Uchiyama, T., Ogasawara, N., Nanba, Y., and Ito, H.,** Conidial germination and appressorial formation of the plant pathogenic fungi on the coverglass or cellophane coated with various lipid components of plant leaf waxes, *Agric. Biol. Chem.*, 43(2), 383, 1979.
17. **Woloshuk, C. P., Sisler, H. D., and Vigil, E. L.,** Action of antipenetrant, tricyclazole on appressoria of *Pyricularia oryzae, Physiol. Plant Pathol.*, 22, 245, 1983.
17a. **Howard, R. J. and Ferrari, M. A.,** Role of melanin in appressoria function, *Exp. Mycol.*, 13, 403, 1989.
18. **Wheeler, M. H. and Bell, A. A.,** Melanins and their importance in pathogenic fungi, in *Current Topics in Medical Mycology*, Vol. 2, McGinnis, M. R., Eds., Springer-Verlag, New York, 1988, chap. 10.
19. **Swan, G. A.,** Structure, chemistry, and biosynthesis of the melanins, *Fortschr. Chem. Org. Naturst.*, 31, 521, 1973.
20. **Nicolaus, R. A. and Piattelli, M.,** Progress in the chemistry of natural black pigments, *Rend. Accad. Sci. Fis. Mat. Naples*, 32, 83, 1965.
21. **Piattelli, M., Fattorusso, E., Nicolaus, R. A., and Magno, S.,** The structure of melanins and melanogenesis. V. *Ustilago* melanin, *Tetrahedron*, 21, 3229, 1965.
22. **Bell, A. A. and Wheeler, M. H.,** Biosynthesis and functions of fungal melanins, *Annu. Rev. Phytopathol.*, 24, 411, 1986.
23. **Iwasaki, S., Nozoe, S., and Okuda, S.,** Isolation and structural elucidation of a phytotoxic substance produced by *Pyricularia oryzae* Cavara, *Tetrahedron Lett.*, 45, 3977, 1969.
24. **Iwasaki, S., Muro, H., Sasaki, K., Nozoe, S., and Okuda, S.,** Isolations of phytotoxic substances produced by *Pyricularia oryzae* Cavara, *Tetrahedron Lett.*, 37, 3537, 1973.
25. **Nukina, M., Sassa, T., Ikeda, M., Umezawa, T., and Tasaki, H.,** Pyriculariol, a new phytotoxic metabolite of *Pyricularia oryzae* Cavara, *Agric. Biol. Chem.*, 45, 2161, 1981.
26. **Araki, F. and Miyagi, Y.,** Effects of fungicides on penetration by *Pyricularia oryzae* as evaluated by improved cellophane method, *J. Pestic. Sci.*, 2, 457, 1977.
27. **Yamaguchi, I., Sekido, S., and Misato, T.,** The effect of nonfungicidal anti-blast chemicals on the melanin biosynthesis and infection by *Pyricularia oryzae, J. Pestic. Sci.*, 7, 523, 1982.
28. **Inoue, S. and Kato, T.,** Mode of rice blast control by chlobenthiazone, *J. Pestic. Sci.*, 8, 333, 1983.
29. **Chida, T. and Sisler, H. D.,** Effect of inhibitors of melanin biosynthesis on appressorial penetration and reductive reactions in *Pyricularia oryzae* and *Pyricularia grisea, Pestic. Biochem. Physiol.*, 29, 244, 1987.
30. **Inoue, S., Uematsu, T., and Kato, T.,** Effects of chlobenthiazone on the infection process by *Pyricularia oryzae, J. Pestic. Sci.*, 9, 689, 1984.
31. **Chida, T. and Sisler, H. D.,** Restoration of appressorial penetration ability by melanin precursors in *Pyricularia oryzae* treated with antipenetrants and in melanin-deficient mutants, *J. Pestic. Sci.*, 12, 49, 1987.
32. **Bustaman, M. and Sisler, H. D.,** Effect of pentachloronitrobenzene, pentachloroaniline, and albinism on epidermal penetration by appressoria of *Pyricularia, Pestic. Biochem. Physiol.*, 28, 29, 1987.

33. **Woloshuk, C. P., Sisler, H. D., Tokousbalides, M. C., and Dutky, S. R.,** Melanin biosynthesis in *Pyricularia oryzae:* site of tricyclazole inhibition and pathogenicity of melanin-deficient mutants, *Pestic. Biochem. Physiol.,* 14, 256, 1980.
34. **Woloshuk, C. P. and Sisler, H. D.,** Tricyclazole, pyroquilon, tetrachlorophthalide, PCBA, coumarine and related compounds inhibit melanization and epidermal penetration by *Pyricularia oryzae, J. Pestic. Sci.,* 7, 161, 1982.
35. **Inoue, S., Maeda, K., Uematsu, T., and Kato, T.,** Comparison of tetrachlorphthalide and pentachlorobenzyl alcohol with chlobenthiazone and other melanin inhibitors in the mechanism of rice blast control, *J. Pestic. Sci.,* 9, 731, 1984.
36. **Bell, A. A., Puhalla, J. E., Tolmsoff, W. J., and Stipanovic, R. D.,** Use of mutants to establish (+)-scytalone as an intermediate in melanin biosynthesis by *Verticillium dahlia, Can. J. Microbiol.,* 22, 787, 1976.
37. **Seto, H. and Yonehara, H.,** Utilization of ^{13}C-^{13}C coupling in structural and biosynthetic studies. VIII. The cyclization pattern of a fungal metabolite, scytalone, *Tetrahedron Lett.,* 5, 487, 1977.
38. **Sankawa, U., Shimada, H., Sato, T., Kinoshita, T., and Yamasaki, K.,** Biosynthesis of scytalone, *Chem. Pharm. Bull.,* 29, 3536, 1981.
39. **Tokousbalides, M. C. and Sisler, H. D.,** Site of inhibition by tricyclazole in the melanin biosynthetic pathway of *Verticillium dahlia, Pestic. Biochem. Physiol.,* 11, 64, 1979.
40. **Wheeler, M. H. and Stipanovic, R. D.,** Melanin biosynthesis in *Thielaviopsis basicola, Exp. Mycol.,* 3, 340, 1979.
41. **Sisler, H. D., Woloshuk, C. P., and Wolkow, P. M.,** Specific chemical regulation of appressorial function, *Tagungser. Akad. Landwirtschaftwiss. DDR (Berlin),* 222, 17, 1984.
42. **Pawalek, J. M. and Korner, A. M.,** The biosynthesis of mammalian melanin, *Am. Sci.,* 70, 136, 1982.
43. **Yamaguchi, I., Sekido, S., Seto, H., and Misato, T.,** Cytotoxic effect of 2-hydroxyjuglone, a metabolite in the branched pathway of melanin biosynthesis in *Pyricularia oryzae, J. Pestic. Sci.,* 8, 545, 1983.
44. **Inoue, S., Kato, T., Jordan, V. W. L., and Brent, K. J.,** Inhibition of appressorial adhesion of *Pyricularia oryzae* to barley leaves by fungicides, *Pestic. Sci.,* 19, 145, 1987.
45. **Okuno, T., Kitamura, Y., and Matsuura, K.,** Mechanism of inhibitory effect of tricyclazole on secondary infection by spores of *Pyricularia oryzae, J. Pestic. Sci.,* 8, 361, 1983.
46. **Shiba, Y. and Nagata, T.,** The mode of action of tricyclazole in controlling rice blast, *Ann. Phytopathol. Soc.,* 47, 662, 1981.
47. **Wheeler, M. H.,** Melanin biosynthesis in *Verticillium dahlia:* dehydration and reduction reaction in cell-free homogenates, *Exp. Mycol.,* 6, 171, 1982.
48. **Wheeler, M. H. and Greenblatt, G. A.,** The inhibition of melanin biosynthetic reactions in *Pyricularia oryzae* by compounds that prevent rice blast disease, *Exp. Mycol.,* 12, 151, 1988.
49. **Wheeler, M. H. and Stipanovic, R. D.,** Melanin biosynthesis and the metabolism of flaviolin and 2-hydroxyjuglone in *Wangeniella dermatitidis, Arch. Microbiol.,* 142, 234, 1985.
50. **Inoue, S., Uematsu, T., Kato, T., and Ueda, K.,** New melanin biosynthesis inhibitors and their structural similarities, *J. Pestic. Sci.,* 16, 589, 1985.
51. **Omata, K., Tomita, H., Nakajima, T., and Natsume, B.,** Design of new melanin biosynthesis inhibitors, *Pestic. Sci.,* 26, 271, 1989.
52. **Butler, M. J., Lazarovits, G., Higgins, V. J., and Lachance, M.-A.,** Partial purification and characterization of a dehydratase associated with the pentaketide melanogenesis pathway of *Phaeococcomyces* sp. and other fungi, *Exp. Mycol.,* 12, 367, 1988.
53. **Tajima, S., Kubo, Y., Furusawa, I., and Shishiyama, T.,** Purification of a melanin biosynthetic enzyme converting scytalone to 1,3,8-trihydroxynaphthalene from *Cochliobolus miyabeanus, Exp. Mycol.,* 13, 69, 1989.
54. **Kubo, Y.,** Melanin biosynthesis in fungi, *Trans. Mycol. Soc. Jpn.,* 27, 487, 1986.
55. **Kubo, Y. and Furusawa, I.,** Localization of melanin in appressoria of *Colletotrichum lagenarium, Can. J. Microbiol.,* 32, 280, 1986.
56. **Kubo, Y., Suzuki, K., Furusawa, I., Ishida, N., and Yamamoto, M.,** Relation of appressorium pigmentation and penetration of nitrocellulose membranes by appressoria of *Colletotrichum lagenarium, Phytopathology,* 72, 498, 1982.
57. **Kubo, Y., Suzuki, K., Furusawa, I., and Yamamoto, M.,** Scytalone as a natural intermediate of melanin biosynthesis in appressoria of *Colletotrichum lagenarium, Exp. Mycol.,* 7, 208, 1983.
58. **Kubo, Y., Furusawa, I., and Yamamoto, M.,** Regulation of melanin biosynthesis during appressorium formation in *Colletotrichum lagenarium, Exp. Mycol.,* 8, 364, 1984.
59. **Kubo, Y., Suzuki, K., Furusawa, I., and Yamamoto, M.,** Melanin biosynthesis as a prerequisite for penetration by appressoria of *Colletotrichum lagenarium*: site of inhibition by melanin-inhibiting fungicides and their action on appressoria, *Pestic. Biochem. Physiol.,* 23, 47, 1985.
60. **Kubo, Y., Furusawa, I., and Shishiyama, J.,** Relation between pigment intensity and penetrating ability in appressoria of *Colletotrichum lagenarium, Can. J. Microbiol.,* 33, 871, 1987.

61. **Suzuki, K., Kubo, Y., Furusawa, I., Ishida, N., and Yamamoto, M.,** Behavior of colorless appressoria in an albino mutant of *Colletotrichum lagenaruim, Can. J. Microbiol.*, 28, 1210, 1982.
62. **Suzuki, K., Furusawa, I., Ishida, N., and Yamamoto, M.,** Protein synthesis during germination and appressorium formation of *Colletotrichum lagenarium apores, J. Gen. Microbiol.*, 124, 61, 1981.
63. **Suzuki, K., Furusawa, I., Ishida, N., and Yamamoto, M.,** Chemical dissolution of cellulose membranes as a prerequisite for penetration from appressoria of *Colletotrichum lagenarium, J. Gen. Microbiol.*, 128, 1035, 1982.
64. **Suzuki, K., Furusawa, I., and Yamamoto, M.,** Role of chemical dissolution of cellulose membranes in the appressorial penetration by *Colletotrichum lagenarium, Ann. Phytopathol. Soc. Jpn*, 49, 481, 1983.
65. **Bell, A. A., Stipanovic, R. D., and Puhalla, J. E.,** Pentaketide metabolites of *Verticillium dahliae:* identification of (+)-scytalone as a natural precursor to melanin, *Tetrahedron*, 32, 1353, 1976.
66. **Stipanovic, R.D. and Bell, A. A.,** Pentaketide metabolites of *Verticillium dehliae*. III. Identification of (−)-3,4-dihydro-3,8-dihydroxy-1(2H)naphthalenone [(−)-vermelone] as a precursor to melanin, *J. Org. Chem.*, 41, 2468, 1976.
67. **Stipanovic, R. D. and Bell, A. A.,** Pentaketide metabolites of *Verticillium dahliae*. II. Accumulation of naphthol derivatives by aberrant-melanin mutant brm-2, *Mycologia*, 69, 164, 1977.
68. **Stipanovic, R. D. and Wheeler, M. H.,** Accumulation of 3,3′-biflaviolin, a melanin shunt product, by tricyclazole treated *Thielaviopsis basicola, Pestic. Biochem. Physiol.*, 13, 198, 1980.
69. **Ishida, N. and Akai, S.,** Electron microscopic observation of cell wall structure during appressorium formation in *Colletotrichum lagenarium, Mycopathol. Mycol. Appl.*, 35, 68, 1968.
70. **Griffiths, D. A. and Campbell, W. P.,** Fine structure of conidial germination and appressorial development in *Colletotrichum atramentarium, Trans. Br. Mycol. Soc.*, 61, 529, 1973.
71. **Politis, D. J. and Wheeler, H.,** Ultrastructural study of penetration of maize leaves by *Colletotrichum graminicila, Physiol. Plant Pathol.*, 3, 465, 1973.
72. **Mercer, P. C., Wood, R. K. S., and Greenwood, A. D.,** Ultrastructure of the parasitism of *Phaseolus vulgaris* by *Colletotrichum lindemuthianum, Physiol. Plant Pathol.*, 5, 203, 1975.
73. **Kozar, F. and Netlitzky,** Studies on hyphal development and appressorium formation of *Colletotrichum graminicola, Can. J. Bot.*, 56, 2234, 1978.
74. **Seifers, D. and Ammon, V.,** Mode of penetration of sycamore leaves by *Gloeosporium platani, Phytopathology*, 70, 1050, 1980.
75. **Mitchel, R. and Sabar, N.,** Autolytic enzymes in fungal cell walls, *J. Gen. Microbiol.*, 42, 39, 1966.
76. **Gull, K. and Trinci, A. P. J.,** Detection of areas of wall differentiation in fungi using fluorescent staining, *Arch. Microbiol.*, 96, 53, 1974.
77. **Polacheck, I. and Rosenberger, R. F.,** *Aspergillus nidulans* mutant lacking -(1,3)-glucan, melanin, and cleistothecia, *J. Bacteriol.*, 132, 650, 1977.
78. **Reyes, F. and Lahoz, R.,** Variations in lysis of walls of *Sclerotinia fructigena* with age of culture, *J. Gen. Microbiol.*, 98, 607, 1977.
79. **Potgieter, H. J. and Alexander, M.,** Susceptibility and resistance of several fungi to microbial lysis, *J. Bacteriol.*, 91, 1526, 1966.
80. **Bloomfield, B. J. and Alexander, M.,** Melanins and resistance of fungi to lysis, *J. Bacteriol.*, 93, 1276, 1967.
81. **Kuo, M. J. and Alexander, M.,** Inhibition of the lysis of fungi by melanins, *J. Bacteriol.*, 94, 624, 1967.
82. **Bull, A. T.,** Chemical composition of wild and mutant *Aspergillus nidulans* cell walls: the nature of polysaccharides and menalin constituents, *J. Gen. Microbiol.*, 63, 75, 1970.
83. **Bull, A. T.,** Inhibition of polysaccharases by melanin: enzyme inhibition in relation to mycolysis, *Arch. Biochem. Biophys.*, 137, 345, 1970.
84. **Okuno, T., Matsuura, K., and Furusawa, I.,** Recovery of appressorial penetration by some melanin precursors in *Pyricularia oryzae* treated with tricyclazole and in a melanin deficient mutant, *J. Pestic. Sci.*, 8, 357, 1983.
85. **Omura, S.,** The antibiotic cerulenin, a novel tool for biochemistry as an inhibitor of fatty acid synthesis, *Bacterio. Rev.*, 40, 681, 1976.
86. **Kubo, Y., Katoh, M., Furusawa, I., and Shishiyama, J.,** Inhibition of melanin biosynthesis by cerulenin in appressoria of *Colletotrichum lagenarium, Exp. Mycol.*, 10, 301, 1986.
87. **Yaegashi, H. and Yamada, M.,** Possible genetic factor conditioning the size of blast lesions induced on rice leaves with *Pyricularia oryzae, Ann. Phytopathol. Soc. Jpn.*, 50, 664, 1984.
88. **Crawford, M. S., Chumley, F. G., Weaver, C. G., and Valent, B.,** Characterization of the meterokaryotic and vegetative diploid phases of *Magnaporthe grisea, Genetics*, 114, 1111, 1986.
89. **Leung, H., Borromeo, E. S., Bernardo, M. A., and Notteghem, J. L.,** Genetic analysis of virulence in the rice blast fungus *Magnaporthe grisea, Phytopathology*, 78, 1227, 1988.
90. **Parsons, K. A., Chumley, F. G., and Valent, B.,** Genetic transformation of the fungal pathogen responsible for rice blast disease, *Proc. Natl. Acad. Sci., U.S.A.*, 84, 4161, 1987.

91. **Dahiya, J. S. and Rimmer, S. R.,** Accumulation of Flaviolin, 4-hydroxyscytalone and 2-hydroxijuglone in tricyclazole-treated cultures of *Leptosphaeria maculans, Phytochemistry,* 27, 3481, 1988.
92. **Lazarovits, G. and Stoessl, A.,** Tricyclazole induces melanin shunt products and inhibits altersolanol A accumulation by *Alternaria solani, Pestic. Biochem. Physiol.*, 31, 36, 1988.
93. **Fernando, T. and Bean, G.,** Effects of tricyclazole on growth, release of aflatoxin, and sterol and fatty acid content, by *Aspergillus* isolates, *Trans. Br. Mycol. Soc.*, 87, 445, 1986.
94. **Hashimoto, N. and Yamaguchi, I.,** in press.

Chapter 7

ANTIFUNGAL AGENTS WITH TARGET SITES IN STEROL FUNCTIONS AND BIOSYNTHESIS

Wolfram Köller

TABLE OF CONTENTS

I. INTRODUCTION

The science of "steroidology", initiating with the discovery of cholesterol as the major sterol found in mammals, has captured significant scientific attention and recognition. Between 1910 and 1985, no less than 15 Nobel Prizes have been awarded to scientists working in this field of biomolecules. The apparent scientific "success" is not surprising in light of the remarkable diversity of sterol functions, ranging from the maintenance of membrane integrity over the production of bile acids to the synthesis of steroid hormones. With the advancement of our knowledge and understanding, the attempts to interfere with sterol metabolism and functions have been increasingly successful. The field of "applied steroidology" reflects the diverse functions of sterols and ranges from interference with contraception over antifungal agents to hypocholesterolemic drugs developed to lower the cholesterol levels in human blood.

The first class of antifungal drugs with a mode of action related to fungal sterols was the polyene macrolides, with nystatin and amphotericin B as the most widely known representatives. These antifungal antibiotics, developed in the 1950s, were restricted to the treatment of human mycotic diseases and had no impact on disease control in agriculture. It was 20 years later that the two disciplines merged. The development of antifungal agents with a specific target site in the biosynthesis of fungal sterols was initiated in the late 1960s with the discovery of antifungal activities inherent to azole derivatives. From the very beginning, this class of azoles was developed for both the control of fungal diseases of plants and the treatment of human and animal mycoses. A second class of commercial fungicides with a sterol-related mode of action is the morpholines, first restricted to the use as fungicides in agriculture and, more recently, extended into the field of antimycotic drugs. A third class of commercial "steroidogenic" antimycotic agents, the allylamines, is still restricted to its application in medicine.

The success, advancement, and impact of antifungal agents with target sites in fungal sterol biosynthesis have been tremendous, and the biochemistry of the various classes of inhibitors has been described in numerous articles. One recently published book was entirely dedicated to this subject,[1] and sterol biosynthesis inhibitors comprise substantial parts of one additional book and two recent conference proceedings.[2-4] The success of sterol biosynthesis inhibitors has also triggered basic research on the function and biosynthesis of fungal sterols and on the identification of novel target enzymes and lead structures of inhibitors. Moreover, the subject has become a clearly interdisciplinary task involving company and university research dedicated to the development of agricultural fungicides, antimycotic agents, hypocholesterolemic drugs, plant growth regulators, and even anticancer agents. Interdisciplinary research, however, is still in its infant stage, and the perspective of an improved collaboration appears promising with regard to the development of novel antifungal agents.

Numerous excellent and comprehensive review articles on the biochemistry of particular

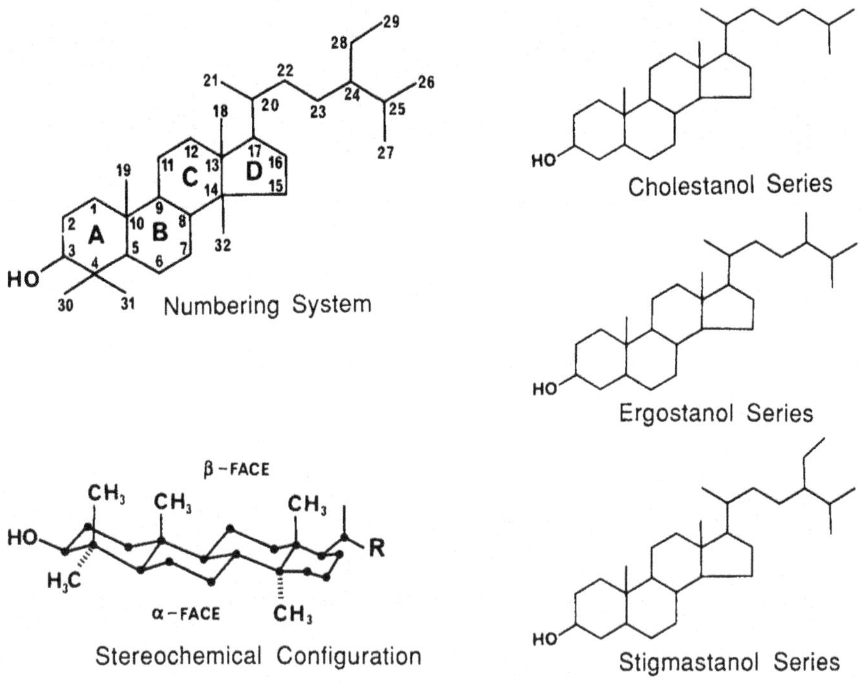

FIGURE 1. Structural features and nomenclature of sterols.

sterol biosynthesis inhibitors have been published over the years, and the immediate need for repetition is not apparent. Instead, an attempt will be made similar to Burden et al.[4a] to summarize the entire field of sterol biosynthesis inhibitors, with emphasis on existing fungicides and antimycotic drugs, but also on novel and undeveloped lead structures, and on opportunities of interdisciplinary collaboration. Naturally, this task requires a summary of fungal sterol chemistry and biochemistry first. The sterol biosynthesis inhibitors will then be described step by step guided by the route of biosynthesis steps. Considering this broad arena, it will not be possible to reference an exhaustive list of original research articles. The goal can only be to provide access to the recent literature, and to emphasize new discoveries, developments, and trends in this field of research.

II. FUNGAL STEROLS

A. STRUCTURES AND DISTRIBUTION

The principles of sterol nomenclature are relatively simple. They are based on a numbering system and three series of sterol prototypes, the cholestanol (no substitution at C-24), ergostanol (C-24 methyl), and stigmastanol (C-24 ethyl) series (Figure 1).[5,6] In a simplified nomenclature, this designation not only assigns the substitution at C-24, but also the stereochemistry of the hydroxy-group at position 3 (β-configuration) and the *trans*-fusion of ring A and ring B. Both of these configurations are preserved in natural sterols. The sterol nomenclature assigns the position of double bonds within this framework of basic rules. For example, the systematic name for ergosterol (Figure 2) would be ergosta-5,7,22-trienol. Excluding the two stereochemical implications described above, the systematic name for ergosterol would change to 5α-ergosta-5,7,22-trien-3β-ol. Many sterols including ergosterol are substituted at the C-24 position of the side chain and, thus, occur as two stereoisomers (Figure 2). In general terms, the stereochemical configuration is characteristic for certain classes of organisms. It is α for the plant sterols campesterol, sitosterol, and

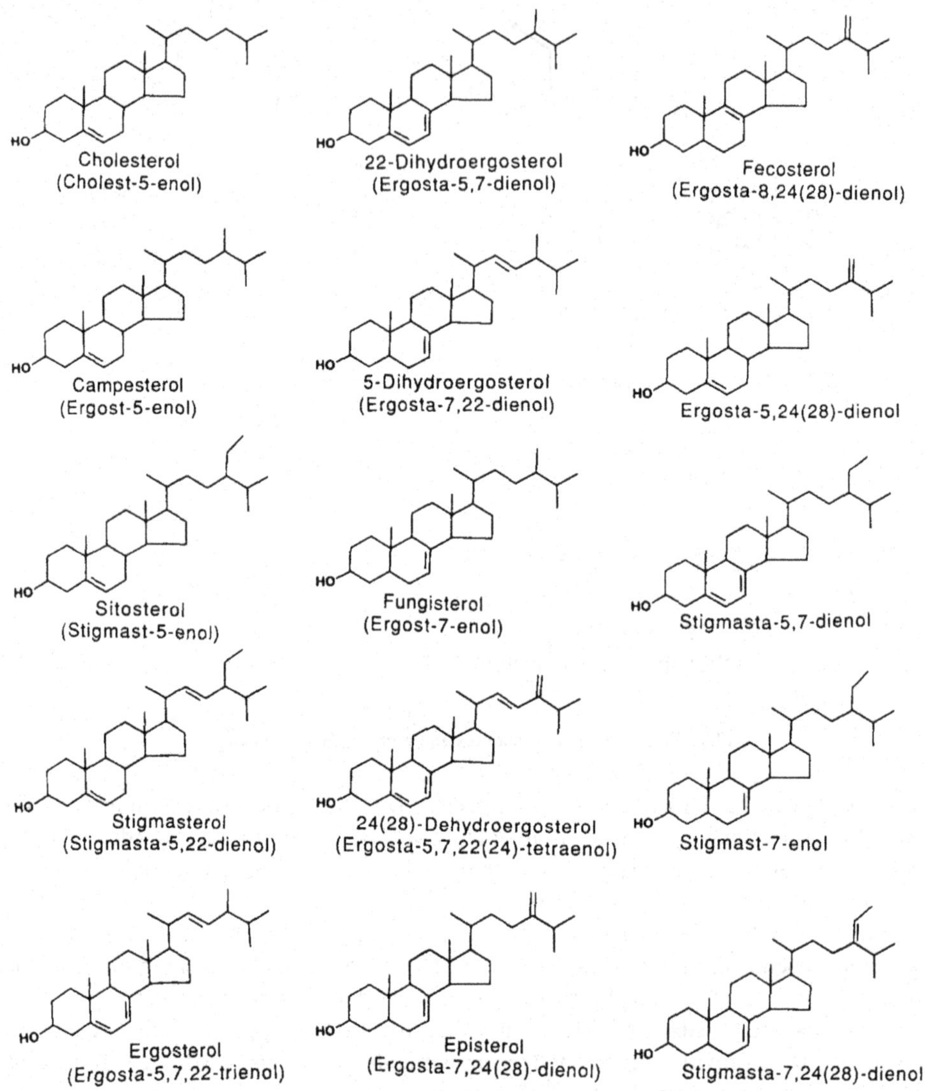

FIGURE 2. Sterol structures.

stigmasterol, and β for ergosterol.[7] In many cases, the systematic names are substituted by trivial names. Some of these trivial names had been assigned with the first description of the respective sterol; the name ergostérine (ergosterol) was proposed in 1889 for the sterol isolated from sclerotia of *Claviceps purpurea* (ergot).[8]

Cholesterol is the main sterol found in mammals; campesterol, sitosterol, and stigmasterol are most prominent in the plant kingdom (see Chapter 8), whereas ergosterol is the characteristic sterol in the higher fungi, the Ascomycetes, Basidiomycetes, and Fungi imperfecti (Figure 2).[7,9-11] Although not entirely free of controversies,[12] it is generally accepted that ergosterol is not present in the lower fungi, the Chytridiomycetes, Hypochytriomycetes, and Oomycetes.[7,9-11,13,14] The Oomycetes can be divided into two groups, those that synthesize or do not synthesize sterols.[14,15] The most prominent sterol found in sterol-synthesizing Oomycetes is fucosterol [stigmasta-5,24(28)-dienol].[10] The economically most important Oomycetes (the plant pathogenic "downy mildews", *Phytophthora* spp. and *Phythium* spp.) belong to the Perenosporales, a group that synthesizes squalene but not sterols and, thus, lacks essential enzymes of sterol biosynthesis.[15-17] Sterols are not an absolute requirement for the vegetative growth of these organisms. A novel group of lipids, the phytophthorols, were isolated from *Phytophthora cactorum*, but not from a sterol-synthesizing oomycete, and a function of these long-chain esters in vegetative growth similar to sterols has been proposed.[18] For both groups of Oomycetes, however, sterols are obligatory for the sexual formation of oospores.[14,16,17] Since vegetative growth and asexual reproduction are the preferred target sites for antifungal agents, it is not surprising that commercial sterol biosynthesis inhibitors are essentially inactive against the Perenosporales. Thus, the higher fungi will be clearly emphasized in this overview.

The initially simple picture of different sterols predominant in the different classes of eukaryotes has gained in complexity since analytical methods such as capillary-GC, HPLC, mass spectrometry, or spectroscopical methods became more refined and powerful.[20-22] For example, no less than 44 different sterols were identified in a survey from 12 genera of the plant family Cucurbitacae.[23] Twenty-two different sterols alone were present in *Cucurbita maxima*, and the relative sterol composition was variable, depending on the developmental stage of the plant and the specific tissues or organs analyzed.[24] Cholesterol, once thought to be a sterol found only in mammals, has since been identified in numerous plants.[25]

Regardless of the great number of sterols present in plants, sterols containing a $\Delta^{5,7}$-diene system, the characteristic feature of ergosterol, are restricted to the higher fungi. This does not imply that all fungal sterols must feature this structural element, and that the diversity of fungal sterols is less complex than that in the plant kingdom. Lösel[10] listed 20 main and 38 minor different sterols identified from fungal organisms. This list is considerably longer, if all sterols from fungi with impaired pathways of sterol biosynthesis, be it by mutation or by chemical inhibition, would be included. Some of the more prominent fungal sterols are shown in Figure 2. The structural sterol diversity is partly due to the lower fungi; but even for higher fungi, the structural diversity remains pronounced, although ergosterol comprises the major sterol in most cases.[10] However, exceptions exist. Ergosterol was totally absent from several powdery mildew fungi; here, ergosta-5,24(28)-dienol was identified as the main sterol (Figure 2).[26] Similar to the powdery mildews, rust uredospores isolated from flax,[27] french bean,[28,29] wheat,[30] and pine[31,32] were also deplete of ergosterol. The rust sterols identified were stigmast-7-enol, stigmasta-5,7-dienol, and stigmasta-7,24(28)-dienol (Figure 2). In addition, cholesterol was found in uredospores of wheat stem rust fungus, and the proportion of cholesterol increased with germination.[30] Based on the similarity between rust and plant sterols, they were initially assumed to be of plant origin.[30] However, evidence for an independent synthesis has been presented,[29,30,32] and the fact that several sterol biosynthesis inhibitors used as fungicides in agriculture provide excellent rust control[33-36] is in support of an independent biosynthetic pathway of rust sterols. A functional reason for the absence of ergosterol from powdery mildew and rust fungi, both obligate plant parasites, has not yet been presented.

In Ascomycetes, Basidiomycetes, and Imperfect fungi, sterols of the stigmastanol series similar to the rust sterols are rare.[10] Exceptions, however, appear to exist. Stigmast-5-enol was isolated from mycelium of *Botrytis cinerea*.[37] In addition to stigmast-5-enol, stigmasta-5,22-dienol was reported to be present in *Pyricularia oryzae, Botrytis cinerea, Fusarium culmorum*, and *Pyrenophora teres*.[38] Since the configuration at the C-24 position has not been determined, it remains unclear whether these stigmastanols are identical with sitosterol and stigmasterol. Furthermore, the results could not be confirmed by others.[38a]

The sterols found in addition to ergosterol in most of the sterol-synthesizing fungi belong to the ergostanol series (Figure 2). In filamentous and dimorphic organisms, ergosta-5,7-dienol, ergosta-7,22-dienol, brassicasterol, and fungisterol are frequently identified as additional sterol components.[5,10,11] In yeasts, 24(28)-dehydroergostanols such as fecosterol, episterol, and 24(28)-dehydroergosterol often comprise large portions of the sterol pool.[9]

The pattern of sterol by-products is not distinctly different for ascomycetes, basidiomycetes, and the imperfect fungi.[9-11] It can even vary from strain to strain of one and the same fungal species. For example, four different sterols, ergosterol, 22-dihydroergosterol, 5-dihydroergosterol, and fungisterol, have been isolated in variable proportions from wild-type isolates of *Ustilago maydis*,[39-41] a basidiomycete extensively used for mode of action studies of sterol biosynthesis inhibitors. A similar diversity from strain to strain of *Saccharomyces cerevisiae* has been summarized by Rattray.[9]

The functional reason for the existence of a number of sterols in addition to ergosterol remains largely unknown. It should be mentioned, however, that at least some of the structural diversity might be related to the developmental and nutritional stage of the analyzed fungus, and to the analytical procedures employed for sterol extraction and preparation. For example, the exclusive free sterol extracted from exponentially growing *U. avenae* was ergosterol;[42,43] two additional sterols, ergosta-5,7-dienol and 24-methylenedihydrolanosterol, were identified when growth approached the linear phase.[43] Both additional sterols are precursors of ergosterol, and both biosynthetic conversions, as described later, require oxygen. Thus, the decreased supply of intracellular oxygen in fungal cultures at a late stage of growth, which has been described to affect the biosynthesis of neutral lipids in cultures of *Lipomyces starkeyi*,[44] might serve as a simple explanation for the appearance of these sterols. The example would imply a biosynthetic rather than functional reason for the presence of sterols in addition to ergosterol. Procedures of sterol analysis might sometimes contribute to an overemphasis of sterol diversity. Sterols occur as free sterols, but also as sterol derivatives, in fungal cells most abundantly esterified to a fatty acid.[9,45] In *S. cerevisiae*, sterol esters were found to be localized in lipid bodies and are thus segregated from the pool of free sterols.[46,47] The proportion of esters can be variable ranging from trace amounts to half of the total lipid.[9] A great part of the analytical work on fungal sterols has been derived from saponified lipid samples. Sterol esters are hydrolyzed during this procedure, and the relative distribution of sterol structures present in the two separate pools of sterols has not been analyzed. Evidence mainly derived from yeasts, however, indicates that the sterols present in the ester fraction are more diverse in structure than the free sterols. It has been suggested that sterol esterification in yeast provides a mechanism to sequester excess sterols in order to maintain an essential amount of free sterols, but also to segregate suboptimal sterols and sterol precursors.[45,48,49] For example, sterol precursors accumulating under the action of the two sterol biosynthesis inhibitors trifluperidol and AY-9944 in *S. cerevisiae* were shown to be primarily esterified.[50,51] Although less experimental data are available for mycelial fungi, the principle of a higher degree of structural diversity of sterols segregated into the ester fraction appears to be of general validity.[10] Regardless of factors that might overemphasize the degree of sterol diversity in higher fungi, ergosterol can be accompanied or entirely substituted by other sterols. Both the functions and biosynthetic implications for this diversification remain largely unknown.

B. BIOSYNTHESIS — ACETATE TO OXIDOSQUALENE

The route of sterol biosynthesis is part of the highly conserved isoprenoid pathway. The successive condensation of C_5 carbon units leads directly to terpenoids such as gibberelic acids (C_{20}), lanosterol (C_{30}), and carotenoids (C_{40}). In addition, intermediates of the pathway are utilized for the synthesis of cytokinines, haem, chlorophyll, ubiquinones, and plastoquinone.[52-54] The biosynthesis of fungal sterols as part of the isoprenoid pathway has been extensively reviewed in the past,[11,54-59] and the treatise of biosynthetic aspects will be mainly restricted to the principles relevant to the mode of action of existing and future antifungal agents.

The initial segment of sterol biosynthesis as outlined in Figure 3 is governed by the formation of the key intermediate isopentenyl diphosphate. 3-Hydroxy-3-methyl-glutaryl-

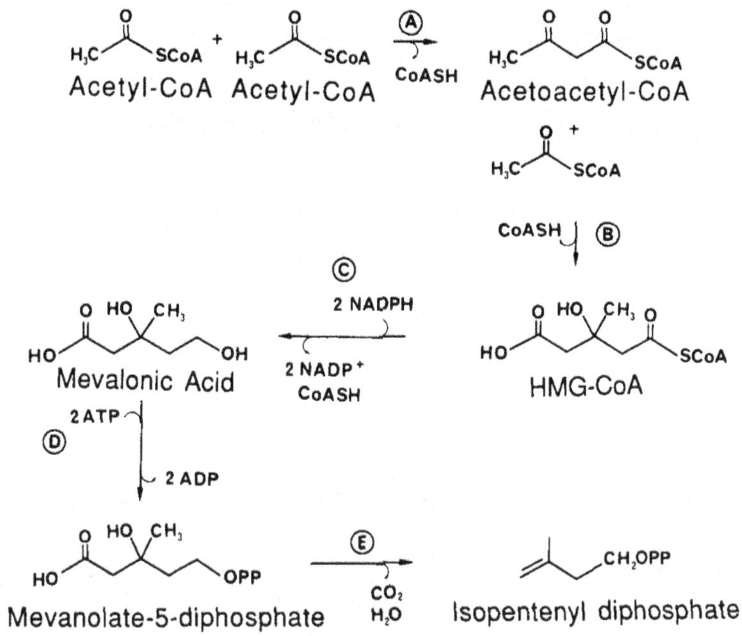

FIGURE 3. Biosynthesis of sterols: condensation of acetate to C_5 units. A, acetoacetyl-CoA tiolase; B, HMG-CoA synthase; C, HMG-CoA reductase; D, mevanolate kinase and mevanolate-5-phosphate kinase; E, mevanolate-5-diphosphate decarboxylase.

CoA (HMG-CoA), a C_6 unit, is formed from the condensation of three acetyl-CoA molecules. These reactions are catalyzed by the enzymes acetoacetyl-CoA thiolase and HMG-CoA synthase. HMG-CoA is then reduced to mevalonic acid, an intermediate central to the isoprenoid pathway. It is not surprising that the respective enzyme, HMG-CoA reductase, serves as a prime target for the regulation of the entire pathway of sterol synthesis. This aspect will be discussed later. The formation of isopentenyl diphosphate, the C_5 building block of all isoprenoids, is accomplished by the decarboxylation of mevanolate-5-diphosphate (Figure 3).

With the exception of HMG-CoA reductase, all enzymes involved in the formation of isopentenyl diphosphate are localized in the cytosol. In rat liver, HMG-CoA reductase is a transmembrane glycoprotein localized in the smooth and rough endoplasmic reticulum, but not in the Golgi and plasma membranes.[60] In addition to this microsomal location, the enzyme has been found in rat liver peroxisomes.[61,62] A dual localization of HMG-CoA reductase in microsomal and mitochondrial membranes was also reported for *S. cerevisiae*.[63] Although the functional relevance for this intracellular separation of the two isoenzymes has not yet been presented, the identification of two separate genes for HMG-CoA reductase in yeast is in support of this observation.[64] A microsomal localization of HMG-CoA reductase has also been reported for the filamentous fungus *Aspergillus nidulans*.[65] However, a soluble cytoplasmic form was found to accompany the membrane bound enzyme. It remains to be clarified whether this difference among fungi is also reflected on the genomic level. The amino acid sequence of the HMG-CoA reductase from yeast as deduced from the cDNA sequence was very similar to the mammalian enzyme,[66] but distinctly different from the first plant enzyme analyzed.[67]

Mevalonic acid, the product of the HMG-CoA reductase action, is phosphorylated twice and finally decarboxylated to isopentenyl diphosphate, the C_5 building block in sterol synthesis.[52,54] The subsequent sequence of reactions involved in the condensation of C_5 units is outlined in Figure 4 and has been summarized recently.[68] Isopentenyl diphosphate (IPP)

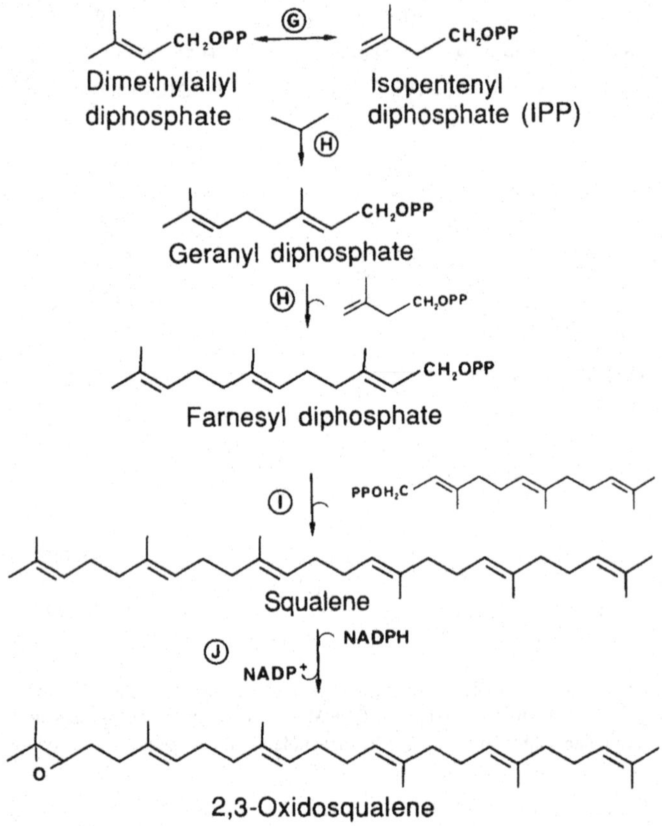

FIGURE 4. Biosynthesis of sterols: condensation of C_5 to C_{30} units. G, IPP-isomerase; H, prenyl transferase; I, squalene synthase; J, squalene epoxidase.

is isomerized to dimethylallyl diphosphate, serving as a donor of IPP in the condensation reaction leading to geranyl diphosphate (C_{10}). Farnesyl diphosphate (C_{15}) is formed by a second condensation of IPP.[52-54] Both enzymes from *S. cerevisiae*, IPP isomerase and prenyl transferase, have been extensively characterized.[69,70] The synthesis of squalene (C_{30}), the specific linear precursor for sterols, takes place in two steps. Two molecules of farnesyl diphosphate are condensed to presqualene diphosphate as an intermediate, which is reductively converted to squalene. The corresponding enzyme, squalene synthase, is an integral protein of microsomal membranes and has been purified and characterized from *S. cerevisiae*.[71]

The immediate precursor for the formation of the first sterol structure, 2,3-oxidosqualene, is formed via double-bond epoxidation (Figure 4). The respective enzyme, squalene epoxidase, is the first enzyme in sterol synthesis that requires oxygen.[54,55] As described later, this enzyme is the target for a novel class of antimycotic drugs, the allylamines (see Section IV.C). The enzyme has been purified from rat liver microsomes and requires, for full activity, NADPH, FAD, and a soluble protein factor.[72] The particulate enzyme from *Candida albicans* was similar except that NADH instead of NADPH was the preferred cosubstrate.[73] Additional differences between the mammalian and fungal enzyme are the requirements for a cytosolic protein factor and the modulation of activity by fatty acids.[74]

C. BIOSYNTHESIS — FORMATION AND MODIFICATION OF STEROLS

The formation of 2,3-oxidosqualene described above involves ten different enzymatic steps, starting out from acetyl-CoA. One additional step is required to yield the first sterol

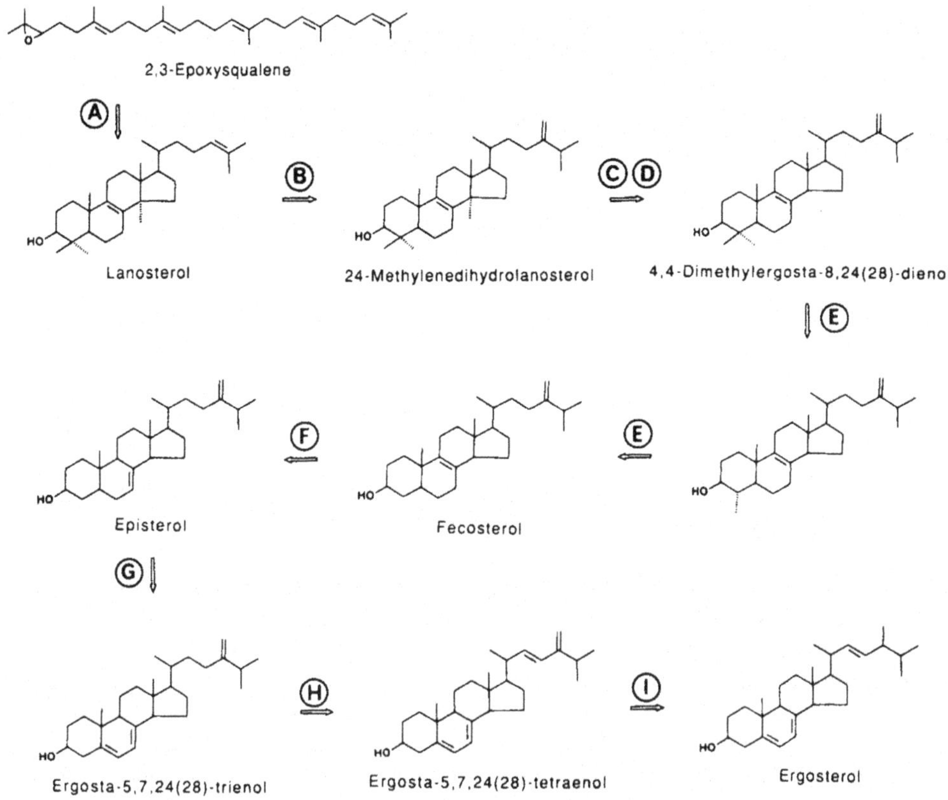

FIGURE 5. Biosynthesis of sterols: cyclization and modification. A, 2,3-oxidosqualene-lanosterol cyclase; B, Δ^{24}-sterol methyltransferase; C, 14α-demethylase; D, Δ^{14}-reductase; E, 4-demethylase; F, Δ^{7}-Δ^{8}-isomerase; G, Δ^{5}-desaturase; H, Δ^{22}-desaturase; I, $\Delta^{24(28)}$-reductase.

ring structure lanosterol. Although the "sterol goal" has been accomplished at this step, the conversion of lanosterol to the final end product ergosterol is again a complex multistep process involving no less than ten distinct enzymatic steps (Figure 5). Based on the lipid character of the intermediate products, all enzymes involved in these steps of sterol modifications are localized in microsomal membranes.[75] Recent findings, however, might indicate that the intracellular localization could be more diverse. For example, rat liver peroxisomes were shown to convert mevalonic acid to cholesterol and, thus, would require the entire set of particulate sterol-converting enzymes.[61,62]

The steps of ergosterol synthesis in fungi involve the methylation of the C-24 side-chain position, three demethylations at the C-4 and C-14 position, and several rearrangements of double-bonds, both in the nucleus and side chain. The individual steps of sterol interconversion do not follow a definite, rigid pattern. The sequence of different enzymatic steps and, thus, the affinity of distinct enzymes for certain sterol intermediates can be slightly different from fungus to fungus. Since the structures of sterol intermediates become increasingly similar during the later steps of double-bond rearrangements, this variability is not surprising. Small structural alterations of enzymes might lead to small changes of K_m-values for the various related substrates, and thus to the reversion of individual steps. One major difference among fungi, however, exists. As described later in greater detail, the respective substrate for the side-chain methylation is lanosterol for most fungi, with *S. cerevisiae* as one of the undisputed exceptions. In this case, the side-chain methylation is accomplished subsequent to the removal of the methyl groups at the C-4 and C-14 position. Because this biosynthetic reversion appears an exception rather than the rule, the overall scheme presented

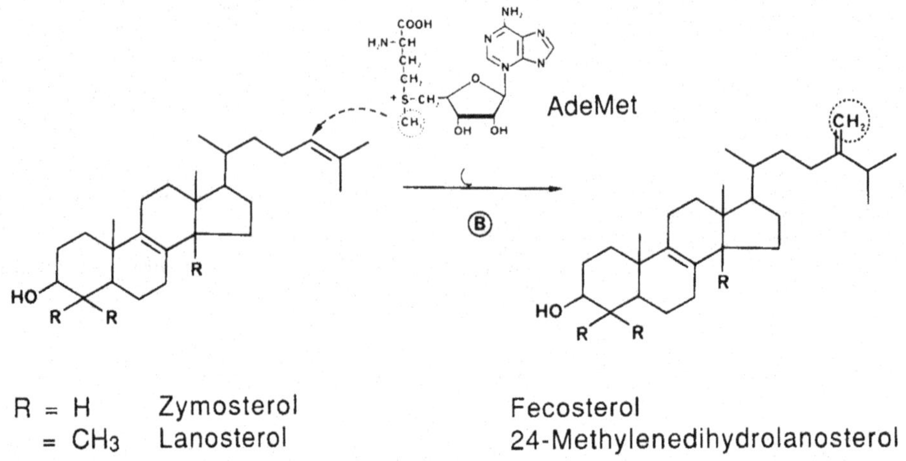

FIGURE 6. Reaction pathway of the Δ^{24}-sterol methyltransferase.

in Figure 5 is restricted to the sequence of steps recognized as the most characteristic pathway in the majority of fungi.[56]

1. Formation of Lanosterol

The cyclization of 2,3-oxidosqualene to lanosterol in mammals and fungi (step A in Figure 5) is catalyzed by the enzyme 2,3-oxidosqualene-lanosterol cyclase. Interestingly, the cyclization product in plants, cycloartenol, is slightly different from lanosterol, although the enzyme reaction is very similar to the formation of lanosterol (Chapter 8). The cyclization of 2,3-oxidosqualene is one of the most intriguing enzymatic reactions, involving no less than ten distinct reaction centers.[54,56] The reaction is initialized by an electrophilic opening of the epoxide ring. The following steps of ring closures and backbone rearrangements proceed through conformationally rigid carbocationic intermediates.[76,77] As discussed later (see Section IV.C), cationic transition-state analogues have been shown to act as powerful cyclase inhibitors. The respective enzyme is microsome bound and has been solubilized and partially purified from hog liver,[78] but not yet from a fungal organism.

2. Side-Chain Methylation

Cholesterol and ergosterol are primarily distinguished by the C-24 methyl group present in ergosterol (Figure 2). This additional substituent is the only carbon in ergosterol not derived from acetate. The methyl donor for this transmethylation is *S*-adenosylmethionine (AdoMet) (Figure 6). The respective enzyme, Δ^{24}-sterol methyltransferase, has been purified and characterized from yeast.[56] The sterol structure essential for the enzyme reaction is the $\Delta^{24,25}$-double bond. Due to the positive charge at the sulfur atom of AdoMet, the methyl group is added electrophilically to the C-24 position, thus generating an electrophilic center at the neighboring C-25 position. This partial positive charge is saturated by a hydrogen shift from C-24 to C-25 and leads to the formation of the $\Delta^{24,28}$-double bond.[56] As discussed later (see Section V.A), the presence of a carbocationic high-energy intermediate at the C-25/C-24 position has been successfully utilized to design transition-state analogues as inhibitors of the Δ^{24}-sterol methyltransferase.

Lanosterol, the initial sterol structure formed in sterol biosynthesis, is widely accepted as the sterol substrate for the transmethylation reaction in most fungi, leading to 24-methylenedihydrolanosterol (Figure 6).[56] One of the few exceptions to this rule is *S. cerevisiae*. Here, sterols with methyl groups in C-4 and C-14 position are not converted by the respective enzyme. For example, *S. cerevisiae*, treated with an inhibitor of the C-14 demethylation, accumulated primarily lanosterol and not 24-methylenedihydrolanosterol.[79] Zymosterol (cholesta-8,24-dienol) is the most likely candidate for the side-chain methylation in yeast (Figure 6). This sterol was both the best *in vitro* substrate for the yeast Δ^{24}-sterol methyltransferase and also accumulated in mutants defective in the enzyme.[80-82] In yeast, Δ^{24}-sterol methyltransferase was found to be localized in microsomes, mitochondria, and a "floating lipid layer" with very low density.[83] An interesting controversy surrounding the most suitable substrate for the methyltransferase has not yet been resolved for *Candida albicans*. Mutants with defects in the C-14 demethylation or cells treated with specific inhibitors of the C-14 demethylation step accumulated primarily 4α-methyl and 4,4-dimethylsterols methylated at C-24.[84,85] These results indicated that, in contrast to *S. cerevisiae*, the methyltransferase present in *C. albicans* was similar to most other fungi and accepted methylated sterols as the most suitable substrates. However, based on sterol labeling studies with [*Me*-^{14}C]methionine, Ryder[86] concluded that C-24 methylation occurred subsequent to the nuclear demethylation. This suggestion was recently confirmed by *in vitro* assays with the solubilized *Candida* enzyme.[87] Similar to yeast, zymosterol was the best substrate *in vitro*, and methylated sterols such as lanosterol were not active as substrates.

3. C-14 Demethylation

According to the different substrates for the Δ^{24}-sterol methyltransferase in *S. cerevisiae* and other sterol-producing fungi, the removal of the C-32 atom protruding into the α-face of the sterol ring system (Figure 1) proceeds either from lanosterol or 24-methylenedihydrolanosterol (Figure 5). The demethylation of lanosterol in yeast resembles closely the situation found in mammals.[56,88] The 14α-demethylases (14α-DMs) belong to the superfamily of cytochrome P-450 monooxygenases. As described below, this class of enzymes is functionally diverse, and the nomenclature became increasingly confusing. Often, several names were used for identical cytochrome P-450 monooxygenases. The first full-length amino acid sequence deduced from cDNA was published in 1983,[89] and progress has been rapid. The alignment of amino acid sequences has been the basis for a new and unified nomenclature. The superfamily of cytochrome P-450 monooxygenases consists of, to date, 11 families.[90-93] According to this nomenclature, the name for fungal 14α-DMs is P450 LI.[93] In the following description, however, trivial (e.g., 14α-DM) rather than systematic names will be used.

The class of cytochrome P-450 monooxygenases (P450s) has been extensively reviewed[94-97] and is also described in Chapter 9. Regardless of the functional diversity of P450s, all enzymes share common mechanistical and structural features. The prototype reaction is the hydroxylation of lipophilic substrates (S–H):

$$S\text{–}H + O_2 + NADPH + H^+ \rightarrow S\text{–}OH + NADP^+ + H_2O$$

The enzyme requires molecular oxygen for activity; one oxygen atom is reduced to water, and one appears in the hydroxylated product. The prosthetic group involved in this hydroxylation reaction is heme iron. In contrast to other heme proteins, the heme of P450s is not covalently bound to the protein. Four of the six coordination sites of the central Fe(III) are occupied by the pyrrole nitrogens of the heme system. The fifth coordination site always involves a cystein thiolate and represents the attachment of the heme to the protein. The peptide sequence around this cystein residue has been found to be highly preserved throughout the different classes of P450s.[90] In the native stage, the sixth bond is weak and often is occupied by water. All P450s are characterized by an absorbance maximum at 420 nm in their oxidized form (soret peak). After reduction of the heme iron to Fe(II), the sixth coordination site is normally occupied by molecular oxygen. Carbon monoxide readily competes with this oxygen binding and leads to a dead end configuration. The soret band of this CO-P450 complex is shifted to 450 nm. The name P450 (pigment 450) was derived from this spectral shift.

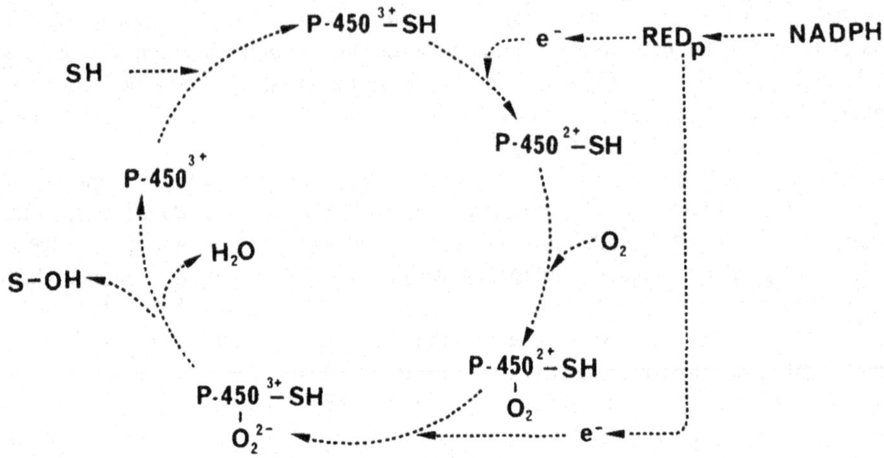

FIGURE 7. Reaction cycle of cytochrome P-450 monooxygenases.

In contrast to other cytochromes that function in electron transfer, P450 has to be regarded an enzyme and, thus, is distinguished by a substrate binding site. Binding of a substrate in the native stage of P450 leaves insufficient space for the nucleophile that normally occupies the sixth coordination site of the central Fe(III). This change from a hexacoordinate to a pentacoordinate geometry of the complex is accompanied by a transition of the five d-orbital electrons of the Fe(III) from low to high spin, a transition that shifts the Soret peak to approximately 390 nm. Normally, the shift is measured as a substrate-induced difference spectrum. A fully high-spin spectrum is indicative for a tight and specific fit of a substrate into the binding site.

The catalytic cycle of cytochrome P-450-catalyzed hydroxylations is shown in Figure 7. The cycle is initiated by the binding of substrate. The first electron is transferred, and the central iron is reduced to Fe(II), a stage that binds molecular oxygen to the sixth ligand position. Oxygen is activated for an internal rearrangement of electrons leading to the original Fe(III) stage and a second electron is transferred. The two oxygen atoms are separated: one is inserted into the substrate and one is reduced to water. A new reaction cycle can start after the product is released from the enzyme. The primary donor of both electrons utilized through the reaction cycle is NADPH, with NADPH-cytochrome P-450 reductase functioning as the normal electron carrier between P450 and NADPH.[96,98]

Two basic biological functions were assigned to the cytochrome P-450 superfamily present in mammals. One group, primarily localized in liver tissues, is involved in the metabolism of foreign aliphatic compounds such as drugs and secondary plant products. This class is distinguished by a tremendous capacity to bind and modify scores of chemicals. This multisubstrate character is mediated by both the multiplicity of different forms of P450s and the overlapping substrate specificities of the various enzymes.[94-96] The second group of P450s is involved in the biosynthesis of various compounds and, thus, is highly substrate specific. The best investigated fungal P450s belong to the group with high substrate specificities.[99] Most important with regard to the scope of this chapter are the enzymes that hydroxylate and modify sterol structures, the steroidogenic P450s. They are involved in numerous steps in the biosynthesis of sterols, steroid hormones, and bile acids.[100-103] Representative examples for these P450–catalyzed reactions are illustrated in Figure 8 and are also described in Chapter 9.

Oxidative modifications catalyzed by cytochrome P-450 monooxygenases occur around the sterol nucleus, the methyl groups attached to the nucleus, and the side chain. Some of these sterol conversions involve the prototype P450 reaction, a single hydroxylation. The

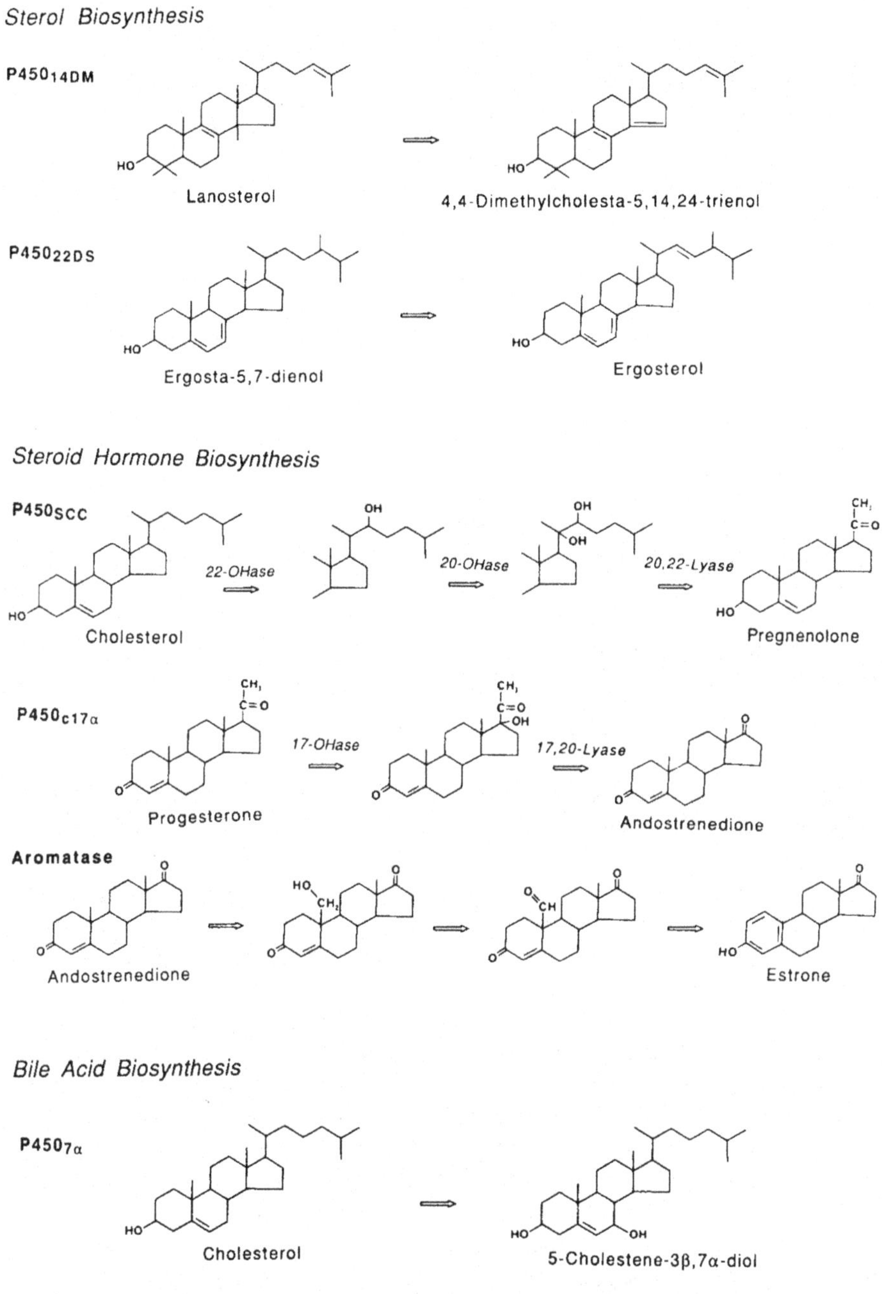

FIGURE 8. P450-catalyzed reactions in sterol synthesis and metabolism.

first enzymatic step leading from cholesterol to bile acids (cholesterol-7α-hydroxylase) exemplifies this reaction type (Figure 8).[103] Many of the P450 oxygenations, however, are more complex and involve a series of successive hydroxylations, but also the cleavage of C–C bonds. The key enzymes of steroid hormone biosynthesis in mammals, the $P450_{scc}$ leading to C_{21} steroids, $P450_{c17\alpha}$ leading to C_{19} steroids, and the aromatase, are representatives of these more complex reaction types (Figure 8). The P450s involved in the removal of the C-14 methyl group of sterol precursors, the 14α-demethylases (14α-DMs), share mechanistical similarities with the complex steroidogenic P450 monooxygenases. The substrates for 14α-DMs are lanosterol in *S. cerevisiae* and mammals (Figure 8), 24-methylenedihydrolanosterol in most higher fungi (Figure 5), and obtusifoliol in plants (Chapter 8).

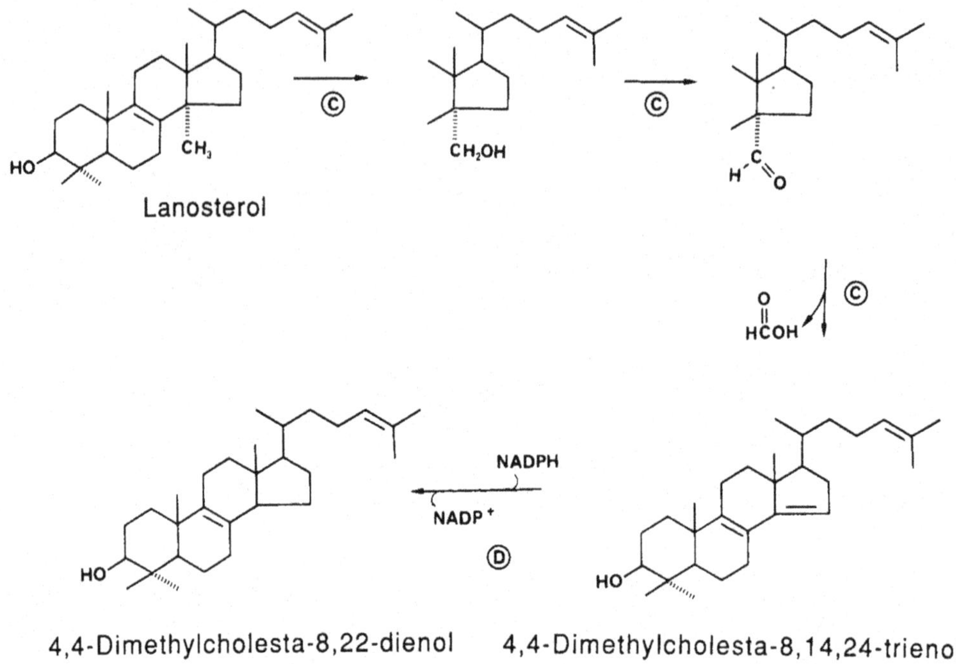

FIGURE 9. Sequence of C-14 demethylation steps. C, 14α-demethylase; D, Δ^{14}-reductase.

As described later (see Section V.B), fungal 14α-DMs represent the currently most important target for fungicides and antimycotic drugs. Almost all of the enzymatic and structural details of fungal 14α-DMs have been elucidated for the enzyme isolated from *S. cerevisiae*.[104] The spectroscopical properties of the 14α-DM purified from yeast microsomes were typical for a cytochrome P-450 monooxygenase.[105] The substrate lanosterol was shown to bind near the heme group; however, the spin state of the enzyme-substrate complex was not completely high-spin, suggesting that the bonding of the sixth coordination site was not completely abolished by the substrate. The mechanistical steps involved in the demethylation reaction are illustrated in Figure 9 and were mainly deduced from work with a reconstituted system, consisting of the lanosterol 14α-DM and the purified NADPH-cytochrome P-450 reductase.[106-109] In intact microsomes and in the presence of oxygen and a NADPH-regenerating system, lanosterol is converted to 4,4-dimethylcholesta-8,24-dienol. However, the reconstituted system converted lanosterol to 4,4-dimethylcholesta-8,14,24-trienol. These two reaction products demonstrated that the entire demethylation reaction consists of two distinct steps, the removal of the methyl group leading to a Δ^{14}-sterol, and the subsequent reduction of the double bond by a separate NADPH-dependent Δ^{14}-reductase. The demethylation sequence is catalyzed in three successive steps. Two successive hydroxylations convert the methyl group to an alcohol and aldehyde. Formic acid is then eliminated in the third step, and the Δ^{14}-double bond is formed. Under normal conditions, the oxidized intermediates are not accessible to chemical analysis, most likely explained by an increased binding affinity of these oxidized intermediates to the enzyme. The mechanistical details of the individual demethylation steps are virtually the same for the lanosterol 14α-DM isolated from liver microsomes,[110-115,115a] and the enzymatic properties of the two enzyme systems are almost identical.

The substrate specificity of the lanosterol 14α-DMs isolated from yeast microsomes is narrow and depends on stringent structural elements present in lanosterol. For example, the binding of dihydrolanosterol (lanosterol without the Δ^{24}-bond) was diminished by a factor of 4.[104] Furthermore, the Δ^8-double bond was shown to be an essential feature. The Δ^7-analogue of lanosterol was a poor substrate; a Δ^6-double bond or the lack of a double bond in ring B abolished the binding of the respective sterols to the enzyme.[116] As mentioned above, the substrate for the 14α-DM demethylases in most other fungi is 24-methylenedihydrolanosterol and, thus, slightly different from lanosterol (Figure 5). The impact of this side-chain alteration on the binding affinity to the two different fungal 14α-DMs has not yet been elucidated. The respective enzyme in microsomes isolated from *C. albicans* has been purified,[117] but potential differences with regard to the binding of lanosterol vs. 24-methylenedihydrolanosterol have not been studied with any fungal enzyme. It was noted, however, that the specific activity of the reconstituted enzyme derived from *C. albicans* was 20-fold lower than that reported for a similar system derived from *S. cerevisiae*, when tested with lanosterol as the substrate.[117] In addition to *C. albicans, C. tropicalis* was shown to produce a 14α-DM immunologically related to the yeast enzyme.[118,119] Regardless of the progress made, the lanosterol 14α-DM from *S. cerevisiae* remains the fungal 14α-DM best characterized in biochemical terms.

In contrast to our still limited knowledge regarding the precise geometry of the active sites of the two types of fungal 14α-DMs, the primary structure deduced from cloned DNAs has been determined for *S. cerevisiae* (530 amino acids)[120,121] and *C. tropicalis* and *C. albicans*.[122,123] Sequence alignment revealed a 516-amino acid overlap with 67% identical and 23% conservatively replaced amino acids. A similar preserved amino acid sequence (62% amino acid homology) was reported between the enzymes from *S. cerevisiae* and *C. albicans*.[123] According to the recommended nomenclature of the P450 superfamily, all three 14α-DMs are cytochromes P-450 L1A1.[92,93]

The high degree of sequence similarity does not automatically imply identical active site structures and, thus, an identical substrate specificity. For example, Lindberg and Negishi[124] investigated this question with two P450s with either coumarin or testosterone as substrates. Regardless of the dissimilar substrates, the two respective enzymes differed in only 11 out of 494 amino acids. The change of a single amino acid to the corresponding amino acid present in the related P450 reversed the substrate specificity. Thus, slight sequence alterations can exert a drastic impact on the geometry of the active site. Evidence, however exists that the 14α-DM present in *C. tropicalis* is not exclusively active toward the endogenous substrate 24-methylenedihydrolanosterol. A lanosterol 14α-DM null mutant of *S. cerevisiae* was restored to aerobic growth upon complementation with the 14α-DM gene from *C. tropicalis*.[125] This result implies that the mutant was able to functionally utilize the foreign 14α-DM to restore ergosterol biosynthesis. Recent advances in cloning techniques, site-directed mutagenesis, and expression of P450s[120-128] have opened the way to explore the relationship between amino acid sequences and substrate specificities, and rapid progress can be expected.

The primary structure of the second component of the yeast 14α-DM, the NADPH-cytochrome P-450 reductase, was deduced from the nucleotide sequence of its cloned DNA.[129] Although functional domains were preserved, the yeast enzyme showed only 33% similarity with mammalian reductases. Reductase null mutants of *S. cerevisiae* should be expected to lack an essential component of the 14α-DM system and, thus, to accumulate lanosterol. However, results obtained with a respective mutant generated by gene disruption were unexpected. The respective yeast strain showed aerobic growth and produced ergosterol, suggesting that alternate routes of electron flow to the 14α-DM must exist.[130] The nature of the alternative component providing electrons remains to be elucidated.

4. C-4 Demethylation

In contrast to the C-14 methyl group, the two methyl groups at C-4 are removed as carbon dioxide. Furthermore, a cytochrome P-450 appears not to be involved in this reaction.

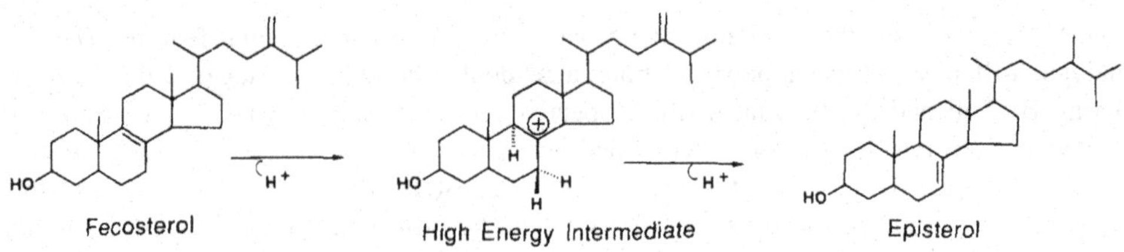

FIGURE 10. Mechanism of Δ^7-Δ^8-isomerization.

Although a reaction sequence for the removal of the two methyl groups has been established,[56,88] many details regarding the organization and identity of enzymes involved remain to be resolved. It is commonly suggested that the demethylation is initiated by the oxidation of the 4α-methyl to a carboxyl group via a hydroxy and aldehyde intermediate. These oxidative steps require oxygen, NADPH, a cyanide sensitive factor, and cytochrome b_5. Decarboxylation of the carboxyl sterol is preceded by the NAD-dependent oxidation of the 3-hydroxy to a 3-keto group. Inversion of the remaining methyl group in β-position takes place to the α-orientation, and the decarboxylation cycle starts again. The NADPH-dependent reduction of the 3-keto group to the original 3β-group comprises the last step of the C-4 demethylation sequence. In contrast to the 14α-DM, little progress has been made in the enzymatic characterization of this important step in sterol biosynthesis.

5. Rearrangement of Double Bonds

The product of the demethylation steps at the C-14 and C-4 position is fecosterol in most fungi (Figure 5) and zymosterol (cholesta-8,24-dienol) in *S. cerevisiae*. The next sequence of biosynthetic reactions is a series of nuclear double-bond rearrangements.[56] The individual steps of these nuclear modifications are well established in mammalian systems.[88] The sequence starts with the isomerization of the Δ^8- to a Δ^7-double bond, the oxidative introduction of the Δ^5-double bond, and the reduction of the Δ^7-double bond. This last step is not required in ergosterol biosynthesis with its $\Delta^{5,7}$ conjugated system. However, Δ^5-analogues of ergosterol have been reported to occur in fungi (see Section II.A), indicating the existence of a similar Δ^7-reductase in fungi.

The membrane-bound Δ^8-Δ^7-isomerase from rat liver and *S. cerevisiae* requires neither oxygen nor cofactors for full activity.[81,131] In yeast, both zymosterol and fecosterol (with fecosterol as the better substrate) but not lanosterol were converted by the respective enzyme. The enzymatic isomerization reaction is reversible, with the Δ^7-isomer being predominant at equilibrium as shown with *in vitro* systems derived from liver and yeast.[81,131] The equilibrium might be more in favor of the Δ^8-isomer in some cases. Although rare, ergosta-5,8,22-trienol (lichesterol) has been reported to accompany ergosterol in some fungi.[10] *Xanthoria perietina*, a fungus found in the symbotic association of lichens,[132] and the mushroom *Flammulina velutipes*[133] are examples for lichesterol-containing organisms. The Δ^8-Δ^7-equilibrium of the HCl-catalyzed isomerization was shown to be nearly 1,[134] and slight variations in enzyme structures might explain a shift in the equilibrium of the Δ^8-Δ^7-isomerase reaction. Studies with enzyme preparations derived from plants and analogues of high-energy intermediates suggested the existence of a carbocationic intermediate in the isomerase-catalyzed reaction according to the scheme presented in Figure 10 (see also Chapter 8).[135,136] Although the stereospecificity of the C-7 hydrogen loss in fungi is opposite from plants and rats,[56,135] the fungal isomerization is likely to involve a similar high-energy and carbocationic intermediate.[137]

The Δ^5-double bond is introduced subsequent to the Δ^8-Δ^7-isomerization (Figure 5). In rat liver, the Δ^5-desaturase requires oxygen, NADH or NADPH, a cyanide-sensitive factor, cytochrome b_5, and, thus, conditions similar to the sterol 4-methyl oxydase described above.[56]

The precise details of the Δ^5-desaturase reaction are still under debate. The mechanism could be a dehydrogenation reaction similar to the desaturation of fatty acids, or a hydroxylation-dehydration reaction with a 3,5- or 3,6-diol as an intermediate.[56] Based on a mutant study with *C. albicans*, evidence was presented for the Δ^5-desaturase-catalyzed formation of a sterol hydroxylated at C-6.[138]

The biosynthesis of ergosterol is finalized with the introduction of the Δ^{22}- and the reduction of the $\Delta^{24(28)}$-double bond. In contrast to the Δ^5-desaturation, the Δ^{22}-desaturation is catalyzed by a typical cytochrome P-450 system ($P450_{22DS}$).[139] The desaturation mechanism might proceed via a hydroxylation/dehydration mechanism or short-lived radical intermediates without hydroxylation.[140] The stereospecific reduction of the $\Delta^{24(28)}$-double bond is accomplished by an NADPH-dependent reductase.[56]

D. REGULATION OF STEROL BIOSYNTHESIS

The cholesterol content of mammalian cells is tightly regulated at the level of sterol uptake from exogenous sources, esterification, and synthesis.[141,142] One of the key regulatory enzymes in the pathway of *de novo* cholesterol biosynthesis is HMG-CoA reductase (see Section II.C.1). Thus, regulation occurs at an early step of the isoprenoid pathway (Figure 3). Oxysterols such as 25-hydroxycholesterol, probably formed from exogenous cholesterol bound to low density lipoprotein, comprise important modulators of this tight regulation.[141,142] It seems clear that the exogenous supply of cholesterol mediates the down-regulation of HMG-CoA reductase activity and, consequently, endogenous sterol synthesis. The reductase level is primarily controlled at the level of protein turnover (exogenous sterols enhance degradation)[143] and gene transcription (accumulating sterols reduce the amount of mRNA transcribed).[144] An octanucleotide sequence localized in the reductase promoter region has been identified to confer the sterol-mediated repression of the reductase gene,[144] a process that was shown to require protein synthesis.[145] Similar sterol regulatory elements have been identified in the 5′-flanking region of the gene coding for HMG-CoA synthase (Figure 3), an enzyme normally coregulated with HMG-CoA reductase in sterol-mediated responses.[146] The current data suggest that the genes are controlled by a conditional positive mechanism. In the absence of sterols, transcription is stimulated by the binding of regulatory proteins to the regulatory elements. In the presence of sterols, the regulatory proteins are inactivated and transcription rates decrease.[146]

Whether a similar sterol-mediated control of HMG-CoA reductase is operative in fungi has not yet been clarified. Self-regulation of fungal sterol biosynthesis and feedback inhibition of HMG-CoA reductase by ergosterol have been reported for *S. cerevisiae* and *Aspergillus nidulans*.[147-150] However, exogenous ergosterol was also reported to suppress the activity of yeast acetoacetyl-CoA synthase and HMG-CoA synthase, with no effect on HMG-CoA reductase.[151,152] Mammalian cells contain a single gene coding for HMG-CoA reductase, whereas two genes have been identified in yeast.[64] The mutation of either one of these genes had only slight effects on the viability of the respective strains, but cells containing mutant alleles of both genes were inviable.[153] Viability was rescued, however, by the expression of the HMG-CoA reductase gene from hamster.[66] Thus, the mammalian enzyme can functionally replace both yeast enzymes. Moreover, sequence homology was not only observed for the catalytic site, but also for the membrane spanning regions involved in the regulation of reductase turnover.[154] Regardless of the striking similarities between the genes present in mammals and yeast, the potentials for a transcriptional control of the fungal reductase similar to mammalian cells, and the corresponding modulators of gene activation and down-regulation still remain to be elucidated.

E. FUNCTION OF STEROLS

Sterols are known to be essential for all eukaryotes, either synthesized *de novo* from acetate or taken up from the environment. As described above, the first sterol structure formed from the cyclization of oxidosqualene is lanosterol in vertebrates and fungal organisms, and this precursor undergoes substantial structural modifications before the end products are reached. Moreover, the products of the sterol pathway are structurally diverse, depending on the class of organisms. Both the extensive biosynthetic modifications and the structural diversity suggest a precise function of sterols and a profound structure-function relationship. The question, why ergosterol is the best-suited sterol in most fungi, has been treated in recent review articles[11,155] and comprises the subject of the following section.

1. Membrane Components

Sterols have been long known as membrane components predominantly localized in the plasma membranes of sterol-requiring organisms.[156-158] Sterols have long been known to interact with phospholipids and to maintain optimal membrane fluidity and integrity under changing environmental conditions. A suboptimal sterol composition of membranes can lead to an increased permeability to nonelectrolytes and, thus, to a severe "leakiness" of membrane-enclosed compartments.[156-158] One of the main functions of biological membranes, the confinement of substances against a concentration gradient would no longer be warranted. This "bulk" function of sterols is highly dependent on the sterol structure. Bloch postulated that sterols, in order to fulfill the bulk function, must have a 3β-hydroxyl group, a nucleus planar at the α-face (Figure 1), and an alkyl side chain.[156] Obviously, these structural requirements are reflected in the evolution of enzymes needed to convert lanosterol to functionally more appropriate sterol structures.

An alternative system for studies on biological sterol functions has been *S. cerevisiae*, and most of our current understanding of sterol functions in fungi has been derived from studies with this organism. Under anaerobic conditions, growth of *S. cerevisiae* is strictly dependent on exogenous sterols and unsaturated fatty acids, both of which require oxygen for synthesis. Structural alterations of the exogenously supplied sterols and their effects on anaerobic growth provided a powerful tool for investigations. A "classical" study employing this technique revealed the following relative importance of structures that distinguish ergosterol from cholesterol in mammals: free hydroxy at C-3 $>$ 24β-methyl-Δ^{22} $>$ 24β-methyl $>$ $\Delta^{5,7}$-diene system.[159] Lanosterol did not support growth, and cholesterol was less efficient than ergosterol.

Clearly, the *in vivo* systems employed in studies on the impact of sterols on growth and development of sterol-deprived organisms have their limitations. For example, the modulation of the physical state of biological membranes by sterols not only has implications on the structural integrity of the lipid bilayer, but also on a broad range of cellular processes, including the activity of membrane-bound enzymes. These effects on enzyme activities can be explained by alterations of the overall physical lipid environment of membrane-bound enzymes, but also by direct enzyme sterol interactions.[157,158] Both effects can be of importance at the same time, as exemplified by the impact of two distinct cholesterol pools on the function of the acetylcholine receptor in membranes from *Torpedo californica*.[160] The importance of specific sterol-enzyme interactions is not always viewed uniformly, as recently summarized for the fungal enzyme chitin-synthetase.[161] Initial experiments with polyene antibiotics had indicated a decisive role of ergosterol in chitin synthesis, but more recent results have left some open questions. Regardless of some conflicting results, however, little doubt exists that sterols not only affect the physical state of membranes, but also are of importance for activity of various membrane-bound enzymes, either indirectly or through direct sterol-enzyme interactions. These more specific and regulatory functions of sterols, in addition to their bulk functions in membranes, as first proposed by Dahl et al. in 1980,[162] have received increasing attention in more recent studies.

2. Regulatory Functions

As discussed in recent reviews,[163,164] the technical difficulty to grow yeast under strict anaerobic conditions has led to numerous conflicting results, in particular with regard to the potency of lanosterol to satisfy the bulk sterol functions, and the importance of the 24-methyl group of ergosterol, a structural feature that distinguishes ergosterol from cholesterol. It could be demonstrated that traces of oxygen were still sufficient to allow small amounts of ergosterol to be synthesized. The combination of anaerobic growth with 2,3-iminosqualene, an inhibitor of the 2,3-oxidosqualene cyclase (see Section IV.C), proved to be a powerful tool that allowed to avoid the interference of endogenous sterol traces with exogenously applied sterols. Studies with various yeast mutants with defects in early steps of sterol biosynthesis provided alternatives to this system.[165] Recent progress has been made employing sterol auxotrophs (such as GL7 and FY3) with biosynthetic defects prior to lanosterol and, most often, an additional defect in heme biosynthesis. The latter mutation warrants that exogenous sterols are taken up under aerobic conditions.[166,167]

Three independent research groups employing these more refined study systems were able to demonstrate essential functions of fungal sterols clearly distinct from the bulk functions in membranes, and the conclusions drawn from many of the earlier results might have to be reinterpreted in some cases. Nes and associates[168,169] (anaerobic growth + 2,3-iminosqualene) were able to show that cholesterol alone could not support growth unless small amounts of ergosterol were present. This "regulatory" function of ergosterol was strictly dependent on the 24β-methyl group. Neither a 24α-methyl nor a 24β-ethyl group could substitute for this function, indicating an essential role for both the bulk and the stereochemistry of the C-24 substituent. Although the molecular mechanism of this regulatory function inherent to the 24β-methyl group remains to be elucidated, the findings reported by Pinto and associates[168,169] answered the question of why an energy-consuming side-chain methylation and a stereospecific reduction of the 24-methylene group had emerged during evolution. The need for this structural feature had not been fully explained from studies on the bulk function of ergosterol in fungal membranes.

Parks and his group[170-172] reached similar conclusions with regard to a dual function of ergosterol. Highly purified cholestanol was not able to support growth of sterol-deficient mutants, unless small amounts of ergosterol were present. Apparently, cholestanol was able to satisfy bulk membrane functions, but not a "sparking" function of ergosterol. However, the 24-methyl group found by others to be mandatory[168,169] was the least important sparking feature of ergosterol. Rather, the Δ^5-double bond was the most important structural requirement.[172] It is interesting to note, however, that the side-chain methylation was also shown to comprise a mandatory step for the expression of sparking sterol functions in plant cell cultures.[173]

In a series of related experiments, Bloch and co-workers[174,175] demonstrated that poor growth of the yeast mutant GL7 mediated by exogenous cholesterol could be increased by supplements with small quantities of ergosterol. This "sterol synergism" was accompanied by a faster rate of oleic acid incorporation into phospholipids and an increased methylation of phosphatidylethanolamine to phosphatidylcholine. Regardless of the slowed-down growth of the sterol-deficient mutant GL7 when supplemented with cholesterol,[174] even a slower growth was not entirely consistent with the absolute requirement for sparking or "regulatory" amounts of ergosterol observed by others. Nes and Dhanuka[176] presented an answer to this open question. Despite the acknowledged inability of the GL7 mutant to synthesize bulk amounts of sterols, small and sparking quantities of ergosterol could be synthesized from acetate. The authors suggested the possibility of two different compartments for ergosterol synthesis, one of which was not affected by the mutations.

The lines of evidence described above strongly suggest at least two different functions of ergosterol. The bulk function in membranes, the maintenance of an optimal membrane fluidity, is similar to the function of cholesterol in mammalian cells.[156-158] This particular role is not strictly dependent on the characteristic structural features of ergosterol, and other sterols such as cholesterol can at least partly secure this function. Ergosterol, however, appears most suited. For example, mutants of *S. cerevisiae* with a block in the side-chain methylation had more ordered membranes than wild-type strains.[177] It should be pointed out, however, that the physical properties of membranes are influenced by both the sterol structure and the phospholipid composition. This interaction is a complicating factor in the interpretation of some results, since fungi have been shown to adjust their phospholipid composition as a response to not only the quantity, but also the structure of membrane sterols.[178] Conflicting results still exist with regard to the bulk qualities of sterols with a 14α-methyl group protruding into the α-face. As described later, one of the most important groups of antifungal agents is targeted against the 14α-DM (see Section V.B) and leads to the replacement of ergosterol by 14α-methyl sterol precursors. Model studies have indicated the decreased effectiveness of these sterols in maintaining membrane integrity.[156-158] In yeast, lanosterol was unsuitable to trigger growth of a sterol auxotroph strain.[179] In contrast, the viability of yeast mutants defective in their lanosterol 14α-demethylase was not always critically affected, and 14α-methyl sterols were shown to satisfy the bulk but not the sparking function of ergosterol.[180] Furthermore, it has been proposed that a 14α-DM mutation was lethal unless a second defect in the Δ^5-desaturase was present.[180] This latter mutation, however, does not always appear mandatory for the viability of 14α-DM mutants.[181,182] It remains to be clarified, whether the "dual compartment" hypothesis for the synthesis of small amounts of regulatory ergosterol proposed by Nes and Dhanuka[176] can serve as an explanation for these differing reports. It must be emphasized, however, that viable 14α-DM mutants have only been described for yeasts. In an attempt to isolate similar mutants for mycelial fungi, Ziogas et al.[183] were unsuccessful and concluded that a block of the 14α-DM was lethal for filamentous fungi.

Regardless of some questions left unanswered, no doubt exists that ergosterol exerts regulatory functions, and studies on the molecular aspects of these regulatory, sparking, or synergistic roles have been initiated. For example, the growth stimulation of the yeast mutant GL7 by ergosterol was preceded by an increased synthesis of phospholipids and an altered metabolism of inositol phosphates.[184] In mammalian cells, metabolism of inositol phosphates is known to be involved in regulatory processes,[185] but so far, little information is available for fungal organisms. Ergosterol at hormonal levels was also involved in the activation of a protein kinase with functions in cell cycle control of *S. cerevisiae*.[186] The GL7 mutant does not die upon starvation for ergosterol but rather accumulates in the G1 state. Ergosterol appears to be required to exit the G1 phase and modulates, at the same time, the activity of kinases.[164] This kinase activation was not dependent on *de novo* protein synthesis, indicating a role of the sparking amounts of ergosterol at the level of enzyme regulation rather than at the level of gene transcription.[164]

Evidence exists that the importance of the 24β-methyl group for growth regulation is not restricted to the yeast *S. cerevisiae*. However, ergosterol does not always serve a dual function. Nes and co-workers[187,188] reported that *Gibberella fujikuroi*, a filamentous ascomycete, produced small amounts of ergosta-7,22-dienol and ergosta-7-enol in addition to ergosterol. Data from inhibition and supplementation experiments suggested that the regulatory function in mycelial growth was restricted to the minor sterols, while ergosterol played the bulk role in membranes. Regardless of the functional discrimination of sterols, the 24β-methyl group was an essential structural feature for the regulatory sterol function. Inhibition of the C-24 transmethylation step led to the severe inhibition of mycelial growth and, at the same time, to the induction of asexual reproduction.[187,188] Similar sterol-regulated developmental steps seem to be more widespread and deserve further attention. For example, the 14α-DM inhibitor flusilazole was shown to induce the formation of appressoria, specialized infection structures crucial for the invasion of plants by plant pathogenic fungi, in germinating conidia of the ascomycete *Venturia inaequalis*, the causal agent of apple scab.[189] The induction of asexual reproduction or the induction of appressoria formation, both triggered by interference with sterol biosynthesis, is a drastic developmental response of fungal organisms.

Considering these effects on fungal development, a direct action of fungal sterols or specific sterol metabolites at the level of gene transcription cannot be excluded. The diversity of the regulatory effects of fungal sterols indicates that considerable work remains to be done to clarify the different molecular levels of regulatory activities inherent to fungal sterols. The various levels most likely comprise effects on the modulation of enzyme activities, the activation of regulatory protein kinases, and a direct involvement in gene expression. Complicating might be the finding that *S. cerevisiae* inherited the genetic competence to reduce the Δ^{24}-double bond of sterol precursors, a key step in the synthesis of mammalian cholesterol but not required for the synthesis of C-24 methylated fungal sterols.[190] Undoubtedly, cholesterol is synthesized in amounts too small to account for any bulk functions. It might, however, have its role in thus far undefined regulatory functions.

Regulatory roles of sterols are not exclusively reported for ergosterol-producing fungi. A successful example to differentiate between the sterol requirements for vegetative growth and more specific events such as asexual reproduction was presented for the sterol-producing oomycete *Saprolegnia ferax*.[19] This lower fungus is devoid if ergosterol and produces cholesterol, cholesta-5,24-dienol, ergosta-5,24(28)-dienol, and stigmasta-5,24(28)-dienol. 25-Azasterol prevented the reduction and alkylation of the Δ^{24}-bond. Under the action of this inhibitor, no cholesterol and stigmasta-5,24(28)-dienol were formed, and only traces of ergosta-5,24(28)-dienol, the main sterol in nontreated mycelium, were identified. This severe interference with sterol biosynthesis had only a slight and transient effect on mycelial growth. However, in an all-or-none mechanism, oosporogenesis was completely prevented. Thus, only the sexual reproduction but not the vegetative growth was severely affected by an altered sterol pattern. This pattern appears dissimilar from *S. cerevisiae* and other ascomycetes and might be related to the formation of steroid hormones, which are known to be involved in the sexual reproduction of some but not all oomycetes.

3. Steroid Hormones

Steroid hormones are synthesized from sterols as their immediate precursors[100,101] and comprise one of the major classes of regulatory compounds. The biochemistry and molecular biology of steroid hormone action are best understood in mammals[191-193] and insects.[194] The existence of specific steroid hormones in plants (bassinolides) acting as plant growth regulators has been established more recently.[195] Little overall information is available on the occurrence and functions of steroid hormones in fungal organisms. By far the best-known example for fungal steroid hormones is the involvement of antheridiol and ooginiol (Figure 11) in the sexual reproduction of the oomycetous water mold *Achlya*.[196] Evidence for antheridiol-regulated changes in the synthesis of specific proteins has been presented,[197-200] and a steroid-binding protein sharing similarities with steroid receptors found in other eukaryotes has been described.[201] Furthermore, three cDNA clones constructed from mRNA isolated from hormone-induced cultures were shown to be transcriptionally down-regulated by antheridiol.[202] These combined results provide clear evidence that the steroid hormones act at the level of gene regulation.

The sex steroid hormones of *Achlya* were shown to derive from fucosterol,[14,196] and the inhibition of sporulation of *Saprolegnia ferax* by the chemical interference with fucosterol metabolism was discussed as evidence for the existence of similar steroid hormones in this sterol-producing Oomycete.[19] However, fucosterol (and cholesterol) accumulated by the "sterol-deficient" Oomycete *Phytophthora cactorum* appeared to trigger sexual reproduction without being transformed to steroids,[14,203] implying that the importance of steroid hormones and the evolutionary history of the Oomycetes might not be uniform.

The action of steroid hormones in fungal organisms is not restricted to their role in

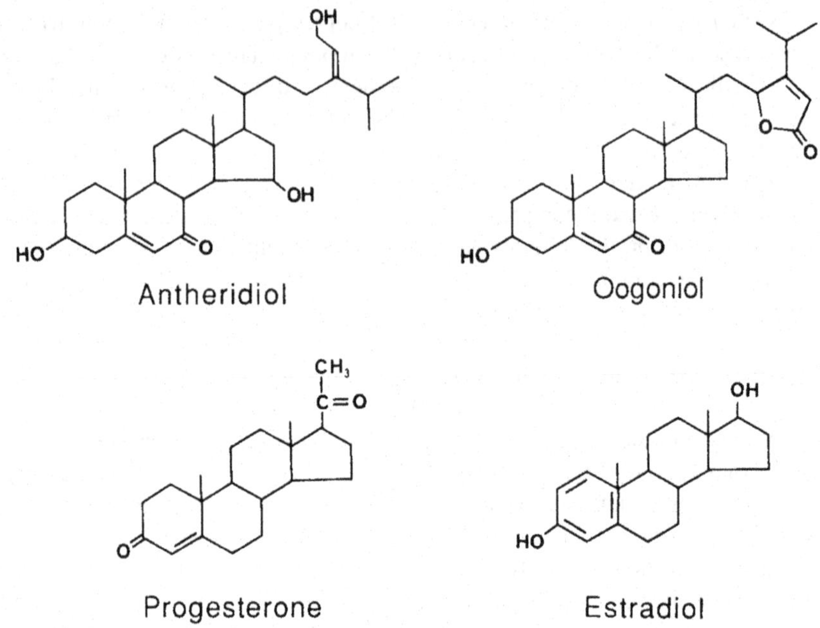

FIGURE 11. Steroid hormones with activities on fungal organisms.

sexual reproduction. Several fungi have been shown to contain steroid binding proteins that recognize mammalian steroid hormones.[204] Since most of these examples are human fungal pathogens, the identified steroid receptors might represent a highly specialized function in host-pathogen interaction rather than a ubiquitous characteristic of fungal organisms. Indeed, the conversion of the dimorphic human pathogen *Paracoccidioides brasiliensis* from its nonpathogenic mycelial to its pathogenic yeast form was triggered by hormonal concentrations of female estrogen hormones, with estradiol (Figure 11) showing the highest activity. This biological activity, paralleled by the existence of highly specific cytosolic estrogen receptors, could serve as an explanation for a remarkable resistance of males to this pathogen.[204]

Evidence exists, however, that fungal steroid binding proteins can also serve as receptors for endogenous fungal steroid hormones. *Candida albicans*, an opportunistic yeast frequently infecting patients receiving glucocorticoids, was shown to contain a high affinity binding protein for corticosterone that also binds progesterone (Figure 11),[204] and a separate binding protein for estradiol.[205] Some of these and other steroid hormones were shown to trigger the transition from the yeast to the pathogenic hyphal form in *C. albicans*,[206] although the effects were not dramatic. The synthesis of similar endogenous steroid hormones was indicated by the presence of a lipid-extractable substance released by *C. albicans* into the medium, which was able to displace corticosterone from the fungal binding site.[207] Thus far, neither the structure nor the biological activity of this putative steroid hormone has been identified. A high-affinity receptor for progesterone was also identified in *Trichophyton mentagrophytes*, a dermatophyte responsible in skin infection.[204] Progesterone was shown to inhibit the growth of the fungus in a dose-dependent fashion. However, the fungus overcame the inhibitory effect of progesterone after 24 to 48 h of exposure, an escape caused by the induced oxidative metabolism of the hormone to more polar metabolies.[208] Binding studies with the less inhibitory metabolites to the steroid receptor strengthened the hypothesis that binding to the receptor accounted for the inhibition of growth.

A high affinity binding protein for estradiol was also identified in *S. cerevisiae*,[204] a steroid hormone receptor apparently not involved in any host-pathogen relationships. Interestingly, no biological effects of estradiol have been identified thus far.[204] However, a growth factor more hydrophilic than ergosterol and probably derived from ergosterol has been isolated from a sterol auxotroph of *S. cerevisiae*, and although the structure has not yet been elucidated, a steroid nature cannot be excluded.[209] In summary, increasing evidence for the existence of fungal steroid hormones with various biological activities and the presence of steroid-binding proteins has evolved in the past. Furthermore, fungal organisms have been long known for their biochemical competence to synthesize and modify steroid hormones.[210] The widespread occurrence and general importance of these putative fungal hormones remain to be substantiated. A more complete understanding might have broad implications for the identification of novel target sites for the control of fungal diseases.

III. INHIBITORY INTERFERENCE WITH STEROL FUNCTIONS

A. POLYENE MACROLIDE ANTIBIOTICS

The polyene macrolide class of drugs represents the first example of antifungal agents interfering with sterols in fungal pathogens. Initiated with the discovery of nystatin in 1950,[211] a secondary product isolated from *Streptomyces* spp., some 200 different polyenes have been described in subsequent years.[212] The most severe drawback inherent to this class of compounds is a substantial toxicity to mammals, and amphotericin B is the only drug currently used in the treatment of systemic mycoses.[213] None of the polyene macrolides has been developed for the control of fungal plant diseases.

Chemically, all macrolide polyenes are amphipatic compounds, characterized by a lactonized ring of carbon atoms containing a hydrophobic part of hydroxyl groups. Differences are the number of carbon atoms within the ring system (12 to 37), the position and number of hydroxyl groups (6 to 14), and the presence or absence of glycosidically bound mycosamine. Three important examples of polyene macrolides are illustrated in Figure 12. The three-dimensional structure of amphotericin B, the clinically most important representative, was clarified by X-ray analysis.[214] The molecule is rod-shaped with an opposing hydrophobic and hydrophilic face, and the length of 2.1 nm is similar to the length of one phospholipid molecule. Both of these molecular features are determinants for the mode of action of amphotericin B. Mode of action studies were initiated soon after the discovery of the polyene macrolide antibiotics and have been summarized in several recent reviews.[215-221]

1. Mode of Action

A relationship between sterol functions in plasma membranes and the mode of action of polyene macrolides was first suggested on the basis of their biological activity spectrum. They are effective against eukaryotic but not prokaryotic organisms, with the exception of those prokaryotes that contain sterols. This indirect evidence was soon substantiated by more conclusive findings. Polyenes were shown to form complexes with sterols added to growth media, and sterols antagonized the fungicidal action of the inhibitors.[222] Furthermore, *Acholeplasma* spp. unable to synthesize sterols were insensitive to the inhibitors unless exogenous sterols were incorporated into their membranes.[223] The main initial effects of polyene macrolides on sensitive organisms are the release of small ions and other cell constituents.[215-221] The structural requirements for sterols to render cells permeable toward ions was a 3β-hydroxyl group, a planar ring system, and a side chain at C-17.[224] The degree of membrane destruction and the nature of compounds released from treated cells, however, are dependent on the nature of the inhibitor. Filipin, a macrolide too toxic to be clinically important, causes release of constituents up to small proteins, whereas the larger polyenes such as nystatin render cell membranes primarily permeable toward K^+.[225]

The loss of K^+ and an associated uptake of protons and, thus, acidification of the treated cells were originally proposed as the main mode of action of polyene macrolides.[225] However,

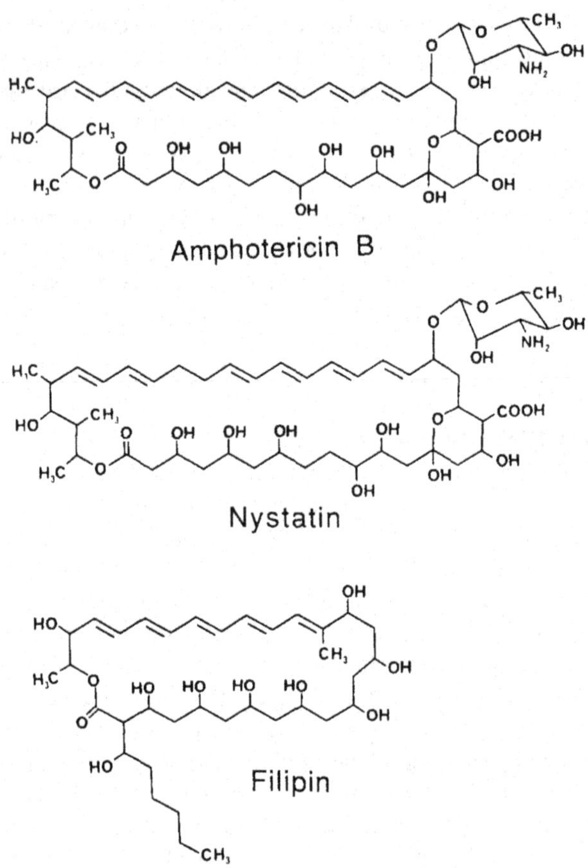

FIGURE 12. Structures of polyene macrolide antibiotics.

this model considering the release of K^+ as the major effect was revised later. For example, the fermentation of maltose by *Saccharomyces cerevisiae* was more sensitive to nystatin and amphotericin B than glucose fermentation.[226] Maltose fermentation was not protected against the inhibition by potassium ions, a protective action frequently observed. The fact that the maltose transport systems are known to be driven by a proton gradient, together with more direct evidence for proton permeability of treated cells, led to the conclusion that the primary effect of the compounds was the increased permeability of protons rather than potassium ions.[226] A proton gradient across the plasma membranes of fungi is maintained by plasma membrane ATPases, and this proton gradient is known to be involved in the active transport of numerous exogenous compounds and in the maintenance of the intracellular pH.[227,228] The different observations with regard to ions and cell constituents released from treated cells might be explained by the concentration dependency of the disorganizing effect of macrolides on membranes.[229] While little doubt exists that the polyene antibiotics are interacting with sterols, and that the efflux of small monovalent ions from treated cells is one of the initial events, the question of whether and to what extent these effects are suitable to explain the antifungal activity is still under discussion.

The initial efforts in the development of molecular models were focused on the specifics of sterol antibiotic interactions, and on the explanation of specificities of some of the macrolides toward ergosterol-containing fungal cells. Polyene macrolides, although water soluble, form aggregates in water, and the antibiotic-sterol interactions in solutions were mostly studied in alcohol-water solvent systems. Under these conditions, a clear interaction of polyene macrolides with sterols could be demonstrated. Furthermore, a selective interaction of amphotericin B with ergosterol over cholesterol was demonstrated in solutions, but also in membrane systems.[215-221] This selectivity was reversed with filipin and most likely explains the high toxicity of this polyene antibiotic toward mammalian cells.[215-221] In a propanol-water solvent system, the specific interaction of amphotericin B with ergosterol was found to be related to the aggregation state of the inhibitor.[230] Ergosterol did not react with monomers of amphotericin B, but rapid complex formation took place in the presence of high concentrations of dimers. The formation of loose complexes did occur, however, in the presence of of only traces of the dimer and thus conditions that prohibited the reaction with cholesterol. Different conformational states of complexes between ergosterol and amphotericin B also related to the preference of the inhibitor for ergosterol over cholesterol were also observed in liposomes.[231,232] It is likely that only one of these conformations is responsible for the drug-induced permeability changes of membranes.

The models proposed to explain the permeability changes are based on numerous studies employing primarily artificial membranes and various experimental approaches.[215-221] In summary, amphotericin B and nystatin increase the permeability to ions and other substances smaller in size than glucose. These properties indicated the presence of an aqueous pore through the membrane with an inner radius limited to 0.4 nm. Such a pore spanning the membrane would consist of eight antibiotic-sterol units, with the hydrophilic portion of the macrolide in the inside and a hydrophobic exterior complexed with sterols. In this model, the mycosamine residue of the inhibitor and the sterol hydroxyl group would be situated toward the water-lipid interface. Since the size of the larger macrolides is similar to the depth of a single membrane layer, a full pore can be formed by two half-pores spanning each of the two phospholipid layers. However, only one half pore is formed when the macrolides are administered from one side of the membrane, and this half-pore was shown to induce permeability changes, provided the acyl chains of the phospholipids were no longer than 18 carbon atoms. The permeabilization properties of this half-pore, which might be in a dynamic stage and might shuttle up and down, relate more closely to the changes observed with intact cells and, thus, comprise the more adequate model. As critically discussed by Bolard,[218] the "half-pore model" is consistent with most but not all of the experimental data reported, and additional effects not related to pore formation might contribute to the disorganization and permeabilization of membranes.

Filipin is smaller in size than nystatin and amphotericin B and lacks of hydrophilic carbohydrate residue. Rather than forming pores, filipin-sterol complexes are thought to disorganize the membrane structure by forming stacked aggregates located in the inner and hydrophobic core of the bilayer. This difference is supported by freeze-fracture analysis of fungal cells after treatment with filipin or amphotericin B. While filipin is leading to the formation of either pits or protrusions on the two fracture faces of plasma membranes,[42] both faces remain relatively smooth after treatment with amphotericin B. Only protein particles are more clustered, leaving slightly elevated and particle-free areas in the membrane.[233]

Since the clinically used inhibitors are members of the "pore-forming" group, most of the mode of action studies have been done with amphotericin B and nystatin. Although compelling evidence for the interaction of these antibiotics with ergosterol in fungal plasma membranes exists, the question of how this ergosterol interaction relates to the death of a fungus is not fully understood. In contrast to model membranes, a living fungus is surrounded by a rigid cell wall, which has to be penetrated before the inhibitors reach their target membranes. Moreover, membrane-bound enzymes are often functionally dependent on sterols and could be directly affected by the antibiotics with their affinity to ergosterol and probably ergosterol binding sites at proteins. Evidence exists that both of these aspects are involved in the fungicidal mode of action of nystatin and amphotericin B. For example, fungal cell walls have been shown to contain binding sites for nystatin and amphotericin B, and these binding sites can greatly affecting the inhibitory activity.[216,218] As described by Gale,[216] *Candida albicans* becomes increasingly resistant in the stationary culture phase. This sensitivity change, named "phenotypic resistance" by Gale,[216] appeared to be mediated by an antibiotic binding, long-chain β-glucan predominant in stationary cultures. The interaction of amphotericin B with cell wall components could be decreased with synergistically acting cell wall-degrading enzyme preparations.[234] In addition to this direct interaction with cell walls, amphotericin B at subinhibitory doses has also been shown to decrease the mannan content of outer cell wall layers, leading to loss in immunological reactivity.[235] Cell wall interactions with nystatin and amphotericin B have also been described for *S. cerevisiae*.[236,237]

Once the macrolides have diffused through the wall, they can exert their action on the plasma membrane. The original model considered the inhibitor-mediated K^+ leakage as the primary event leading to the death of fungal cells,[225] and the molecular events leading to K^+ leakage could be explained with the half-pore model described above. However, the results of more recent studies are not entirely in support of this model. There is increasing evidence that the fungicidal effect of the larger polyene macrolides is not directly related to K^+ leakage. For example, amphotericin B was shown to act fungistatic at low concentrations, accompanied by a reversible leakage of K^+. To the contrary, at high concentrations the antibiotic exerted fungicidal effects. Since the intracellular K^+ concentration had no effect on the distinction between fungistatic and fungicidal effects, neither of them appeared to be explained by the leakage of K^+.[238] Spin-probe analysis of amphotericin B and nystatin-treated membranes revealed phase transitions to less fluid membranes and were not in support of the pore model.[219] These alterations of membrane properties could alter the activity of fungal plasma membrane ATPases and could explain the leakage of small ions.[219] A direct inhibitory effect of macrolide antibiotics on fungal chitin synthetase has been proposed, but as critically discussed by Lösel,[10] a major contribution of this effect to the inhibition of fungal growth appears unlikely.

Macrolide-induced lipid peroxidation was recently considered to contribute to the fungicidal effect.[239,240] Protoplasts and whole cells of *C. albicans* were protected from the lethal action of amphothericin B under oxygen-free conditions and by exogenous catalase and superoxide dismutase. To the contrary, ascorbic acid acting as a prooxidant augmented the killing of fungal cells. However, neither the protective effect of active oxygen exclusion nor the enhancing action of ascorbic acid had any effect on the leakage of K^+. In summary, there is now increasing evidence that the formation of half pores leading to K^+ leakage is not the exclusive mode of action of the polyene macrolides. Lipid peroxidation resulting in severe and overall damage of membranes might be functionally involved. How this oxidative lipid damage is triggered by the inhibitors, and how specific interactions of the antibiotics with ergosterol can be linked to these events still need to be elucidated. The detailed understanding of all factors involved in the interaction of polyene antibiotics with either fungal or human erythrocyte plasma membranes will have positive impact on the current clinical efforts to minimize the toxic side effects of amphotericin B and semisynthetic derivatives.[241-243]

2. Mechanism of Resistance

The spectrum of fungi affected by the polyene antibiotics including amphotericin B is relatively broad. According to Scholer and Polak,[244] fungal species with fungistatic MIC values for amphotericin B around 1 μg/ml or less are considered "highly sensitive". Some fungal species have MIC values of 10 μg/ml and were defined "insensitive". Phenotypes of normally "sensitive" fungal species were considered resistant, if their MIC values are around or above this "insensitivity" range. Resistant phenotypes originate from spontaneous mutations, and their number in unexposed fungal populations is extremely low.[244] Under the long-term selection force of the inhibitor, however, the resistant phenotypes can increase in frequency. Although practical resistance has not created substantial problems in the past, resistant *Candida* spp. have been isolated from patients under long-term treatment with amphotericin B or nystatin.[244] A second source for phenotypes resistant to polyene antibiotics is the mutagenesis of fungal propagules and subsequent selection on antibiotic-containing media. Most studies on the mechanism of fungal resistance to polyene antibiotics have been done with such laboratory mutants.

As described above, the larger polyene macrolides such as nystatin and amphotericin B bind most effectively to ergosterol and less to other sterols. Based on this property, a logical "strategy" for a fungal organism would be to alter the structure of sterols incorporated into the membranes. Indeed, the majority of resistant phenotypes fall into this category.[244,245] The departure from ergosterol is most often mediated by a mutation in later steps of sterol biosynthesis, and the mutagenesis and selection of fungi on nystatin are frequently the method of choice to obtain mutants with such defects for studies on sterol functions and biosynthesis.[165] The biosynthesis of ergosterol dictates that sterol precursors are gradually modified toward the ergosterol structure, and it is not surprising that the level of resistance is usually higher for mutants with defects in earlier steps of sterol synthesis.[244] For example, a polyene-resistant mutant of *C. albicans* contained >98% methylsterols and no ergosterol.[246] The qualitative change of sterol structures, however, is not the exclusive mechanisms of resistance. Quantitative changes, and most commonly the reduction of sterol contents, were frequently reported to cause similar effects.[183,244,245] A reducted amount of ergosterol in membranes might indicate an adaptation of phospholipid structures and compositions to these conditions and, consequently, a decreased membrane-disturbing effect of the inhibitors.[218]

Ziogas et al.[183] analyzed both the ergosterol content and composition of *Aspergillus nidulans* mutants selected on pimaricin and found a correlation between the resistance level and the ergosterol content of 17 mutants, with low ergosterol contents leading to elevated levels of resistance. The relatively poor correlation coefficient, however, indicated that the mechanism of resistance for some mutants was not related to the ergosterol content, and the authors concluded that mechanisms of resistance unrelated to sterols must exist. The degradation of nystatin by an induced enzyme system has been reported for some dermatophytic fungi,[247] but this mechanism has received little attention and confirmation in more recent studies. Considering the diverse factors that influence the sensitivity of fungal organisms to polyene macrolides (e.g., lipid peroxidation and the interaction with cell wall components), a diversity of the molecular mechanisms causing resistance would not be surprising. One of the characterized examples departing from a sterol-related expression of resistance is a nystatin-resistant mutant of *S. cerevisiae* with unimpaired sterol metabolism.[248] Differences, however, were observed in the composition of cell walls, and these cell wall modifications were discussed as a potential mechanism of resistance.

Both the sterol and cell wall composition of fungal species have evolved in an "optimizing" process, and a departure from this optimum is often accompanied by impaired properties of mutants that have gained in resistance to the macrolide inhibitors. Indeed, both the growth rate and virulence of resistant phenotypes were frequently reduced. Scholer and Polak[244] considered this resistance penalty as the major reason for the extremely few reported failures of chemotherapy with amphotericin B. The authors discussed this fitness penalty in relation to an increased vulnerability of resistant phenotypes to the host's immunoresponses. The emergence of problems, however, might occur with immunocompromised patients under long-term treatment with the antibiotics, and a clear understanding of the mechanisms of resistance would aid the prediction of resistance risks under these conditions and the development of counteractive measures.

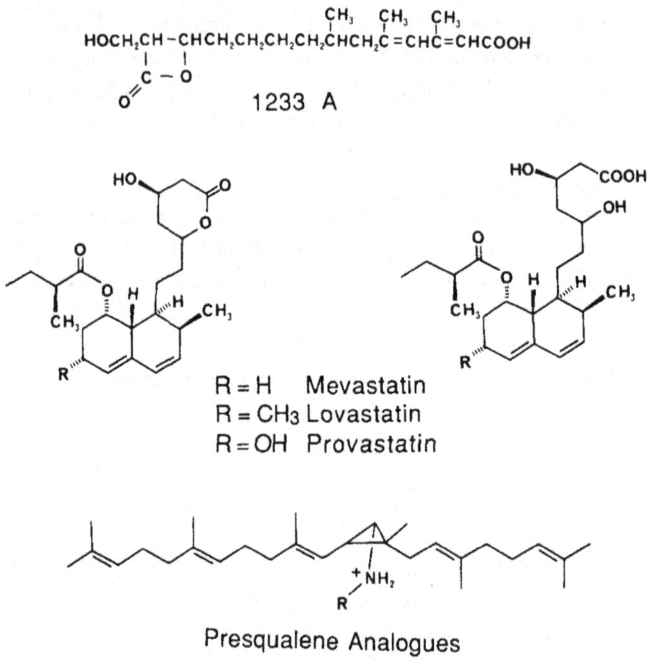

FIGURE 13. Inhibitors of presqualene biosynthetic steps.

IV. INHIBITORS OF STEROL BIOSYNTHESIS: ACETATE TO LANOSTEROL

With the exception of squalene epoxidase, all other enzymes involved in sterol biosynthesis prior to lanosterol (see Section II.B) are not targets for developed antifungal agents. They have received, however, increasing attention as targets for hypocholesterolemic drugs.[249] With respect to prospects of future antifungal inhibitors, the early enzymes in the biosynthetic pathway of sterols are both attractive and unattractive. They are attractive, because not only sterols, but also other compound classes are branching from the isoprenoid pathway,[52-54] and the effect of an ''early enzyme'' inhibitor could be expected to be multiple. The potential disadvantage lies in the highly conserved nature of the early part in isoprenoid synthesis, and inhibitors could be too nonselective to be useful as systemic fungicides and drugs. Small differences, however, in the active site structures of enzymes converting identical substrates are often sufficient to warrant a high degree of selectivity of inhibitors, and comparative studies with identical enzymes isolated from different sources could evaluate the potential for selectivity optimizations, once inhibitor lead structures were identified. Furthermore, the cloning of an increasing number of enzyme genes will greatly facilitate future target evaluations. It will be possible to compare more and more protein sequences of identical enzymes present in different classes of organisms, and enzymes difficult to isolate will become readily available via the expression of the genes in transformed organisms. In addition to the direct inhibition of enzymes, an interference with regulatory mechanisms can also be considered as target for antifungal agents. For these reasons, not only existing antifungal agents but also some inhibitor ''leads'' will be described in the following section.

A. ACETATE TO SQUALENE

The β-lactone 1233A (L-659,699) (Figure 13), an antibiotic originally isolated from *Cephalosporium*, was identified as a specific inhibitor of HMG-CoA synthase (step B in Figure 3) prepared from rat liver.[250,251] Acetoacetyl-CoA thiolase and HMG-CoA reductase were not affected. A completely different class of antibiotics produced by several fungi have been identified as powerful competitive inhibitors of HMG-CoA reductase (step C in Figure 4).[252] Of the structurally similar compounds, mevastatin (compactin, ML-236B), lovastatin (mevinolin, monacolin K, MB-520B), and pravastin (CS-514, SQ 31000) have received considerable attention as promising hypocholesterolemic drugs (Figure 13).[252-254,254a] They all exist in a lactone and open hydroxy acid form, which can be isolated as the sodium salt. As discussed by Bach and Lichtenthaler,[255-257] the open acid form resembles the mevaldyl moiety of mevaldyl-CoA thiohemiacetal, a proposed intermediate in enzyme catalysis. As expected, mevastatin competes with HMG-CoA and CoASH, but not NADPH. Either part of the inhibitor, the 3,5-dihydrovalerate and the polyhydronaphtyl moiety, is a poor inhibitor even when applied in combination. Based on this lack of inhibition, Nakamura and Abeles[258] concluded that the inhibitor binds to both the HMG-CoA site and to an additional site close to the active site.

Most of the biochemical work has been focused on HMG-CoA reductase derived from liver, but mevastatin has also been identified as inhibitor of plant and yeast HMG-CoA reductases.[255-257] This *in vitro* activity was also expressed *in vivo*. Mevastatin was shown to inhibit plant growth,[256] and mevastatin and lovastatin inhibited the growth of several yeasts with MIC values as low as 0.1 μg/ml.[259] The inhibitor potency of mevastatin is not highly selective and similar for HMG-CoA reductases isolated from liver, plant, and yeast, with K_i values of 0.5, 2.2, and 3.5 n*M*, respectively.[255] The results of an interesting and successful approach to substitute the polyhydronaphthyl moieties present in the antibiotics by fully synthetic residues have been summarized by Lee,[254] and a similar approach might be suitable to increase the inhibitor selectivity toward HMG-CoA reductases present in fungi. Interesting are the sensitivity changes to mevastatin when each of the two genes coding for HMG-CoA reductase was disrupted separately.[64] When HMG1 contributing >80% of the wild-type's activity was disrupted, the mutants became more sensitive to the inhibitor. The effect was less pronounced when HMG2 was disrupted. Although the reason for this sensitivity shift remains to be clarified, the example demonstrates that the techniques of "gene additions" and "gene disruptions" will become a powerful tool for the evaluation of novel targets for antifungal agents.

Inhibitors of enzymes subsequent to HMG-CoA reductase have received less attention in the past. Noteworthy are fluorinated mevanolate analogues as inhibitors of mevanolate-5-bisphosphate decarboxylase (step F in Figure 4),[260] and 3-(fluoromethyl)-3-butenyl diphosphate as a suicide substrate for IPP-isomerase (step G in Figure 4).[68,261] These substrates, however, are relatively small and leave little room for the chemical modification of these inhibitor leads. This restriction might be different for analogues of carbocationic intermediates described as inhibitors of squalene synthase (step I in Figure 4). Squalene is formed from two molecules of farnesyldiphosphate in two distinctive steps,[54] and the inhibitors synthesized by Poulter et al.[262] are ammonium analogues of the intermediate presqualene diphosphate (Figure 13).

B. SQUALENE EPOXIDASE

Squalene epoxidase (step J in Figure 4) is the only prelanosterol enzyme inhibited by a new class of antimycotic drugs, the allylamines. The antifungal properties of naftifine (Figure 14), the first representative of this class, were reported in 1981.[263] Since then allylamines have been the subject of extensive chemical programs,[264,265,265a] and terbinafine as well as more recent derivatives such as SDZ-87-469 (Figure 14) with improved activities against *Candida* spp. are in different stages of clinical testing and development.[241] Thus far, the allylamines remain restricted to the chemotherapy of mycotic diseases, and derivatives for the control of plant pathogenic fungi have not yet been described.

The mode of action of allylamines have been thoroughly investigated, starting with the

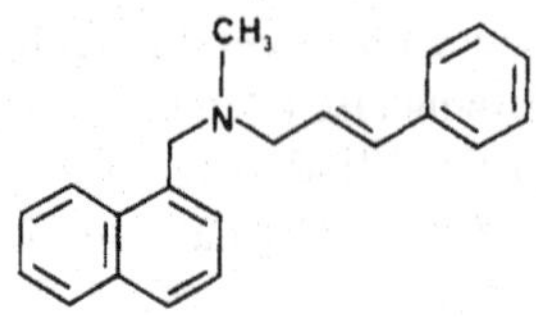

Naftifine

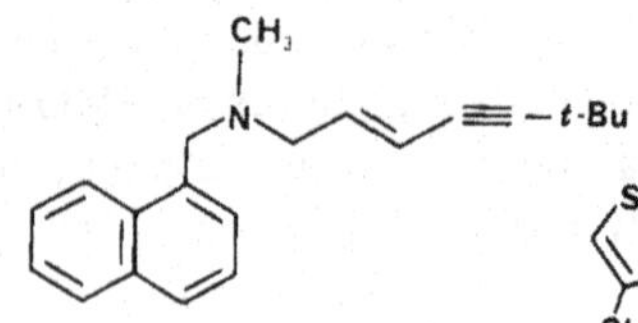

Terbinafine

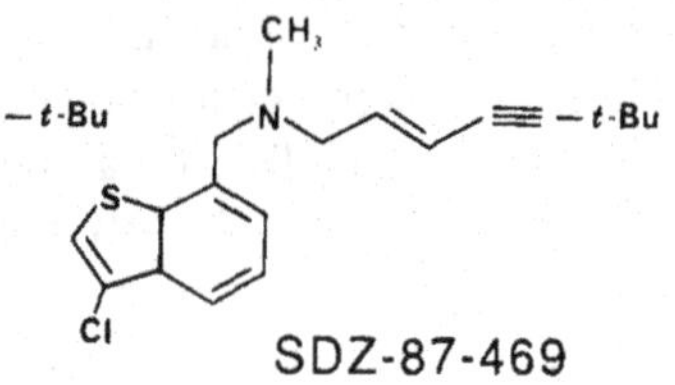

SDZ-87-469

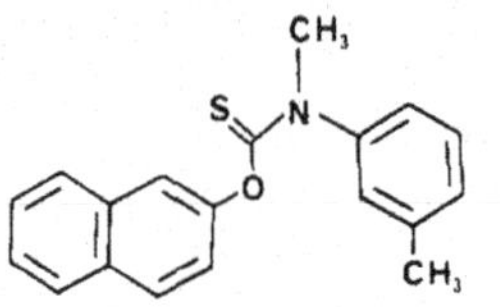

Tolnaftate

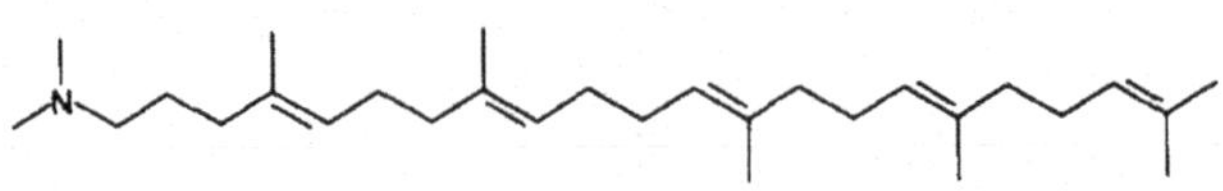

2-Aza-2,3-dihydrosqualene

FIGURE 14. Inhibitors of squalene epoxidase.

monitoring of sterol biosynthesis in the presence of [^{14}C]-acetate and proceeding to inhibitor assays in cell-free systems. First indications for the inhibition of sterol biosynthesis at the point of squalene epoxidation were reported in 1982.[266,267] The results of an increasing number of biochemical studies mainly done by Ryder and co-workers have been described in detail in several reviews,[268-275] and the key features of allylamine antimycotics will only be summarized. Naftifine and terbinafine are most active against dermatophytes such as *Trichophyton rubrum* and other filamentous fungi, with less pronounced activity toward *Candida* spp. Both naftifine and terbinafine have been identified as relatively exclusive inhibitors of squalene epoxidase without major side effects on other metabolic processes in fungi, with terbinafine performing as the more efficient inhibitor in all systems studied. In mammals, both naftifine and terbinafine interact with hepatic cytochrome P-450 monooxygenases involved in detoxification, but not with P450 systems active in steroid hormone and prostacyclin synthesis.[276]

The high degree of target specificity in fungi is best illustrated by the inhibitory potency of terbinafine in different assay systems derived from *C. albicans*. The I_{50} values obtained with the incorporation of acetate into squalene in whole cells, the incorporation of mevalonic acid into squalene in a cell-free system, and with *in vitro* assays employing squalene epoxidase were all 0.008 ng/ml (0.03 μM).[273] This high degree of correlation indicates that the relatively low activity toward *Candida* spp. when compared to filamentous fungi is not explained by a reduced uptake or rapid detoxification of the inhibitor. It remains to be resolved whether a differential sensitivity of the target enzyme or other ergosterol-related effects are an explanation for the higher sensitivity of filamentous fungi. Indications exist that sterol-related secondary effects might be involved. Regardless of an almost complete block of squalene epoxide synthesis, *C. albicans* continued to produce ergosterol for several hours, whereas synthesis was immediately blocked in dermatophytes.[273] It should be emphasized that the function of sterols in fungi has been mainly elucidated for yeasts such as *Saccharomyces cerevisiae* (see Section II.E). The mode of action studies described above has been performed with *Candida* spp. in the yeast form of this dimorphic fungus. The filamentous form has indeed been shown to be more susceptible to allylamines,[273] and an increased understanding of sterol functions in filamentous fungi will be required to finally elucidate the cause of allylamine selectivity.

The consequence of squalene epoxidase inhibition is not only the depletion of the ergosterol pool, but also the strong accumulation of squalene. This precursor accumulation does not diminish the incorporation of precursors into squalene and thus leads to squalene levels as high as 5% of the cell dry weight.[271-273] Most likely, squalene is incorporated into membranes and, thus, is responsible for the fungicidal effects of allylamines. For example, the viability of *T. mentagrophytes* was reversely related to the amount of accumulating squalene, whereas the ergosterol content was only slightly decreased.[273]

Most interesting are comparative *in vitro* studies with squalene epoxidases derived from *C. albicans* and rat liver.[276] The properties of these enzymes are similar, although differences with regard to cofactor requirements exist (see Section II.B). Both naftifine and terbinafine are powerful inhibitors of the fungal enzyme, with K_i values of 1.1 and 0.03 μM, respectively. The inhibition is reversible and noncompetitive in character, indicating the binding of allylamines to regulatory sites of the terminal oxidase rather than competing with squalene as the substrate. This conclusion, however, might be premature. As discussed below, the inhibition of 2,3-oxidosqualene sterol cyclases by compounds designed as high-energy intermediates and thus active site-directed inhibitors followed similar noncompetitive kinetics, and these unexpected findings might be related to the hydrophobic character of both the substrates and the inhibitors and thus experimental conditions that complicate the interpretation of "classical" inhibition kinetics. The squalene epoxidase isolated from rat liver was almost insensitive to the inhibitors; the K_i value for terbinafine was higher by a factor of over 3000.[273] The enzyme isolated from guinea pig liver was more sensitive, but was still over 100-fold less sensitive than the fungal enzyme. Interesting is the finding that the inhibition of mammalian enzymes followed competitive kinetics, indicating a weak binding of allylamines into the squalene binding domain of the active site.[273] Apparently, the fungal and mammalian enzymes are strikingly different with respect to the binding of the inhibitor. This difference might be explained by either the presence or absence of a regulatory binding site leading to noncompetitive or competitive inhibition, or by different physicochemical properties of the membrane-bound enzymes that make the active site more or less accessible to the substrate and the inhibitor. Regardless of the precise explanation for the different inhibitor kinetics, the difference between the fungal and mammalian enzyme exemplifies that a very high degree of inhibitor selectivity can be achieved by utilizing differences of functionally identical enzymes. Inhibitor data derived from *in vitro* assays are yet not available for the squalene epoxidase from plants. Radish root elongation was inhibited at high concentration of naftifine,[277] but this relatively insensitive response might have been caused by insufficient uptake and translocation of the inhibitor. Substantial side effects on other enzymes such as mammalian P450 monooxygenases have not been reported yet.[278]

The chemistry of specific inhibitors of fungal squalene epoxidases is not restricted to allylamines. The thiocarbamate antifungal agents tolnaftate (Figure 14) and tolciclate had been in use for decades for the topical treatment of dermatophyte skin infections, with little activity against *Candida* spp. Both compounds have been identified as inhibitors of fungal squalene epoxidase and, thus, as "analogues" of allylamines.[279,280] In contrast to naftifine and terbinafine, however, the lack of activity against *Candida* spp. could be explained by the presence of a penetration barrier in *Candida* cell walls. Similar to the allylamines, tolnaftate and tolciclate were highly selective inhibitors of the fungal enzyme with a low inhibitory potency when tested with squalene epoxidase derived from rat liver.[273]

This specificity, however, can be reversed for some aza-analogues of squalene originally designed as inhibitors of 2,3-oxidosqualene-lanosterol cyclase (see following section). With 2-aza-2,3-dihydrosqualene (Figure 14), the rat liver enzyme was indeed more effectively inhibited than the squalene epoxidase derived from *C. albicans*.[281] The I_{50} value for the fungal epoxidase was 32.9 μM, and 2.4 μM for the epoxidase derived from rat liver. This reversion of specificities in fungal and mammalian systems as a consequence of the inhibitor structures represents an example for subtle active site differences leading to different inhibitor responses. Although allylamine resistance has not been reported for target pathogens, Orth et al.[281a,b] provided evidence for a decreased affinity of terbinafine to the target for a mutant of *Ustilago maydis* with a high level of resistance.

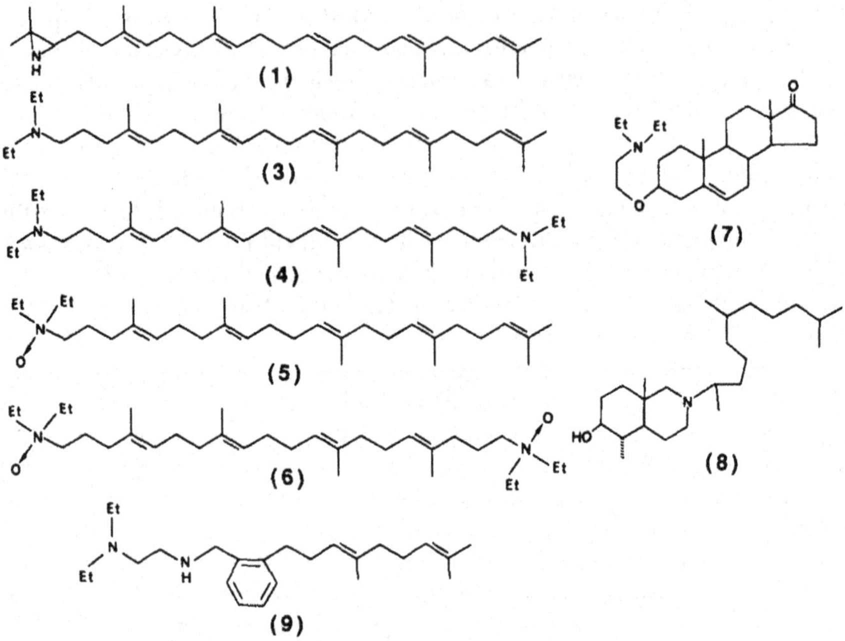

FIGURE 15. Inhibitors of 2,3-oxidosqualene-sterol cyclase.

C. 2,3-OXIDOSQUALENE-STEROL CYCLASE

While the allylamines as inhibitors of squalene epoxidase were discovered via the random synthesis of naftifine, the discovery of its antifungal activity, and the subsequent optimization of the lead structure, considerable emphasis has been placed on the rational design of inhibitors of the next following enzymatic step catalyzed by 2,3-oxidosqualene-(lano)sterol cyclase, yielding lanosterol in fungi and mammals (step A in Figure 5) and cycloartenol in plants (Chapter 8). None of these inhibitor structures has been reported to be under development as antimycotic drugs or fungicides, but the design efforts have revealed interesting lead structures and have identified potential differences between the fungal, plant, and mammalian cyclase. The formation of the first sterol structure from 2,3-oxidosqualene is initiated by an electrophilic opening of the epoxide yielding a partial positive charge at C-2, and then proceeds through several carbocationic intermediates (see Section II.C.1). The strategy of inhibitor design concepts has been based on these high-energy carbocationic intermediates stabilized during the course of enzyme catalysis. Most of the inhibitors are analogues of these high-energy intermediates by mimicking positive charges at various positions of the squalene/sterol molecule, "freezing" the route of catalysis once the inhibitor is bound into the active site of the enzyme.

Positive charges under physiological pH conditions can be most easily accomplished by the introduction of basic nitrogen atom at the position of an expected carbocationic intermediate, and several potent inhibitors of this type have been described for the 2,3-oxidosqualene-sterol cyclase catalyzing the formation of lanosterol in mammalian and fungal systems and cycloartenol in plants. Unfortunately, most of the inhibitors have only been tested with the enzymes isolated from liver and plant tissue (see Chapter 8), but not with fungal cyclases. Antifungal agents, however, have to be based on putative differences of the enzyme present in fungi in order to minimize toxic side effects to both mammals and plants. Therefore, the discussion of this class of inhibitors will be focused on the present knowledge regarding fungal enzymes.

The first cyclase inhibitor reported to mimic the carbocationic intermediate at the C-2 position of squalene was 2,3-iminosqualene (**1**) (Figure 15).[282] This analogue of the high-

TABLE 1
Inhibitor Action of Aza Derivatives on Oxidosqualene-Sterol Cyclases

	I_{50} (μM)[b]			
Compound[a]	**Rat liver**	**Pea seedlings**	***S. cerevisiae***	***C. albicans***
1	0.4 (0.2)[c]	0.2	10	0.14[c]
2	7.5 (4.8)[c]	1.3	10	3.3[c]
3	3.2	0.55	14 (12.5)[d]	
4	3.0		6 (7.5)[d]	
5	1.5	0.15	14 (14.0)[d]	
6	1.5	0.15	4 (3.0)[d]	
7	1.9	0.2	0.1	
8	2.0[e]	1.0[e]		
9	20.8			1.9

[a] For inhibitor structures (except **2**) see Figure 15; **2** is 2-aza-2,3-dihydrosqualene (Figure 14).
[b] Data from Cattel et al.[284] unless stated otherwise.
[c] Jolidon et al.[285]
[d] Balliano et al.[286]
[e] Taton et al.[292] (plant enzyme derived from maize embryo tissue).

energy intermediate was shown to inhibit the growth of *S. cerevisiae* and *Gibberella fujikuroi*.[168,169,187,188] For the latter fungus, however, the inhibition of mycelial growth was only transient, and the inhibitor was slowly cyclisized to the corresponding epiminolanosterol.[283] A second inhibitor of the fungal cyclase, 2-aza-2,3-dihydrosqualene [(**2**); for structure see Figure 14], with its additional action on squalene epoxidase, has already been described in the previous section. This compound, together with various derivatives including 2,3-iminosqualene (**1**), has been evaluated in a comparative study involving the cyclases from rat liver, pea seedlings, and *S. cerevisiae*.[284] Some of these and additional inhibitors have also been tested with the cyclase derived from *C. albicans*.[285] The structures of representative inhibitors and corresponding inhibitor data obtained with the three enzyme sources are presented in Figure 15 and Table 1. The azasqualene derivatives are reversible inhibitors of cyclases derived from all three sources. However, the highest activity was always observed for the plant enzyme, whereas the yeast cyclase was least inhibited. 2,3-Iminosqualene (**1**) appears as an exception. The *C. albicans* cyclase was considerably more sensitive than the yeast enzyme, although no *in vivo* activity was noticed with concentrations >0.3 m*M*.[285]

The inhibitory potency toward the yeast enzyme could be slightly increased by the introduction of a second *N,N*-diethylaza group (**4**). Interesting was the comparison of the inhibitor kinetics for *N,N*-diethylazasqualene (**3**) and its *N*-oxide (**5**).[286] The former aza derivative appeared as a noncompetitive inhibitor and, thus, follows kinetics similar to those observed for other cyclase inhibitors,[287] but also, as described above, for the allylamines as inhibitors of squalene epoxidase. These results were surprising, since all of these inhibitors were designed as analogues of high-energy intermediates binding into the catalytic site and, thus, should act as competitive inhibitors. Moreover, the validity of the design concept and the action as active site-directed inhibitors have been largely confirmed in structure-activity relationships,[284,286,287] regardless of the apparently noncompetitive inhibition type. However, the limitations of kinetic studies with a complex anisotropic system, as discussed by Duriatti et al.,[287] might serve as an explanation for the unexpected kinetic data. Indeed, while bis-*N,N*-diethylazasqualene (**4**) appeared to inhibit the yeast enzyme noncompetitively, the *N*-oxide (**5**) followed competitive kinetics.[286] The authors discussed this difference as an indication for differential accessibilities of the enzyme's active site.

Both *N,N*-diethylazasqualene (**3**) and bis-*N,N*-diethylazasqualene (**4**) inhibited growth of *S. cerevisiae* at concentrations of 60 and 15 μM, respectively.[286] Sterol analysis revealed that inhibition of growth was accompanied by a sharp increase in 2,3-oxidosqualene, indicating a highly specific inhibition of the oxidosqualene cyclase. It should be mentioned, however, that azasqualene and bis-azasqualene derivatives exerted an additional and nonspecific effect on membranes, as indicated by their antibacterial activity and their effect on liposomes.[288] Although superior enzyme inhibitors *in vitro*, the *N*-oxide analogues (**5**) and (**6**) were far less active *in vivo*, indicating a restricted permeability of these particular inhibitors.

One of the longer known cyclase inhibitors is 3β-(β-diethylaminoethoxy)andost-5-en-17-one (**7**) (Figure 15), also known as U18666A,[289] and its methyl derivative.[290] For the latter inhibitor, the reduction of the $\Delta^{24(28)}$-double bond was identified as a second (and more sensitive) target site when sterols from treated yeast cells were analyzed.[290] This alternative mode of action in fungi has not been investigated for other cyclase inhibitors. Among the inhibitors tested by Cattel et al.[284] in a cell-free assay system, U18666A was by far the most powerful inhibitor of the fungal cyclase. In contrast to the azasqualene analogues, the fungal enzyme was more sensitive than the respective enzymes from plants and liver (Table 1). This reversion of inhibitor sensitive indicates substantial differences between the plant, mammalian, and fungal cyclases. Based on the limited number of tested inhibitors, it could be speculated that the fungal enzyme exhibits a preference for cyclic, "sterol-like" inhibitors, whereas "squalene-like" compound are more easily bound to the plant and animal cyclases. This suggestion is substantiated by recent results obtained with a more "cyclic" azasqualene derivative (**9**) (Figure 15).[285] The inhibitor was highly active as inhibitor of the *C. albicans* enzyme (Table 1) and was active *in vivo*. The differential structural requirements might also serve as an alternative explanation for the differences observed for the inhibitory effects of *N*-(1-*n*-dodecyl)-heterocycles.[291] Sevéral of these compounds were powerful inhibitors of the liver cyclase, but were virtually inactive when tested with the yeast enzyme. The authors postulated that the dodecyl chain was positioned over a region outside the active site of the liver enzyme, a region not present in the fungal cyclase.

All epoxisqualene-cyclase inhibitors described above were designed to mimic the carbocation postulated to occur during the epoxy opening at the C-2 position. During the ring closure of oxidosqualene to lanosterol, an additional carbocationic intermediate is likely to occur at the C-8 position of the forming sterol skeleton. The 8-azadecalin derivative (**8**) mimics the carbocationic high-energy intermediate at this position and was recently identified as a powerful *in vitro* and *in vivo* inhibitor of oxidosqualene cyclization in mammals.[292,293] No data on the antifungal performance of this interesting inhibitor are yet available.

D. REGULATION OF HMG-CoA REDUCTASE

Mammalian cells can satisfy their sterol requirements through *de novo* biosynthesis and from nutrients. Thus, it is not surprising that sterol biosynthesis is tightly regulated, with HMG-CoA reductase as one of the key regulatory elements (see Section II.D). Oxygenated sterols are involved in both the repression of gene transcription and enhancement of enzyme turnover (see Section II.D). Oxysterol-binding proteins have been purified, and the sequence deduced from cDNA revealed similarities with proteins that bind to DNA.[294] A surprisingly large number of oxygenated sterols have been identified as mediators of HMG-CoA reductase down-regulation, involving hydroxy and keto functions at the various ring positions and the side chain of cholesterol, but also oxygenated lanosterol derivatives.[141,142,295-298] In many cases, it remains unknown whether the synthetic oxysterols exert their regulatory activities as such or subsequent to metabolic conversion. For example, Panini et al.[299] concluded from inhibitor studies that endogenously formed oxylanosterols play a precursor role in the formation of the authentic regulatory oxysterols. A link between the C-14 demethylation of lanosterol and the formation of endogenous HMG-CoA modulators has also been suggested by others investigating inhibitors of the 14α-DM.[300-302] Lanost-8-en-32-aldehyde, an intermediate in the 14α-DM catalyzed demethylation, has been identified as a putative regulator of HMG-CoA reductase in mammalian cell cultures.[302]

Although indications for the sterol-mediated feedback regulation of fungal HMG-CoA reductases exist,[147-150] the involvement of oxygenated sterols in this process as well as the cellular level of potential regulatory functions has not been clarified. Interesting is the striking structural similarity between the mammalian HMG-CoA reductase and the two yeast enzymes, and the observation that the mammalian enzyme was fully functional in the replacement of both yeast enzymes.[66,154] The two genes present in *S. cerevisiae* were recently shown to be regulated by heme.[303] The expression of the HMG1 gene was stimulated, and the expression of the HMG2 gene was repressed by heme. Furthermore, the uptake of exogenous sterols by yeast maintained under anaerobic conditions is abolished by heme synthesis.[166] Since oxygen is required in both heme and sterol synthesis, the regulation of the two yeast genes by heme might reflect a mechanism to monitor intracellular sterol levels. This explanation is appealing for facultative anaerobic organisms like yeast, but does not exclude a second level of regulation directly involving sterols, or oxidized sterol metabolites. A "feedback" inhibition of HMG-CoA reductase activity as a result of lanosterol accumulation has been reported for *S. cerevisiae*,[304] but a potential oxylanosterol nature of the actual modulator has not been investigated. Regardless of lacking details, little doubt exists that HMG-CoA reductase also serves as a key regulatory enzyme of isoprenoid and, thus, sterol biosynthesis in fungi. Interference with these regulatory events might become an interesting target for novel antifungal agents, as it is a target for hypocholesterolemic agents in drug research.

V. INHIBITORS OF STEROL BIOSYNTHESIS: LANOSTEROL TO ERGOSTEROL

A. SIDE-CHAIN METHYLATION

The side-chain methylation at the C-24 position is an intriguing target for antifungal agents. It is this structural feature that differentiates ergosterol and other fungal enzymes from cholesterol. Furthermore, compelling evidence exists that the 24-methyl group not only is an important feature for membrane integrity, but also serves as a regulatory element (see Section II.E.2). Inhibition of the C-24 methylation step has indeed been shown to exert drastic effects on reproductive events in fungal organisms.[188] Surprisingly, no specific inhibitors of the fungal Δ^{24}-sterol methyltransferase are yet available for the control of fungal diseases. However, interesting lead structures and inhibitor design strategies have been described.

During the methylation reaction, two carbocations at the C-25 and C-24 position have been postulated to occur as intermediates (see Section II.C.2). The presence of these carbocationic high-energy intermediates has provided the basis for the design of inhibitors that act as analogues of these intermediates. Both the introductions of positive charges in the C-25 and C-24 position have been a successful approach, and a variety of derivatives have been found to act as powerful *in vivo* and *in vitro* inhibitors of the plant Δ^{24}-cycloartenol methyltransferase.[305,306]

A similar approach was also successful for the fungal methyltransferase. 24,25-Epiminolanosterol has been mentioned earlier as a fungal side-chain methylation inhibitor with antifungal activities toward *Gibberella fujikuroi*.[188] Some additional inhibitor structures are given in Figure 16. According to the most suitable substrate of the transferase in *S. cerevisiae*, zymosterol, 25-aza-24,25-dihydrozymosterol (**1**) was found to be the best enzyme inhibitor among several similar azasterol derivatives.[307] Treatment of cells with (**1**) led to inhibition of yeast growth and the accumulation of several sterols not methylated in the C-24 position.

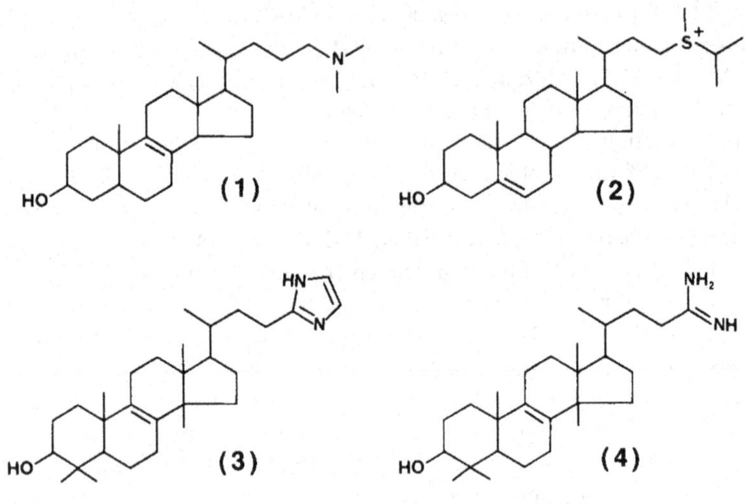

FIGURE 16. Inhibitors of Δ^{24}-sterol methyltransferase.

25-Aza-24,25-dihydrolanosterol, the C-4 and C-14 methylated derivative, was less active by a factor of approximately 10.[307] Ator et al.[87] have extended the design approach to several interesting structures with strong inhibitor activity toward the methyl transferase derived from *Candida albicans* [compounds (**2**) to (**4**) in Figure 16]. Although the best enzyme substrate for the *Candida* transferase was, like in yeast, zymosterol and not lanosterol, the lanosterol analogues (**3**) and (**4**) were equally inhibitory. Interesting is the analysis of inhibitor kinetics of these methylase inhibitors. According to Oehlschlager et al.,[307] (**1**) was competitive to *S*-adenosyl-methionine (SAM) and noncompetitive to zymosterol. With the *Candida* transferase, the inhibitor kinetics were even more complex and discussed as indication for the binding of the inhibitors into the sterol binding site, followed by binding of a second inhibitor into the SAM site.[87] The postulated binding of the inhibitors into the SAM binding site, however, might indicate the possibility that the inhibitors lack specificity and might act on methylating enzymes other than the sterol methyltransferases. This lack of specificity has been reported for the antibiotic sinefungin,[82] a more generally active methyltransferase inhibitor.

B. C-14 DEMETHYLATION

In the late 1960s the outstanding activity of pyridines, pyrimidines, imidazoles, and triazoles was discovered by independent research groups. Structures of this class of inhibitors are shown in Figure 18. Buthiobate (**4**), triarimol (**1**), miconazole (**9**), and clotrimazole (**12**) were the first developmental compounds derived from these synthesis programs. The only common feature of these compounds was the presence of a heteroaromate containing at least one nitrogen atom with a free pair of electrons, a structural feature that later was shown to comprise the mechanistical link between these inhibitors. Initially, the compounds were not viewed as biochemically related. As summarized by Büchel,[308,309] the rationale for the synthesis of early azole derivatives under his direction was based on a "carbocation hypothesis" and, thus, was not at all related to fungal sterol biosynthesis. Ragsdale and Sisler recognized structural similarities between triparanol (Figure 21), a developmental hypocholesterolemic drug under investigation since 1959,[310] and the pyrimidine fungicide triarimol (**1**) and presented the first evidence that the inhibitor site of this pyrimidine derivative was located in fungal sterol biosynthesis.[311] As described below (see Section V.C), triparanol has since been identified as an inhibitor of fungal sterol biosynthesis, although with a target site different from triarimol. A full account of the triarimol mode of action study was

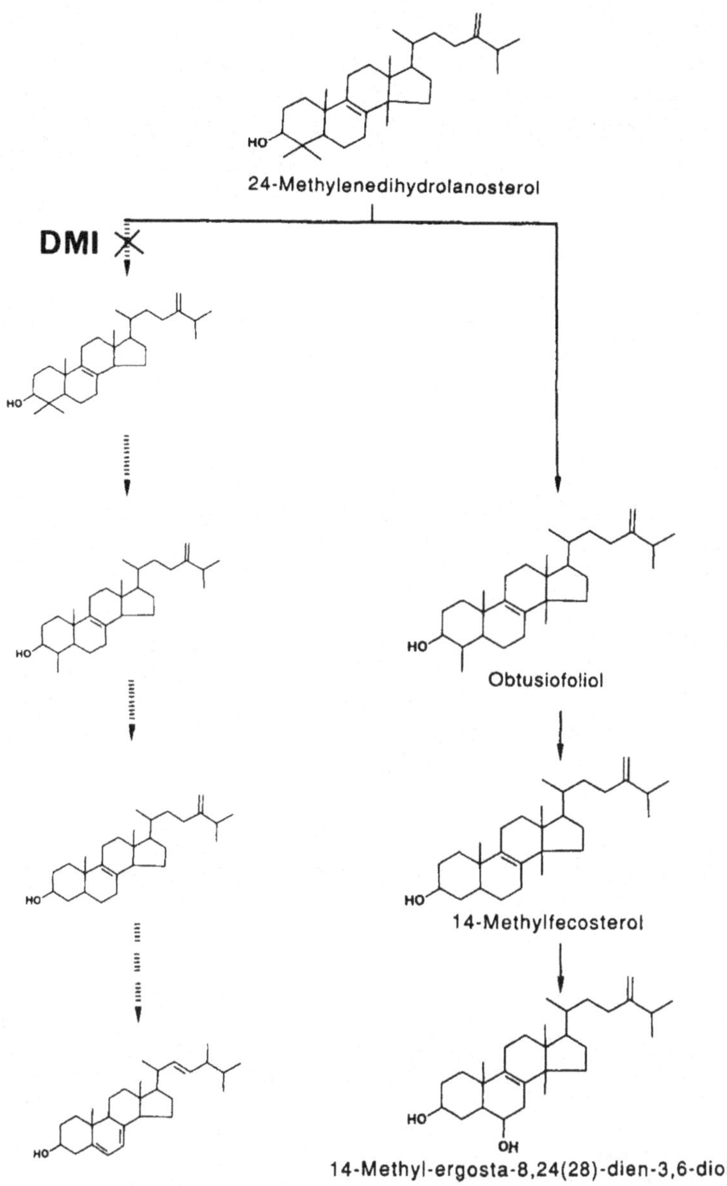

FIGURE 17. Sterol biosynthesis under the action of sterol demethylation inhibitors.

published in 1975.[41] Treatment of *Ustilago maydis* with triarimol led to a dose- and time-dependent accumulation of 24-methylenedihydrolanosterol and, after prolonged treatment, to small amounts of obtusifoliol and 14-methyl fecosterol (Figure 17). This drastic change in fungal sterol composition indicated a specific inhibition of the fungal 14α-DM (see Section II.C.3). Obtusifoliol and 14-methyl fecosterol, the two 14-methylated sterols normally not found in fungi, were thought to originate from the slow demethylation of accumulating 24-methylenedihydrolanosterol at the C-4 position. C-14 methyl sterols with double bonds other than Δ^8 have rarely been identified, indicating that these Δ^8-precursors are not accepted as substrates for enzymes active in the late steps of sterol biosynthesis. The original mode of action scheme suggested by Ragsdale was soon verified for the pyridine buthiobate (**4**),[312] the triazole triadimefon (**21**),[313] and the imidazole miconazole (**9**),[85] and it became clear that these structurally diverse compounds shared a common mode of action as inhibitors of fungal

14α-DMs. The biochemical term DMI, which stands for sterol demethylation inhibitor, has been introduced to describe this entire class of antifungal agents.

The class of DMIs comprises by far the most extensively investigated group of inhibitors with antifungal properties. The reason for this tremendous and ongoing attention is obvious: DMIs are highly potent antifungal agents; the structural flexibility and, thus, patentability of derivatives proved to be tremendous; the biochemical knowledge gathered over the last 15 years was substantial enough to guide the synthesis and evaluation of new derivatives; and both sectors of antifungal chemotherapy were covered. The usefulness of DMIs for both agriculture and medicine became apparent during the early phase of research and development. Whereas buthiobate and triarimol were under development as agricultural fungicides, clotrimazole and miconazole were found to be most useful as topical antifungal drugs. Although research continued to involve all four possible heteroaromates initially discovered as potent essential groups, the azole structures became more successful than the pyridines and pyrimidines. Within the group of azole compounds, more imidazoles have been introduced as antimycotics, and more triazoles are in use as fungicides in agriculture. This distinction, however, is not exclusive (Table 2).

Research and development of DMIs has been ongoing for over 20 years. Zirngibl[314] listed over 800 patents filed for azole fungicides until 1982, and the number has since clearly surpassed 1000. Considering this tremendous attention it is not surprising that numerous commercial products have been introduced to both fields of application. In their review article published in 1987, Scheinpflug and Kuck[34] listed 23 DMIs used in agriculture. Fromtling[315] listed, in 1988, 13 antifungal agents available for clinical use, and 7 more in late stages of development. More structures have since been added to this arsenal of DMIs, and a prediction of 25 compounds eventually marketed in each field of application appears conservative.

The literature published on DMIs reflects the tremendous attention this class of antifungal agents has received. Comprehensive and excellent review articles have been published on the technical and biological aspects of DMIs used in agriculture[33,34,36] and medicine,[314-318] but also on the mode of action of DMIs.[217,219,319-333] Considering the wealth of available information, the treatise of DMIs in this chapter can only be a structured summary with emphasis on more recent developments and findings. The structures of important DMIs are presented in Figure 18, and additional information, including access to respective biochemical literature, is compiled in Table 2.

1. Primary Mode of Action

The initial methods employed in mode of action studies were either the analysis of fungal sterols via GLC and GLC-MS, or the incorporation of radioactive sterol precursors, normally acetate, into sterols, followed by separation of sterols by TLC. This set of methods has been applied to numerous DMIs, and departures from the mode of action scheme proposed by Ragsdale[41] are rare (Figure 17). Ebert et al.[39] identified, after prolonged treatment of *U. maydis* with etaconazole, a DMI closely related to propiconazole (**25**), the accumulation of 14-methyl ergosta-8,24(28)-dien-3,6-diol, in addition to 24-methylenedihydrolanosterol, obtusifoliol, and 14-methyl fecosterol. This hydroxysterol originating from 14-methyl fecosterol was not detected after 4 h of treatment, but comprised 2% after 13 h and 7% after 44 h of treatment. The 3,6-diol was also detected in a 14α-DM-deficient mutant of *C. albicans*,[138] in *S. cerevisiae* treated with fluconazole (**39**),[360] in *C. albicans* treated for 24 h with itraconazole (**38**) and ketoconazole (**15**) in the presence of ^{14}C-acetate,[361] and in *Aspergillus fumigatus* treated with itraconazole.[355] Considering the less hydrophobic character of the 3,6-diol, it could have frequently escaped analysis, and the accumulation of the 3,6-diol might be a more widespread phenomenon. Vanden Bossche et al.[361] speculated that the C-6 hydroxylation might represent a reaction of the stressed cell toward the "detoxification"

TABLE 2
List of DMI Fungicides and Antimycotic Drugs

Compound	Common name	Year of description	Company	Ref.
Pyridines and pyrimidines (Agriculture)				
1	Triarimol	1970	Eli Lilly	322, 324
2	Fenarimol	1975	Eli Lilly	322, 324
3	Nuarimol	1975	Eli Lilly	322, 324
4	Buthiobate	1975	Sumitomo	325, 334
5	Pyrifenox	1986	Maag	335
Imidazoles (Agriculture)				
6	Imazalil	1972	Janssen	324, 332
7	Prochloraz	1977	FBC/Schering	324, 332
8	Triflumidol	1983	Nippon Soda	336
Imidazoles (Medicine)				
9	Miconazole	1969	Janssen	324, 332
10	Isoconazole	1969	Janssen	324, 332
11	Econazole	1969	Janssen	324, 332
12	Clotrimazole	1972	Bayer	332, 337
13	Bifonazole	1972	Bayer	337, 338
14	Butoconazole	1978	Syntex	316
15	Ketoconazole	1979	Janssen	324, 332
16	Tioconazole	1979	Pfizer	324
17	Oxiconazole	1979	Siegfried	339
18	Fenticonazole	1981	Recordati	316
19	Sulconazole	1984	Syntex	316
20	Omoconazole	1988	Siegfried	340
Triazoles (Agriculture)				
21	Triadimefon	1973	Bayer	324, 326
22	Triadimenol	1977	Bayer	324, 326
23	Bitertanol	1978	Bayer	324, 332
24	Diclobutrazole	1979	ICI	324, 332
25	Propiconazole	1979	Janssen/Ciba-Geigy	324, 327
26	Penconazole	1983	Janssen/Ciba-Geigy	337
27	Diniconazole	1983	Sumitomo	341
28	Flutriafol	1983	ICI	342, 343
29	Flusilazole	1984	Du Pont	344, 345
30	Hexaconazole	1986	ICI	346
31	Tebuconazole	1986	Bayer	38
32	Myclobutanil	1986	Rohm & Haas	347
33	Cyproconazole	1986	Sandoz	348
34	Furconazole	1988	Rhone Poulenc	349
35	CGA 169374	1988	Ciba-Geigy	350
Triazoles (Medicine)				
36	Vibunazole	1980	Bayer	351, 352
37	Terconazole	1983	Janssen	353
38	Itraconazole	1984	Janssen	354, 355
39	Fluconazole	1885	Pfizer	356, 357
40	ICI 195-739	1987	ICI	358, 359

of accumulating precursors. More recent evidence suggested the involvement of a C-6 hydroxylation in the introduction of the $\Delta^{5,6}$-double bond.[138,360] According to this proposal, the Δ^5-desaturase would accept 14-methyl fecosterol as an artificial substrate. However, the subsequent dehydration step following hydroxylation would not be accomplished under these structural constraints of the artificial substrate.

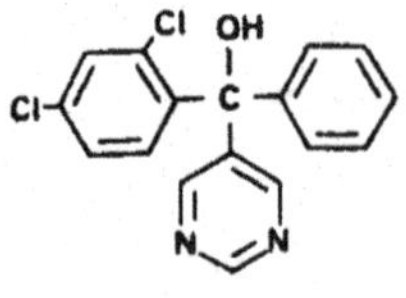

(1) Triarimol

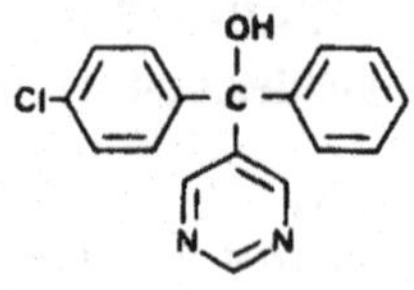

(2) Fenarimol

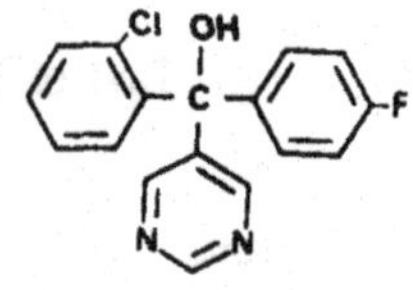

(3) Nuarimol

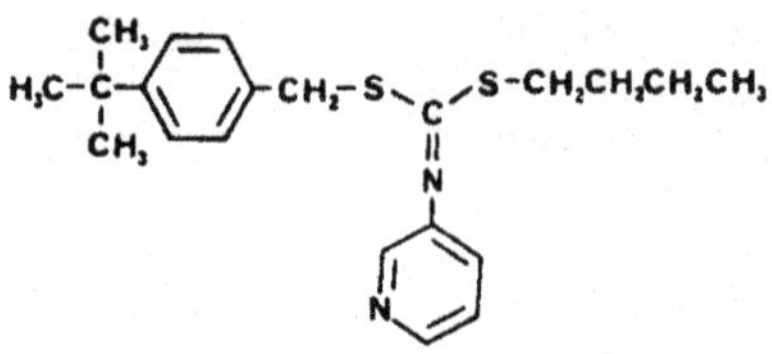

(4) Buthiobate

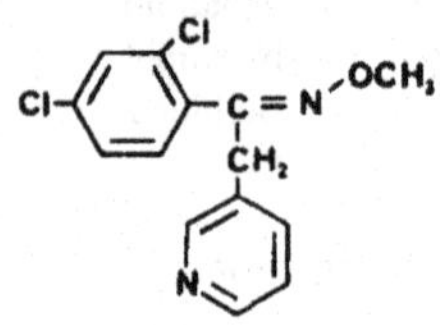

(5) Pyrifenox

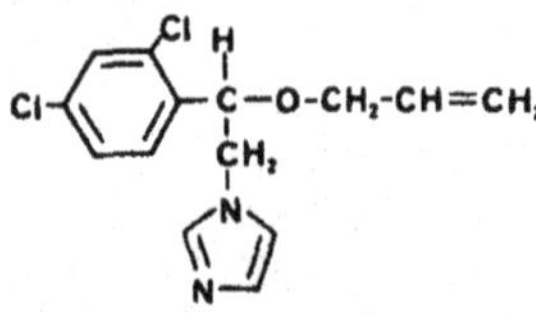

(6) Imazalil

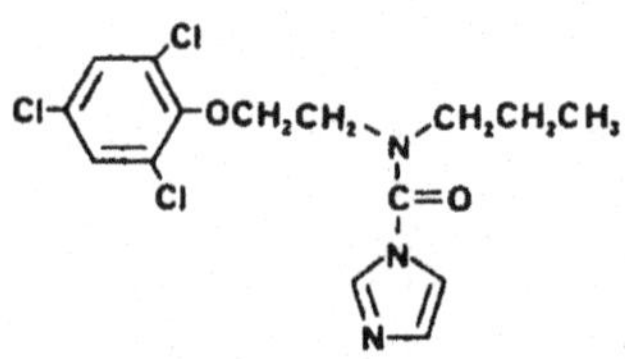

(7) Prochloraz

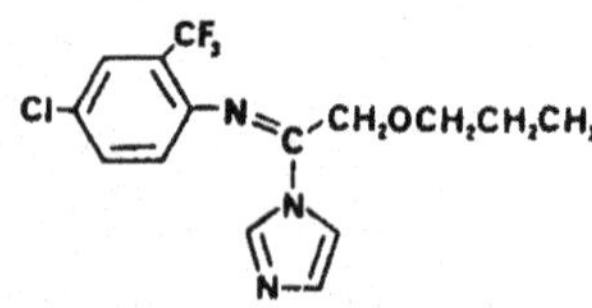

(8) Triflumidol

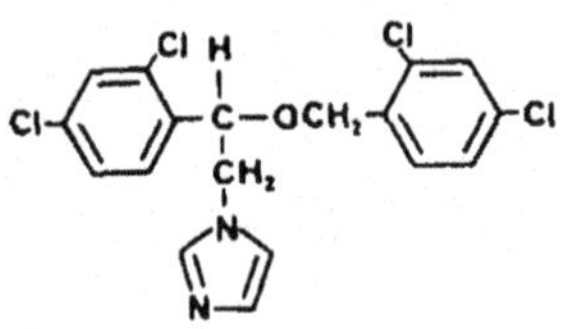

(9) Miconazole

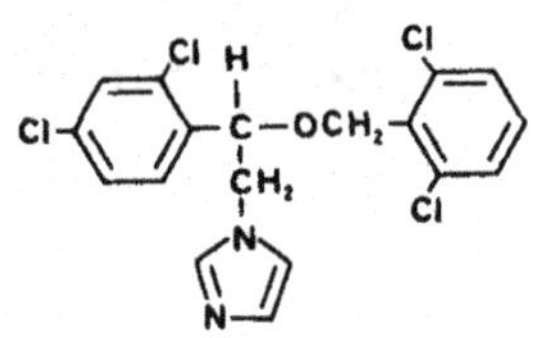

(10) Isoconazole

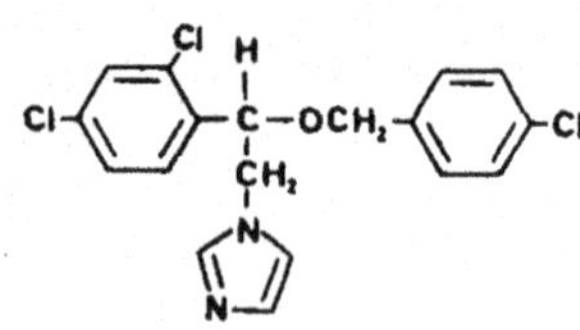

(11) Econazole

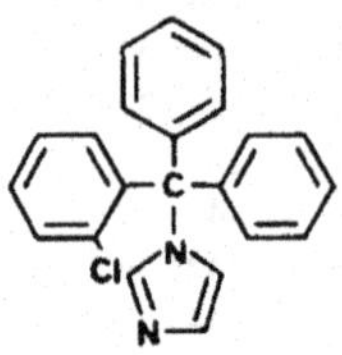

(12) Clotrimazole

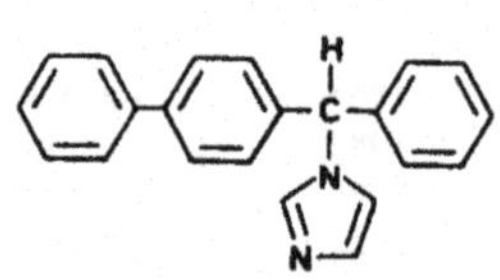

(13) Bifonazole

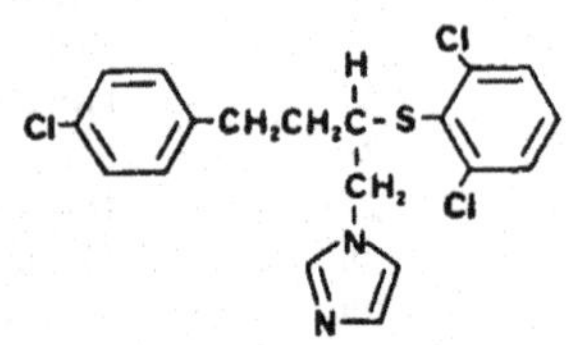

(14) Butoconazole

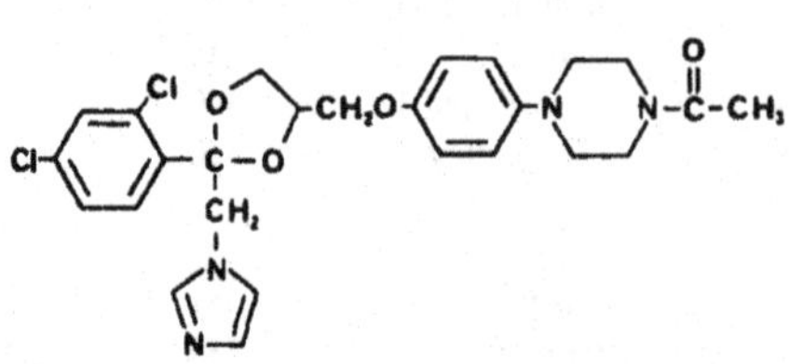

(15) Ketoconazole

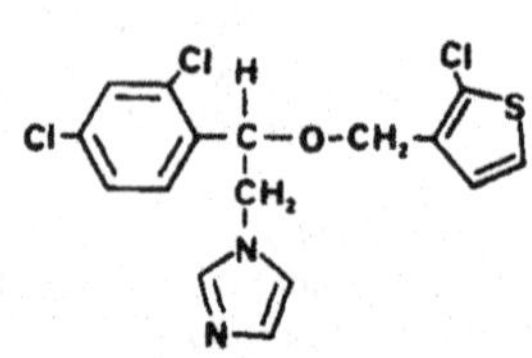

(16) Tioconazole

FIGURE 18. Structures of sterol demethylation inhibitors.

Puzzling are data presented by Berg et al.,[351] who reported a high content of ergosta-8,24(28)-dien-3,6-diol in cultures of *C. albicans*. The authors claimed that the properties of this C-14 demethylated 3,6-diol were identical with 14-methyl ergosta-8,24(28)-dien-3,6-diol described by Ebert et al.,[39] and this important discrepancy with regard to the absence or presence of the 14-methyl group remains to be clarified. Interestingly, Berg and co-

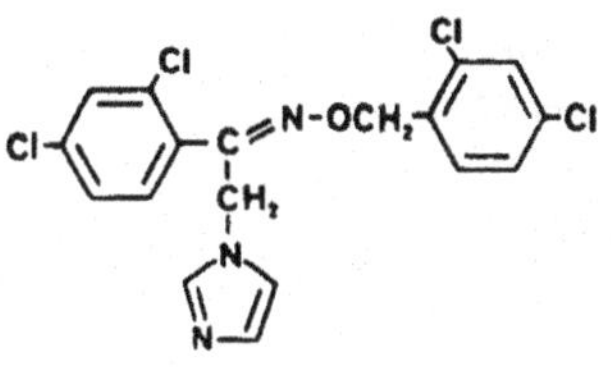

(17) Oxiconazole

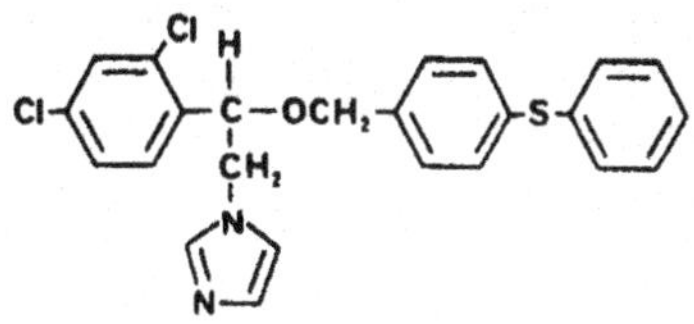

(18) Fenticonazole

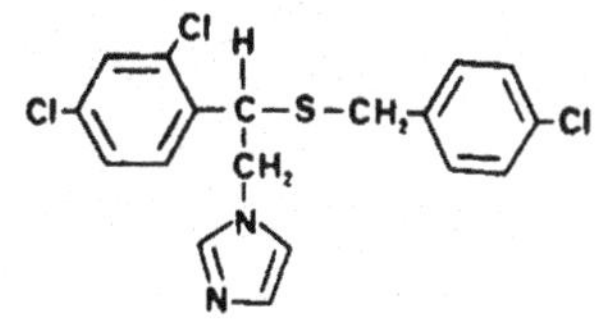

(19) Sulconazole

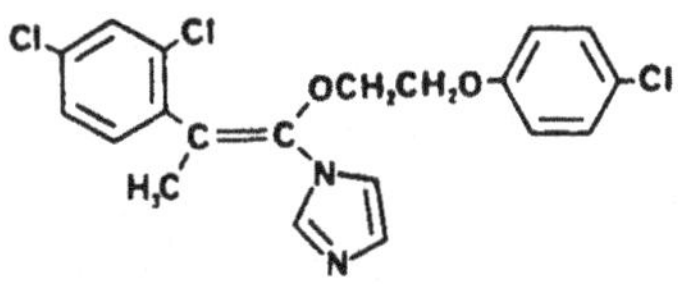

(20) Omoconazole

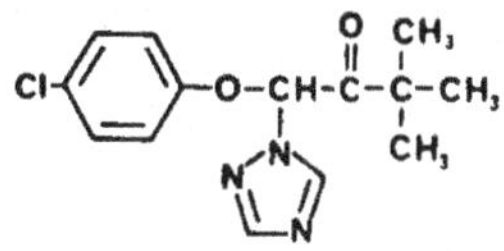

(21) Triadimefon

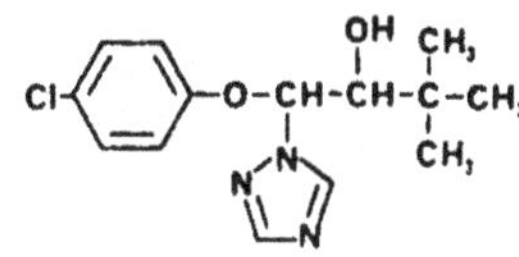

(22) Triadimenol

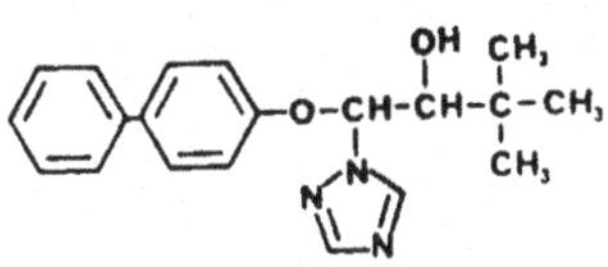

(23) Bitertanol

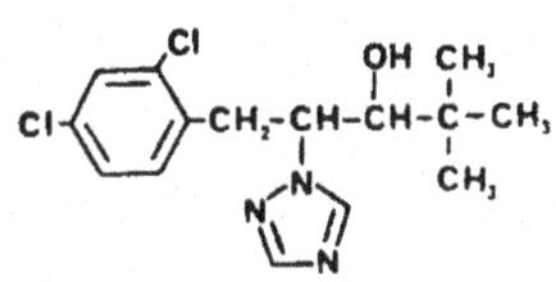

(24) Diclobutrazole

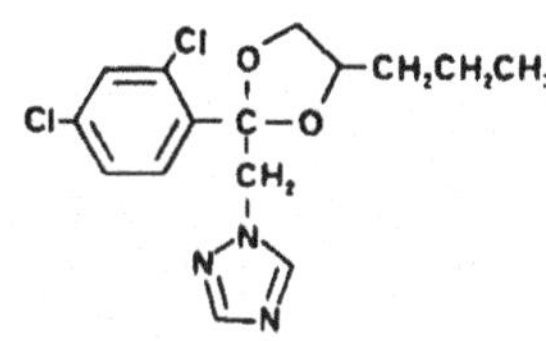

(25) Propiconazole

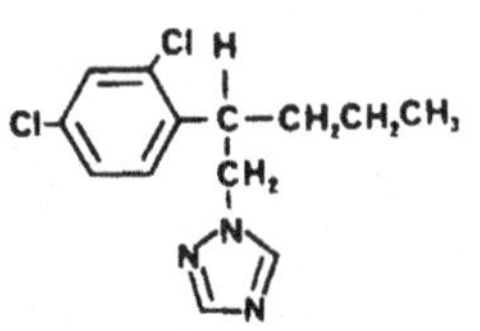

(26) Penconazole

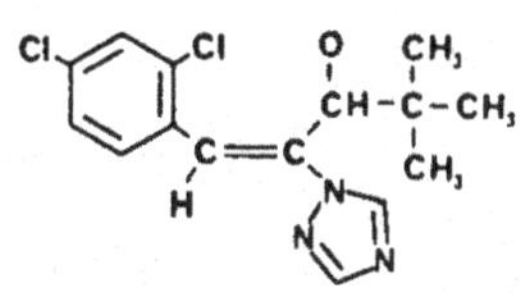

(27) Diniconazole

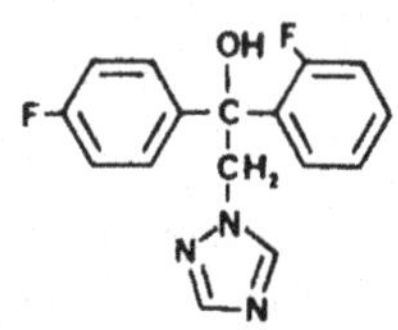

(28) Flutriafol

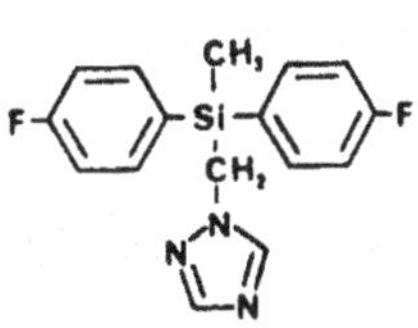

(29) Flusilazole

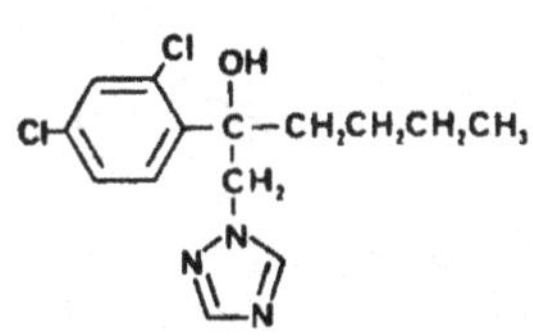

(30) Hexaconazole

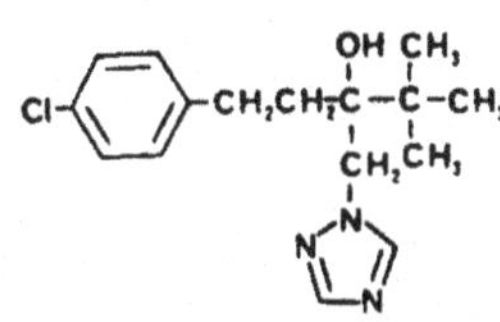

(31) Tebuconazole

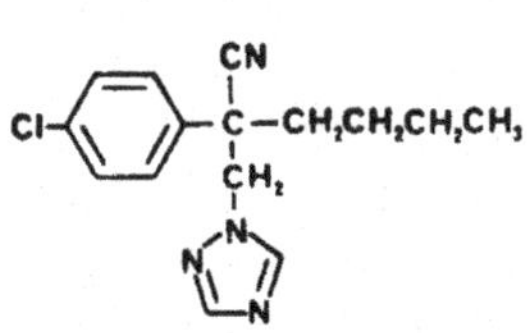

(32) Myclobutanil

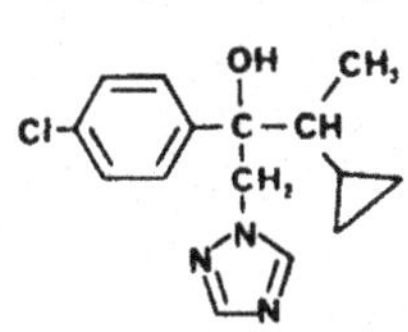

(33) Cyproconazole

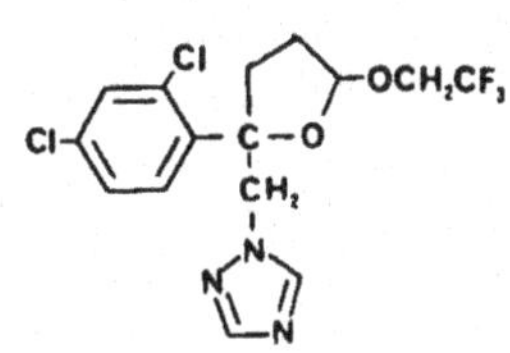

(34) Furconazole (LS 840606)

FIGURE 18 (continued).

workers departed not only at this point from the widely accepted inhibitor scheme presented in Figure 17. In their recent mode of action studies on vibunazole (**36**), tebuconazole (**31**), and fluconazole (**39**),[38,351] $\Delta^{5,8}$ as well as Δ^5 (plant) sterols were reported to accumulate under the action of the DMIs, in addition to the normally reported C-14 methyl sterols. For example, ergosta-5,8-dienol was reported to accumulate in *Trichophyton mentagrophytes* treated with fluconazole, and the level of 24-methylenedihydro-lanosterol, the normally

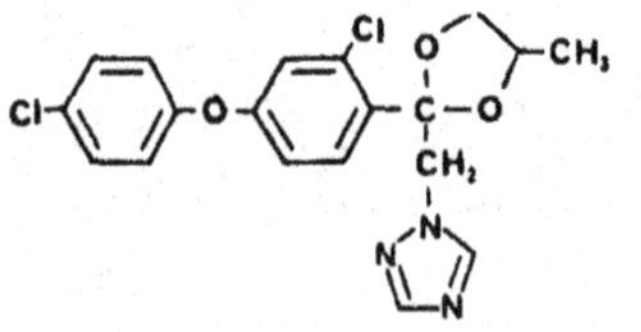

(35) CGA 169274

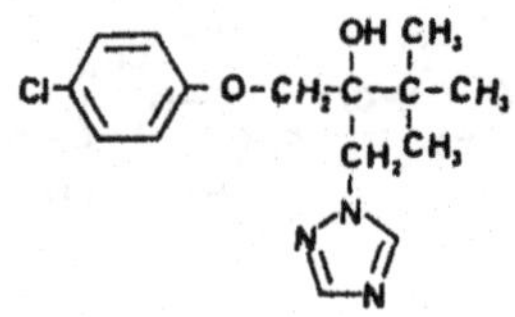

(36) Vibunazole

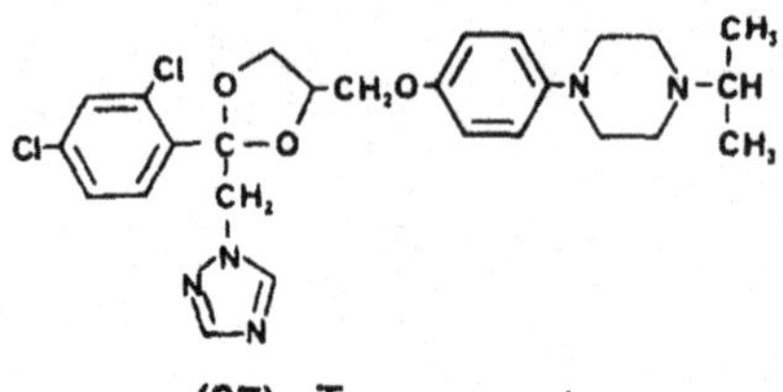

(37) Terconazole

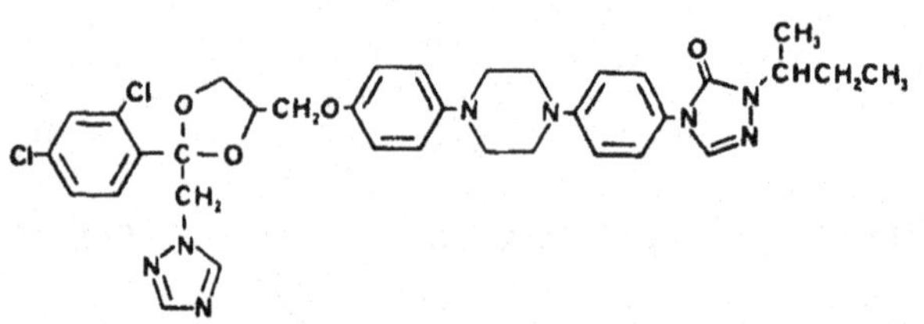

(38) Itraconazole

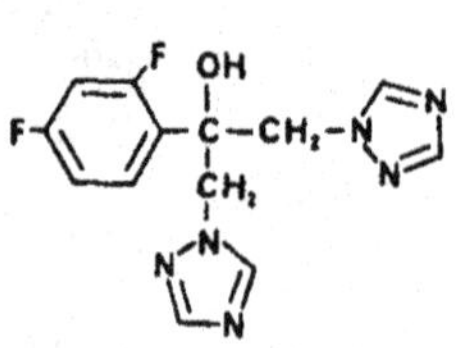

(39) Fluconazole

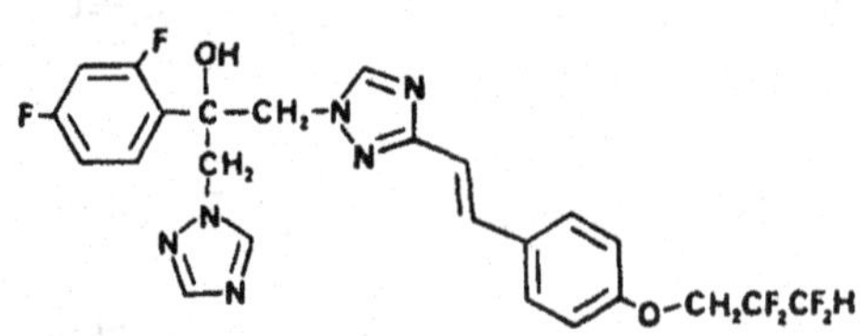

(40) ICI 195-739

FIGURE 18 (continued).

accumulating precursor, was only slightly elevated.[351] In contrast, "normal" precursor accumulation was reported for fluconazole-treated *S. cerevisiae*,[360] and fluconazole inhibited the C-14 demethylase in a cell-free system derived from *C. albicans*.[362] Based on the accumulation of unusual $\Delta^{5,8}$ and Δ^{5} sterols, Berg et al.[351] speculated that the Δ^{8}-Δ^{7}-isomerization might proceed via Δ^{8} reduction, C-7 hydroxylation, and dehydration to yield the Δ^{7}-double bond, a biosynthetic scheme that departs considerably from the widely accepted mechanism of the Δ^{8}-Δ^{7}-isomerization reaction involving a carbocationic high-energy intermediate (see Section II.C.5). Some DMIs such as tebuconazole (**31**) were assumed to interfere with the C-7 hydroxylation, in addition to their activity as 14α-DM inhibitors. Unfortunately, little information on the experimental conditions and on analytical sterol data was described by Berg et al.,[38,351] and these interesting speculations require more studies to gain in validity.

Regardless of some unusual and preliminary findings described above, the vast majority of DMIs follows the scheme presented in Figure 17 and leads to the accumulation of artificial C-14 methylated sterols. An additional effect is the rapid depletion of the ergosterol pool. The time course of events is exemplified in Table 3 for the treatment of *U. maydis* with triarimol.[41] The inhibitory effect, the strong accumulation of 14α-methyl sterols, became

TABLE 3
Time Course of Triarimol Action on Sterols of *Ustilago maydis*

Sterols	(% of nontreated control)			
Time after treatment	**1 h**	**3.5 h**	**5.5 h**	**9.5 h**
Desmethyl sterols	63	26	18	5
14α-Methyl sterols	900	4,500	12,800	1,800
Total sterols	87	94	90	99

Calculated from Ragsdale;[41] *U. maydis* was treated with 2 mg/l triarimol.

apparent as early as 1 h after treatment and continued over the entire period. The second effect, the disappearance of ergosterol, was equally drastic. The effect of triarimol on the reproduction of *U. maydis* was only observed after 3.5 h, a time span required for one cell cycle. At this time, the pool of desmethyl sterols was diminished to 26% of the nontreated control. A similar rapid depletion of ergosterol was observed after the treatment of *U. avenae* with triadimenol.[43] Considering one reproductive cycle over this period of inhibitor action and a complete block of ergosterol synthesis, ergosterol synthesized prior to inhibitor action should comprise roughly 50% of the sterols extracted, provided the pool of ergosterol is stable. However, the content is significantly lower (Table 3), indicating that ergosterol is actively degraded once the C-14 demethylation step is blocked. This ergosterol turnover has not yet received adequate attention, and it remains unknown whether ergosterol degradation represents normal ergosterol turnover, or a turnover increased by the accumulation of 14α-methyl precursors. Furthermore, the dynamics of ergosterol degradation have not yet been investigated. Naturally, the pattern of authentic sterols and sterol precursors is not only determined by the accumulation of 14α-methyl precursors, but also by the turnover of the two sterol pools. The chemical analysis of sterol patterns subsequent to treatment of DMIs provides a static picture but no data on the dynamics of sterol turnover. This metabolic turnover might be different for different fungi and might explain the variable levels of total sterols identified in fungi treated with DMIs. For example, the total sterol content was only slightly affected in *U. maydis* treated with triarimol (Table 3),[41] but it was more than doubled in *Cercospora arachidicola* and *Cercosporidium personatum* treated with propiconazole.[363]

Once the mode of action of DMIs as inhibitors of the fungal 14α-DM had been determined by the analysis of sterol patterns of DMI-treated fungi, the experimental procedures became more sophisticated and advanced. The first analysis of DMI potencies in a cell-free system composed of microsomes derived from *S. cerevisiae*, a NADPH-regenerating system, and either mevalonic acid or lanosterol as substrates of ergosterol synthesis was presented by Kato in 1977.[364] Buthiobate (**4**) and triarimol (**1**) were shown to inhibit the 14α-DM at concentrations around 10 μM. The demethylation in the C-4 position was not inhibited at a tenfold higher concentration, indicating a specific interaction of the inhibitors with the cytochrome P-450-catalyzed demethylation at the C-14 position. The biochemical characterization of fungal 14α-DMs as typical cytochrome P-450 monooxygenase has been described in detail above (see Section II.C.3). Basic amines, but also pyridine and azole derivatives, had long been known to exhibit a strong affinity to other cytochrome P-450 monooxygenases, characterized by a difference spectrum with an absorbance peak around 430 nm and a trough around 390 nm (Type II spectrum).[104,365] The imidazoles miconazole (**9**) and ketoconazole (**15**)[366] and the pyridine buthiobate (**4**)[334] were the first DMIs shown to cause similar spectral changes when incubated with microsomes or a reconstituted 14α-DM derived from *S. cerevisiae*. Buthiobate formed a 1:1 complex with the yeast 14α-DM.[334]

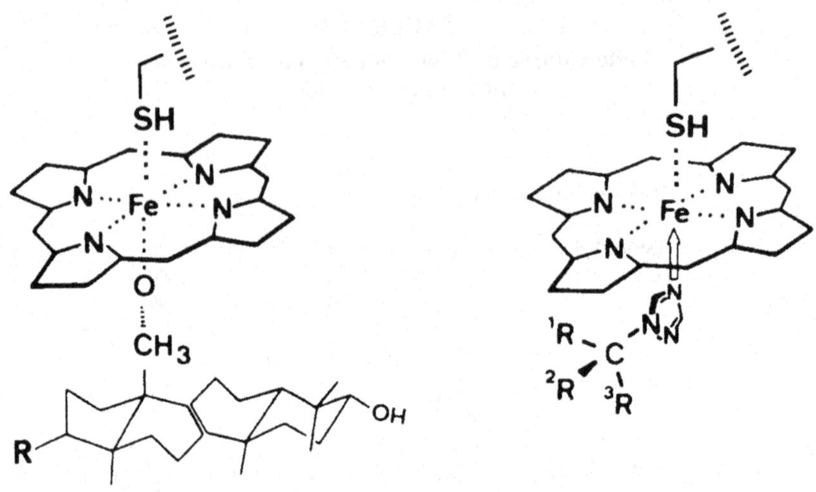

FIGURE 19. Molecular model of DMI action.

This high affinity was also observed for triadimefon (**21**), triadimenol (**22**), ketoconazole (**15**), and itraconazole (**38**).[367,368] Although slight differences between triazoles and imidazoles were observed, the spectral changes caused by the binding of the inhibitors to the 14α-DM were indicative for the interaction of a basic nitrogen with the sixth coordination site of the P450 heme iron. Differences were apparent when the 14α-DM-inhibitro complexes were treated with CO. Whereas buthiobate, triadimefon, and triadimenol were readily displaced by CO, this was not easily achieved with ketoconazole and itraconzaole,[367,368] indicating a tighter fit of the two latter DMIs. These spectroscopical studies have led to a refined model of inhibitor action.[104,324,332] DMIs interact with two different sites at the 14α-DM (Figure 19). The hydrophobic substituents bind to the active site normally occupied by the enzyme substrate lanosterol or 24-methylenedihydrolanosterol. Thereby, the basic nitrogen of the essential heteroaromate is positioned toward the center iron of the porphyrin system and blocks the binding of the oxygen molecule that would normally be activated and transferred to the C-32 carbon of 14α-methyl precursor (see Section II.C.3). Surprising and not in accord with this widely accepted model is the non-competitive nature of 14α-DM inhibiton by ICI 153,066.[368a] The scheme of inhibitor action implies that the first hydroxylation step within the sequence of the C-14 demethylation reaction is prevented. This model is in agreement with the sterol structures extracted from DMI-treated fungi. Only C-14 methyl sterols, but not C-32 hydroxy and carbonyl intermediates, have thus far been identified.

The amazing structural flexibility of DMIs (Figure 18) implies that the lanosterol binding site can be successfully occupied by a variety of structures. Numerous structure-activity studies including sophisticated computer modeling have been published in the past.[336,340,341,343,347,358] Within one series of structures, this approach has yielded valuable information with regard to "optimal inhibitor fits". The class of DMIs as a whole, however, appears more variable, and the structural requirements for a fit into the lanosterol binding site can only be described in general terms. In most cases, two of the substituents are hydrophobic, with at least one halogenated phenyl, and one substituent is frequently distinguished by a hydrophilic group. Apparently, not all of these structures exhibit equally high affinities to the lanosterol binding site. The fact that buthiobate, triadimefon, and triadimenol, but not ketoconazole and itraconazole, could be readily displaced by CO indicates that ketoconazole and itraconazole are more tightly bound.[367,368]

The Type II spectral changes observed upon binding of DMIs to 14α-DM present in

TABLE 4
***In Vitro* Activity of DMIs as Inhibitors of Fungal 14α-DMs**

DMI	*Saccharomyces cerevisiae* I_{50} (n*M*) Cell free	ΔA-titration	*Candida albicans* I_{50} (n*M*) Cell free	ΔA-titration
(**1**) Triarimol	10,000[a]			
(**4**) Buthiobate	300[b]			
(**6**) Imazalil		45[c]		50[c]
(**7**) Prochloraz	10,000[a]			
(**9**) Miconazole	90[d]	86[c]	90[d]	76[c]
			200[e]	
			350[f]	
(**10**) Isoconazole		48[g]		37[g]
(**11**) Econazole		56[g]	100[f]	53[g]
			90[e]	
(**12**) Clotrimazole		32[c]	100[f]	36[c]
			120[e]	
(**13**) Bifonazole		960[c]		1,170[c]
(**14**) Butoconazole			60[e]	
(**15**) Ketoconazole	70[d]	23[c]	70[d]	30[c]
	50[h]		60[f]	
	60[i]		50[e]	
(**16**) Tioconazole	45[g]		60[f]	37[g]
			50[i]	
(**17**) Oxiconazole		41[g]		44[g]
(**19**) Sulconazole		56[g]		53[g]
(**21**) Triadimefon	10,000[a]			
	200[i]			
(**24**) Diclobutrazole	3,000[a]			
(**25**) Propiconazole		68[c]		61[c]
(**26**) Penconazole		46[c]		19[c]
(**27**) Diniconazole[*S*(+)]	70[j]	140[j]		
(**28**) Flutriafol		50[k]		
(**29**) Flusilazole	90[l]			
(**37**) Terconazole		55[g]		41[g]
(**38**) Itraconazole	60[i]	28[g]		31[c]
(**39**) Fluconazole	50[h]	260[c]		249[c]

a Gadher et al.[369]
b Aoyama et al.[334]
c Vanden Bossche et al.[370]
d Willemsens et al.[371]
e Pye and Marriott.[372]
f Marriott.[373]
g Vanden Bossche et al.[330]
h Shaw et al.[362]
i Yoshida and Aoyama.[368]
j Yoshida et al.[374]
k Baldwin et al.[402]
l Trzaskos and Henry.[344]

microsomes from *S. cerevisiae* and *C. albicans* could be applied to spectrophotometrically titrate the enzyme with DMIs and, thus, to determine I_{50} values for inhibitor binding. This method, together with sterol biosynthesis in cell-free systems, has yielded a great number of *in vitro* inhibitor data. A list of I_{50} values determined with a variety of methods is presented in Table 4. Considering the differences of experimental methods, the results are amazingly similar. Only the concentrations reported by Gadher et al.[369] for triarimol, prochloraz,

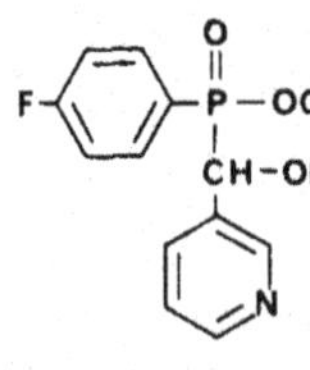

(41)

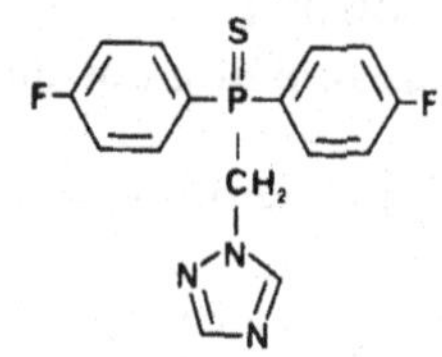

(42)

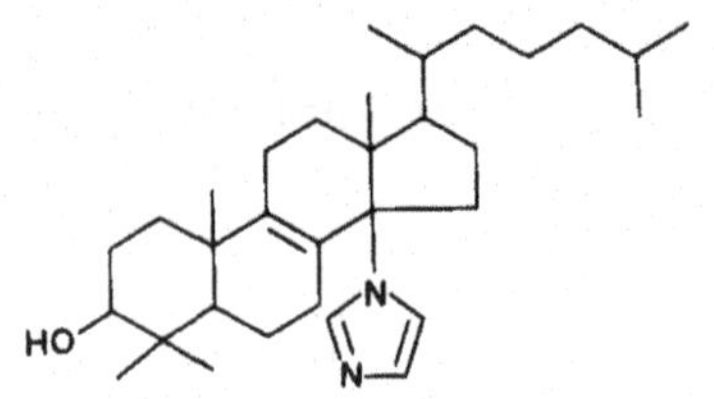

(43)

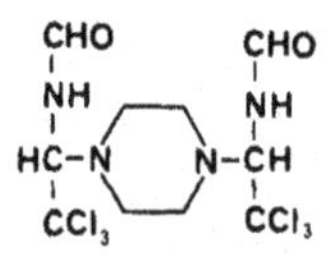

(44) Triforine

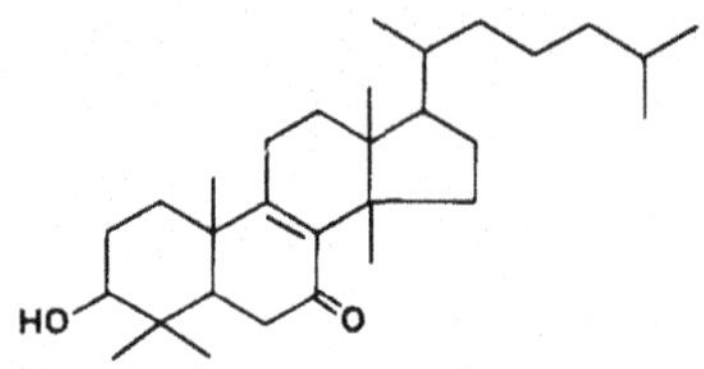

(45)

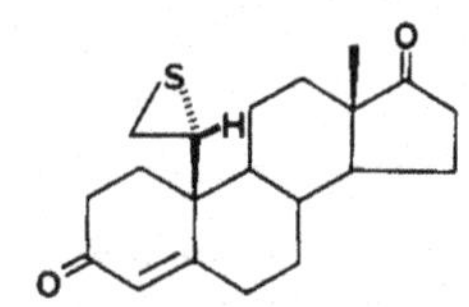

(46)

FIGURE 20. Structures of unusual demethylation inhibitors.

triadimefon, and diclobutrazole appear high when compared to data found by others (Table 4). In general, the I_{50} values for the *in vitro* inhibition of 14α-DMs derived from *S. cerevisiae* and *C. albicans* are centered around 100 n*M*, with little differences between DMIs and the fungal assay system. Bifonazole appears to represent one of the exceptions, with I_{50} values approximately 20 times higher than "normal". The K_m of the yeast 14α-DM for lanosterol as the substrate is 6000 n*M*,[104] indicating that the binding affinity of most DMIs is higher by a factor of approximately 100.

Regardless of the structural flexibility of DMIs, it is surprising that the basic chemistry has not reached a higher level of diversification. From the class of developed DMIs, only flusilazole (**29**) departs from the normal chemistry and contains silicium instead of carbon as the central atom (Figure 18). A modification of this central inhibitor part has also been accomplished with phosphinates and phosphine sulfides (Figure 20). Sasaki et al.[375] reported that phosphinates substituted with pyridine acted as DMIs, with (**41**) being the most efficient inhibitor. This concept was later extended to triazole phosphine sulfides (**42**).[376] Despite the chemical departure from normal DMI structures, the essential groups of silicium and phosphorous containing DMIs remained heteroaromatic. This restriction was even conserved with lanosterol analogues modified at the C-32 position (**43**) (Figure 20).[377] Among several lanosterol derivatives, the imidazole-modified structure was found to be the most effective DMI and fungal growth inhibitor. However, activity was also pronounced with free and substituted amino groups.[377] The potential suitability of a nonaromatic nitrogen to inhibit the 14α-DM has been recognized before. Triforine (**44**), a piperazine derivative in use as a fungicide in agriculture, was one of the early examples for a DMI mode of action (Figure 20).[378] An interesting departure from both the chemistry of DMIs and the precise interaction with the 14α-DM was described for the lanosterol analogue 7-oxo-24,25-dihydrolanosterol (**45**) (Figure 20).[379] Different from the coordination of basic groups with the porphyrin iron, the inhibitor apparently blocked the flow of electrons to the cytochrome oxyferro intermediate.

The restriction of essential groups to pyridines, pyrimidines, and azoles for the vast majority of DMIs is surprising, since the chemistry of known cytochrome P-450 inhibitors is considerably more diverse.[365] Interesting with regard to the antifungal DMIs is a comparison with some known aromatase inhibitors. The 14α-DM and the aromatase are mechanistically almost identical enzymes (Figure 8). Both are cytochrome P-450 monooxygenases, and both demethylate at a sterol ring structure via successive hydroxylations and elimination of formic acid (see Section II.3.C). Only the position of the methyl group is different: C-14 for the 14α-DM and C-19 for the mammalian aromatase. It is thus not surprising that some of the DMI antifungal agents containing heteroaromates as essential groups act as potent aromatase inhibitors. It even has been possible to largely increase the aromatase specificity by slight chemical modifications of existing DMIs.[380] This aspect is described in detail in Chapter 9. The opposite situation, the suitability of structurally modified aromatase inhibitors to act as DMIs, has not yet been reported. For the group of aromatase inhibitors, the chemistry of essential groups is much more diverse and includes residues such as acetylene, nitrils, sulfhydryl, and azides in the C-10 position of androgens.[380] One of the most potent inhibitors, 10-thiirane substituted androgens (**46**) (Figure 20), showed competitive aromatase inhibition with a K_i of 1 nM.[381,382] Difference spectra were similar to 14α-DMs interacting with DMIs and indicated that the sulfur atom of the thiiram was bound to the heme iron.[382] It remains to be determined how successful an adaptation of this chemistry could be for the design of novel inhibitors of fungal 14α-DMs.

Such design efforts would be greatly facilitated by the knowledge of the precise active site geometry of fungal 14α-DMs. The only P450 thus far crystallized and structurally analyzed is cytochrome P-450$_{cam}$, an enzyme responsible for the hydroxylation of camphor in the soil bacterium *Pseudomona putida*.[383,384] In prokaryotes, P450s are soluble enzymes and not membrane-bound like all P450s in eukaryotes. Although the P450$_{cam}$ shares sequence homologies with several eukaryotic P450s,[385] and although the genes of fungal 14α-DMs have been isolated (see Section II.C.3), a similarly refined three-dimensional structural analysis would only be possible if the fungal 14α-DM gene could be modified toward the expression as soluble enzyme in a transformed microorganism. This engineered enzyme should remain unchanged with regard to the binding of lanosterol and DMIs. Until this goal has not been accomplished, the design of novel DMIs will have to rely on semiempirical approaches.

2. Stereochemistry and Activity

Regardless of the apparent flexibility of DMIs, the fit into the active site of fungal 14α-DMs is greatly influenced by stereochemical constraints. An examination of the DMI structures presented in Figure 18 reveals that almost all DMIs are chiral compounds with one or more asymmetric carbon centers. A stereochemical discrimination was first observed with triadimefon (**21**), a chiral compound that undergoes biological (and chemical) reduction to the four isomers of triadimenol (**22**) (Figure 18).[386-388] The stereochemical impact on the activity of DMIs has been described in a recent review,[389] and the discussion will be restricted to a brief summary and more recent findings. Two factors have emerged as a complication of stereochemical activity discriminations. Azole and pyrimidine compounds very similar to DMIs have been developed as plant growth regulators (PGRs) rather than fungicides. The primary mode of action of these PGRs is the inhibition of *ent*-kaurene oxidation and, thus, gibberellin biosynthesis.[389-392] However, not only PGRs but also DMIs such as triadimenol exert effects on plant growth and development, and increasing evidence suggests that C-14 demethylation of obtusifoliol rather than the inhibition of gibberellin synthesis is the primary reason for this effect.[393-397] This aspect will be described in greater detail in Chapter 8.

Interesting with regard to stereochemical considerations is a comparison of triadimenol (**22**), diclobutrazole (**24**), and paclobutrazole. The latter azole lacks the 2-chlorine present in dichlobutrazole and was developed as a PGR.[389] It has, in addition, fungicidal activity.[395] The *S,R*-enatiomer of triadimenol is the most potent DMI, followed by the *R,R*-form.[389,395] This ranking is entirely different for dichlobutrazole and paclobutrazole, with a ranking of *R,R* > *S,S*.[389,395] This example demonstrates that the optimal fit into the active site of fungal (and plant) 14α-DMs is substantially affected by stereochemical implications, and that slight chemical modifications can have drastic effects on the stereochemical discrimination. The stereochemical dissimilarity of the most potent enantiomers of triadimenol and paclobutrazole suggests that the slight change from –O– to $-CH_2-$ reverses the most optimal positioning of the two hydrophobic substituents within the lanosterol binding site.[389] It should be mentioned that the enantiomer of paclobutrazole most active as an inhibitor of *ent*-kaurene oxidation is *S,S*, and thus opposite to the most active DMI enantiomer.[394,395]

The dioxolano-azoles ketoconazole (**15**), propiconazole (**25**), CGA 169374 (**35**), terconazole (**37**), and itraconazole (**38**) are mixtures of four isomers.[389] A stereochemical discrimination of fungicidal activities had first been reported for etaconazole, a derivative closely related to propiconazole.[389] This study has been extended to propiconazole with similar results. The 2*S*-isomers were the most active inhibitors with little difference between the 2*S*,4*S*- and the 2*S*,4*R*-isomers.[398] Considerable information has also been compiled for the vinylazole diniconazole (**27**), and the closely related PGR uniconazole.[389] The compounds exist as two geometrical isomers, and activity was highly restricted to the *E*-isomer.[399] These *E*-isomers contain, in addition, one asymmetric center, and the fungicidal activity was considerably higher for the *R*-configuration.[399] The same selectivity was observed in cell-free studies with the 14α-DM prepared from *S. cerevisiae*.[374] The inhibitory action and the inhibitor binding were clearly more pronounced for the *R*-enantiomer of diniconazole. The relation between lanosterol and the lowest energy conformer of this enantiomer was recently described by Katagi.[341] The opposite enantiomer with *S*-configuration was responsible for the PGR activity of both diniconazole and uniconazole.[389,400] Identical stereochemical constraints have been reported for the triapenthenol, a PGR closely related to uniconazole and diniconazole.[391] The *Z*-isomer of the vinyl-imidazole omoconazole (**20**) (*E*-configuration with regard to the dichlorophenyl and imidazole residues) was found to be more active as an antifungal agent, and the toxicity to mice was lower compared to the less active *E*-isomer.[340]

Surprisingly little information is currently available on the impact of stereochemical configuration on activities of ethylazoles with their mostly two isomers.[389] The *S*(–)-isomer of tebuconazole (**31**) and the (–) isomer of vibunazole (**36**) were reported to exert considerably higher levels of antifungal activities.[38,351,352] Similar results were reported for hexaconazole (**30**).[400a]

The numerous examples indicate that stereochemical activity discrimination is, most likely, a widespread phenomenon among the DMIs. The apparent flexibility of the lanosterol binding site allowing to accommodate structurally diverse DMIs is superimposed by stereochemical restrictions. Commercial DMIs are composed of mixtures of two or more isomers, with one being considerably more active. This discrimination would call for stereospecific synthetic routes, which could result in lower application rates and reduced side effects in both plants and mammals. An alternative improvement would be the reduction of stereochemical discrimination via structural modifications toward an improved fit of all isomers. This goal appears feasible. All four isomers of cyproconazole (**33**) were more or less equally effective,[401] and a similar lack of discriminatory binding to fungal 14α-DM has been reported for the two isomers of flutriafol (**28**),[402] although the (–) isomer was more active *in vivo*.[400a]

3. Secondary Modes of Action

The activity of DMIs acting as powerful inhibitors of P450 systems is clearly not restricted to fungal 14α-DMIs. Many of the DMIs have been identified as inhibitors of various cytochrome P-450 monooxygenases in mammals, and the systematic application of these findings has led to more specific DMIs, but also to novel applications of similar inhibitor structures in medicine. This aspect is described in great detail in Chapter 9. In fungal organisms, P450s other than 14α-DM are not or only slightly affected. The Δ^{22}-side-chain desaturation might be an exception. Ragsdale[41] found after treatment of *U. maydis* with triarimol indications for a secondary effect on the Δ^{22}-desaturase, like the 14α-DM, a cytochrome P-450 monooxygenase (see Section II.C.3). Similar results have been reported for diniconazole.[399] To the contrary, buthiobate (**4**) was not active as an inhibitor of this desaturation when tested in a cell-free system derived from yeast.[334] Cell-free assays with a larger variety of DMIs are required to decide whether this secondary effect is based on direct enzyme inhibition or on factors related to precursor accumulation and growth inhibition.

Side effects on fungal enzymes other than P450s have been reported with increasing frequency. For example, the interference of clotrimazole (**12**), triadimenfon (**21**), and bifonazole (**13**) with fungal HMG-CoA reductases has been discussed as a secondary mode of action.[304,338] However, the mechanisms of enzyme inhibition were distinctly different. HMG-CoA reductase appeared to become down-regulated as a consequence of lanosterol accumulation under the action of clotrimazole and triadimefon. In contrast, bifonazole had a clear direct inhibitory action on fungal HMG-CoA reductase. Unfortunately, inhibitor kinetics have not yet been presented, and it remains unknown whether bifonazole acts as an active site-directed inhibitor or interferes with HMG-CoA reductase by binding to a regulatory site.

Bifonazole was also shown to cause the irregular deposition of chitin in cell walls of *Candida albicans*.[403] This phenomenon of disturbed chitin deposition was also shown for *Ustilago maydis* and *Penicillium italicum* treated with imazalil,[404] and for *Aspergillus* spp. treated with itraconazole.[405] However, bifonazole had no pronounced and direct effects on the activity of chitin synthase,[406] and the reason for the disturbance of chitin deposition remains largely unknown. Explanations for DMI-mediated cell wall alterations might derive from the recent identification of two different chitin synthases in *Saccharomyces cerevisiae*, encoded by separate genes. Studies aided by molecular biology techniques have revealed that one isozyme is responsible for septum formation and cell division. A repair function has been assigned to the second and more abundant isozyme.[407] Ultrastructural studies with budding yeasts such as *S. cerevisiae*,[408] *C. albicans*,[408,409] *Ustilago* spp.,[404,410] and *Taphrina deformans*[411] treated with DMIs strongly suggest a functional imbalance between these potential isozymes. Treated cells cause the formation of chains and clusters of interconnected cells. The budding process is strongly impaired, and both septa and peripheral cell walls become thicker and structurally altered. Cells frequently have incomplete septa, or septa are completely lacking. Moreover, the release of protoplasts from DMI-treated *Candida* cells was greatly impaired.[412] Similar alterations at the cell wall have been reported for DMI-treated mycelial fungi.[404,405,409,413,414] As discussed by Sancholle et al.,[411] strong circumstantial evidence suggests that these pronounced effects on the cell wall formation are a consequence of 14α-methyl sterol accumulation and thus indirect. Apparently, the regulatory role of ergosterol on the intracellular targeting and activity of chitin synthase(s) is severely altered when ergosterol is replaced. A more detailed understanding of these regulatory mechanisms at the level of cell wall synthesis might lead to antifungal compounds that interfere directly rather than indirectly with these events crucial for fungal growth and proliferation.

In addition to cell wall alterations, an indirect effect of DMIs on the composition of fatty acids has been frequently reported and discussed.[322-327,332] Major changes comprise an increased pool of free fatty acids and a higher degree of unsaturation in both the pool of free and phospholipid fatty acids. It is unlikely that DMIs interfere directly with the respective enzymes.[415] Enzymes active in fatty acid metabolism are more likely modulated by the replacement of ergosterol. Alternatively, the organism could also react in an adaptive and active response to counteract the changing sterol environment. Such adaptive responses have been reported.[178]

The secondary effects of DMIs described above are most likely a consequence of inhibitor action rather than direct effects, and several alternative modes of DMI action suggested in the past might well be explained by the exchange of ergosterol by 14α-methyl precursors. For example, the sterol composition has been shown to be critical for mitochondrial functions in yeast,[416,417] and this requirement could be an explanation for decreased ATP levels in *C. albicans* treated with DMIs.[418] As discussed by Vanden Bossche,[332] the direct interaction of DMIs with fungal 14α-DM and changes in sterol composition has also been related to an unbalanced generation and metabolism of destructive active oxygen. Indeed, lipid peroxidation has been suggested as a side effect of DMI action.[419] The peroxidation of phospholipids and, thus, the destruction of biological membranes as a secondary effect could also be the reason for membrane damages as indicated, for example, by the release of K^+ ions from DMI-treated *C. albicans*.[420]

An alternative cause for membrane damages is the direct interaction of some DMIs with membranes. These direct effects are dependent on the chemical structure of particular DMIs and require relatively high dosages. For example, Sud and Feingold found that anaerobically grown *S. cerevisiae* devoid of sterol synthesis could be inhibited by miconazole (**9**) and clotrimazole (**12**), but not ketoconazole (**15**). However, MIC values were higher by a factor of 32 for clotrimazole (**12**), but not ketoconazole (**15**). However, MIC values were higher by a factor of 32 for clotrimazole, 130 for miconazole, and >500 for ketoconazole when compared to aerobically grown yeast.[421,422] The differential influence of miconazole and ketoconazole on membranes, as indicated by these different biological activities, was confirmed in detailed physicochemical investigations.[423,424] Direct membrane damage was shown to result in leakage of ions and small molecules,[415,425] and to inhibitory effects of membrane-bound enzymes. Examples for the latter effect are mitochondrial and plasma membrane ATPases.[423,426,428] Since resistance of *S. cerevisiae* to miconazole was reflected in resistance of mitochondrial targets,[427] a direct rather than a DMI mode of action of miconazole appears to be operative at least in this case. Direct effects on membrane properties are somewhat similar to the mode of action of macrolide polyenes described above and have been widely investigated for the antimycotic imidazole DMIs.[219,316,324] Similar membrane-damaging effects of triazole DMIs used in agriculture have been described for propiconazole (**25**) and penconazole (**26**) when applied at high concentrations.[429]

4. Mechanism of Growth Inhibition and Fungal Specificity

Little doubt exists that all DMIs are efficient inhibitors of fungal 14α-DMs. However, the direct membrane-damaging effects described above have been related to the fungicidal and thus cell-killing activity of imidazole antimycotics.[219,316,324] Effects on membranes are rapid, and they are also responsible for the activity of some of the DMIs on Gram-positive bacteria.[430] Many of the imidazole DMIs are in use as topical agents for the treatment of fungal skin infections, and doses high enough to achieve fungicidal effects are likely to occur under these conditions. Thus, it is not surprising that in overviews on antimycotic DMIs, the fungicidal effect is often emphasized over the 14α-DM inhibitory activity as the primary mode of action.[219,316] Direct and fungicidal effects have been described for miconazole (**9**) and many of the structurally similar DMIs such as isoconazole (**10**), econazole (**11**), tioconazole (**16**), fenticonazole (**18**), sulconazole (**19**).[316] However, the high inhibitor concentrations necessary to exert direct and membrane-damaging effects are unlikely to be reached for orally applied antimycotic drugs such as ketoconazole (**15**), itraconazole (**38**), or fluconazole (**39**). Direct membrane-damaging effects would not necessarily warrant fungal specificity and would lead to high mammalian toxicity. As described above, ketoconazole was clearly different from miconazole with regard to direct and fungicidal effects on membranes, and the DMI activity has to be regarded as the antifungal primary target site. Interesting is the comparison of various *in vitro* inhibitor data. While ketoconazole's I_{50} values determined for the 14α-DM from yeast and *C. albicans* were around 0.05 μM (Table 3), 100 μM was required to inhibit NADH oxidation in mitochondria from *C. albicans* and succinate oxidation by rat liver mitochondria.[431] The inhibition of 14α-DMs is also the most likely primary effect of DMIs used in agriculture. Most of these fungicides are systemic compounds,[33,34] and direct interference with membranes would lead to pronounced phytotoxic side effects.

As described above, the primary DMI action leads to the accumulation of 14α-methyl sterol precursors and the depletion of the desmethyl sterol pool. The precursors are incorporated into the plasma membrane and eventually replace ergosterol.[432] The primary action on sterol synthesis stops fungal growth and reproduction, but this effect is fungistatic in character and thus different from the fungicidal action observed after direct membrane damage at high concentrations. A good example is results reported for a DMI-resistant strain of *U. avenae*.[43] At concentrations fully inhibitory to the sensitive parent strain, proliferation of the resistant phenotype was also inhibited during an initial response. Similar to the sensitive strain, ergosterol was almost completely replaced by 14α-methyl sterols within 6 h. However, the inhibitory action was only transient. Growth resumed after 12 h, while the precursor content declined and ergosterol levels increased. Thus, the viability and potential for full recovery from inhibitor action were not affected even after the almost complete saturation of membranes with 14α-methyl sterols for several hours.

The inhibition of fungal 14α-DMs and, consequently, the replacement of sterols are clearly the primary event of DMI action. How this event is translated into the curtailment of fungal growth and reproduction is less well understood. As described earlier (see Section II.E), 14α-methyl sterols are less suited to serve as "bulk" sterols in membranes. In model studies with liposomes, the replacement of ergosterol with lanosterol was shown to reduce membrane fluidity,[320] and this reduced fluidity might also be the reason for the strong and regular aggregation of membrane particles (proteins) observed after treatment of *U. avenae* with triadimenol.[433,434] Furthermore, leakage of glucose from membrane vesicles was increased when ergosterol was replaced by lanosterol,[324] and Sancholle et al.[435] concluded that the release of phosphate from cells of *Taphrina deformans* treated with propiconazole (**25**) was caused by the inhibitory action of the DMI rather than by direct membrane damage. Accumulating sterol precursors not only affect the permeability of membranes, but also the active transport of nutrients such as amino acids.[436] Some indication exists that the indirect effect of DMI action on fatty acid saturation is an important contribution to growth inhibition. For example, oleic acid antagonized the inhibition of *C. albicans* growth by several imidazole DMIs but not the inhibition of ergosterol synthesis.[415]

Drastic alterations of intracellular membrane structures and, ultimately, membrane destructions have been reported for DMI-treated fungal cells. For example, Hippe[410] observed after treatment of *U. avenae* with triadimefon (**21**) and nuarimol (**3**) for 6 h an increased number of structurally altered mitochondria, and an increased abundance of endoplasmic reticulum, vacuoles, and lipid bodies. Extensive membrane lysis and destruction after 16 h of treatment indicated fungitoxic rather than fungistatic effects. These data are not always easy to interpret, since the DMI concentrations applied are often extremely high and unbalanced with regard to fungistatic effects. They could indicate membrane damage caused directly by the DMI or, as described above, by the peroxidative destruction of phospholipids as a secondary effect on the balance of active oxygen. Numerous 14α-DM-deficient mutants of various fungi have been described to be viable (see Section II.E), and the mere replacement of ergosterol by 14α-methyl sterols cannot be regarded lethal and destructive. However, the growth rates of most of these mutants are severely hampered. Interesting is the comparison of a 14α-DM-deficient mutant of *U. maydis* described by Walsh and Sisler[437] with the respective wild-type strain. Ergosterol was completely absent in the mutant, and the sporidial doubling time was reduced from 2.5 to 6.3 h. Thus, the total replacement of ergosterol by 14α-methyl sterols was neither sufficient to abolish viability nor to completely prevent reproduction. The most drastic effect was morphological in character. Sporidia were multicellular, highly branched and irregular in shape, and identical with wild-type sporidia treated with fenarimol. While the slower growth might be explained by the decreased suitability of 14α-methyl sterols in "bulk" membrane functions, the drastic effect on septa formation and budding of cells is more likely explained by the broad regulatory functions of ergosterol receiving increasing attention (see Section II.E.2). This would imply that the depletion of the ergosterol pool as a consequence of DMI action is at least as important as the accumulation of 14α-methyl sterols.

Similar regulatory functions of ergosterol in the control of fungal cell differentiation are most likely involved in the demonstrated effects of DMIs on developmental processes. For example, many of the azole DMIs have the potential to inhibit the conversion of the yeast phase to pseudo or true hyphae in *C. albicans* at concentrations low enough to suggest a specific rather than fungitoxic interference.[438-442] Studies with a polyene-resistant *C. albicans* mutant unable to form hyphae and wild-type strains treated with various concentrations of clotrimazole demonstrated that the inhibition of morphogenesis was clearly related to the increased levels of 14α-methyl sterols.[443] Although not entirely free of conflicts for all pathogens, morphogenetic transformations are regarded important in disease initiation by medically important pathogens.[444,445] Other effects of DMIs on host-pathogen interactions have been reported. For example, ketoconazole was by far the most effective DMI blocking the binding of the steroid hormone corticosterone (Figure 11) to a high-affinity binding protein present in *C. albicans*.[204] Effects of DMIs on crucial events in host-pathogen interactions as a consequence of their primary action might explain the sometimes remarkable discrepancies between *in vitro* and *in vivo* activities reported for some of the DMIs developed as antifungal drugs.[324,332,338,446]

Effects of DMIs on fungal differentiation have also been reported for plant pathogenic fungi. In general, the initial steps of host plant invasion (spore germination, appressoria formation, and surface penetration) are not prevented by DMIs.[447] The lack of activity has often been explained by the presence of endogenous storage sterols utilized during these early developmental steps. This might be correct in many cases, but spores of *Botrytis cinerea* and *Venturia inaequalis* were shown to synthesize sterols immediately after the onset of germination.[448] Furthermore, germ tube elongation of *V. inaequalis* conidia was shown to be sensitive to the DMI flusilazole (**29**), although the sensitivity drastically increased during the course of germination.[189] Interestingly, flusilazole also induced the formation of appresoria, a crucial step of early cell differentiation during the course of plant infection.[189] A second crucial developmental step in the disease cycle of plant pathogenic fungi is the initiation of sporulation. Imazalil was shown to act as a powerful antisporulant of *Penicillium italicum* at a concentration not sufficient to inhibit vegetative growth.[449]

Some of the side effects reported for DMIs appear to reflect stress conditions rather than specific consequences of DMI action. For example, the adherence of *C. albicans* to epithelial cells was not only inhibited by sublethal concentrations of miconazole but also by 5-fluorocytosine.[450] Likewise, plant host reactions resembling typical responses of resistance cultivars were not only noticed after treatment with DMIs,[451] but also with several nonrelated fungicides.[452] Clearly, studies concerned with side effects of DMIs should be comparative and should include inhibitors with target sites other than sterol biosynthesis in order to differentiate DMI-specific and more general stress-related effects. Regardless of these constraints, the interference of DMIs with regulatory functions of sterols deserves more attention in the future. Although the current state of knowledge with regard to regulatory sterol functions (see Section II.E) is still rather limited, more research efforts in this field might lead to novel and specific targets in the control of fungal diseases, and DMIs (and other sterol biosynthesis inhibitors) could serve as valuable investigative tools.

The examples discussed above indicate that both the modification of "bulk" membrane properties by accumulating 14α-methyl sterols and the disturbance of regulatory sterol function by ergosterol depletion are responsible for the fungistatic inhibition of fungal growth and development. Considering that the intrinsic activity of DMIs as inhibitors of 14α-DMs from *S. cerevisiae* and *C. albicans* is, with a few exceptions, very similar (Table 4), the *in vivo* activity should also be expected to be comparable. However, this is not always the case. DMIs are long known to act relatively discriminatory with respect to sterol-producing fungi.[324,326]

First attempts to explain the molecular mechanisms responsible for these discriminatory *in vivo* activities have been presented for ketoconazole (**15**) and itraconazole (**38**).[355,453-455] Both DMIs have almost identical *in vitro* activities as inhibitors of 14α-DMs (Table 4). For itraconazole, this intrinsic activity was closely reflected in the *in vivo* activity toward *C. lusitaniae* and *Aspergillus fumigatus*. In contrast, ketoconazole was 100 times (*A. fumigatus*) or even 1000-fold (*C. lusitaniae*) less active. Several hypotheses to explain this discrepancy have been tested and discussed by Vanden Bossche et al.[355] The authors noted that the effects of experimental conditions such as growth media and length of treatment on ergosterol biosynthesis was much more pronounced for ketoconazole than for itraconazole. Uptake was higher for itraconazole than for ketoconazole, but physicochemical data and calculations suggested that itraconazole might be more prone to a plasma membrane localization. Similarly, a triazole DMI (ICI 153066) related to flutriafol (**28**) was shown to be more permeable to the mycelial rather than the yeast form of *C. albicans*,[456] a growth response largely influenced by the media conditions. Overall, however, differences with regard to uptake levels were rejected as a full explanation for the differences between ketoconazole and itraconazole.[355] Furthermore, drastically different rates of inhibitor detoxification were not responsible for discriminatory activities.

The chemical difference between the imidazole ketoconazole and the triazole itraconazole appears to comprise a more likely explanation. The pK_a values of imidazole DMIs range from 6.3 to 6.7[330] and determine a very sensitive protonation status within the physiological pH range. In contrast, triazoles with their pK_a of <2 are unprotonated under all physiological conditions. Considering the molecular model of 14α-DM inhibition (Figure 19), it is not surprising that the protonated form a imidazole DMI was a much weaker inhibitor when tested with microsomes from *S. cerevisiae*. While the I_{50} values of ketoconazole varied from 200 n*M* at pH 5 to 30 m*M* at pH 7.5, the corresponding values for itraconazole were low and stable over the entire pH range.[355] Measurements of intracellular pH values in fungal cultures under various growth conditions led Vanden Bossche et al. to the conclusion that this parameter contributes substantially to the differential *in vivo* activities observed for the two DMIs.

Interesting with regard to the discriminatory activity of DMIs is a comparison of *in vitro* data derived from cell-free assays (*S. cerevisiae* microsomes) with *in vivo* ED_{50} values obtained with a wild-type isolate of *U. avenae*, a basidiomycete not related to yeast but very sensitive to DMIs.[457] For most of the DMIs tested, the correlation between the two sensitivity values is unexpectedly high, with *in vivo* sensitivities of *U. avenae* being approximately twofold lower than the yeast I_{50} values would suggest (Table 5). Exceptions are imazalil and itraconazole with considerably lower *in vivo* activities toward *U. avenae*, and propiconazole and diniconazole, where the *in vivo* activities are higher than the intrinsic activities determined with the yeast 14α-DM. Both the good correlations and the departures are not related to the imidazole or triazole nature of the inhibitors. Several explanations are possible. The selectivity of particular DMIs could be based on a differential interaction of the inhibitor with the 14α-DMs present in *S. cerevisiae* and *U. avenae*. Unfortunately, cell-free enzyme assays have thus far only been developed for *S. cerevisiae* and *Candida* species. However, *in vivo* and *in vitro* data derived from three *Candida* spp. indicate that drastic differences with regard to the binding affinities of DMIs to fungal 14α-DMs might be involved in selective activities. The I_{50} values for miconazole (**9**) as determined with cell-

TABLE 5
Comparison of *In Vitro (S. cerevisiae)* and *In Vivo (U. avenae)* Activities of Imidazole and Triazole DMIs

DMI	*S. cerevisiae* I_{50} (n*M*)[a]	*U. avenae* ED_{50} (n*M*)	ED_{50}/I_{50}
(**6**) Imazalil (imidazole)	45	265	5.9
(**38**) Itraconazole (triazole)	28	103	3.6
(**12**) Clotrimazole (imidazole)	32	88	2.8
(**37**) Terconazole (triazole)	55	125	2.3
(**15**) Ketoconazole (imidazole)	23	50	2.1
(**26**) Penconazole (triazole)	46	88	1.9
(**9**) Miconazole (imidazole)	86	144	1.7
(**10**) Isoconazole (imidazole)	48	70	1.5
(**13**) Bifonazole (imidazole)	960	1320	1.4
(**25**) Propiconazole (triazole)	68	20	0.2
(**27**) Diniconazole (triazole)	140	11	0.08

[a] Data are taken from Table 4 (ΔA-titration).
[b] Calculated from Köller and Wubben.[457]

free systems derived from *C. krusei*, *C. albicans*, and *C. pseudotropicalis* were 1900, 200, and 26 n*M*, relating to MIC values of 12.5, 12.5, and 0.05 μg/ml.[372] The *in vivo* difference between *C. krusei* and *C. pseudotropicalis* reflected the differences observed *in vitro*, but the *C. albicans* data indicate that other factors must be involved.

These additional selectivity factors could be related to the influence of other microsomal proteins on the DMI sensitivity of 14α-DMs operative in living cells. For example, the *in vitro* I_{50} values determined for ketoconazole and itraconazole with isolated microsomes from *C. albicans* were both 30 n*M* (Table 4). When the same fungus was preincubated with the inhibitors for 4 h and the effect on the short-term incorporation (1 h) of acetate into desmethyl and 14α-methyl sterols was analyzed, I_{50} values for the inhibition of ergosterol synthesis were 4 and 0.55 n*M* for ketoconazole and itraconazole, respectively.[355] Thus, the target appeared more sensitive *in situ* than *in vitro*. When *C. albicans* was incubated for 24 h in stand cultures in the presence of tioconazole and ^{14}C-acetate, 50% ergosterol biosynthesis was achieved at 21 n*M* and, thus, at a concentration close to the corresponding *in vitro* activity (Table 4). However, the inhibition of both ergosterol synthesis and fungal growth trailed off at about 60% and stayed constant up to a 1000-fold higher concentration.[457a] Full inhibition of growth and sterol synthesis was only achieved at fungicidal concentrations. Although the incorporation of radioactivity into the 3,6-diol pool was not determined in this study, the target response did not follow saturation kinetics.

Similar trailing effects have never been reported for dose-response curves obtained in cell-free systems (Table 4). The discrepancies between *in vitro* and *in situ* inhibition of the 14α-DM indicate that other regulatory processes might be involved in determining the intracellular binding affinities of DMIs to 14α-DM, and some evidence for a more complex *in vivo* situation does exist. For example, Trzaskos and Henry found that the binding constant for flusilazole determined with purified rat liver 14α-DM was tenfold lower than for the enzyme localized in intact microsomes.[344] The authors concluded that binding of the DMI to other microsomal P450s isozymes "protected" the 14α-DM from inhibition. Alternatively, DMI binding affinities could also be influenced by the protein-protein interaction with redox partners. For example, the disruption of the gene of the yeast NADPH cytochrome P-450 reductase, which is thought to serve as the main electron delivery system in the P450 cycle (see Section II.C.3), had no effect on the capacity of *S. cerevisiae* to synthesize ergosterol,[130] indicating that alternative redox partners must exist. The gene disruption, however, made the transformant 200-fold more sensitive to ketoconazole.[130] Although the alternative electron transferring component remains unknown, the example demonstrates that changes apart from the 14α-DM can affect the binding affinities of DMIs. For the rat liver 14α-DM, the involvement of cytochrome b_5 in electron transfer was not apparent from reconstitution experiments, but it might be involved in intact microsomes.[111,112] Indeed, the presence of cytochrome b_5 has been reported to affect the K_m and V_{max} of other microsomal cytochrome P-450 monooxygenases.[458] Similarly, the complex of cytochrome P-450_{scc} with its redox partners in mitocondrial membranes was sensitively affected by the amounts of available substrate.[459] These examples indicate that DMI binding affinities are not only influenced by the intrinsic properties of the fungal 14α-DM itself, but also by protein-protein interactions with other membrane components involved in the demethylation cycle.

Insensitivity or "natural tolerance" of fungi to DMIs is not necessarily related to the *in vivo* response of the respective 14α-DMs. Weete and Wise[460] investigated three ergosterol-producing fungi with drastically different sensitivities to propiconazole (**25**), *Mucor rouxii, Rhizopus stolonifer,* and *Aspergillus ochraceus.* The ED_{50} values for these three pathogens were 80, 8, and 1 μg/ml, respectively. Regardless of the insensitive growth response of *M. rouxii*, the 14α-DM was relatively sensitive, and both ergosterol depletion and accumulation of 14α-methyl precursors were observed at propiconazole concentrations too low to achieve appreciable effects on growth inhibition. In contrast to the sensitive *A. ochraceus*, propiconazole had no significant effect on the accumulation of free fatty acids, the increase in linoleic acid, and the incorporation of acetate into triglycerides in *M. rouxii* and *R. stolonifer*. Only with *M. rouxii* propiconazole imposed an increased incorporation of acetate into the pool of sterol esters. The authors concluded that the mechanism of tolerance experienced with *M. rouxii* most likely resides in a sterol requirement less sensitive to structural alterations of sterols. The results might also indicate that the accumulation of free fatty acids comprises a substantial role in the mechanism of fungal growth inhibition, as suggested by Georgopapadaku et al.[415] All examples described above indicate that the intrinsic activities of DMIs as inhibitors of fungal 14α-DMs are by no means a good indicator for the sensitivity of fungal organisms. Considering I_{50} values in the range of 50 to 100 n*M* (Table 4), the *in vivo* activities should be expected to be centered around 0.015 to 0.030 μg/ml. Such low sensitivities have been observed (see Table 5), but they are considerably higher for most fungi investigated. The intracellular pH as a determinant of imidazole binding affinities, the interaction of fungal 14α-DMs with microsomal redox partners involved in the demethylation cycle, and the functional role of sterols in their interplay with other lipids can contribute to the *in vivo* activities of these inhibitors. More intensive studies on the molecular understanding of factors involved in mandating the *in vivo* sensitivities of fungi could lead to DMIs with an increased spectrum of antifungal activities. They also could facilitate the elucidation of the mechanisms involved in fungal resistance to DMIs.

5. Mechanism of Resistance

The development of practical resistance to DMIs is, thus far, different for compounds used in agriculture and medicine. In agriculture, practical resistance has developed for several powdery mildews, but not yet for various other plant diseases controlled with DMIs.[461-463] The development of practical resistance is related to a broad and continuous range of DMI sensitivities found in wild-type populations of plant pathogens. For example, the sensitivity distribution of *Venturia inaequalis* isolates to flusilazole (**32**) was log-normal in shape, and ED_{50} values ranged from 0.8 to 170 ng/ml.[464] Under the selection force of a DMI fungicide, the frequency of less sensitive isolates will increase over time, and the levels of disease control might decrease.[461-463]

The distribution of sensitivities in unexposed wild-type populations of human pathogens has received less attention in the past. Sensitivities to itraconazole were determined for a large number of *Candida* isolates.[465] Unfortunately, the sensitivity distribution was only presented as a whole for the different *Candida* spp. tested in this study, and the range of sensitivities (MIC values from 0.04 to 25 μg/ml) represents both the variation within the population of one species and the species selectivity of itraconazole. Thus, it is not surprising that the sensitivity distribution appears discontinuous in character. Interestingly, an identical presentation of data yielded a log-normal distribution for amphotericin B (MIC values from 0.19 to 6.25 μg/ml), indicating that the activity of this drug is less species-specific. The most extensive data base on the sensitivity variation of human pathogens to itraconazole (**38**) has been described by van Cutsem et al.[466] and was summarized by Grant and Clissold.[467] The data clearly indicate that the itraconazole sensitivities of individual isolates was greatly variable for a large number of different pathogens. For example, the MIC values for over 1000 clinical isolates of *C. albicans* varied from 0.001 to 10 μg/ml; two isolates required 100 μg/ml for complete growth inhibition. Considering these broad DMI sensitivities, little doubt exists that population responses reported for plant pathogens could also occur with human pathogens.

Thus far, reports on DMI-resistant fungal phenotypes that resisted a therapy with DMIs are rare. This promising experience might be related to the restriction of most of the previous DMIs to topical applications at high fungicidal concentrations. As discussed above, the fungicidal mode of action is different from the fungistatic action and less specific with regard to target sites. However, it is not easy to predict whether resistance will never develop to DMIs in their fungitoxic mode. This would depend on the individual sensitivities of the directly affected target sites and, thus, on the contribution of particular targets to the overall fungicidal activity. As shown for *S. cerevisiae* with mitochondrial functions sensitive to miconazole, alterations of this target site could confer resistance.[427] Also interesting are findings with a miconazole-resistant isolate of *C. glabrata*.[468] Resistance was reflected in the effect of miconazole on oxygen consumption of intact cells but not when mitochondrial functions were tested *in vitro*. This discrepancy could be caused by a reduced uptake of the inhibitor into resistant cells. Indeed, the inhibition of ergosterol synthesis was much more sensitive in protoplasts than in intact cells, indicating that the cell wall might serve as a penetration barrier. The restriction of uptake levels would lead to resistance expression, regardless of the number and nature of target sites affected by the inhibitor. Whether such mechanisms of resistance could become important under practical conditions depends entirely on the pathogenicity of respective mutants, a criterion not yet investigated.

More likely is the development of resistance to DMI antimycotics applied in their fungistatic mode for the treatment of deep and systemic fungal diseases, in particular in cases of long-term or even prophylactic therapies of immunocompromised patients. This would be similar to the development of resistance of plant pathogens challenged by DMIs systemically translocated within plants. It is not surprising that DMI-resistant phenotypes of *C. albicans* have been isolated from patiens relapsing after long-term treatment with ketoconazole (**15**).[245,330,469] Whether the selection of resistant phenotypes capable of resisting chemotherapy will impose more widespread problems in the future will depend on the pathogenic fitness of resistant phenotypes, but also on the genetics and dynamics of fungal populations. In theory, this applies to both fields of DMI use, and since the epidemiology of plant pathogens is much better understood than for human pathogens, a closer cooperation among mycologists engaged in either of the fields would be helpful to provide predictive tools for the development of DMI resistance under clinical conditions.

For plant pathogenic fungi, DMI resistance has been correlated with impaired pathogenic fitness, although several exceptions have been reported.[333,470-472] Similarly, a DMI-resistant phenotype of *C. albicans* was shown to be less pathogenic than sensitive strains.[469] However, it remains to be determined whether this reduced fitness is always a consequence of resistance. Moreover, the genetics of resistance remains diverse, ranging from the presence of one resistance gene to polygenic control.[333,470] The understanding of the molecular mechanism(s) of fungal resistance to this important group of fungicides and antimycotic drugs would be of great value for the prediction of pathogenic fitness parameters and, thus, resistance risks, and for the design of countermeasures and management strategies. Unfortunately, they remain far from clear. The biochemistry of DMI resistance has been studied with both laboratory mutants and phenotypes isolated from plants or patients under ketokonazole treatment.[245,330,333] Several models have been suggested in the past, and these models and findings will be discussed in relation to the mechanisms contributing to the "natural tolerance" of fungi to DMIs, as described above.

a. Reduced Uptake of DMIs

The best investigated molecular model of DMI resistance originated from studies with mutants of *Aspergillus nidulans* and *Penicillium italicum*. Uptake of fenarimol and various triazole DMIs into sensitive strains proceeded fast.[473,474] After 10 min of incubation, the intracellular inhibitor concentrations in a sensitive isolate of *P. italicum* was 4 to 100 times higher than the concentration in the medium, depending on the respective DMI.[475] However, the initial accumulation was only transient and was followed by a rapid decline of cell-bound inhibitors.[473-475] The initial phase of rapid uptake was completely lacking in DMI-resistant mutants, and levels of uptake remained low from the beginning. In both cases, however, inhibitors of respiration and thus ATP synthesis lead to rapid uptake of the inhibitors.[473,474] These results have been discussed as evidence for an energy-dependent efflux system, which is induced in sensitive isolates upon inhibitor treatment, but resides as a constitutive system in resistant isolates. In sensitive but nonresistant isolates the target site is thought to be saturated during the initial phase of high inhibitor concentrations.[473,474] Although the effect of inhibitors of ATP production was not studied, similar uptake patterns were observed with sensitive and resistant isolates of *C. albicans* and the triazole DMI ICI 153066.[436] Orthovanadate as an inhibitor of the plasma membrane ATPase was shown to act as an inhibitor of DMI efflux, indicating a role of this proton pump in the molecular mechanism of active inhibitor excretion.[476] This aspect, however, remains largely unexplored. For a highly resistant isolate of *P. italicum*, the differences in initial uptake levels are not entirely suited to explain the levels of resistance.[475] For example, resistance factors of 42 for bitertanol and 54 for fenarimol were reflected by three to four times higher intracellular inhibitor concentrations during the initial phase of incubation.

Interesting with regard to the biphasic uptake of pyrimidine and triazole DMIs into sensitive isolates is a comparison with the imidazole imazalil (**6**). Siegel and Solel[477] reported an uptake pattern similar to fenarimol into a sensitive and resistant strain of *A. nidulans*. However, when a loosely bound proportion of imazalil was removed, the uptake differences were considerably less pronounced. Similarly, no uptake difference of imazalil and the lack of initial DMI accumulation was observed for *P. italicum* isolates.[475] Imazalil exerted, in addition, by far the lowest level of resistance. However, preincubation of a sensitive strain with imazalil prevented the initial phase of fenarimol influx, and this result was discussed as evidence that imazalil was capable of inducing the active efflux system.[478] As discussed above, the difference between triazoles and imidazoles is primarily the pK_s and, thus, the status of protonation at physiological pH values. Indeed, the uptake of the imidazole ketoconazole (**15**) was much more affected by the pH than that of the triazole itraconazole (**38**),[355] and similar pH dependencies have been reported for imazalil.[479]

Uptake of DMIs is normally measured as the amount of inhibitor retained by fungal cells subsequent to a series of washings. This procedure does not allow conclusions with regard to the cellular inhibitor localization. They could be bound to the cell wall, retained within the plasma membrane, or bound to intracellular structures including the target site. In particular, the amount of DMI bound to the cell wall can be substantial. In exponentially growing cultures of *C. albicans*, 75% of miconazole (**9**) was found in the cell wall, whereas 60% was localized in the microsomal fraction during the stationary phase.[480] Moreover, the cell wall was discussed as a penetration barrier restricting the uptake of miconazole into a DMI-resistant strain of *C. glabrata*.[468] DMI uptake was also shown to be affected by the lipid composition of fungal cells. Hitchcock et al.[456] related a higher phospholipid/free sterol ratio in mycelial cultures of *C. albicans* with increased uptake of a triazole DMI. A similar relationship was observed with a set of sensitive and DMI-resistant isolates grown in the yeast form, with resistant strains showing a lower phospholipid/sterol ratio and reduced uptake.[481] Moreover, a mutant of *C. albicans* with high contents of 14α-methyl sterols was almost impermeable to the DMI.[245,246] Boiron et al.[482] suggested that the uptake of ketoconazole (**15**) into *C. albicans* was an active and energy-requiring process at low inhibitor concentrations. In contrast to other systems, metabolic inhibitors did decrease rather than increase the amount of the DMI taken up at these low concentrations.

In summary, a restricted uptake of DMIs into resistant fungal cells has been related to a constitutive efflux system not present in sensitive strains, to differences with regard to sterol contents, and to cell walls acting as penetration barriers. Limited information is still available on the nature of putative DMI transport systems and the cellular localization of inhibitors retained by fungal cells. With respect to a reduced inhibitor uptake as a mechanism of resistance, only the proportion of inhibitor that actually reaches the target site is of interest. In particular, the binding of DMIs to fungal cell walls can be substantial, and more work with protoplasts is required to elucidate the precise nature of DMI uptake and efflux mechanisms.

b. Removal of DMIs from the Target Site

Hippe[483] employed electron microscopic autoradiography to determine the intracellular localization of triadimenol within strains of *U. avenae* sensitive and resistant to this DMI. The content of triadimenol was lower in sensitive than in resistant cells. More distinct, however, was a pronounced accumulation of the inhibitor in vacuoles of resistant cells, accompanied by a decreased cytoplasmatic concentration. This difference was discussed as evidence for the removal of the inhibitor from the target site via segregation into vacuoles. Unfortunately, the cells had been pretreated with triadimenol for 15 h prior to the incubation with the radioactive label. The triadimenol concentration chosen was inhibitory to the sensitive but not the resistant strain and, consequently, the membrane sterols at the onset of incubation with the radioactive inhibitor were altered in sensitive but not resistant cells. Considering the effects of 14α-sterols on membrane properties including potential effects on transport systems and membrane permeability, the observed differences with regard to intracellular inhibitor localization might reflect the differences in membrane composition rather than the differential sensitivity to DMIs. Theoretically, the relief of the target from a DMI could also be accomplished by an enhanced inhibitor degradation. Thus far, however, no indications exist that DMI resistance can be based on this option.

c. Modifications of Sterol Synthesis and Requirements

Numerous mutants of yeasts and dimorphic fungi growing in the yeast form have been reported to remain vital albeit a defective 14α-DM and, thus, high levels of 14α-methyl sterols. Two of these mutants, *U. maydis*[437] and *C. albicans*,[246] have been described above. It is not surprising that the lack of an active target site of DMIs rendered high levels of DMI resistance. The structure and properties of a mutated 14α-DM ($P450_{SG1}$) and the interaction of the defect enzyme with DMIs have been investigated for the nystatin-resistant yeast mutant *S. cerevisiae* SG1.[121,484,485] A single nucleotide change from Gly-310 to Asp-310 led to drastic changes in the heme environment and substrate binding site. Spectral properties and computer modeling indicated a histidine residue bound to the sixth coordination site of the heme iron. The mutated enzyme showed no significant interaction with lanosterol and, thus, no 14α-DM activity.

The DMIs ketoconazole (**15**) and diniconazole (**27**) but not buthiobate (**4**) were bound to the mutated enzyme. However, the binding characteristics of diniconazole were clearly different from the native 14α-DM and thus substantiated the structural modification of the lanosterol binding site. Thus far, the viability of 14α-DM mutants has only been documented for yeasts. An attempt to obtain similar mutants for mycelial fungi was unsuccessful,[183] indicating that the mutation might be lethal under these conditions. Considering these severe penalties it appears highly unlikely that the loss of a functional DMI target site would comprise a mechanism of DMI resistance under practical conditions. Only an active site modification that would render a relatively unimpaired binding of the substrate, but a substantially decreased affinity to DMIs, could lead to vital and resistant mutants.

A mechanism of DMI resistance related to a mutation in sterol metabolism remote form the 14α-DM has been suggested for *S. cerevisiae*. Watson et al.[360] found that a mutant deficient in Δ^5-desaturation was resistant to fluconazole (**39**). Resistance was not related to a decreased sensitivity of the target site. The mutant accumulated even higher levels of 14α-methyl precursors when treated at a sublethal fluconazole concentration. 14-Methyl fecosterol was the main sterol found in the treated mutant, whereas high levels of 14α-methyl-ergosta-8,24(28)-dien-3,6-diol accumulated in the wild-type strain. The authors concluded that the 3,6-diol had more severe membrane-damaging effects, and that 14-methyl fecosterol was much more suited to provide a functional sterol. Circumstantial evidence indicates that this mechanism of resistance is rare and might be restricted to *S. cerevisiae*. The replacement of $\Delta^{5,7}$- by Δ^7-sterols as an indication for a mutation of the Δ^5-desaturase has not yet been reported for other DMI-resistant fungi. Resistant isolates of *U. avenae*,[42,43] *U. maydis*,[486] *C. albicans*,[245] *P. italicum*,[487] *Botrytis cinerea*,[486] *Cercospora beticola*,[471] and *Cladosporium cucumerinum*[488] all contained $\Delta^{5,7}$-sterols as their main sterol.

An interesting structural change of sterols has been reported for laboratory mutants of *Cercospora beticola*.[471] All but 1 out of 25 DMI-resistant isolates produced ergosterol and ergosta-5,22-dienol, whereas ergosta-5,7-dienol comprised the main sterol in the sensitive parent strain. The accumulation of 14-methylenedihydrolanosterol and obtusifoliol under the action of DMIs was not affected by this change in sterol structure, but the presence of active Δ^{22}-desaturation in resistant isolates might indicate a different membrane composition that compensates for effectively for accumulating precursors. Interesting with regard to this change in number and position of double bonds is the ranking of the sparking function of sterols in *S. cerevisiae*. While the Δ^5-double bond was essential for sparking, the Δ^{22}-double bond was found more efficient than the Δ^7-unsaturation.[172] Thus, the structural change of *C. beticola* sterols might indicate a regulatory rather than a structural adaptation. This interesting aspect deserves further investigation.

In theory, the mechanisms of "natural insensitivity" could also be operative as a mechanism of resistance. The most pronounced difference of the "naturally insensitive" *Mucor rouxii* treated with propiconazole was, as described above, the lack of free fatty acid accumulation and an increased incorporation of sterols into sterol esters.[460] However, sterol precursors accumulated to high levels at propiconazole concentrations far less than those required for growth inhibition. This lack of correlation between growth inhibition and inhibition of sterol biosynthesis was indicative for a less restricted role of sterols in cell functions. Thus far, little evidence exists that a decreased functional demand on sterol structures has been an explanation for resistance. Although differences between sensitive and resistant isolates with regard to the total lipid composition have been reported,[245,488] growth of resistant phenotypes remained sensitive to precursor accumulation and ergosterol depletion. Only higher concentrations of DMIs were required to achieve sterol precursor accumulation in DMI-resistant strains of *U. avenae*,[457] *Candida albicans*,[436] *P. italicum*,[487] and *Cladosporium cucumerinum*.[488]

d. Target Site Modifications

Changes in fungal 14α-DMs affecting the binding of lanosterol and DMIs and, thus, the sensitivity of the target do not necessarily lead to a nonfunctional enzyme as described above for DMI-resistant yeast mutants deficient in ergosterol synthesis. For example, a higher copy number of expressed 14α-DM genes would lead to a higher enzyme content and, thus, to higher DMI doses required to achieve a certain level of inhibition. Indeed, this gene-dosage effect led to ketoconazole resistance of *S. cerevisiae* transformants and has been utilized as a strategy to identify and isolate the 14α-DM gene.[489] Similarly, the expression of the 14α-DM gene from *Candida tropicalis* in *S. cerevisiae* resulted in a transformant less sensitive to ketoconazole.[125] The relation between the number of gene copies and levels of resistance has not yet been determined, and it might be speculated that unpenalized resistance factors achieved via gene dosage effects might not be high. To the contrary, the disruption of the *S. cerevisiae* gene for NADPH-cytochrome P-450 reductase generated a transformant 200 times more sensitive to ketoconazole without any impairment of ergosterol synthesis.[130] This sensitivity change indicated that alternatives must exist to provide the reduction equivalents required during the demethylation cycle of lanosterol, and that the interaction of the 14α-DM with redox partners has, most likely, impact on the sensitivity to DMIs. In the reported case the transformant became more sensitive, but the opposite scenario, a change toward resistance mediated by the interaction of 14α-DM with alternative redox partners, might also be feasible.

The possibility of target site changes leading to resistance, either through increased copy numbers or through structural modifications without impact on substrate binding, has not been widely tested. The sensitivity of 14α-DMs isolated from three DMI-resistant isolates of *C. albicans* varied slightly when tested in cell-free assays, but the variation did not correlate with resistance.[436,490] No indication for a reduced sensitivity of the DMI target was also observed for a resistant laboratory isolate of *U. avenae*. Here, the expression of resistance was shown to be induced rather than constitutive.[43] During the initial phase of growth inhibition, 14α-methyl sterols accumulated similar to the sensitive counterpart, indicating a sensitive target under normal growth conditions. Pulse-chase experiments with lipid precursors revealed that the target recovered from the inhibitor action while growth resumed. This process required protein synthesis,[491] indicating that either the DMI was actively removed from the target, or that the 14α-DM became less sensitive as a consequence of newly synthesized proteins.

A reduced affinity of DMIs to cytochrome P-450 was reported for a resistant but greatly impaired strain of *C. albicans* (B41628) isolated from a patient with muscocutaneous candidosis who relapsed after treatment with ketoconazole.[469] The resistance factors determined for miconazole (**9**), ketoconazole (**15**), and itraconazole (**38**) were 50, 25, and 5, respectively.[469] These resistance factors were reflected in cell-free inhibitor binding tests with microsomal fractions. The I_{50} values of P450 binding determined for the resistant and sensitive isolate differed by a factor of 69 for miconazole, 17 for ketoconazole, and 3 for itraconazole.[361] A factor of 10 was determined for fluconazole.[492] This result indicated that resistance was caused by a reduction of inhibitor binding. The CO difference spectrum of the reduced P450 isolated from the resistant isolate showed a slightly different maximum absorbance and might reflect a structural change.[361] Interesting is the observation that the reduction of DMI binding was diminished when the isolate was subcultured in drug-free medium.[492] It would be interesting to test whether the decreased DMI binding is induced and dependent on protein synthesis similar to the *U. avenae* isolate described above.

e. Conclusion

In summary, a decreased uptake of inhibitor mediated by either a constitutive efflux system or an altered lipid or cell wall composition, the segregation of the inhibitor into vacuoles, the adaptation of sterol structures toward reduced membrane disturbing effects, and a reduced binding of the inhibitor to the target site have been suggested as mechanisms of resistance. This variety of models might indicate that several options exist for a fungus to acquire DMI resistance. The variety of options might be reflected by the genetic control of DMI resistance, ranging from a single gene over polygenic control. Since many of the molecular models are still in an infant stage of experimental refinement, this conclusion appears premature. Clearly, more work is required before more conclusive and elaborate models will emerge.

C. Δ^{14}-REDUCTION AND Δ^{8}-Δ^{7}-ISOMERIZATION

The C-14 demethylation of lanosterol or 24-methylenedihydrolanosterol is followed by the reduction of the Δ^{14}-double bond and the Δ^{8}-Δ^{7}-isomerization (see Section II.C.5). Inhibitors for both of these enzymatic steps have been characterized and described in the past. Thus far, only one group, the morpholine fungicides, has been developed to commercial antifungal products. The clarification of the morpholine mode of action indicated that the Δ^{14}-double bond reduction and the Δ^{8}-Δ^{7}-isomerization are mechanistically very similar and thus prone to similar inhibitor principles. Therefore, both enzymes have to be discussed as closely related targets for antifungal inhibitors.

1. Mode of Action of Morpholines

Dodemorph and the closely related tridemorph (Figure 21) are the first representatives of antifungal *N*-substituted morpholine derivatives developed for the control of plant diseases, primarily powdery mildews.[35] These compounds were introduced in 1965 and 1969, respectively. Fenpropimorph and fenpropidin (Figure 21), both agricultural fungicides with an improved spectrum of disease control, were introduced during the early 1980s.[33,493] Although fenpropidin departs from the morpholine structure, it clearly belongs to this group of antifungal compounds. Amorolfine (Figure 21) is closely related to fenpropimorph, but has improved *in vitro* activity against human pathogens in addition to its activity against plant diseases. It therefore represents the first morpholine derivative developed for clinical use.[494-496] The mode of action of morpholines has been outlined in several recent review articles and thus will be summarized only.[497-500,500a]

Since dodemorph and tridemorph were introduced a decade earlier than the phenylpropyl morpholines, most of the initial mode of action work was done with these inhibitors. From 1972 to 1980, several modes of action such as the direct interference with fungal membranes and the inhibition of protein synthesis or respiration, but also impact on lipid and sterol biosynthesis were suggested.[497-499] Kato et al.,[501] investigating *Botrytis cinerea* treated with tridemorph, found a strong accumulation of the Δ^{8}-sterols fecosterol, ergosta-8,22,24(28)-trienol, and ergosta-8,22-dienol (site B in Figure 22) and proposed the Δ^{8}-Δ^{7}-isomerization step as the primary target for this morpholine derivative. To the contrary, Kerkenaar et al.[502] found ignosterol and some other less pronounced Δ^{14}-sterols to accumulate in response to tridemorph treatment of *U. maydis* and suggested the Δ^{14}-reductase (site A in Figure 22) rather than the Δ^{8}-Δ^{7}-isomerase as the major target site for morpholine action. Support for the Δ^{8}-Δ^{7}-isomerization as the primary inhibitor site came from work with *Saprolegnia ferax, B. cinerea, P. expansum,* and *S. cerevisiae*.[503,504] Others working with *P. italicum*[505] and *Pyricularia oryzae*[337] were in support of the Δ^{14}-reductase as the main morpholine target. This apparent discrepancy was resolved by Baloch et al.[40] in their comparative study with *U. maydis* and *S. cerevisiae*. Employing tridemorph, fenpropimorph, and fenpropidin as sterol biosynthesis inhibitors, a careful chemical analysis of accumulating sterols revealed that both enzymes were targets for morpholine action, although the relative ranking was different. While the activity of tridemorph was largely restricted to the Δ^{8}-Δ^{7}-isomerase, fenpropimorph and fenpropidin were found to inhibit the Δ^{14}-reductase in addition to the

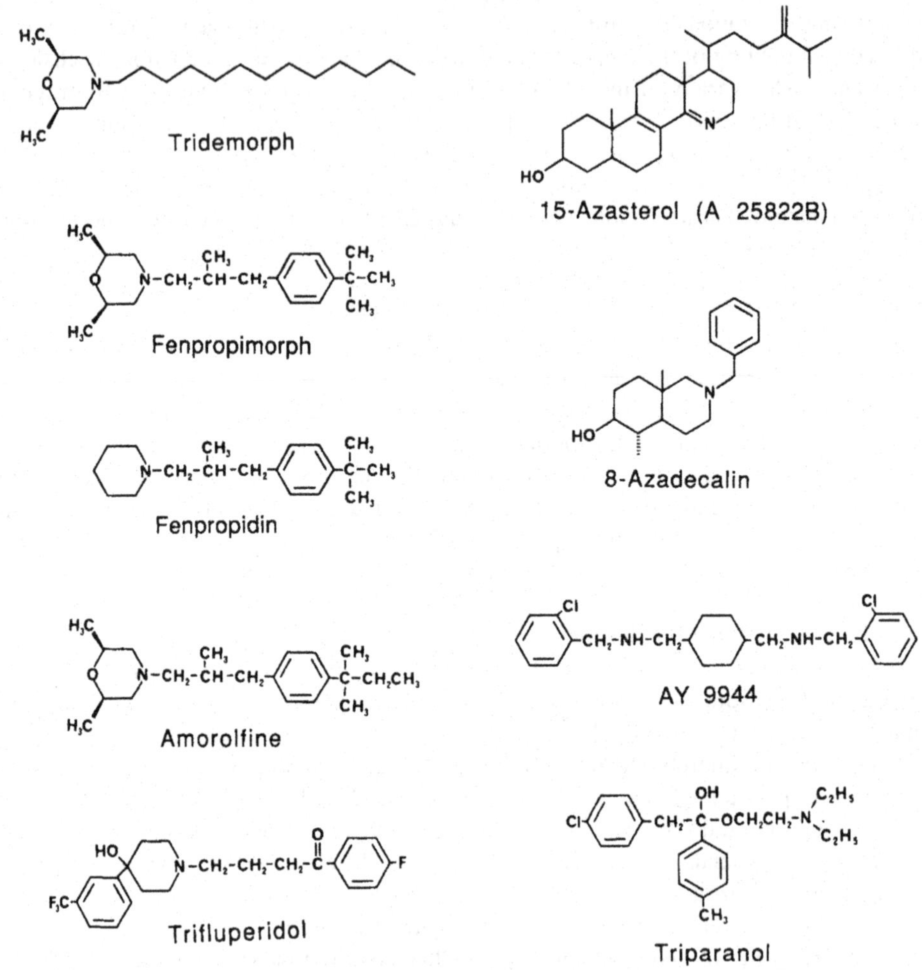

FIGURE 21. Inhibitors of Δ^7-Δ^8-isomerization.

former enzyme. This dual mode of action was confirmed in cell-free assays employing the two enzyme systems derived from *S. cerevisiae* (Table 6).[137] Although differences with regard to inhibitory activities were obvious for the morpholines tested, the Δ^8-Δ^7-isomerase was in all three cases the more sensitive target enzyme, a finding not entirely in agreement with the sterol patterns isolated from treated fungi. Kerkenaar[500] found that the supply of oxygen had considerable impact on the pattern of accumulating sterol precursors and suggested that the morpholines can undergo oxidation to the *N*-oxido derivatives which then tend to be more powerful inhibitors of the Δ^{14}-reductase. This shift of *N*-oxido derivatives toward an increased inhibitory activity of the Δ^{14}-reductase has indeed been demonstrated for fenopropimorph and related cyclic amines tested in respective plant systems.[509] The putative alteration of target enzyme specificities via enzymatic *N*-oxidation might explain the initial conflicting results with regard to the precise morpholine target sites.

The dual morpholine mode of action and a similar relative ranking of tridemorph, fenpropimorph, and fenopropidin were confirmed in a cell-free study with Δ^{14}-reductase and Δ^8-Δ^7-isomerase derived from maize embryos.[506] The inhibitor data for both the fungal and plant system are summarized in Table 6. Since the Δ^8-Δ^7-isomerase inhibition with tridemorph was shown to be pH-dependent[136] and the tests were performed under different pH conditions, the absolute values are not directly comparable. However, the remarkable re-

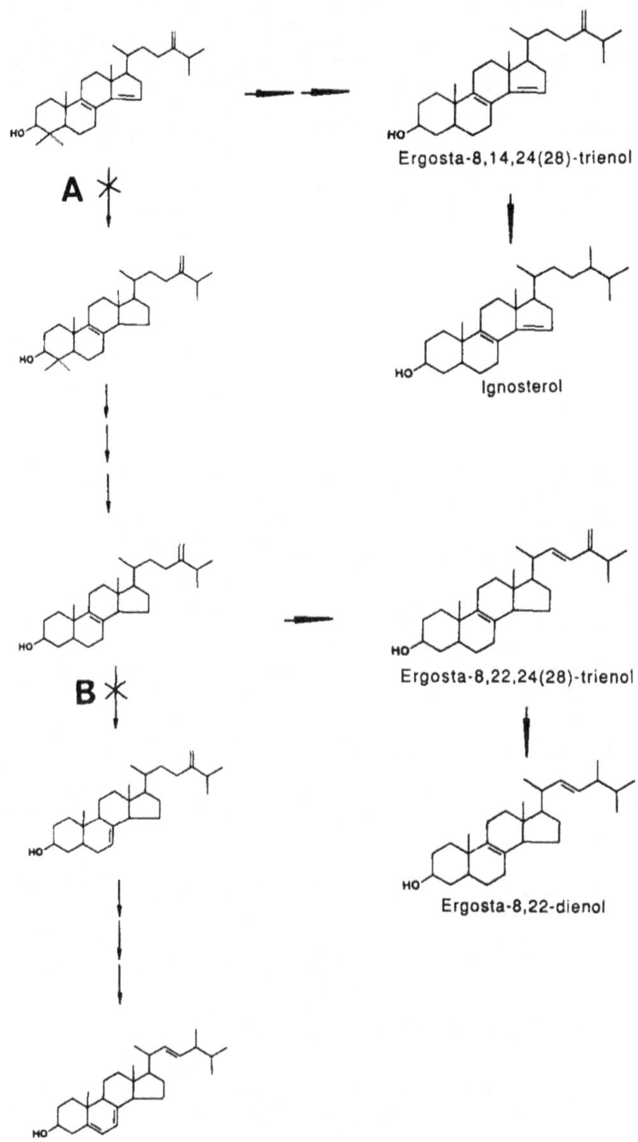

FIGURE 22. Sterol biosynthesis under the action of Δ^7-Δ^8-isomerization inhibitors.

version of inhibitor potencies of fenpropimorph and fenpropidin for the two isomerases from yeast or plant tissue might indicate selectivity differences between the plant and the fungal enzyme. It should be noted that in plants the morpholines also inhibit the cycloeucalenol-obtusifoliol-isomerase, in addition to the Δ^7-Δ^8-isomerase and Δ^{14}-reductase (see Chapter 8).[136,506,507] With a pK_a of 7.5,[136] the morpholines are positively charged under physiological conditions and, thus, have been discussed as analogues of the carbocationic high-energy intermediate shown to occur during the Δ^8-Δ^7-isomerization reaction (Figure 10).[136,507] The fact that both the fungal and plant Δ^{14}-reductases were inhibited provided strong evidence for the existence of a similar carbocationic intermediate during the reduction of the Δ^{14}-double bond.[137,506] The molecular model of morpholine action based on this concept suggests that the inhibitor is bound to the sterol binding site of both enzymes and exerts its inhibitory effect of mimicking the carbocationic high-energy intermediates of both the Δ^8-Δ^7-isomerization and the Δ^{14}-reduction (Figure 23).

TABLE 6
Inhibition of Δ¹⁴-Reductase and Δ⁸-Δ⁷-Isomerase by Analogues of Carbocationic High Energy Intermediates

	I_{50} (μM)			
	Saccharomyces cerevisiae[a]		Maize embryos[b]	
Compound	Δ^{14}-Reductase	Δ^{8}-Δ^{7}-Isomerase	Δ^{14}-Reductase	Δ^{8}-Δ^{7}-Isomerase
Tridemorph	98	0.033	25	0.4
Fenpropimorph	2.3	0.013	0.8	0.4
Fenpropidin	1.8	0.300	3.0	0.08
15-Azasterol	0.002[c]		0.003	0.8
AY 9944			40	0.5
8-Azadecalin			0.08	0.13

[a] Data from Baloch and Mercer.[137]
[b] Data from Taton et al.[506]
[c] From Bottema and Parks.[514]

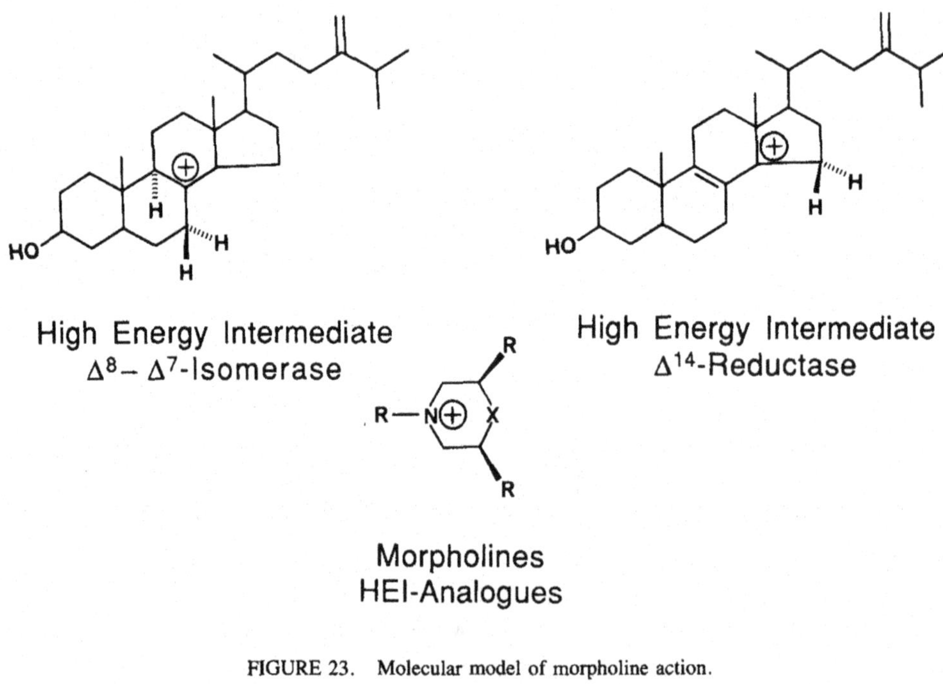

FIGURE 23. Molecular model of morpholine action.

Although data derived from cell-free assays have not been reported for amorolfine, the mode of action has been shown to be very similar. Ignosterol and ergosta-8,14,24(28)-trienol (Figure 22) were the main sterols isolated from treated *C. albicans, Histoplasma capsulatum,* and *Trichopyton mentagrophytes*[339,508] The sterol pattern suggests a preferential inhibition of the Δ^{14}-reductase by amorolfine, although a Δ^{8}-Δ^{7}-isomerization inhibition cannot be excluded from the presented data.

The 2,6-dimethyl morpholine derivatives tridemorph, fenpropimorph, and amorolfine (Figure 21) are stereoisomers, and the impact of stereochemistry on the inhibitory activity has been extensively investigated. All three morpholine derivatives exist as *cis* and *trans* form with regard to the methyl substituents. For tridemorph, both *cis* and *trans* were equally active.[493] For fenpropimorph[493] and amorolfine,[494] the *cis*-form was shown to be clearly more active, and the respective common names are restricted to the *cis*-isomers. In addition to the 2,6-dimethylmorpholine moiety, fenpropimorph and amorolfine are distinguished by an additional asymmetric center in the propyl interchain. In both cases, the *S*-enantiomers were identified as the more active inhibitors.[493,494] Conflicting results have been reported with regard to the optical rotations inherent to the *S*-isomers of fenpropimorph and amorolfine.[493,494] The *cis*- and *S*-preference of fenpropimorph has been confirmed in cell-free assays of the Δ^8-Δ^7-isomerization;[136,137] no data are yet available for the Δ^{14}-reductase.

Similar to the class of DMIs discussed above, the primary event of morpholine action is the inhibition of specific steps in sterol biosynthesis and, consequently, the accumulation of sterol precursors replacing the normally synthesized sterols. The translation of these events into fungal growth inhibition is less well understood. The morphology of *U. maydis* and *Penicillium italicum* treated with fenpropimorph was indistinguishable from that observed after DMI treatment.[404] Apparently, the inhibition of cell budding, induction of excessive hyphal branching, and disturbance of chitin deposition are not dependent on the chemical nature of accumulating sterol precursors. Thus, the question of how the accumulation of artificial sterols is interrelated with the inhibition of fungal growth might be similar for the DMIs and the morpholines. Interference with the "bulk" function of sterols in fungal membranes has been reported for *S. cerevisiae* and other fungi treated with fenpropimorph.[510] *S. cerevisiae* accumulated ergosta-8,14,24(28)-trienol as the major sterol, resulting in a greater fluidity of plasma membranes. On the other hand, the arrest of cell proliferation in the unbudded G_1 phase was linked to the depletion of ergosterol, indicating the involvement of regulatory functions.[510a] Sterol precursor-containing membranes were also found to be more permeable to glucose. Whether the interference with bulk functions of sterols comprises the exclusive or major effect on fungal growth and reproduction, and how the inhibition of ergosterol synthesis interferes with the regulatory functions of sterols will have to be established in future studies.

Only a few side effects in addition to sterol biosynthesis inhibition have been reported for the morpholines. Polak[508] found a strong accumulation of squalene along with sterol precursors in *T. mentagrophytes* treated with amorolfine. It remains unknown whether this effect represents a direct inhibitory side effect or a regulatory event due to the disturbance of sterol functions. An interesting direct interference of morpholines with cholesterol oxidases has been reported by Kerkenaar.[500] These enzymes are either involved in the synthesis of steroid hormones or, in microorganisms, the oxidative degradation of sterols. Two distinct steps are involved, the oxidation of the 3-hydroxyl group and a subsequent Δ^5-Δ^4-isomerization. The latter step resembles the Δ^8-Δ^7-isomerization and thus is thought to be prone to inhibition by morpholines.[500] Thus far, no indication exists that this side effect has impact on the antifungal activity of morpholines.

Little information is currently available on fungal resistance to morpholines. Although laboratory resistance to morpholines has been reported for several fungi,[498,499] problems with practical resistance have not yet surfaced.[463] Studies on mechanisms of morpholine resistance employing laboratory mutants are lacking. They would provide not only insight into the potentials of resistance development, but they also might contribute to open questions regarding the precise mode of action.

2. Mode of Action of Additional Lead Structures

Several compounds with inhibitory activities toward the fungal Δ^{14}-reductase and Δ^8-Δ^7-isomerase all contain cyclic and noncyclic nitrogens (Figure 21) and, thus, might well fall into the category of analogues of carbocationic high-energy intermediates. 15-Azasterol (A 25822B) (Figure 21) is a potent antimycotic antibiotic produced by *Geotrichum flavobrunneum*. At sublethal concentrations, the inhibitor led to the strong accumulation of ignosterol (Figure 22) in *S. cerevisiae*[511,512] and *U. maydis*,[513] indicating a specific inhibition of the Δ^{14}-reductases. This strong inhibitory potency was confirmed in a cell-free assay system, and a K_i of 2 n*M* was determined.[514] This value is almost identical with the I_{50} for the respective enzyme isolated from maize tissue (Table 6).[506] It is apparent from these data that 15-azasterol is a 1000-fold more powerful Δ^{14}-reductase inhibitor than the morpholines fenpropidin and fenpropimorph (Table 6). The inhibitor structure closely resembles the substrate of fungal Δ^{14}-reductases and, thus, can be expected to act as a specific inhibitor. Indeed, the activity as an inhibitor of the plant Δ^8-Δ^7-isomerase was less by a factor of almost 300 (Table 6). In contrast to *U. maydis*, the antifungal activity was only transient for *S. cerevisiae*, most likely mediated by a deactivating modification of the inhibitor.[515] A series of various azasterol mimics has been identified as another class of fungal Δ^{14}-reductase inhibitors.[500a]

In contrast to 15-azasterol, the hypocholesterolemic drug AY 9944 (Figure 21) was identified as a relatively specific inhibitor of the Δ^8-Δ^7-isomerase derived from maize embryos with low activity toward the plant Δ^{14}-reductase (Table 6). With the partly purified Δ^8-Δ^7-isomerase solubilized from rat liver microsomes, AY 9944 was shown to act as a competitive inhibitor.[131] This type of inhibition is not immediately apparent from the chemical structure of the inhibitor. Although the growth of *S. cerevisiae* was not significantly affected by the drug, a pronounced alteration of the sterol pool has been reported.[51] Most pronounced was an elevated level of sterol esters, and within this ester pool, the fraction of Δ^8-sterols was simultaneously increased. This indicated an inhibition of the Δ^8-Δ^7-isomerase, although the inhibitory activity *in vivo* was counteracted by the segregation of accumulating sterol precursors. The effects of AY 9944 on fungi other than *S. cerevisiae* have not yet been investigated.

In cell-free assays of both the Δ^8-Δ^7-isomerase and Δ^{14}-reductase derived from plant tissue, a benzyl-substituted 8-azadecalin (Figure 21) was identified as an equally powerful inhibitor for both enzymes (Table 6).[506] Unfortunately, no data are available on the effects of these derivatives on fungal organisms. The azadecalins (see Chapter 8) represent an interesting group of inhibitors derived from the rational design of inhibitors based on concepts of enzyme catalysis.[516,517] The inhibitors are thought to interfere with carbocationic intermediates at the 8- and 14-position of the respective enzyme substrates (Figure 23). Similar carbocationic intermediates occur during the ring opening of cycloeucalenol in plants yielding obtusifoliol (see Chapter 8). Consequently, azadecalins act, like morpholines, as inhibitors of this plant isomerase not present in fungal organisms.[506] Moreover, a carbocationic intermediate with a similar location within the sterol structure is thought to be generated during the cyclization of epoxysqualene, and 8-azadecalin derivatives are also inhibitors of oxidosqualene-sterol cyclases (see Section IV.C; Table 1 and Figure 15).

Although no data derived from cell-free assay systems are yet available, evidence exists that the hypocholesterolemic drugs trifluperidol and triparanol can both be considered inhibitors of fungal Δ^8-Δ^7-isomerase. In yeast, trifluperidol (Figure 21) caused the accumulation of Δ^8-sterols in both the free and esterified sterol pool, with preference for the esterified sterols.[50] Leroux et al.[518] listed piperalin, a powdery mildew fungicide with structural similarities to trifluperidol, as an inhibitor of fungal Δ^8-Δ^7-isomerases. Triparanol (Figure 21) was shown to inhibit the growth of *S. cerevisiae*, causing the accumulation of Δ^8-sterols.[519] In rat hematoma cells, triparanol also acted as an inhibitor of the Δ^8-Δ^7-isomerase, but only at high concentrations. At lower concentrations, the reduction of both the Δ^7 and $\Delta^{24(25)}$-double bond was inhibited.[520] Both of these enzymatic steps are lacking in the synthesis of ergosterol.

In summary, both the Δ^{14}-reductase and Δ^8-Δ^7-isomerase are proven targets for commercial antifungal inhibitors. More lead structures derived from rational inhibitor design based on the analogue concept of carbocationic high-energy intermediates or from drugs originally developed as hypocholesterolemic agents have been described, and optimization with regard to the demands of fungal enzymes as well as increased *in vivo* activities appears possible.

TABLE 7
Sterol Biosynthesis Inhibitors

Enzyme	Inhibitors	Figure
HMG-CoA synthase	β-Lactones	13
HMG-CoA reductase	Mevastatin, lovastatin	13
Squalene synthase	Presqualene analogues	13
Squalene epoxidase	Allylamines[a]	14
Oxidosqualene-sterol cyclase	Azasqualene derivatives	15
Δ^{24}-Lanosterol methyltransferase	25-Azasterols	16
14α-Demethylase	Pyridines,[a] pyrimidines[a]	18
	Azoles,[a] lanosterol analogues	
Δ^{14}-Reductase	Morpholines,[a] 15-azasterol, 8-azadecalin	21
4-Demetylase	None	
Δ^{8}-Δ^{7}-Isomerase	Morpholines,[a] 8-azadecalin, AY 9944, triparanol, trifluperidol	21
Δ^{5}-Desaturase	None	
Δ^{22}-Desaturase	DMI side effect (?)	
$\Delta^{24(28)}$-Reductase	23-Azacholesterol	

[a] Commercial product.

VI. FUTURE PERSPECTIVES

The pathway of fungal sterol biosynthesis has emerged as the currently most important pathway of antifungal agents in both fields of application, agriculture and medicine. Four individual target enzymes, the squalene epoxidase, the 14α-demethylase, and the Δ^{14}-reductase in combination of Δ^{8}-Δ^{7}-isomerase, are inhibited by three different classes of established fungicides and antimycotic drugs (Table 7). Lead structures and inhibitor design concepts for more numerous enzymes involved in the conversion of acetate to ergosterol have been described over the last decade (Table 7). In particular, the concept of designing analogues of high-energy carbocationic intermediates has yielded promising lead structures. Some of these lead structures with antifungal potencies have been described above. Additional experimental information derived from plant enzymes is described for numerous inhibitors in Chapter 8 and recent reviews,[516,517] and chances to extend this design concept to fungal enzymes appear possible and promising.

Inhibitors of enzymes involved in the prelanosterol part of the sterol biosynthetic route have been described. Many of the enzyme substrates, however, are relatively small and water-soluble molecules, and inhibitors might lend little room for structural modifications. HMG-CoA reductase with its complex coenzyme binding sites and putative regulatory functions comprises a clear exception. Relatively unique and hydrophobic enzyme substrates appear in the synthesis and modification of squalene, and it is not surprising that the allylamines as inhibitors of the squalene epoxidase have been developed to antifungal drugs. The antifungal activities of these inhibitors are not only based on the inhibition of ergosterol biosynthesis, but also on the accumulation of lipophilic and membrane-disturbing precursors due to the lack of a pronounced feedback inhibition.

Within the postlanosterol pathway, the structural modifications generate sterols increasingly similar to the end products. Thus, early steps such as the 14α-demethylation appear most promising to yield a disturbing precursor accumulation. Inhibitors of the C-4 demethylation should have similar effects, but no inhibitor structures have yet been found. This is not surprising since the enzymology of this important demethylation step remains largely unknown. The same applies to the introduction of the Δ^{5}-double bond. Inhibitors of the latter enzyme might lead to a severe disturbance of regulatory rather than bulk functions of sterols (see Section II.E.2). The same applies to the side-chain methylation at the C-24 position and the reduction of the $\Delta^{24(28)}$-double bond. Several inhibitors of the Δ^{24}-lanosterol methyltransferase have been described (Table 7). Less information is available on inhibitors of the $\Delta^{24(28)}$-reductase. 23-Azacholesterol has been reported to inhibit this step at lower concentrations than those required to block the C-24 methylation.[521] Similarly, 3β-(dimethylaminoethoxy)androst-5-en-17-one has been identified as an inhibitor more active toward the $\Delta^{24(28)}$-reduction rather than the cyclization of oxidosqualene. With the exception of a putative DMI side effect, no inhibitors have been described for the Δ^{22}-desaturase. In this case, however, it could be argued that the *in vivo* effect of a putative Δ^{22}-desaturase inhibitor would not be great. Neither the bulk nor the regulatory functions of sterols appear to rely substantially on this structural feature, and vital fungal genotypes lacking the Δ^{22}-double bond have been described.[471]

Overall, the sterol biosynthetic pathway leaves room for more and improved future antifungal inhibitors. Furthermore, a more thorough understanding of the regulatory roles inherent to fungal sterols could lead to inhibitors that interfere directly with the regulatory binding sites at crucial enzymes or receptors. Future discoveries in this field of applied mycology appear likely, and the progress made in the understanding of fungal "steroidology" will trigger these developments.

REFERENCES

1. **Berg, D. and Plempel, M.,** *Sterol Biosynthesis Inhibitors — Pharmaceutical and Agrochemical Aspects,* Ellis Horwood, Chichester, 1988.
2. **Haug, G. and Hoffmann, H.,** *Chemistry of Plant Protection, Vol. 1, Sterol Biosynthesis, Inhibitors and Anti-Feeding Compounds,* Springer-Verlag, Berlin, 1986.
3. **Fromtling, R. A.,** *Recent Trend in the Discovery, Development and Evaluation of Antifungal Agents,* J. R. Prous Science Publishers, 1987.
4. **Georgiev, V. St.,** *Antifungal Drugs,* New York Academy of Sciences, New York, 1988.

4a. **Burden, R. S., Cooke, D. T., and Carter, G. A.,** Inhibitors of sterol biosynthesis and growth in plants and fungi, *Phytochemistry,* 28, 1791, 1989.

5. **Magalith, P. Z.,** *Steroid Microbiology,* Charles C Thomas, Springfield, IL, 1986, chap. 2.
6. IUPAC-IUB Joint Commission on Biochemical Nomenclature (JBCN), The nomenclature of steroids — recommendations 1989, *Eur. J. Biochem.,* 186, 429, 1989.
7. **Nes, W. R.,** Structure-function relationships for sterols in *Saccharomyces cerevisia,* in *Ecology and Metabolism of Plant Lipids,* Fuller, G. and Nes, W. D., Eds., American Chemical Society, Washington, D.C., 1987, 252.
8. **Tauret, C.,** Sur un noveau principle immédiate de l'ergot de seigle, l'ergostérine, *C.R. Acad.,* 108, 98, 1889.
9. **Rattray, J. B. M.,** Yeasts, in *Microbial Lipids,* Vol. 1, Ratledge, C. and Wilkinson, S. G., Eds., Academic Press, London, 1988, 555.
10. **Lösel, D. M.,** Fungal lipids, in *Microbial Lipids,* Vol. 1, Ratledge, C. and Wilkinson, S. G., Eds., Academic Press, London, 1988, 700.
11. **Weete, J. D.,** Structure and function of sterols in fungi, *Adv. Lipid Res.,* 23, 115, 1989.
12. **Warner, S. A., Sovocool, G. W., and Domnas, A. J.,** Sterols of selected species of Oomycetes and Hypochytridiomycetes, *Mycologia,* 75, 285, 1983.
13. **Weete, J. D., Fuller, M. S., Huang, M. Q., and Gandhi, S.,** Fatty acids and sterols of selected Hyphochytriomycetes and Chytridiomycetes, *Exp. Mycol.,* 13, 183, 1989.
14. **Nes, W. D.,** Biosynthesis and requirement for sterols in the growth and reproduction of oomycetes, in *Ecology and Metabolism of Plant Lipids,* Fuller, G. and Nes, W. D., Eds., American Chemical Society, Washington, D.C., 1987, 304.
15. **Berg, L. and Patterson, G. W.,** Phylogenetic implications of sterol biosynthesis in the oomycetes, *Exp. Mycol.,* 10, 175, 1986.
16. **Nes, W. D.,** Control of fungal physiology by polycyclic isopentenoids and its phylogenetic implications, in *Isopentenoids in Plants — Biochemistry and Function,* Nes, W. D., Fuller, G., and Tsai, L.-S., Eds., Marcel Dekker, New York, 1984, 267.

17. **Nes, W. D., Le, P. H., Patterson, G. W., and Kerwin, J. L.,** A comparison of cycloartenol and lanosterol biosynthesis and metabolism by the oomycetes, *Experientia,* 42, 556, 1986.
18. **Nes, W. D.,** Phytophthorols-Novel lipids produced by *Phytophthora cactorum, Lipids,* 23, 9, 1989.
19. **Nes, W. D. and Le, P. H.,** Regulation of sterol biosynthesis in *Saprolegnia ferax* by 25-azasterol, *Pestic. Biochem. Physiol.,* 30, 87, 1988.
20. **Nes, W. D. and Parish, E. J., Eds.,** *Analysis of Sterols and Other Biologically Significant Steroids,* Academic Press, San Diego, 1989.
21. **Nes, W. R.,** A comparison of methods for the identification of sterols, *Methods Enzymol.,* 111, 3, 1985.
22. **Patterson, G. W.,** Chemical and physical methods in the analysis of plant sterols, in *Isopentenoids in Plants — Biochemistry and Function,* Nes, W. D., Fuller, G., and Tsai, T.-S., Eds., Marcel Dekker, New York, 1984, 293.
23. **Akihisa, T., Shimizu, N., Ghosh, P., Thakur, S., Rosentein, F. U., Tamura, T., and Matsumoto, T.,** Sterols of the cuburbitaceae, *Phytochemistry,* 26, 1693, 1987.
24. **Fenner, G. P., Patterson, G. W., and Lusby, W. R.,** Development regulation of sterol biosynthesis in *Curcurbita maxima* L., *Lipids,* 24, 271, 1989.
25. **Heftmann, E.,** Metabolism of cholesterol in plants, in *Isopentenoids in Plants — Biochemistry and Function,* Nes, W. D., Fuller, G. and Tsai, T.-S., Eds., Marcel Dekker, New York, 1984, 487.
26. **Loeffler, R. S. T., Butters, J. A., and Hollomon, D. W.,** The lipid contents of powder mildews and their sensitivity to fungicides, in *Proc. 1984 British Crop Prot. Conf.,* BCPC Publications, Croydon, England, 1984, 911.
27. **Jackson, L. L. and Frear, D. S.,** Lipids of rust fungi. II. Stigmast-7-enol and stigmasta-7,24(28)-dienol in flax rust uredopores, *Phytochemistry,* 7, 651, 1968.
28. **Lin, H.-K., Langenbach, R. J., and Knoche, H. W.,** Sterols of *Uromyces phaseoli* uredospores, *Phytochemistry,* 11, 2319, 1972.
29. **Lin, H.-K. and Knoche, H. W.,** Origin of sterols in uredospores of *Uromyces phaseoli, Phytochemistry,* 13, 1795, 1974.
30. **Novak, R., Kim, W. K., and Rohringer, R.,** Sterols of healthy and rust-infected primary leaves of wheat and of non-germinated and germinated uredospores of wheat stem rust, *Can. J. Bot.,* 50, 185, 1972.
31. **Carmack, C. L., Weete, J. D., and Kelley, W. D.,** Hydrocarbons, fatty acids and sterols of *Cronartium fusiforme* aeciospores, *Physiol. Plant Pathol.,* 8, 43, 1976.
32. **Weete, J. D. and Kelley, W. D.,** Fatty acids and sterols of *Cronartium fusiforme* basidiospores, *Lipids,* 12, 398, 1977.
33. **Kuck, K. H. and Scheinpflug, H.,** Biology of sterol-biosynthesis inhibiting fungicides, in *Chemistry of Plant Protection,* Vol. 1, Haug, G. and Hoffmann, H., Eds., Springer-Verlag, Berlin, 1986, 65.
34. **Scheinpflug, H. and Kuck, K. H.,** Sterol biosynthesis inhibiting piperazine, pyridine, pyrimidine and azole fungicides, in *Modern Selective Fungicides — Properties, Applications and Mechanisms of Action,* Lyr, H., Ed., John Wiley & Sons, New York, 1987, 173.
35. **Pommer, K.-H.,** Morpholine fungicides, in *Modern Selective Fungicides — Properties, Applications and Mechanisms fo Action,* Lyr, H., Ed., John Wiley & Sons, New York, 1987, 143.
36. **Schulz, U. and Scheinpflug, H.,** Sterol biosynthesis inhibiting fungicides: antifungal properties and application in cereals, in *Sterol Biosynthesis Inhibitors — Pharmaceutical and Agrochemical Aspects,* Berg, D. and Plempel, M., Eds., Ellis Horwood, Chichester, 1988, 211.
37. **Flesch, P., Gwinner, L., Müller, S., and Simon, P.,** Isolierung und Identifizierung von Sterinen im Metabolismus des Pilzes *Botrytis cinerea, Z. Naturforsch,* 38c, 207, 1983.
38. **Berg, D., Born, L., and Büchel, K.-H.,** HWG1608 — chemistry and biochemistry of a new azole fungicide, *Pflanzenschutz Nachr. (Bayer),* 40, 111, 1987.
38a. **Loeffler, R. S. and Hayes, A. L.,** Sterols of the plant pathogenic fungi *Botryis cinerea* and *Pyrenophora teres, Phytochemistry,* 29, 3423, 1990.
39. **Ebert, E., Gaudin, J., Muecke, W., Ramsteiner, K., and Vogel, C.,** Inhibition of ergosterol biosynthesis by etaconazole in *Ustilago avenae, Z. Naturforsch.,* 38c, 28, 1982.
40. **Baloch, R. I., Mercer, E. I., Wiggins, T. E., and Baldwin, B. C.,** Inhibition of ergosterol biosynthesis in *Saccharomyces cerevisiae* and *Ustilago maydis* by tridemorph, fenpropimorph and fenpropidin, *Phytochemistry,* 23, 2219, 1984.
41. **Ragsdale, N.,** Specific effects of triarimol on sterol biosynthesis in *Ustilago maydis, Biochim. Biophys. Acta,* 380, 81, 1975.
42. **Hippe, S. and Köller, W.,** Ultrastructure and sterol composition of laboratory strains of *Ustilago avenae* resistant to triazole fungicides, *Pestic. Biochem. Physiol.,* 26, 209, 1986.
43. **Smith, F. D. and Köller, W.,** The expression of resistance of *Ustilago avenae* to the sterol demethylation inhibitor triadimenol is an induced response, *Phytopathology,* 80, 584, 1990.
44. **Uzuka, Y., Naganuma, T., Tanaka, K., and Suzuki, K.,** Relation between neutral lipid accumulation and the growth phase in the yeast *Lipomyces strakeyi,* a fat producing yeast, *Agric. Biol. Chem.,* 49, 851, 1985.

45. **Parks, L. W.,** Metabolism of sterols in yeast, *CRC Crit. Rev. Microbiol.*, 301, 1978.
46. **Clausen, M. K., Christiansen, K., Jensen, P. K., and Behnke, O.,** Isolation of lipid particles from baker's yeast, *FEBS Lett.*, 43, 176, 1974.
47. **Schaffner, G. and Matile, Ph.,** Structure and composition of baker's yeast lipid globules, *Biochem. Physiol. Pflanz.*, 176, 659, 1981.
48. **Parks, L. W., Lewis, T. A., Low, C., and Haeckler, K.,** Effect of intracellular sterol concentration on sterol esterification, in *The Metabolism, Structure and Function of Plant Lipids*, Stumpf, P. K. and Nes, W. D., Eds., Plenum Press, New York, 1987, 57.
49. **Lewis, T. A., Rodriguez, R. J., and Parks, L. W.,** Relationship between intracellular sterol content and sterol esterification and hydrolysis in *Saccharomyces cerevisiae, Biochim. Biophys. Acta,* 921, 205, 1987.
50. **Sobus, M. T., Holmlund, C. E., and Wittaker, N. F.,** Effects of the hypercholesteremic agent trifluperidol on the sterol, steryl ester, and fatty acid metabolism of *Saccharomyces cerevisiae, J. Bacteriol.*, 127, 1310, 1977.
51. **Pereira, R., Holmlund, C. E., and Wittaker, N. F.,** The effect of AY-9944 on yeast sterol and sterol ester metabolism, *Lipids,* 18, 545, 1983.
52. **Gray, J. C.,** Control of isoprenoid biosynthesis in higher plants, *Adv. Bot. Res.*, 14, 25, 1987.
53. **Ratledge, C. and Wilkinson, S. G.,** Terpenoid lipids, in *Microbial Lipids,* Vol. 1, Ratledge, C. and Wilkinson, S. G., Eds., Academic Press, New York, 1988, 55.
54. **Coolbear, T. and Threlfall, D. R.,** Biosynthesis of terpenoid lipids, in *Microbial Lipids,* Vol. 2, Ratledge, C. and Wilkinson, S. G., Eds., Academic Press, New York, 1989, 115.
55. **Weete, J. D.,** *Lipid Biochemistry of Fungi and Other Organisms,* Plenum Press, New York, 1980.
56. **Mercer, E. I.,** The biosynthesis of ergosterol, *Pestic. Sci.*, 15, 133, 1984.
57. **Harrison, D. M.,** The biosynthesis of triterpenoids and steroids, *Nat. Prod. Rep.*, 2, 525, 1985.
58. **Caspi, E.,** The mode of incorporation of C-2 hydrogen atoms of mevalonic acid into protosterols and sterols, *Tetrahedron,* 42, 3, 1986.
59. **Faust, J. R., Trzaskos, J. M., and Gaylor, J. L.,** Cholesterol biosynthesis, in *Biology of Cholesterol,* Yeagle, P. L., Ed., CRC Press, Boca Raton, FL, 1988, 19.
60. **Reinhart, M. P., Billheimer, J. T., Faust, J. R., and Gaylor, J. L.,** Subcellular localization of the enzymes of cholesterol biosynthesis and metabolism in rat liver, *J. Biol. Chem.*, 262, 9649, 1987.
61. **Krisans, S. K., Pazirandeh, M., and Keller, G. A.,** Localization of 3-hydroxy-3-methylglutaryl-coenzyme A reductase in rat liver peroxisomes, in *Peroxisomes in Biology and Medicine,* Fahimi, H. D. and Sies, H., Eds., Springer-Verlag, Berlin, 1987, 40.
62. **Thompson, S. L., Burrows, R., Laub, R. J., and Krisans, S. K.,** Cholesterol synthesis in rat liver peroxisomes, *J. Biol. Chem.*, 262, 17420, 1987.
63. **Williams, S., Crabbe, T. B., Haslam, J. M., and Quain, D. E.,** The control of sterol biosynthesis in *Saccharomyces cerevisiae, Biochem. Soc. Trans.*, 14, 1086, 1986.
64. **Basson, M. E., Thorsness, M., and Rine, J.,** *Saccharomyces cerevisiae* contains two functional genes encoding 3-hydroxy-3-methylglutaryl-coenzyme A reductase, *Proc. Natl. Acad. Sci. U.S.A.*, p. 5563, 1986.
65. **Evans, J. L. and Gealt, M. A.,** The 3-hydroxy-3-methylglutaryl-coenzyme A reductase of *Aspergillus nidulans, Exp. Mycol.*, 132, 1988.
66. **Basson, M. E., Thorsness, M., Finer-Moore, J., Strout, R. M., and Rine, J.,** Structural and functional conservation between yeast and human 3-hydroxy-3-methylglutaryl-coenzyme A reductases, the rate-limiting enzyme of sterol biosynthesis, *Mol. Cell. Biol.*, 8, 3797, 1988.
67. **Learned, R. M. and Fink, G. R.,** 3-Hydroxy-3-methylglutaryl-coenzyme A reductase from *Arabidopsis thaliana* is structurally distinct from the yeast and animal enzymes, *Proc. Natl. Acad. Sci. U.S.A.*, 86, 2779, 1989.
68. **Poulter, C. D.,** Isopentenyl diphosphate to squalene — enzymology and inhibition, in *Biochemistry of Cell Walls and Membranes in Fungi,* Kuhn, P. J., Trinci, A. P. J., Jung, M. J., Goosey, M. W., and Copping, L. G., Eds., Springer-Verlag, Berlin, 1990, 169.
69. **Anderson, M. S., Muehlbacher, M., Street, I. P., Proffitt, J., and Poulter, C. D.,** Isopentenyl diphosphate:dimethylallyl diphosphate isomerase — an improved purification of the enzyme and isolation of the gene from *Saccharomyces cerevisiae, J. Biol. Chem.*, 15, 19169, 1989.
70. **Anderson, M. S., Yarger, J. G., Burck, C. L., and Poulter, C. D.,** Farnesyl diphosphate synthase — molecular cloning, sequence, and expression of an essential gene from *Saccharomyces cerevisiae, J. Biol. Chem.*, 264, 19176, 1989.
71. **Sasiak, K. and Rilling, H. C.,** Purification to homogeneity and some properties of squalene synthase, *Arch. Biochem. Biophys.*, 260, 622, 1988.
72. **Ono, T., Nakazono, K., and Kosaka, H.,** Purification and partial characterization of squalene epoxidase from rat liver microsomes, *Biochim. Biophys. Acta,* 709, 84, 1982.
73. **Ryder, N. S. and DuPont, M.-C.,** Properties of a particulate squalene epoxidase from *Candida albicans, Biochim. Biophys. Acta,* 794, 466, 1984.

74. **Ryder, N. S.,** Squalene epoxidase — enzymology and inhibition, in *Biochemistry of Cell Walls and Membranes in Fungi,* Kuhn, P. J., Trinci, A P. J., Jung, M. J., Goosey, M. W., and Copping, L. G., Eds., Springer-Verlag, Berlin, 1990, 189.
75. **Gaylor, J. L.,** Membrane-bound enzymes of cholesterol biosynthesis from lanosterol, in *Membranes and Transport,* Vol. 1, Martonosi, A. N., Ed., Plenum Press, New York, 1982, 249.
76. **Van Tamelen, E. E.,** Bioorganic characterization and mechanism of the 2,3-oxidosqualene-lanosterol conversion, *J. Am. Chem. Soc.,* 104, 6480, 1982.
77. **Johnson, W. S., Lindell, S. D., and Steele, J.,** Rate enhancement of biomimetric polyene cyclizations by a cation-stabilizing auxiliary, *J. Am. Chem. Soc.,* 109, 5852, 1987.
78. **Duriatti, A. and Schuber, F.,** Partial purification of 2,3-oxidosqualene-lanosterol cyclase from hog liver, *Biochem. Biophys. Res. Commun.,* 151, 1378, 1988.
79. **Nakanishi, S., Nishio, T., Yabusaki, Y., Fujisaki, S., and Katsuki, H.,** Preparation of C^{14}-labeled and unlabeled sterol intermediates from yeast using metabolic inhibitors, *J. Biochem.,* 96, 1665, 1984.
80. **Moore, J. T. and Gaylor, J. L.,** Isolation and purification of an S-adenosylmethionine:Δ^{24}-sterol methyltransferase from yeast, *J. Biol. Che.,* 244, 6334, 1969.
81. **Yabusaki, Y., Nishino, T., Ariga, N., and Katsuki, H.,** Studies on Δ^{8}-Δ^{7} isomerization and methyl transfer of sterols in ergosterol biosynthesis of yeast, *J. Biochem.,* 85, 1531, 1979.
82. **McCammon, M. T., McLean-Bowen, C. A., and Parks, L. W.,** Genetic, cellular, and biochemical analysis of the yeast Δ^{24}-sterol methyltransferase, in *Biochemistry of S-Adenosylmethionine and Related Compounds,* Usdin, E., Borchardt, R. T., and Creveling, C. R., Eds., Macmillan, London, 1982, 499.
83. **McCammon, M. T., Hartmann, M.-A., Bottema, C. D. K., and Parks, L. W.,** Sterol methylation in *Saccharomyces cerevisiae, J. Bacteriol.,* 157, 475, 1984.
84. **Pierce, A. M., Pierce, H. D., Unrau, A. M., and Oehlschlager, A. C.,** Lipid composition and polyene resistance of *Candida albicans* mutants, *Can. J. Biochem.,* 56, 135, 1978.
85. **Vanden Bossche, H. D., Willemsens, G., Cools, W., Lauwers, W. F., and LeJeune, L.,** Biochemical effects of miconazole on fungi. II. Inhibition of ergosterol biosynthesis in *Candida albicans, Chem. Biol. Interact.,* 21, 59, 1978.
86. **Ryder, N. S.,** Effect of allylamine antimycotic agents on fungal sterol biosynthesis measured by sterol side-chain methylation, *J. Gen. Microbiol.,* 131, 1595, 1985.
87. **Ator, M. A., Schmidt, S. J., Adams, J. L., and Dolle, R. E.,** Mechanism and inhibition of Δ^{24}-sterol methyltransferase from *Candida albicans* and *Candida tropicalis, Biochemistry,* 28, 9633, 1989.
88. **Schroepfer, G. F.,** Sterol biosynthesis, *Annu. Rev. Biochem.,* 51, 555, 1982.
89. **Tarr, G. E., Black, S. D., Fujita, V. S., and Coon, M. J.,** Complete amino acid sequence and predicted membrane topology of phenobarbital-induced cytochrome P-450 (isoenzyme 2) from rat liver microsomes, *Proc. Natl. Acad. Sci. U.S.A.,* 80, 6552, 1983.
90. **Nebert, D. W. and Conzales, F. J.,** P-450 genes: structure, evolution, and regulation, *Annu. Rev. Biochem.,* 56, 945, 1987.
91. **Gonzales, F. J.,** The molecular biology of cytochrome P450s, *Pharmacol. Rev.,* 40, 243, 1988.
92. **Nebert, D. W., Adesnik, M., Coon, M. J., Estabrook, R. W., Gonzales, F. J., Guengerich, F. P., Gunsalus, I. C., Johnson, E. F., Kemper, B., Phillips, I. R., Sato, R., and Waterman, M. R.,** The P450 gene superfamily. Recommended nomenclature, *DNA,* 6, 1, 1987.
93. **Nebert, D. W., Nelson, D. R., Adesnik, M., Coon, M. J., Estabrook, R. W., Gonzales, F. J., Guengerich, F. P., Gunsalus, I. C., Johnson, E. F., Kemper, B., Levin, W., Phillips, I. R., Sato, R., and Waterman, M. R.,** The P450 superfamily: updated listing of all genes and recommended nomenclature for the chromosomal loci, *DNA,* 8, 1, 1989.
94. **Oritz de Montellano, P. R., Ed.,** *Cytochrome P-450 — Structure, Mechanism, and Biochemistry,* Plenum Press, New York, 1986.
95. **Guengerich, F. P., Ed.,** *Mammalian Cytochrome P-450,* Vol. 1 and 2, CRC Press, Boca Raton, FL, 1987.
96. **Black, S. D. and Coon, M. J.,** P-450 cytochromes: structure and function, *Adv. Enzymol.,* 60, 35, 1987.
97. **Oritz de Montellano, P. R.,** Oxygen activation and transfer, in *Cytochrome P-450 — Structure, Mechanism, and Biochemistry,* Oritz de Montellano, P. R., Ed., Plenum Press, New York, 1986, 217.
98. **Peterson, J. A. and Prough, R. A.,** Cytochrome P-450 reductase and cytochrome b5 in cytochrome P-450 catalysis, *Cytochrome P-450 — Structure, Mechanism, and Biochemistry,* Oritz de Montellano, P. R., Ed., Plenum Press, New York, 1986, 89.
99. **Käppeli, O.,** Cytochromes P-450 of yeasts, *Microbiol. Rev.,* 50, 244, 1986.
100. **Jefcoate, C. R.,** Cytochrome P-450 enzymes in sterol biosynthesis and metabolism, in *Cytochrome P-450 — Structure, Mechanism, and Biochemistry,* Oritz de Montellano, P. R., Ed., Plenum Press, New York, 1986, 387.
101. **Hall, P. F.,** Cytochromes P-450 and the regulation of steroid synthesis, *Steroids,* 48, 133, 1987.
102. **Masters, B. S. S., Muerhoff, and Okita, R. T.,** Enzymology of extrahepatic cytochromes P-450, in *Mammalian Cytochromes,* Vol. 1, Guengerich, F. P., Ed., CRC Press, Boca Raton, FL, 1987, 107.

103. **Danielson, H. and Wikvall, K.,** Biosynthesis of bile acids, in *Cholesterol 7α-Hydroxylase (7α-Monooxygenase)*, Fears, R. and Sabine, J. R., Eds., CRC Press, Boca Raton, FL, 1986, 9.
104. **Yoshida, Y.,** Cytochrome P450 of fungi: primary target for azole antifungal agents, in *Current Topics in Medical Mycology*, Vol. 2, McGinnis, M. R., Ed., Springer-Verlag, New York, 1988, 388.
105. **Yoshida, Y. and Aoyama, J.,** Yeast cytochrome P450 catalyzing lanosterol 14α-demethylation. I. Purification and spectral properties, *J. Biol. Chem.*, 259, 1655, 1984.
106. **Aoyama, J., Yoshida, Y., and Sato, R.,** Yeast cytochrome P450 catalyzing lanosterol 14α-demethylation. II. Lanosterol metabolism of purified $P450_{DM}$ and by intact microsomes, *J. Biol. Chem.*, 259, 1661, 1984.
107. **Aoyama, Y., Yoshida, Y., Sonoda, Y., and Sato, Y.,** Deformylation of 32-oxo-24,25-dihydrolanosterol by the purified cytochrome $P\text{-}450_{14DM}$ (lanosterol 14α-demethylase) from yeast. Evidence confirming the intermediate step of lanosterol 14α-demethylation, *J. Biol. Chem.*, 264, 18502, 1989.
108. **Aoyama, J. and Yoshida, Y.,** Evidence for the contribution of a sterol 14-reductase to the 14α-demethylation of lanosterol by yeast, *Biochem. Biophys. Res. Commun.*, 134, 659, 1986.
109. **Aoyama, J., Yoshida, Y., Sonoda, Y., and Sato, R.,** Metabolism of 32-hydroxy-24,25-dihydrolanosterol by purified cytochrome $P\text{-}450_{14DM}$ from yeast, *J. Biol. Chem.*, 262, 1239, 1987.
110. **Trzaskos, J. M., Bowen, W. D., Shafiee, A., Fisher, R. T., and Gaylor, J. L.,** Cytochrome P-450-dependent oxidation of lanosterol in cholesterol biosynthesis, *J. Biol. Chem.*, 259, 13402, 1984.
111. **Paik, Y.-K., Trzaskos, J. M., Shafiee, A., and Gaylor, J. L.,** Microsomal enzymes of cholesterol biosynthesis from lanosterol, *J. Biol. Chem.*, 259, 13413, 1984.
112. **Trzaskos, J. M., Kawata, S., and Gaylor, J. L.,** Microsomal enzymes of cholesterol biosynthesis, *J. Biol. Chem.*, 261, 1451, 1986.
113. **Trzaskos, J. M., Fisher, R. T., and Favata,** Mechanistic studies of lanosterol C-32 demethylation, *J. Biol. Chem.*, 261, 16937, 1986.
114. **Shafiee, A., Trzaskos, J. M., Paik, Y.-K., and Gaylor, J. L.,** Oxidative demethylation of lanosterol in cholesterol biosynthesis: accumulation of sterol intermediates, *J. Lipid Res.*, 27, 1, 1986.
115. **Fischer, R. T., Stam, S. H., Johnson, P. R., Ko, S. S., Magolda, R. L., Gaylor, J. L., and Trzaskos, J. M.,** Mechanistic studies of lanosterol 14α-methyl demethylase: substrate requirements for the component reactions catalyzed by a single cytochrome P-450 isozyme, *J. Lipid Res.*, 30, 1621, 1989.
115a. **Fischer, R. T., Trzaskos, J. M., Magolda, R. L., Ko, S. S., Brosz, C. C., and Larsen, B.,** Lanosterol 14α-demethylase-isolation and characterization of the third metabolically generated oxidative demethylation intermediate, *J. Biol. Chem.*, 266, 6124, 1991.
116. **Aoyama, J., Yoshida, Y., Sonoda, Y., and Sato, R.,** Role of the 8-double bond of lanosterol in the enzyme-substrate interaction of cytochrome $P\text{-}450_{14DM}$ (lanosterol 14α-demethylase), *Biochim. Biophys. Acta*, 1001, 196, 1989.
117. **Hitchcock, C. A., Dickinson, K., Brown, S. B., and Evans, E. G.,** Purification and properties of cytochrome P-450-dependent 14α-sterol demethylase from *Candida albicans*, *Biochem. J.*, 263, 573, 1989.
118. **Loper, J. C., Chen, C., and Dey, C. R.,** Gene engineering in yeast for biodegradation: immunological cross-reactivity among cytochrome P-450 system proteins of *Saccharomyces cerevisiae* and *Candida tropicalis*, *Hazard. Waste Hazard. Mater.*, 2, 131, 1985.
119. **Sanglard, D., Käppli, O., and Fiechter, A.,** The distinction of different types of cytochromes P450 from the yeasts *Candida tropicalis* and *Saccharomyces uvarum*, *Arch. Biochem. Biophys.*, 251, 176, 1986.
120. **Kalb, V. F., Woods, C. W., Turi, T. G., Dey, C. R., Sutter, T. R., and Loper, J. C.,** Primary structure of the P450 lanosterol demethylase gene from *Saccharomyces cerevisiae*, *DNA*, 6, 529, 1987.
121. **Ishida, N., Aoyama, Y., Hatanaka, R., Oyama, Y., Imajo, S., Ishiguro, M., Oshima, T., Nakazato, H., Noguchi, T., Maitra, U. S., Mohan, V. P., Sprinson, D. B., and Yoshida, Y.,** A single amino acid substitution converts cytochrome $P\text{-}450_{14DM}$ to an inactive form, cytochrome $P450_{SG1}$: complete primary structures deduced from cloned DNAs, *Biochem. Biophys. Res. Commun.*, 155, 317, 1988.
122. **Chen, C., Kalb, V. F., Turi, T. G., and Loper, J. C.,** Primary structure of the cytochrome P450 lanosterol 14α-demethylase gene from *Candida tropicalis*, *DNA*, 7, 617, 1988.
123. **Lai, M. H. and Kirsch, D. R.,** Nucleotide sequence of cytochrome P450 L1A1 (lanosterol 14α-demethylase) from *Candida albicans*, *Nucleic Acid Res.*, 17, 804, 1989.
124. **Lindberg, R. L. P. and Negishi, M.,** Alteration of mouse cytochrome $P450_{coh}$ substrate specificity by mutation of a single amino-acid residue, *Nature*, 339, 632, 1989.
125. **Chen, C., Turi, T. G., Sanglard, D., and Loper, J. C.,** Isolation of the *Candida tropicalis* gene for P450 lanosterol demethylase and its expression in *Saccharomyces cerevisiae*, *Biochem. Biophys. Res. Commun.*, 146, 1311, 1987.
126. **Anderson, S., Davis, D. L., Dahlbäck, H., Jörnvall, H., and Russell, D. W.,** Cloning, structure, and expression of the mitochondrial cytochrome P-450 sterol 26-hydroxylase, a bile acid biosynthetic enzyme, *J. Biol. Chem.*, 264, 8222, 1989.
127. **Furuya, H., Shimizu, T., Hirano, K., Hatano, M., Fujii-kuriyama, Y., Raag, R., and Poulos, T. L.,** Site-directed mutageneses of rat liver cytochrome $P\text{-}450_d$: catalytic activities toward benzphetamine and 7-ethoxycoumarin, *Biochemistry*, 28, 6848, 1989.

128. **Imai, Y. and Nakamura, M.**, Point mutations at threonine-301 modify substrate specificity of rabbit liver microsomal cytochrome P-450 (laurate (ω-1)-hydroxylase and testosterone 16α-hydroxylase), *Biochem. Biophys. Res. Commun.*, 158, 717, 1989.
129. **Yabusaki, Y., Murakami, H., and Ohkawa, H.**, Primary structure of *Saccharomyces cerevisiae* NADPH-cytochrome P450 reductase deduced from nucleotide sequence of its cloned gene, *J. Biochem.*, 103, 1004, 1988.
130. **Sutter, T. R. and Loper, J. C.**, Disruption of the *Saccharomyces cerevisiae* gene for NADPH-cytochrome P450 reductase causes increased sensitivity to ketoconazole, *Biochem. Biophys. Res. Commun.*, 160, 1257, 1989.
131. **Paik, Y.-K., Billheimer, J. T., Magolda, R. L., and Gaylor, J. L.**, Microsomal enzymes of cholesterol biosynthesis from lanosterol — solubilization and purification of steroid 8-isomerase, *J. Biol. Chem.*, 261, 6470, 1986.
132. **Lenton, J. R., Goad, L. J., and Goodwin, T. W.**, Sterols of *Xanthoria parientina:* evidene for two sterol 'pools' and the identification of a novel C_{28} triene, ergosta-5,8,22-trien-3-β-ol, *Phytochemistry*, 12, 1135, 1973.
133. **Yokokawa, H. and Mitsuhashi, T.**, The sterol composition of mushrooms, *Phytochemistry*, 20, 1349, 1981.
134. **Gaylor, J. L., Delwiche, C. V., and Swindell, A. C.**, Enzymatic isomerization (Δ^8-Δ^7) of intermediates of sterol biosynthesis, *Steroids*, 8, 353, 1966.
135. **Benveniste, P.**, Sterol biosynthesis, *Annu. Rev. Plant Physiol.*, 37, 275, 1986.
136. **Taton, M., Benveniste, P., and Rahier, A.**, Mechanism of inhibition of sterol biosynthesis enzymes by *N*-substituted morpholines, *Pestic. Sci.*, 21, 269, 1987.
137. **Baloch, R. I. and Mercer, E. I.**, Inhibition of sterol Δ^8-Δ^7-isomerase and Δ^{14}-reductase by fenpropimorph, tridemorph and fenpropidin in cell-free enzyme systems from *Saccharomyces cerevisiae*, *Phytochemistry*, 26, 663, 1987.
138. **Shimokawa, O., Kato, Y., Kawano, K., and Nakayama, H.**, Accumulation of 14α-methylergosta-8,24(28)-dien-3β,6α-diol in 14α-demethylation mutants of *Candida albicans:* genetic evidence for the involvement of 5-desaturase, *Biochim. Biophys. Acta*, 1003, 15, 1989.
139. **Hata, S., Nishino, T., Katsuki, H., Aoyama, Y., and Yoshida, Y.**, Characterization of Δ22-desaturation in ergosterol biosynthesis of yeast, *Agric. Biol. Chem.*, 51, 1349, 1987.
140. **Oritz de Montellano**, Cytochrome P-450 catalysis: radical intermediates, *Trends Pharmacol. Sci.*, 10, 354, 1989.
141. **Rudney, H. and Sexton, R. C.**, Regulation of cholesterol biosynthesis, *Annu. Rev. Nutr.*, 6, 245, 1986.
142. **Goldstein, J. L. and Brown, M. S.**, Regulation of the mevanolate pathway, *Nature*, 343, 425, 1990.
143. **Jingami, H., Brown, M. S., Goldstein, J. L., Anderson, R. W. G., and Luskey, K. L.**, Partial deletion of membrane-bound domain of 3-hydroxy-3-methylglutaryl coenzyme A reductase eliminates sterol-enhanced degradation and prevents formation of crystalloid endoplasmic reticulum, *J. Cell Biol.*, 104, 1693, 1987.
144. **Osborne, T. F., Gil, G., Goldstein, J. L., and Brown, M. S.**, Operator constitutive mutation of 3-hydroxy-3-methylglutaryl-coenzyme A reductase promoter abolishes protein binding to sterol regulatory elements, *J. Biol. Chem.*, 263, 3380, 1988.
145. **Trzaskos, J. M., Jonas, M., and Chen, H. W.**, Sterol-mediated suppression of HMG-CoA reductase mRNA levels in cultured cells requires protein synthesis, *Biochem. Biophys. Res. Commun.*, 161, 267, 1989.
146. **Smith, J. R., Osbornes, T. F., Brown, M. S., Goldstein, J. L., and Gil, G.**, Multiple sterol regulatory elements in promoter for hamster 3-hydroxy-3-methylglutaryl-coenzyme A synthases, *J. Biol. Chem.*, 263, 18480, 1988.
147. **Boll, M., Löwel, M., Still, J., and Berndt, J.**, Sterol biosynthesis in yeast 3-hydroxy-3-methylglutaryl-coenzyme A reductase as a regulatory enzyme, *Eur. J. Biochem.*, 54, 435, 1975.
148. **Quain, D. E. and Haslam, J. M.**, The effects of catabolic derepression on the accumulation of steryl esters and the activity of β-hydroxy-β-3methylglutaryl-CoA reductase in *Saccharomyces cerevisiae*, *J. Gen. Microbiol.*, 111, 343, 1979.
149. **Bard, M. and Downing, J. F.**, Genetic and biochemical aspects of yeast sterol regulation involving 3-hydroxy-3-methylglutaryl-coenzyme A reductase, *J. Gen. Microbiol.*, 125, 415, 1981.
150. **Pinto, W., Lozano, R., and Nes, W. R.**, Inhibition of sterol biosynthesis by ergosterol and cholesterol in *Saccharomyces cerevisiae*, *Biochim. Biophys. Acta*, 836, 89, 1985.
151. **Trocha, P. and Sprinson, D.**, Localization and regulation of early enzymes of sterol biosynthesis in yeast, *Arch. Biochem. Biophys.*, 174, 45, 1976.
152. **Servouse, M. and Karst, F.**, Regulation of early enzymes of ergosterol biosynthesis in *Saccharomyces cerevisiae*, *Biochem. J.*, 240, 541, 1986.
153. **Basson, M. E., Thorsness, M., and Rine, J.**, Identifying mutations in duplicated functions in *Saccharomyces cerevisiae:* genetic analysis of an early step, *Genetics*, 117, 645, 1987.

154. **Sengstag, C., Stirling, C., Schekman, R., and Rine, J.,** Genetic and biochemical evaluation of eucaryotic membrane topology: multiple transmembrane domains of *Saccharomyces cerevisiae* 3-hydroxy-3-methylglutaryl coenzyme A reductase, *Mol. Cell. Biol.*, 10, 672, 1990.
155. **Vanden Bossche, H.,** Importance and role of sterols in fungal membranes, in *Biochemistry of Cell Walls and Membranes in Fungi*, Kuhn, P. J., Trinci, A P. J., Jung, M. J., Goosey, M. W., and Copping, L. G., Eds., Springer-Verlag, Berlin, 1990, 135.
156. **Bloch, K.,** Sterol structure and membrane function, *CRC Crit. Rev. Biochem.*, 14, 47, 1983.
157. **Yeagle, P. L.,** Cholesterol and the cell membrane, *Biochim. Biophys. Acta*, 822, 267, 1985.
158. **Yeagle, P. L.,** Cholesterol and the cell membrane, in *Biology of Cholesterol*, Yeagle, P. L., Ed., CRC Press, Boca Raton, FL, 1988, 121.
159. **Nes, W. R., Sekula, B. C., Nes, W. D., and Adler, J. H.,** The functional importance of structural features of ergosterol in yeast, *J. Biol. Chem.*, 253, 6218, 1978.
160. **Leibel, W. S., Firestone, L. L., Legler, D. C., Braswell, L. M., and Miller, K. W.,** Two pools of cholesterol in acetylcholine receptor-rich membranes from *Torpedo*, *Biochim. Biophys. Acta*, 897, 249, 1987.
161. **Lösel, D. M.,** Functions of lipids: specialized roles in fungi and algae, in *Microbial Lipids*, Vol. 2, Ratledge, C. and Wilkinson, S. G., Eds., Academic Press, London, 1989, 367.
162. **Dahl, J. S., Dahl, C. E., and Bloch, K.,** Sterols in membranes: growth characteristics and membrane properties of *Mycoplasma capricolum* cultured on cholesterol and lanosterol, *Biochemistry*, 19, 1467, 1980.
163. **Nes, W. R.,** Structure-function relationships for sterols in *Saccharomyces cerevisiae*, in *Ecology and Metabolism of Plant Lipids*, Fuller, G. and Nes, W. D., Eds., American Chemical Society, Washington, D.C., 1987, 253.
164. **Dahl, C. and Dahl, J.,** Cholesterol and cell function, in *Biology of Cholesterol*, Yeagle, P. L., Ed., CRC Press, Boca Raton, FL, 1988, 147.
165. **Parks, L. W., Bottema, C. D. K., Rodriguez, R. J., and Lewis, T. A.,** Yeast sterols: yeast mutants as tools for the study of sterol metabolism, *Methods Enzymol.*, 111, 333, 1985.
166. **Lewis, T. A., Taylor, F. R., and Parks, L. W.,** Involvement of heme biosynthesis in control of sterol uptake by *Saccharomyces cerevisiae*, *J. Bacteriol.*, 163, 199, 1985.
167. **Nes, W. R., Dhanuka, I. C., and Pinto, W. J.,** Evidence for facilitated transport in the absorption of sterols by *Saccharomyces cerevisiae*, *Lipids*, 21, 102, 1986.
168. **Pinto, W. J., Lozano, R., Sekula, B. C., and Nes, W. R.,** Stereochemically distinct roles for sterols in *Saccharomyces*, *Biochem. Biophys. Res. Commun.*, 112, 47, 1983.
169. **Pinto, W. J. and Nes, W. R.,** Stereochemical specificity for sterols in *Saccharomyces cerevisiae*, *J. Biol. Chem.*, 258, 4472, 1983.
170. **Rodriguez, R. J., Taylor, F. R., and Parks, L. W.,** A requirement for ergosterol to permit growth of yeast sterol auxotrophs on cholestanol, *Biochem. Biophys. Commun.*, 106, 435, 1982.
171. **Bottema, C. D. K., Rodriguez, R. J., and Parks, L. W.,** Influence of sterol on yeast plasma membrane properties, *Biochim. Biophys. Res. Commun.*, 813, 313, 1985.
172. **Lorenz, R. T., Casey, W. M., and Parks, L. W.,** Structural discrimination in the sparking function of sterols in the yeast *Saccharomyces cerevisiae*, *J. Bacteriol.*, 171, 6169, 1989.
173. **Goad, L. J.,** Application of sterol synthesis inhibitors to investigate the sterol requirements of protozoa and plants, *Biochem. Soc. Trans.*, 18, 63, 1990.
174. **Ramgopal, M. and Bloch, K.,** Sterol synergism in yeast, *Proc. Natl. Acad. Sci. U.S.A.*, 80, 712, 1983.
175. **Kawasaki, S., Ramgopal, M., Chin, J., and Bloch, K.,** Sterol control of the phosphatidylethanolamine-phosphatidylcholine conversion in the yeast mutant GL7, *Proc. Natl. Acad. Sci. U.S.A.*, 82, 5715, 1985.
176. **Nes, W. R. and Dhanuka, I. C.,** Inhibition of sterol synthesis by Δ^5-sterols in a sterol auxotroph of yeast defective in oxidosqualene cyclase and cytochrome P-450, *J. Biol. Chem.*, 263, 11844, 1988.
177. **Lees, N. A., Bard, M., Kemple, M. D., Haak, R. A., and Kleinhans, F. W.,** ESR determination of membrane order parameter in yeast sterol mutants, *Biochim. Biophys. Acta*, 553, 469, 1979.
178. **Low, C., Rodriguez, R. J., and Parks, L. W.,** Modulation of yeast plasma membrane composition of a yeast sterol auxotroph as a function of exogenous sterol, *Arch. Biochem. Biophys.*, 240, 530, 1985.
179. **Buttke, T. M. and Bloch, K.,** Comparative responses of the yeast mutant strain GL7 to lanosterol, cycloartenol, and cyclolaudenol, *Biochem. Biophys. Res. Commun.*, 92, 229, 1980.
180. **Taylor, F. R., Rodriguez, R. J., and Parks, L. W.,** Requirement for a second sterol biosynthetic mutation for viability of a sterol C-14 demethylation defect in *Saccharomyces cerevisiae*, *J. Bacteriol.*, 155, 64, 1983.
181. **King, D. J., Wiseman, A., Kelly, D. E., and Kelly, S. L.,** Differences in the cytochrome P-450 enzymes of sterol C-14 demethylase mutants of *Saccharomyces cerevisiae*, *Curr. Genet.*, 10, 261, 1985.
182. **Bard, M., Lees, N. D., Barbuch, R. J., and Sanglard, D.,** Characterization of a cytochrome P-450 deficient mutant of *Candida albicans*, *Biochem. Biophys. Res. Commun.*, 147, 794, 1987.
183. **Ziogas, B. N., Sisler, H. D., and Lusby, W. R.,** Sterol content and other characteristics of pimaricin-resistant mutants of *Aspergillus nidulans*, *Pestic. Biochem. Physiol.*, 20, 320, 1983.

184. **Dahl, J. S. and Dahl, C. E.**, Stimulation of cell proliferation and phosphoinosite metabolism in *Saccharomyces cerevisiae* GL7 by ergosterol, *Biochem. Biophys. Res. Commun.*, 133, 844, 1985.
185. **Shears, S. B.**, Metabolism of the inositol phosphates produced upon receptor activation, *Biochem. J.*, 260, 313, 1989.
186. **Dahl, C. E., Biemann, H.-P., and Dahl, J. S.**, A protein kinase antigenically related to $pp60^{v\text{-}src}$ possibly involved in yeast cell cycle control: positive *in vivo* regulation by sterol, *Proc. Natl. Acad. Sci. U.S.A.*, 84, 4012, 1987.
187. **Nes, W. D. and Heupel, R. C.**, Physiological requirement for biosynthesis of multiple 24β-methyl sterols in *Gibberella fujikuroi*, *Arch. Biochem. Biophys.*, 244, 211, 1986.
188. **Nes, W. D., Hanners, P. K., and Parish, E. J.**, Control of fungal sterol C-24 transalkylation: importance to developmental regulation, *Biochem. Biophys. Res. Commun.*, 139, 410, 1986.
189. **Smith, F. D. and Köller, W.**, The effect of flusilazole on the germination of conidia of sensitive and resistant isolates of *Venturia inaequalis*, *Phytopathology*, 80, 1000, 1990.
190. **Xu, S. and Nes, W. D.**, Biosynthesis of cholesterol in the yeast mutant Erg 6, *Biochem. Biophys. Res. Commun.*, 155, 509, 1988.
191. **Ringold, G., Ed.**, *Steroid Hormone Action*, Alan R. Liss, New York, 1988.
192. **Clark, C. R., Ed.**, *Steroid Hormone Receptors — Their Intracellular Localisation*, Ellis Horwood, Chichester, 1987.
193. **Gronemeyer, H., Ed.**, *Affinity Labelling and Cloning of Steroid and Thyroid Hormone Receptors*, Ellis Horwood, Chichester, 1988, 1.
194. **Svoboda, J. A. and Thompson, M. J.**, Variability in steroid metabolism among phytophagous insects, in *Ecology and Metabolism of Plant Lipids*, Fuller, G. and Nes, W. D., Eds., American Chemical Society, Washington, D.C., 1987, 176.
195. **Meudt, W. J.**, Chemical and biological aspects of brassinolide, in *Ecology and Metabolism of Plant Lipids*, Fuller, G. and Nes, W. D., Eds., American Chemical Society, Washington, D.C., 1987, 53.
196. **Margalith, P. Z.**, *Steroid Microbiology*, Charles C Thomas, Springfield, IL, 1986, chap. 5.
197. **Brunt, S. A. and Silver, J. C.**, Steroid-hormone induced changes in secreted proteins in the filamentous fungus *Achlya*, *Exp. Cell Res.*, 163, 22, 1986.
198. **Brunt, S. A. and Silver, J. C.**, Cellular localization of steroid hormone-regulated proteins during sexual development in *Achlya*, *Exp. Cell Res.*, 165, 306, 1986.
199. **Horton, J. S. and Horgen, P. A.**, Synthesis of an antheridiol-inducible polypeptide during sexual morphogenesis of *Achlya ambisexualis* E87, *Can. J. Biochem. Cell Biol.*, 63, 355, 1985.
200. **Brunt, S. A. and Silver, J. C.**, Steroid hormone-regulated basic proteins in *Achlya ambisexualis*, *Exp. Mycol.*, 11, 65, 1987.
201. **Riehl, R. M. and Toft, D. O.**, Analysis of the steroid receptor of *Achlya ambisexualis*, *J. Biol. Chem.*, 259, 15324, 1984.
202. **Horton, J. S. and Horgen, P. A.**, Molecular cloning of cDNAs regulated during steroid-induced sexual differentiation in the aquatic fungus *Achlya*, *Exp. Mycol.*, 13, 263, 1989.
203. **Nes, W. D. and Stafford, A. E.**, Evidence for metabolic and functional discrimination of sterols by *Phytophthora cactorum*, *Proc. Natl. Acad. Sci. U.S.A.*, 80, 3227, 1983.
204. **Feldman, D.**, Evidence for the presence of steroid hormone receptors in fungi, in *Steroid Hormone Action*, Ringold, G., Ed., Alan R. Liss, New York, 1988, 169.
205. **Das, M. and Datta, A.**, Steroid binding protein(s) in yeast, *Biochem. Int.*, 11, 171, 1985.
206. **Kinsman, O. S., Pitblado, K., and Coulson, C. J.**, Effect of mammalian steroid hormones and luteinizing hormone on the germination of *Candida albicans* and implications for vaginal candidosis, *Mycoses*, 31, 617, 1988.
207. **Loose, D. S., Schurman, D. J., and Feldman, D.**, A corticosteroid binding protein and endogenous ligand in *Candida albicans* indicating a possible steroid receptor system, *Nature*, 293, 477, 1981.
208. **Clemons, K. V., Stover, E. P., Schär, G., Statis, P. A., Chan, K., Tökés, L., Stevens, D. A., and Feldman, D.**, Steroid metabolism as a mechanism of escape from progesterone-mediated growth inhibition in *Trichophyton mentagrophytes*, *J. Biol. Chem.*, 264, 11186, 1989.
209. **Parks, L. W., Rodriguez, R. J., and Low, C.**, An essential fungal growth factor derived from ergosterol: a new end product of sterol biosynthesis?, *Lipids*, 21, 89, 1986.
210. **Mahato, S. B. and Mukherjee, A.**, Steroid transformations by microorganisms, *Phytochemistry*, 23, 2131, 1984.
211. **Hazen, E. L. and Brown, R.**, Two antifungal agents produced by a soil actinomycete, *Science*, 112, 423, 1950.
212. **Ryley, J. F., Wilson, R. G., Gravestock, M. B., and Poyser, J. P.**, Experimental approaches to antifungal therapy, *Adv. Pharmacol. Chemother.*, 18, 49, 1981.
213. **Schaffner, C. P.**, Polyene macrolides in clinical practice: pharmacology and adverse and other effects, in *Macrolide Antibiotics — Chemistry, Biology, and Practice*, Omura, S., Ed., Academic Press, New York, 1984, 457.

214. **Mechlinski, W., Schaffner, C. B., Ganis, P., and Avitabile, G.,** Structure and absolute configuration of the polyene macrolide antibiotic amphotericin B, *Tetrahedron Lett.*, 44, 3873, 1970.
215. **Kerridge, D. and Whelan, W. L.,** The polyene macrolide antibiotics and 5-fluorocytosine: molecular actions and interactions, in *Mode of Action of Antifungal Agents,* Trinci, A. P. J. and Ryley, J. F., Eds., Cambridge University Press, Cambridge, 1984, 343.
216. **Gale, E. F.,** Mode of action and resistance mechanisms of polyene macrolides, in *Macrolide Antibiotics — Chemistry, Biology, and Practice,* Omura, S., Ed., Academic Press, New York, 1984, 425.
217. **Kerridge, D.,** Mode of action of clinically important antifungal drugs, *Adv. Microb. Physiol.*, 27, 1, 1986.
218. **Bolard, J.,** How do the polyene macrolide antibiotics affect cellular membrane properties?, *Biochim. Biophys. Acta,* 864, 257, 1986.
219. **Thomas, A. H.,** Suggested mechanisms for the antimycotic activity of the polyene antibiotics and the *N*-substituted imidazoles, *J. Antimicrob. Chemother.*, 17, 269, 1986.
220. **Medoff, G.,** The mechanism of action of amphotericin, in *Aspergillus and Aspergillosis,* Vanden Bossche, H., Mackenzie, D. W. R., and Cauwenbergh, G., Eds., Plenum Press, New York, 1987, 161.
221. **Kerridge, D.,** Polyene macrolide antibiotics, in *Aspergillus and Aspergillosis,* Vanden Bossche, H., Mackenzie, D. W. R., and Cauwenbergh, G., Eds., Plenum Press, New York, 1987, 147.
222. **Lampen, J. O., Arnov, P. M., and Safferman, R. S.,** Mechanism of protection by sterols against polyene antibiotics, *J. Bacteriol.*, 80, 200, 1960.
223. **Weber, M. M. and Kinsky, S. C.,** Effect of cholesterol on the sensitivity of *Mycoplasma laidlawii, J. Bacteriol.*, 89, 306, 1965.
224. **De Kruijff, B., Gerritsen, V. J., Oerlemans, A., Demel, R. A., and Van Deenen, L. L. M.,** Polyene antibiotic sterol interaction in membranes of *Acholeplasma laidlawii* cells and lecithin liposomes, *Biochim. Biophys. Acta,* 339, 30, 1974.
225. **Lampen, J. O.,** Interference by polyene antibiotics (especially nystatin and filipin) with specific membrane functions, in *Biochemical Studies of Antimicrobial Drugs,* Newton, B. A. and Reynolds, P. E., Eds., Cambridge University Press, Cambridge, 1966, 111.
226. **Palacios, J. and Serrano, R.,** Proton permeability induced by polyene antibiotics: a plausible mechanism for their inhibition of maltose fermentation in yeast, *FEBS Lett.*, 91, 198, 1978.
227. **Serrano, R.,** Structure and function of proton translocating ATPase in plasma membranes, *Biochim. Biophys. Acta,* 947, 1, 1988.
228. **Portillo, F. and Serrano, R.,** Growth control strength and active site of yeast plasma membrane ATPase studied by site-directed mutagenesis, *Eur. J. Biochem.*, 186, 501, 1989.
229. **Cohen, B. E.,** Concentration and time-dependence of amphotericin B induced permeability changes across ergosterol-containing liposomes, *Biochim. Biophys. Acta,* 857, 117, 1985.
230. **Gruda, I. and Dusault, N.,** Effect of the aggregation state of amphotericin B on its interaction with ergosterol, *Biochem. Cell Biol.*, 66, 177, 1988.
231. **Vertrut-Croquin, A., Bolard, J., Chubbert, M., and Gary-Bobo, C.,** Differences in the interaction of the polyene antibiotic amphotericin B with cholesterol- or ergosterol-containing phospholipid vesicles. A circular dichroism and permeability study, *Biochemistry,* 22, 2939, 1983.
232. **Gruda, I. and Bolard, J.,** On the existence of amphotericin B-sterol complex in lipid vesicles and in propanol-water systems, *Biochem. Cell Biol.*, 65, 234, 1987.
233. **Sekiya, T., Yano, K., and Nozawa, Y.,** Effects of amphotericin B and its methyl ester on plasma membranes of *Candida albicans* and erythrocytes as examined by freeze-fracture electron microscopy, *Sabouraudia,* 20, 303, 1982.
234. **Oranusi, N. A. and Trinci, A. P. J.,** Growth of bacteria on chitin, fungal cell walls and fungal biomass, and the effect of extracellular enzymes produced by these cultures on the antifungal activity of amphotericin B, *Microbios,* 43, 17, 1985.
235. **Al-Bassam, T., Poulain, D., Giumelly, P., Lematre, J., and Bonaly, R.,** Chemical and antigenic alterations of *Candida albicans* cell walls related to the action of amphotericin sub-inhibitory doses, *J. Antimicrob. Chemother.*, 15, 263, 1985.
236. **Beezer, A. E.,** Interaction of nystatin with *Saccharomyces cerevisiae:* an example of enthalpy-entropy compensation?, *Microbios,* 39, 117, 1984.
237. **Ellul, H., Beezer, A. E., Brian, A. P. R., Miles, R. J., and Sivayogan, S. R.,** The effect of chemical modifications of *Saccharomyces cerevisiae* on electrophoretic mobility, cell-wall structure and amphotericin B uptake, *Biochim. Biophys. Acta,* 845, 151, 1985.
238. **Brajtburg, J., Medoff, G., Kobayashi, G., and Elberg, S.,** Influence of extracellular K^+ or Mg^{2+} on the stages of the antifungal effects of amphotericin B and fillipin, *Antimicrob. Agents Chemother.*, 18, 593, 1980.
239. **Sokol-Anderson, M. L., Brajtburg, J., and Medoff, G.,** Amphotericin B-induced oxidative damage and killing of *Candida albicans, J. Infect. Dis.*, 154, 76, 1986.
240. **Brajtburg, J., Elberg, S., Kobayashi, G. S., and Medoff, G.,** Effects of ascorbic acid on the antifungal action of amphotericin B, *J. Antimicrob. Chemother.*, 24, 333, 1989.

241. **Saksena, A. K., Girijavallabhan, V. M., Cooper, A. B., and Loebenberg, D.,** Recent advances in antifungal agents, *Annu. Rep. Med. Chem.*, 24, 111, 1989.
242. **Schaffner, C. P.,** Amphotericin B derivatives, in *Recent Trends in the Discovery, Development and Evaluation of Antifungal Agents,* Fromtling, R. A., Ed., J. R. Prous Science, Barcelona, 1988, 595.
243. **Brajtburg, J., Powderly, W. G., Kobayashi, G. S., and Medoff, G.,** Amphotericin B: delivery systems, *Antimicrob. Agents Chemother.*, 34, 381, 1990.
244. **Scholer, H. J. and Polak, A.,** Resistance to systemic antifungal agents, in *Antimicrobial Drug Resistance,* Bryan, L. E., Ed., Academic Press, Orlando, 1984, 393.
245. **Hitchcock, C. A., Russell, N. J., and Barrett-Bee, K. J.,** Sterols in *Candida albicans* mutants resistant to polyene or azole antifungals, and of a double-mutant *C. albicans* 6.4, *CRC Crit. Rev. Microbiol.*, 15, 111, 1987.
246. **Hitchcock, C. A., Barrett-Bee, K. J., and Russell, N. J.,** The lipid composition and permeability to azole of an azole- and polyene-resistant mutant of *Candida albicans, J. Med. Vet. Mycol.*, 25, 29, 1987.
247. **Capek, A. and Simek, A.,** Antimicrobial agents. XII. Relationship between biochemical resistance and microbial degradation of antimycotics, *Folia Microbiol.*, 16, 472, 1971.
248. **Beezer, A. E., Miles, R. J., Park, W.-B., Smith, A. R. W., and Brian, P. R.,** A nystatin-resistant mutant of *Saccharomyces cerevisiae:* isolation and characterization by electron microscopy and chemical analysis of whole cells, cell-walls and protoplasts, *Microbios*, 47, 117, 1986.
249. **Steinberg, D.,** Hypocholesterolemic drugs and lipoprotein metabolism, in *Drugs Affecting Lipid Metabolism,* Paoletti, R., Ed., Springer-Verlag, Berlin, 1987, 69.
250. **Omura, S., Tomoda, H., Kumagai, H., Greenspan, M. D., Yodkovitz, J. B., Chen, J. S., Alberts, A. W., Martin, I., Mochales, S., Monghan, R. L., Chabala, J. C., Schwartz, R. E., and Patchett, A. A.,** Potent inhibitory effect of antibiotic 1233A on cholesterol biosynthesis which specifically blocks 3-hydroxy-3-methylglutaryl coenzyme A synthase, *J. Antibiot.*, 40, 1356, 1987.
251. **Greenspan, M. D., Yodkovitz, J. B., Lo, C.-Y. L., Chen, J. S., Alberts, A., Hunt, V. M., Chang, M. N., Yang, S. S., Thompson, K. L., Chiang, Y.-C. P., Chabala, J. C., Monaghan, R. L., and Schwartz, R. E.,** Inhibition of hydroxymethylglutaryl-coenzyme A synthase by L-659,699, *Proc. Natl. Acad. Sci. U.S.A.*, 84, 7488, 1987.
252. **Endo, A.,** Specific nonsterol inhibitors of HMG-CoA reductase, in *Regulation of HMG-CoA Reductase,* Preiss, B., Ed., Academic Press, Orlando, 1985, 49.
253. **Goto, Y.,** The profile of an HMG-CoA reductase inhibitor, CS-514 (SQ 31,000), in *Drugs Affecting Lipid Metabolism,* Paoletti, R., Ed., Springer-Verlag, Berlin, 1987, 247.
254. **Lee, T.-J.,** Synthesis, SARs and therapeutic potential of HMG-CoA reductase inhibitors, *Trends Pharmacol. Sci.*, 8, 442, 1987.
254a. **Joshua, H., Schwartz, M. S., and Wilson, K. E.,** L-669,262, a potent HMG-CoA reductase inhibitor, *J. Antibiotics,* 44, 366, 1991.
255. **Bach, Th. J. and Lichtenthaler, H. K.,** Mechanisms of inhibition by mevinolin (MK 803) of microsome-bound radish and of partially purified yeast HMG-CoA reductase (EC.1.1.1.34), *Z. Naturforsch.*, 38c, 212, 1985.
256. **Bach, Th. J. and Lichtenthaler, H. K.,** Inhibition by mevinolin of plant growth, sterol formation and pigment accumulation, *Physiol. Plant.*, 59, 50, 1983.
257. **Bach, Th. J. and Lichtenthaler, H. K.,** Plant growth regulation by mevinolin and other sterol biosynthesis inhibitors, in *Ecology and Metabolism of Plant Lipids,* Fuller, G. and Nes, W. D., Eds., American Chemical Society, Washington, D.C., 1987, 109.
258. **Nakamura, C. E. and Abeles, R. H.,** Mode of interaction of β-hydroxy-β-methylglutaryl coenzyme A reductase with strong binding inhibitors: compactin and related compounds, *Biochemistry,* 24, 1364, 1985.
259. **Ikeura, R., Murakawa, S., and Endo, A.,** Growth inhibition of yeast by compactin (ML-236B) analogues, *J. Antibiot.*, 41, 1148, 1988.
260. **Reardon, J. E. and Abeles, R. H.,** Inhibition of cholesterol biosynthesis by fluorinated mevanolate analogues, *Biochemistry,* 26, 4717, 1987.
261. **Poulter, C. D., Muehlbacher, M., and Davis, D. R.,** Isopentenyldiphosphate isomerase. Mechanism of active site directed irreversible inhibition by 3-(fluoromethyl)-3-butenyl diphosphate, *J. Am. Chem. Soc.*, 111, 3740, 1989.
262. **Poulter, C. D., Capson, T. L., Thompson, M. D., and Bard, R. S.,** Squalene synthetase. Inhibition by ammonium analogues of carbocationic intermediates in the conversion of presqualene diphosphate to squalene, *J. Am. Chem. Soc.*, 111, 3734, 1989.
263. **Georgopoulos, A., Petrayni, G., Mieth, H., and Drews, J.,** In vitro activity of naftifine, a new antifungal agent, *Antimicrob. Agents Chemother.*, 19, 386, 1981.
264. **Stütz, A.,** Allylamine derivatives — a new class in antifungal chemotherapy, *Angew. Chem. Int. Ed. Engl.*, 26, 320, 1987.
265. **Stütz, A.,** Synthesis and structure-activity correlations within allylamine antibiotics, *Ann. N.Y. Acad. Sci.*, 544, 46, 1988.

265a. **Nussbaumer, P., Ryder, N. S., and Stütz, A.,** Allylamine antimycotics: Recent trends in structure-activity relationships and synthesis, *Pestic. Sci.*, 31, 437, 1991.

266. **Paltauf, F., Daum, G., Zuder, G., Hogenauer, G., Schulz, G., and Seidl, G.,** Squalene and ergosterol biosynthesis in fungi treated with naftifine, a new antimycotic agent, *Biochim. Biophys. Acta*, 712, 268, 1982.

267. **Ryder, N. S. and Troke, P. F.,** The activity of naftifine as a sterol synthesis inhibitor of *Candida albicans*, in *Current Chemotherapy and Immunotherapy*, Periti, P. and Grassi, G. G., Eds., American Society of Microbiology, Washington, D.C., 1982, 10116.

268. **Petranyi, G., Ryder, N. S., and Stütz, A.,** Allylamine derivatives: new class of synthetic antifungal agents inhibiting fungal squalene epoxidase, *Science*, 224, 1239, 1984.

269. **Ryder, N. S.,** Selective inhibition of squalene epoxidation by allylamine antimycotic agents, in *Microbial Cell Wall Synthesis and Autolysis*, Nombela, C., Ed., Elsevier, Amsterdam, 1984, 313.

270. **Ryder, N. S.,** Biochemical mode of action of the allylamine antimycotic agents naftifine and SF 86-327, in *In Vivo and In Vitro Evaluation of Antifungal Agents*, Iwata, K. and Vanden Bossche, H., Eds., Elsevier, Amsterdam, 1986, 89.

271. **Ryder, N. S.,** Squalene epoxidase as the target of antifungal allylamines, *Pestic. Sci.*, 21, 281, 1987.

272. **Ryder, N. S.,** Mechanism of action of the allylamine antimycotics, in *Recent Trends in the Discovery, Development and Evaluation of Antifungal Agents*, Fromtling, R. A., Ed., J. R. Prous Science Publisher, Barcelona, 1987, 451.

273. **Ryder, N. S.,** Mode of action of allylamines, in *Sterol Biosynthesis Inhibitors — Pharmaceutical and Agrichemical Aspects*, Berg, D. and Plempel, M., Eds., Ellis Horwood, Chichester, 1988, 151.

274. **Ryder, N. S.,** Inhibition of squalene epoxidase and sterol-side chain methylation by allylamines, *Biochem. Soc. Trans.*, 18, 45, 1990.

275. **Ryder, N. S.,** Mechanism of action and biochemical selectivity of allylamine antibiotics, *Ann. N.Y. Acad. Sci.*, 544, 208, 1988.

276. **Ryder, N. S. and DuPont, M.-C.,** Inhibition of squalene epoxidase by allylamine antibiotic compounds, *Biochem. J.*, 230, 765, 1985.

277. **Bach, T. J.,** Selected natural and synthetic enzyme inhibitors of sterol biosynthesis as molecular probes for in vivo studies concerning the regulation of plant growth, *Plant Sci.*, 39, 183, 1985.

278. **Schuster, I.,** Potential of allylamines to inhibit cytochrome P-450, in *Recent Trends in the Discovery, Development and Evaluation of Antifungal Agents*, Fromtling, R. A., Ed., J. R. Prous Science Publisher, Barcelona, 1987, 471.

279. **Barrett-Bee, K. J., Lane, A. C., and Turner, R. W.,** The mode of antifungal action of tolnaftate, *J. Med. Vet. Mycol.*, 24, 155, 1986.

280. **Ryder, N. S., Frank, I., and DuPont, M.-C.,** Ergosterol biosynthesis inhibition by the thiocarbamate antifungal agents tolnaftate and tolciclate, *Antimicrob. Agents Chemother.*, 29, 858, 1986.

281. **Ryder, N. S., DuPont, M.-C., and Frank, I.,** Inhibition of fungal and mammalian sterol biosynthesis by 2-aza-2,3-dihydrosqualene, *FEBS Lett.*, 204, 239, 1986.

281a. **Orth, A. B., Henry, M. J., and Sisler, H. D.,** Mechanism of resistance to terbinafine in two isolates of *Ustilago maysid, Pestic. Biochem, Physiol.*, 37, 182, 1990.

281b. **Orth, A. B. and Sisler, H. D.,** Mode of action of terbinafine in *Ustilago maydis* and characterization of resistant mutants, *Pestic, Biochem, Physiol.*, 37, 53, 1990.

282. **Corey, E. J., Oritz de Montellano, P. R., Lin, K., and Dean, P. D. G.,** 2,3-Iminosqualene, a potent inhibitor of the enzymatic cyclization of 2,3-oxidosqualene to sterols, *J. Am. Chem. Soc.*, 89, 2797, 1967.

283. **Nes, W. D. and Parish, E. J.,** Metabolism of 2(*RS*), 3-epiminosqualene to 24 (*RS*), 25-epiminolanosterol by *Gibberella fujikuroi, Lipids*, 23, 375, 1988.

284. **Cattel, L., Cerutti, M., Delprino, L., Balliano, G., Duriatti, A., and Bouvier-Navé, P.,** The squalene-2,3-epoxide cyclase as a model for the development of new drugs, *Lipids*, 21, 31, 1986.

285. **Jolidon, S., Polak, A.-M., Guerry, P., and Hartman, P. G.,** Inhibitors of 2,3-oxidosqualene lanosterol-cyclase as potential antifungal agents, *Biochem. Soc. Trans.*, 18, 47, 1990.

286. **Balliano, G., Viola, F., Ceruti, M., and Cattel, L.,** Inhibition of sterol biosynthesis in *Saccharomyces cerevisiae* by *N,N*-diethylsqualene and derivatives, *Biochim. Biophys. Acta*, 959, 9, 1988.

287. **Duriatti, A., Bouvier-Nave, P., Benveniste, P., Schuber, F., Delprino, L., Balliano, G., and Cattel, L.,** *In vitro* inhibition of animal and higher plants 2,3-oxidosqualene-sterol cyclases by 2-aza-2,3-dihydrosqualene and derivatives, and by other ammonium-containing molecules, *Biochem. Pharmacol.*, 34, 2765, 1985.

288. **Ceruti, M., Balliano, G., Viola, F., Cattel, L., Gerst, N., and Schuber, F.,** Synthesis and biological activity of azasqualenes, bis-azasqualenes and derivatives, *Eur. J. Med. Chem.*, 22, 199, 1987.

289. **Sexton, R. C., Panini, S. R., Azran, F., and Rudney, H.,** Effects of 3β-[2-(diethylamino)ethoxy]androst-5-en-17-one on the synthesis of cholesterol and ubiquinone in rat intestinal epithelial cell cultures, *Biochemistry*, 22, 5687, 1983.

290. **Field, R. B., Holmlund, C. E., and Whittaker, N. F.,** The effect of the hypocholesteremic compound 3β-(β-dimethylaminoethoxy)androst-5-en-17-one on the sterol and steryl ester composition of *Saccharomyces cerevisiae, Lipids,* 14, 741, 1979.
291. **Mercer, E. I. and Morris, P. K.,** Differences in the inhibitory effects of *N*-(1-*n*-dodecyl)-heterocycles on the 2,3-oxidosqualene lanosterol-cyclase of rat liver and yeast, *Comp. Biochem. Physiol. B,* 80, 341, 1985.
292. **Taton, M., Benveniste, P., and Rahier, A.,** *N*-[(1,5,9)-Trimethyldecyl]-4α,10-dimethyl-8-aza-*trans*-decal-3β-ol, a novel inhibitor of 2,3 oxidosqualene cycloartenol and lanosterol cyclases, *Biochem. Biophys. Res. Commun.,* 138,'764, 1986.
293. **Gerst, N., Duriatti, A., Schuber, F., Taton, M., Benveniste, P., and Rahier, A.,** Potent inhibition of cholesterol biosynthesis in 3T3 fibroblasts by *N*-[(1,5,9)-trimethyldecyl]-4α,10-dimethyl-8-aza-*trans*-decal-3β-ol, a new 2,3-oxidosqualene cyclase inhibitor, *Biochem. Pharmacol.,* 37, 1955, 1988.
294. **Dawson, P. A., Ridgway, N. D., Slaughter, C. A., Brown, M. S., and Goldstein, J. L.,** cDNA cloning and expression of oxysterol-binding protein, an oligomer with a potential leucine zipper, *J. Biol. Chem.,* 264, 16798, 1989.
295. **Wilson, W. K., Wang, K.-S., Kisic, A., and Schroepfer, G. J.,** Concerning the chemical synthesis of 3β-hydroxy-5a-cholet-8(14)en-15-one, a novel regulator of cholesterol metabolism, *Chem. Phys. Lipids,* 48, 7, 1988.
296. **Schroepfer, G. J., Parish, E. J., and Kandutsch, A. A.,** Inhibitors of sterol biosynthesis. Synthesis and activities of ring C oxygenated sterols, *Chem. Phys. Lipids,* 46, 147, 1988.
297. **Parish, E. J., Honda, H., Chitrakorn, S., and Taylor, F. R.,** A facile synthesis of lanost-8-en-3β-ol-24-one (24-ketosterol). An inhibitor of 3-hydroxy-3-methylglutaryl coenzyme A reductase, *Chem. Phys. Lipids,* 48, 255, 1988.
298. **Schroepfer, G. J., Parish, E. J., Tsuda, M., and Kandutsch, A. A.,** Inhibitors of sterol synthesis. Chemical synthesis, properties and effects of 4,4-dimethyl-15-oxygenated sterols on sterol synthesis and on 3-hydroxy-3-methylglutaryl coenzyme A reductase activity in cultured mammalian cells, *Chem. Phys. Lipids,* 47, 187, 1988.
299. **Panini, S. R., Sexton, R. C., Gupta, A. K., Parish, E. J., Chitrakorn, S., and Rudney, H.,** Regulation of 3-hydroxy-3-methylglutaryl coenzyme A reductase activity and cholesterol biosynthesis by oxylanosterol, *J. Lipid Res.,* 27, 1190, 1986.
300. **Marco de la Calle, C., Hwang, W., Pullinger, C. R., and Gibbons, G. F.,** A relationship between the activities of hepatic lanosterol 14α-demethylase and 3-hydroxy-3-methylglutaryl coenzyme A reductase, *Biochem. J.,* 250, 33, 1988.
301. **Favata, M. F., Trzaskos, J. M., Chen, H. W., Fischer, R. T., and Greenberg, R. S.,** Modulation of 3-hydroxy-3-methylglutaryl coenzyme A reductase by azole antimycotics requires lanosterol demethylation, but not epoxylanosterol formation, *J. Biol. Chem.,* 262, 12254, 1987.
302. **Trzaskos, J. M., Favata, M. F., Fischer, R. T., and Stam, S. H.,** *In situ* accumulation of 3β-hydroxylanost-8-en-32-aldehyde in hepatocyte cultures — a putative regulator of 3-hydroxy-3-methylglutaryl coenzyme A reductase activity, *J. Biol. Chem.,* 262, 12261, 1987.
303. **Thorsness, M., Schafer, W., D'Ari, L., and Rine, J.,** Positive and negative transcriptional control by heme of genes encoding 3-hydroxy-3-methylglutaryl coenzyme A reductase in *Saccharomyces cerevisiae, Mol. Cell. Biol.,* 9, 5702, 1989.
304. **Berg, D., Draber, W., von Hugo, H., and Mayer, D.,** The effect of clotrimazole and triadimefon on 3-hydroxy-3-methyl-glutaryl-CoA-reductase (EC 1.1.134): activity in *Saccharomyces cerevisiae, Z. Naturforsch.,* 36c, 798, 1981.
305. **Rahier, A., Génot, J.-C., Schubert, F., Benveniste, P., and Narula, A. S.,** Inhibition of S-adenosyl-L-methionine sterol-C-24-methyltransferase by analogues of a carbocationic ion high-energy intermediate, *J. Biol. Chem.,* 259, 15215, 1984.
306. **Rahier, A., Taton, M., Bouvier-Navé, P., Schmitt, P., Benveniste, P., Schuber, F., Narula, A. S., Cattel, L., Anding, C., and Place, P.,** Design of high energy intermediate analogues to study sterol biosynthesis in higher plants, *Lipids,* 21, 52, 1986.
307. **Oehlschlager, A. C., Angus, R. H., Pierce, H. D., and Srinivasan, R.,** Azasterol inhibition of Δ^{24}-sterol methyltransferase in *Saccharomyces, Biochemistry,* 23, 3582, 1984.
308. **Büchel, K. H.,** The history of azole chemistry, in *Fungicide Chemistry — Advances and Practical Applications,* Green, M. B. and Spilker, D. A., Eds., American Chemical Society, Washington, D.C., 1986, 1.
309. **Büchel, K. H. and Plempel, M.,** The azole story, in *Chronicles of Drug Discovery,* Vol. 2, Bindra, J. S. and Lednicer, D., Eds., John Wiley & Sons, New York, 1983, 235.
310. **Blohm, T. R. and MacKenzie, R. D.,** Specific inhibition of cholesterol biosynthesis by a synthetic compounds (MER-29), *Arch. Biochem. Biophys.,* 85, 245, 1959.
311. **Ragsdale, N. N. and Sisler, H. D.,** Inhibition of ergosterol biosynthesis in *Ustilago maydis* by the fungicide triarimol, *Biochem. Biophys. Res. Commun.,* 46, 2048, 1972.

312. **Kato, T., Tanaka, S., Ueda, M., and Kawase, Y.**, Inhibition of sterol biosynthesis in *Monilinia fructigena*, *Agric. Biol. Chem.*, 39, 169, 1975.
313. **Buchenauer, H.**, Wirkungsmechanismus von Bayleton (Triadimefon) in *Ustilago avenae*, *Pflanzenschutz Nachr. (Bayer)*, 29, 281, 1976.
314. **Zirngibl, L.**, Fifteen years of structural modifications in the field of antifungal monocyclic 1-substituted 1*H*-azoles, *Prog. Drug Res.*, 27, 253, 1983.
315. **Fromtling, R. A.**, Imidazoles as medically important antifungal agents: an overview, *Drugs Today*, 20, 325, 1984.
316. **Fromtling, R. A.**, Overview of medically important antifungal azole derivatives, *Clin. Microbiol. Rev.*, 1, 187, 1988.
317. **Berg, D., Büchel, K. H., Plempel, M., and Regel, E.**, Antimycotic sterol biosynthesis inhibitors, *Trends Pharmacol. Sci.*, 7, 233, 1986.
318. **Saag, M. S. and Dismukes, W. E.**, Azole antifungal agents: emphasis on new triazoles, *Antimicrob. Agents Chemother.*, 32, 1, 1988.
319. **Vanden Boscche, H., Willemsen, G., Marichal, P., Cools, W., and Lauwers, W.**, The molecular basis for the antifungal activities of N-substituted azole derivatives. Focus on R 51 211, in *Mode of Action of Antifungal Agents*, Trinci, A. P. J. and Ryley, J. F., Eds., Cambridge University Press, Cambridge, 1984, 321.
320. **Vanden Bossche, H., Lauwers, W., Willemsen, G., Marichal, P., Cornelissen, F., and Cools, W.**, Molecular basis for the antimycotic and antibacterial activity of N-substituted imidazoles and triazoles: the inhibition of isoprenoid biosynthesis, *Pestic. Sci.*, 15, 188, 1984.
321. **Sisler, H. D., Ragsdale, N. N., and Waterfield, W. F.**, Biochemical aspects of the fungitoxic and growth regulatory action of fenarimol and other pyrimidin-5-ylmethanols, *Pestic, Sci.*, 15, 167, 1984.
322. **Sisler, H. D. and Ragsdale, N. N.**, Biochemical and cellular aspects of the antifungal action of ergosterol biosynthesis inhibitors, in *Mode of Action of Antifungal Agents*, Trinci, A. P. J. and Ryler, J. F., Eds., Cambridge University Press, Cambridge, 1984, 257.
323. **Borgers, M.**, Antifungal azole derivatives, in *The Scientific Basis of Antimicrobial Chemotherapy*, Greenwood, D. and O'Grady, F., Eds., Cambridge University Press, Cambridge, 1985, 133.
324. **Vanden Bossche, H.**, Biochemical targets for antifungal azole derivatives: hypothesis on the mode of action, in *Current Topics in Medical Mycology, Vol. 1*, McGinnis, M. R., Ed., Springer-Verlag, New York, 1985, 313.
325. **Kato, T.**, Sterol-biosynthesis in fungi, a target for broad spectrum fungicides, in *Chemistry of Plant Protection, Vol. 1*, Haug, G. and Hoffmann, H., Eds., Springer-Verlag, Berlin, 1986, 1.
326. **Buchenauer, H.**, Mechanism of action of triazolyl fungicides and related compounds, in *Modern Selective Fungicides — Properties, Applications and Mechanism of Action*, Lyr, H., Ed., Longman Scientific, Harlow, 1987, 205.
327. **Weete, J. D.**, Mechanism of fungal growth suppression by inhibitors of ergosterol biosynthesis, in *Ecology and Metabolism of Plant Lipids*, Fuller, G. and Nes, W. D., Eds., American Chemical Society, Washington, D.C., 1987, 268.
328. **Janssen, P. A. J. and Vanden Bossche, H.**, Mode of action of cytochrome P-450 monooxygenases inhibitors. Focus on azole derivatives, *Arch. Pharm. Chem. Sci. Ed.*, 15, 23, 1987.
329. **Janssen, P. A. J.**, New antifungal agents: a result of target oriented drug design, in *Frontiers in Microbiology*, De Clercq, E., Ed., Martinus Nijhoff, Dordrecht, 1987, 29.
330. **Vanden Bossche, H., Willemsens, G., and Marichal, P.**, Anti-*Candida* drugs — the biochemical basis for their activity, *CRC Crit. Rev. Microbiol.*, 15, 57, 1987.
331. **Vanden Bossche, H. and Janssen, P. A. J.**, Biological activities of modifiers of cytochrome P-450 activities, in *Drug Metabolism: From Molecule to Man*, Bridges, J. W. and Gibson, G. G., Eds., Taylor & Francis, London, 1987, 244.
332. **Vanden Bossche, H.**, Mode of action of pyridine, pyrimidine and azole antifungals, in *Sterol Biosynthesis Inhibitors — Pharmaceutical and Agrichemical Aspects*, Berg, D. and Plempel, M., Eds., Ellis Horwood, Chichester, 1988, 79.
333. **Köller, W.**, Sterol demethylation inhibitors: mechanism of action and resistance, in *Fungicide Resistance in North America*, Delp, C. J., Ed., APS Press, St. Paul, 1988, 79.
334. **Aoyama, Y., Yoshida, Y., Hata, S., Nishino, T., and Katsuki, H.**, Buthiobate: a potent inhibitor for yeast cytochrome P-450 catalyzing 14α-demethylation of lanosterol, *Biochem. Biophys. Res. Commun.*, 115, 642, 1983.
335. **Masner, P. and Kerkenaar, A.**, Effect of the fungicide pyrifenox on sterol biosynthesis in *Ustilago maydis*, *Pestic. Sci.*, 22, 61, 1988.
336. **Nakayama, A., Ikura, K., Katsuura, K., Hashimoto, S., and Nakata, A.**, Quantitative structure-activity relationships, conformational analysis and computer graphics study of triflumizole analogs, fungicidal *N*-(1-imidazole-1-ylalkylidene)anilines, *J. Pestic. Sci.*, 14, 23, 1989.

337. **Berg, D.,** Biochemical mode of action of fungicides — ergosterol biosynthesis inhibitors, in *Fungicide Chemistry — Advances and Practical Application,* Green, M. B. and Spilker, D. A., Eds., American Chemical Society, Washington, D.C., 1986, 25.
338. **Berg, D., Büchel, K. H., Plempel, M., and Regel, E.,** Biochemical characteristics of bifonazole, in *Recent Trends in the Discovery, Development and Evaluation of Antifungal Agents,* Fromtling, R. A., Ed., J. R. Prous Science Publishers, Barcelona, 1988, 313.
339. **Polak-Wyss, A., Lengsfeld, H., and Oesterhelt,** Effect of oxiconazole and Ro 14-4767/002 on sterol pattern in *Candida albicans, J. Med. Vet. Mycol.,* 23, 433, 1985.
340. **Zirngibl, L., Fisher, J., Jahn, U., and Thiele, K.,** Structure-activity relationships of 2-(1H-imidazol-1-yl)vinyl ethers. Route to the new broad-spectrum antifungal agent omoconazole, *Ann. N.Y. Acad. Sci.,* 544, 63, 1988.
341. **Katagi, T.,** Structural studies of the vinyltriazole fungicide diniconazole (ER pure) related to its binding to cytochrome P-450, *J. Agric. Food Chem.,* 36, 344, 1988.
342. **Baldwin, B. C.,** Role of biochemical studies in the discovery of agricultural fungicides, in *In Vitro and In Vivo Evaluation of Antifungal Agents,* Iwata, K. and Vanden Bossche, H., Eds., Elsevier, Amsterdam, 1986, 185.
343. **Marchington, A. F. and Lambros, S. A.,** Computer design of fungicides, in *Modern Selective Fungicides — Properties, Applications and Mechanism of Action,* Lyr, H., Ed., Longman, Harlow, 1987, 325.
344. **Trzaskos, J. M. and Henry, M. J.,** Comparative effects of the azole-based fungicide flusilazole on yeast and mammalian lanosterol 14α-methyl demethylase, *Antimicrob. Agents Chemother.,* 33, 1228, 1989.
345. **Henry, M. J.,** Mode of action of the fungicide flusilazole in *Ustilago maydis, Pestic. Sci.,* 28, 35, 1990.
346. **Shepard, M. C., Noon, R. A., Worthington, P. A., McClellan, W. D., and Lever, B. G.,** Hexaconazole: a novel triazole fungicide, *Proc. 1986 Br. Crop. Prot. Conf. Pests Dis.,* 1, 363, 1986.
347. **Fujimoto, T. T., Quinn, J. A., Egan, A. R., Shaber, S. H., and Ross, R. R.,** Quantitative structure-activity relationship studies of the fungitoxic properties of phenetyl 1,2,4-triazoles, *Pestic. Biochem. Physiol.,* 30, 199, 1988.
348. **Roberson, R. W., Fuller, M. S., and Grabski, C.,** Effects of the demethylase inhibitor, cyproconazole, on hyphal tip cells of *Sclerotium rolfsii, Pestic. Biochem. Physiol.,* 34, 130, 1989.
349. **Zech, B., Gouot, J. M., Merindol, B., and Greiner, P.,** LS 840606 — a new broad-spectrum fungicide, *Proc. 1988 Br. Crop Prot. Conf. Pests Dis.,* 2, 503, 1988.
350. **Ruess, W., Riebli, P., Herzog, J., Speich, J., and James, J. R.,** CGA 169374, a new systemic fungicide with a novel broad-spectrum activity against disease complexes in a wide range of crops, *Proc. 1988 Br. Crop Prot. Conf. Pests Dis.,* 2, 543, 1988.
351. **Berg, D., Büchel, K. H., Krämer, W., Plempel, M., and Scheinpflug, H.,** Mechanistic studies as a tool for the development of new compounds in *Sterol Biosynthesis Inhibitors — Pharmaceutical and Agrochemical Aspects,* Berg. D. and Plempel, M., Eds., Ellis Horwood, Chichester, 1988, 168.
352. **Ritter, W., Holmwood, G., Ahr, H.-J., Detzer, K., Kraatz, U., Plempel, M., Scherling, D., and Siefert, H.-M.,** Vibunazole and its isomers, *Ann. N.Y. Acad. Sci.,* 544, 74, 1988.
353. **Issacson, D. M., Tolman, E. L., Tobia, A. J., Rosenthale, M. E., McGuire, J. L., Vanden Bossche, H., and Janssen, P. A. J.,** Selective inhibition of 14α-desmethyl sterol synthesis in *Candida albicans* by terconazole, a new triazole antimycotic, *J. Antimicrob. Chemother.,* 21, 333, 1988.
354. **Vanden Bossche, H.,** Itraconazole: a selective inhibitor of the cytochrome P-450 dependent ergosterol biosynthesis, in *Recent Trends in the Discovery, Development and Evaluation of Antifungal AGents,* Fromtling, R. A., Ed., J. R. Prous Science Publishers, Barcelona, 1987, 207.
355. **Vanden Bossche, H., Marichal, P., Geerts, H., and Janssen, P. A. J.,** The molecular basis for itraconazole's activity against *Aspergillus fumigatus,* in *Aspergillus and Aspergillosis,* Vanden Bossche, H., MacKenzie, D. W. R., and Cauwenbergh, G., Eds., Plenum Press, New York, 1988, 171.
356. **Marriott, M. S. and Richardson, K.,** The discovery and mode of action of fluconazole, in *Recent Trends in the Discovery, Development and Evaluation of Antifungal Agents,* Fromtling, R. A., Ed., J. R. Prous Science Publishers, Barcelona, 1987, 207.
357. **Richardson, K., Cooper, K., Marriott, M. S., Tarbit, M. H., Troke, P. F., and Whittle, P. J.,** Design and evaluation of a systemically active agent, fluconazole, *Ann. N.Y. Acad. Sci.,* 544, 4, 1988.
358. **Boyle, F. T., Gilman, D. J., Gravestock, M. B., and Wardleworth, J. M.,** Synthesis and structure-activity relationships of a novel antifungal agent, ICI 197, 739, *Ann. N.Y. Acad. Sci.,* 544, 86, 1988.
359. **Barrett-Bee, K., Lees, J., Pinder, P., Campbell, J., and Newboult, L.,** Biochemical studies with a novel antifungal agent, ICI 195,739, *Ann. N.Y. Acad. Sci.,* 544, 231, 1988.
360. **Watson, P. F., Rose, M. E., Ellis, S. W., England, H., and Kelly, S. L.,** Defective sterol C5-6 desaturation and azole resistance: a new hypothesis for the mode of action of azole antifungals, *Biochem. Biophys. Res. Commun.,* 164, 1170, 1989.
361. **Vanden Bossche, H., Marichal, P., Gorrens, J., Bellens, D., Verhoeven, H., Coene, M.-C., Lauwers, W., and Janssen, P. A. J.,** Interaction of azole derivatives with cytochrome P-450 isozymes in yeast, fungi, plants and mammalian cells, *Pestic. Sci.,* 21, 289, 1987.

362. **Shaw, J. T. B., Tarbit, M. H., and Troke, P. F.,** Cytochrome P-450 mediated sterol synthesis and metabolism: differences in sensitivity to fluconazole and other azoles, in *Recent Trends in the Discovery, Development and Evaluation of Antifungal Agents,* Fromtling, R. A., Ed., J. R. Prous Science Publishers, Barcelona, 1987, 207.
363. **Hancock, H. G. and Weete, J. D.,** Effects of triazoles on fungi. IV. Growth and lipids of *Cercospora arachidicola* and *Cercospora personatum, Pestic. Biochem. Physiol.,* 24, 395, 1985.
364. **Kato, T.,** Mechanism of action of denmert, an inhibitor of sterol biosynthesis, *Neth. J. Plant Pathol.,* 83 (Suppl. 1), 113, 1977.
365. **Oritz de Montellano, P. R. and Reich, N. O.,** Inhibition of cytochrome P-450 enzymes, in *Cytochrome P-450 — Structure, Mechanism, and Biochemistry,* Oritz de Montellano, P. R., Ed., Plenum Press, New York, 1986, 273.
366. **Vanden Bossche, H. and Willemsens, G.,** Effect of antimycotics, miconazole and ketoconazole, on cytochrome P-450 in yeast microsomes and rat liver microsomes, *Arch. Int. Physiol. Biochim.,* 90, B218, 1982.
367. **Yoshida, Y. and Aoyama, Y.,** Interaction of azole fungicides with yeast cytochrome P-450 which catalyzes lanosterol 14α-demethylation, in *In Vitro and In Vivo Evaluation of Antifungal Agents,* Iwata, K. and Vanden Bossche, H., Eds., Elsevier, Amsterdam, 1986, 123.
368. **Yoshida, Y. and Aoyama, Y.,** Interaction of azole antifungal agents with cytochrome $P\text{-}450_{14DM}$ purified from *Saccharomyces cerevisiae* microsomes, *Biochem. Pharmacol.,* 36, 229, 1987.
368a. **Barrett-Bee, K., Newboult, L., and Pinder, P.,** Biochemical changes associated with the antifungal action of the triazole ICI 153,006 on *Candida albicans* and *Trichophyton quickeanum, FEMS Microbiol. Lett.,* 79, 127, 1991.
369. **Gadher, P., Mercer, E. I., Baldwin, B. C., and Wiggins, T. E.,** A comparison of the potency of some fungicides as inhibitors of sterol 14-demethylation, *Pestic. Biochem. Physiol.,* 19, 1, 1983.
370. **Vanden Bossche, H., Marichal, P., Gorrens, J., Bellens, D., Verhoeven, H., Coene, M.-C., Lauwers, W., and Janssen, P. A. J.,** Interaction of azole derivatives with cytochrome P-450 isozymes in yeast, fungi, plants and mammalian cells, *Pestic. Sci.,* 21, 289, 1987.
371. **Willemsens, G., Cools, W., and Vanden Bossche, H.,** Effects of miconazole and ketoconazole on sterol synthesis in a subcellular fraction of yeast and mammalian cells, in *The Host Invader Interplay,* Vanden Bossche, H., Ed., Elsevier, Amsterdam, 1980, 691.
372. **Pye, G. W. and Marriott, M. S.,** Inhibition of sterol C14 demethylation by imidazole-containing antifungals, *Sabouraudia,* 20, 325, 1982.
373. **Marriott, M. S.,** Inhibition of sterol biosynthesis in *Candida albicans* by imidazole-containing antifungals, *J. Gen. Microbiol.,* 117, 253, 1980.
374. **Yoshida, Y., Aoyama, Y., Takano, H., and Kato, T.,** Stereo-selective interaction of enantiomers of diniconazole fungicide with purified P-450/14DM from yeast, *Biochem. Biophys. Res. Commun.,* 137, 513, 1986.
375. **Sasaki, M., Kato, T., Yamamoto, S., and Mukai, K.,** Antifungal activity of [hydroxy(pyridin-3-yl-methyl]-phenylphosphinates and related compounds, *J. Pestic. Sci.,* 9, 717, 1984.
376. **Sasaki, M., Takano, H., and Kato, T.,** 1*H*-1,2,4-triazole-1-ylmethyldiphenylphosphine sulfides, a new class of fungicides, *Agric. Biol. Chem.,* 51, 1727, 1987.
377. **Cooper, A. B., Wright, J. J., Ganguly, A. K., Desai, J., Loebenberg, D., and Parmegiani, R.,** Synthesis and antifungal properties of 14-aminomethyl-substituted lanosterol derivatives, *Ann. N.Y. Acad. Sci.,* 544, 109, 1988.
378. **Sherald, J. L. and Sisler, H. D.,** Antifungal mode of action of triforine, *Pestic. Biochem. Physiol.,* 5, 477, 1975.
379. **Aoyama, Y., Yoshida, Y., Sonoda, Y., and Sato, Y.,** 7-Oxo-24,25-dihydrolanosterol: a novel lanosterol 14α-demethylase ($P\text{-}450_{14DM}$) inhibitor which blocks electron transfer to the oxyferro intermediate, *Biochim. Biophys. Acta,* 922, 270, 1987.
380. **Covey, D. F.,** Aromatase inhibitors: specific inhibitors of oestrogen biosynthesis, in *Sterol Biosynthesis Inhibitors — Pharmaceutical and Agrochemical Aspects,* Berg, D. and Plempel, M., Ellis Horwood, Chichester, 1988, 534.
381. **Childers, W. E. and Robinson, C. H.,** Novel 10β-thiiranyl steroids as aromatase inhibitors, *J. Chem. Soc. Chem. Commun.,* p. 320, 1987.
382. **Kellis, J. T., Childers, W. E., Robinson, C. H., and Vickery, L. E.,** Inhibition of aromatase cytochrome P-450 by 10-oxirane and 10-thiirane substituted androgens, *J. Biol. Chem.,* 262, 4421, 1987.
383. **Poulos, T. L., Finzel, B. C., and Howard, A. J.,** High resolution crystal structure of cytochrome P450cam, *J. Mol. Biol.,* 195, 687, 1987.
384. **Poulos, T. L. and Howard, A. J.,** Crystal structures of metyrapone- and phenylimidazole-inhibited complexes of cytochrome $P\text{-}450_{cam}$, *Biochemistry,* 26, 8165, 1987.
385. **Kalb, V. F. and Loper, J. C.,** Proteins from eight eukaryotic cytochrome P-450 families share a segment region of sequence similarities, *Proc. Natl. Acad. Sci. U.S.A.,* 85, 7221, 1988.

386. **Gasztonyi, M.,** The diastereomeric ratio in the triadimenol produced by fungal metabolism of triadimenfon, and its role in fungicidal selectivity, *Pestic. Sci.*, 12, 433, 1981.

387. **Buchenauer, H. and Grossmann, F.,** Fungitoxic properties of the diastereomers of triadimenol and transformation of triadimefon into the triadimenol isomers by fungi, *Z. Pflanzenkr. Pflanzenschutz*, 89, 309, 1982.

388. **Deas, A. H. B. and Clifford, D. R.,** Metabolism of the 1,2,4-triazolylmethane fungicides, triadimefon, triadimenol, and diclobutrazole by *Aspergillus niger*, *Pestic. Biochem. Physiol.*, 17, 120, 1982.

389. **Köller, W.,** Isomers of sterol synthesis inhibitors: fungicidal effects and plant growth regulator activities, *Pestic. Sci.*, 18, 129, 1987.

390. **Graebe, J. E.,** Gibberellin biosynthesis and control, *Annu. Rev. Plant Physiol.*, 38, 419, 1987.

391. **Lürssen, K.,** Triazole plant growth regulators: effects and mode of action, in *Sterol Biosynthesis Inhibitors — Pharmaceutical and Agrochemical Aspects*, Berg, D. and Plempel, M., Eds., Ellis Horwood, Chichester, 1988, 305.

392. **Rademacher, W., Fritsch, H., Graebe, J. E., Sauter, H., and Jung, J.,** Tetcyclacis and triazole-type plant retardants: their influence on the biosynthesis of gibberellins and other metabolic processes, *Pestic. Sci.*, 21, 241, 1987.

393. **Köller, W.,** Plant growth regulator activities of stereochemical isomers of triadimenol, *Physiol. Plant.*, 71, 309, 1987.

394. **Burden, R. S., Clark, T., and Holloway, P. J.,** Effects of ergosterol-inhibiting fungicides and plant growth regulators on sterol composition of barley plants, *Pestic. Biochem. Physiol.*, 27, 289, 1987.

395. **Burden, R. S., Carter, G. A., Clark, T., Cooke, D. T., Croker, S. J., Deas, A. H. B., Hedden, P., James, C. S., and Lenton, J. R.,** Comparative activities of the enantiomers of triadimenol and paclobutrazole as inhibitors of fungal growth and plant sterol biosynthesis, *Pestic. Sci.*, 21, 253, 1987.

396. **Taton, M., Ullmann, P., Benveniste, P., and Rahier, A.,** Interaction of triazole fungicides and plant growth regulators with microsomal cytochrome P-450-dependent obtusifoliol 14-methyl demethylase, *Pestic. Biochem. Physiol.*, 30, 178, 1988.

397. **Rahier, A. and Taton, M.,** Plant sterol biosynthesis inhibitors: the 14-demethylation steps, their enzymology and inhibition, *Biochem. Soc. Trans.*, 18, 1990, 1990.

398. **Ebert, E., Eckhardt, W., Jäkel, K., Moser, P., Sozzi, D., and Vogel, C.,** Quantitative structure activity relationships of fungicidally active triazoles: analogs and stereoisomers of propiconazole and etaconazole, *Z. Naturforsch.*, 44c, 85, 1988.

399. **Takano, H. and Kato, T.,** Biological activity and mode of action of diniconazole, in *Pesticide Science and Biotechnology*, Greenhalg, R. and Roberts, T. R., Eds., Blackwell Scientific, Oxford, 1987, 197.

400. **Takano, H., Oguri, Y., and Kato, T.,** Antifungal and plant growth regulating activities of enantiomers of (*E*)-1(2,4-dichlorophenyl)-4,4-dimethyl-2(1,2,4-triazol-1-yl)-1-peneten-3-ol (S-3308L), *J. Pestic. Sci.*, 373, 1986.

400a. **Worthington, P. A.,** Synthesis of 1,2,4-triazole compounds related to the fungicides flutriatol and hexaconazole, *Pestic. Sci.*, 31, 457, 1991.

401. **Gisi, U., Schaub, F., Wiedmer, H., and Ummel, E.,** SAN 619 F, a new triazole fungicide, *Proc. 1986 Br. Crop Prot. Conf. Pest Dis.*, 1, 33, 1986.

402. **Baldwin, B. C., Wiggins, T. E., Marchington, A. F., and Worthington, P. A.,** The discovery and mode of action of PP450, a new broad spectrum fungicide for cereals, *Meded. Fac. Landbouwwet. Rijksuniv. Gent*, 49/2a, 303, 1984.

403. **Barug, D., Samson, R. A., and Kerkenaar, A.,** Microscopic studies of *Candida albicans* and *Torulopsis glabrata* after *in vitro* treatment with bifonazole, *Drug Res.*, 33, 528, 1983.

404. **Kerkenaar, A. and Barug, D.,** Fluorescence microscope studies of *Ustilago maydis* and *Penicillium italicum* after treatment with imazalil or fenpropimorph, *Pestic. Sci.*, 15, 199, 1984.

405. **Marichal, P., Gorrens, J., and Vanden Bossche, H.,** The action of itraconazole and ketoconazole on growth and sterol synthesis in *Aspergillus fumigatus* and *Aspergillus niger*, *J. Med. Vet. Mycol.*, 22, 13, 1984.

406. **Hector, R. F. and Braun, P. C.,** The effect of bifonazole on chitin synthesis in *Candida albicans*, in *Recent Trends in the Discovery, Development and Evaluation of Antifungal Agents*, Fromtling, R. A., Ed., J. R. Prous Science Publishers, Barcelona, 1987, 369.

407. **Silverman, S. J.,** Similar and different domains of chitin synthases 1 and 2 of *S. cerevisiae*: two isozymes with distinct functions, *Yeast*, 5, 459, 1989.

408. **Yamaguchi, H.,** Morphological aspects of azole action, in *Sterol Biosynthesis Inhibitors — Pharmaceutical and Agrochemical Aspects*, Berg, D. and Plempel, M., Eds., Ellis Horwood, Chichester, 1988, 56.

409. **Borgers, M.,** Ultrastructural correlates of antimycotic treatment, in *Current Topics in Medical Mycology*, Vol. 2, McGinnis, M. R., Ed., Springer-Verlag, New York, 1988, 1.

410. **Hippe, S.,** Electron microscopic investigations on the cytology of *Ustilago avenae* as influenced by the systemic fungicides triadimefon, nuarimol, and imazalil nitrate, *Pestic. Biochem. Physiol.*, 21, 170, 1984.

411. **Sancholle, M., Dargent, R., Weete, J. D., Rushing, A. E., Miller, K. S., and Montant, C.,** Effects of triazoles on fungi. VI. Ultrastructure of *Taphrina deformans, Mycologia,* 80, 162, 1988.
412. **Bastide, M., Jouvert, S., and Bastide, M. J.,** A comparison of the effects of several antifungal imidazole derivatives and polyenes on *Candida albicans, Can. J. Microbiol.,* 28, 1119, 1982.
413. **Rhichmond, D. V.,** Effects of triadimefon on the fine structure of germinating conidia of *Botrytis allii, Pestic. Biochem. Physiol.,* 21, 74, 1984.
414. **Fuller, M. S., Roberson, R. W., and Gisi, U.,** Effects of the sterol demethylation inhibitor, cyproconazole, on hyphal tip cells of *Sclerotium rolfsii.* III. Cell wall cytochemistry, *Pestic. Biochem. Physiol.,* 36, 115, 1990.
415. **Georgopapadaku, N. H., Dix, B. A., Smith, S. A., Freudenberger, J., and Funke, P. T.,** Effect of antifungal agents on lipid biosynthesis and membrane integrity in *Candida albicans, Antimicrob. Agents Chemother.,* 31, 46, 1987.
416. **Parks, L. W., McLean-Bowen, C., Taylor, F., and Hough, S.,** Sterols in yeast subcellular fractions, *Lipids,* 13, 730, 1978.
417. **Astin, A. and Haslam, J.,** The effect of altered membrane sterol composition on oxidative phosphorylation in a heam mutant of *Saccharomyces cerevisiae, Biochem. J.,* 166, 287, 1977.
418. **Odds, F. C., Cheesman, S. L., and Abbott, A. B.,** Suppression of ATP in *Candida albicans* by imidazole and derivative antifungal agents, *J. Med. Vet. Mycol.,* 23, 415, 1985.
419. **Lyr, H.,** Lipid peroxidation: a side effect of sterol demethylation inhibitor fungicides in *Mucor mucedo* (L.) Fres and *Ustilago maydis, Pestic. Biochem. Physiol.,* 32, 197, 1988.
420. **Cope, J.,** Mode of action of miconazole on *Candida albicans:* effect on growth, viability and K^+ release, *J. Gen. Microbiol.,* 119, 245, 1980.
421. **Sud, I. J. and Feingold, D. S.,** Heterogeneity of action mechanisms among antimycotic imidazoles, *Antimicrob. Agents Chemother.,* 20, 71, 1981.
422. **Sud, I. J. and Feingold, D. S.,** Mechanisms of action of the antimycotic imidazoles, *J. Invest. Dermatol.,* 76, 438, 1981.
423. **Vanden Bossche, H., Ruyschaert, J. M., Defrise-Quertain, F., Willemsens, G., Cornelissen, F., Marichal, P., Cools, W., and Van Cutsen, J.,** The interaction of miconazole and ketoconazole with lipids, *Biochem. Pharmacol.,* 31, 2609, 1982.
424. **Brasseur, R., Vandenbosch, C., Vanden Bossche, H., and Ruyschaert, J. M.,** Mode of insertion of miconazole, ketokonazole and deacylated ketokonazole in lipid layers, *Biochem. Pharmacol.,* 32, 2175, 1983.
425. **Arndt, R. and Schulz-Harder, B.,** Effect of miconazole on *Saccharomyces cerevisiae, J. Pharm. Pharmacol.,* 35, 668, 1983.
426. **Portillo, F. and Gancedo, C.,** Mode of action of miconazole on yeasts: inhibition of the mitochondrial ATPase, *Eur. J. Biochem.,* 143, 273, 1984.
427. **Portillo, F. and Gancedo, C.,** Mitochondrial resistance to miconazole in *Saccharomyces cerevisiae, Mol. Gen. Genet.,* 199, 495, 1985.
428. **Dufour, J. P., Boutry, M., and Goffeau, A.,** Plasma membrane ATPase of yeast, *J. Biol. Chem.,* 225, 5735, 1980.
429. **Dahmen, H., Hoch, H. C., and Staub, T.,** Differential effects of sterol inhibitors on growth, cell membrane permeability, and ultrastructure of two target fungi, *Phytopathology,* 78, 1033, 1988.
430. **Sud, I. J. and Feingold, D. S.,** Action of antifungal imidazoles on *Staphylococcus aureus, Antimicrob. Agents Chemother.,* 22, 470, 1982.
431. **Uno, J., Shigematsu, M. L., and Arai, T.,** Primary site of action of ketoconazole on *Candida albicans, Antimicrob. Agents Chemother.,* 21, 912, 1982.
432. **Weete, J. D., Sancholle, M., Tuozé-Soulet, J.-M., Bradley, J., and Dargent, R.,** Effects of triazoles on fungi. III. Composition of plasma membrane-enriched fraction of *Taphrina deformans, Biochim. Biophys. Acta,* 812, 633, 1985.
433. **Hippe, S.,** Hexagonal clustering of intramembrane particles in the plasmalemma of *Ustilago avenae* after fungicide treatment, *Pestic. Biochem. Physiol.,* 24, 161, 1985.
434. **Hippe, S. and Lüth, H.,** A simple physical model for fungicide induced hexagonal clustering of intramembrane particles in the plasmalemma of *Ustilago avenae, J. Theor. Biol.,* 121, 351, 1986.
435. **Sancholle, M., Weete, J. D., and Montant, C.,** Effects of triazoles on fungi. I. Growth and cellular permeability, *Pestic. Biochem. Physiol.,* 21, 31, 1984.
436. **Ryley, J. F., Wilson, R. G., and Barrett-Bee, K. J.,** Azole resistance in *Candida albicans, J. Med. Vet. Mycol.,* 22, 53, 1984.
437. **Walsh, R. C. and Sisler, H. D.,** A mutant of *Ustilago maydis* deficient in sterol C-14 demethylation characteristics and sensitivity to inhibitors of ergosterol biosynthesis, *Pestic. Biochem. Physiol.,* 18, 122, 1982.
438. **Johnson, E. M., Richardson, M. D., and Warnock, D. W.,** Effect of imidazole antifungals on the development of germ tubes by strains of *Candida albicans, J. Antimicrob. Chemother.,* 12, 303, 1983.

439. **Odds, F. C., Cockayne, A., Hayward, J., and Abbott, A. B.,** Effects of imidazole- and triazole derivative antifungal compounds on the growth and morphological development of *Candida albicans* hyphae, *J. Gen. Microbiol.*, 131, 2581, 1985.
440. **Schaude, M., Ackerbauer, H., and Mieth, H.,** Inhibitory effect of antifungal agents on germ tube formation in *Candida albicans*, *Mykosen*, 30, 281, 1987.
441. **Barug, D., De Groot, C., and Samson, R. A.,** Morphology and ultrastructure of *Candida albicans* after *in vitro* treatment with bifonazole, in *Recent Trends in the Discovery, Development and Evaluation of Antifungal Agents*, Fromtling, R. A., Ed., J. R. Prous Publishers, Barcelona, 1987, 353.
442. **Marichal, P., Gorrens, J., Van Cutsem, J., and Vanden Bossche, H.,** Culture media for the study of the effects of azole derivatives on germ tube formation and hyphal growth of *Candida albicans*, *Mykosen*, 26, 76, 1985.
443. **Shimokawa, O., Kato, Y., and Nakayama, H.,** Accumulation of 14-methyl sterols and defective hyphal growth in *Candida albicans*, *J. Med. Vet. Mycol.*, 24, 327, 1986.
444. **Shepherd, M. G.,** Morphogenetic transformation of fungi, in *Current Topics in Medical Mycology*, Vol. 2, McGinnis, M. R., Ed., Springer-Verlag, New York, 1988, 278.
445. **Borgia, G., Tallarino, A., Crowell, J., Lambiase, A., Cicciarello, S., Reynaud, L., and Piazza, M.,** Mycelium-yeast transition in *Histoplasma capsulatum*. Early ultrastructural changes, *Mycoses*, 32, 584, 1989.
446. **Plempel, M., Berg, D., Büchel, K.-H., and Ritter, W.,** Experimental antimycotic activity of azole derivatives — experience, knowledge and questions, in *Sterol Biosynthesis Inhibitors — Pharmaceutical and Agrochemical Aspects*, Berg, D. and Plempel, M., Ellis Horwood, Chichester, 1988, 349.
447. **Scheinpflug, H.,** The pathogenesis of plant diseases: the effect of modern fungicides, in *Fungicide Chemistry — Advances and Practical Applications*, Green, M. B. and Spilker, D. A., Eds., American Chemical Society, Washington, D.C., 1986, 73.
448. **Pontzen, R. and Scheinpflug, H.,** Effects of triazole fungicides on sterol biosynthesis during spore germination of *Botrytis cinerea*, *Venturia inaequalis*, and *Puccinia graminis* f. sp. *tritici*, *Neth. J. Plant Pathol.*, 95 (Suppl. 1), 151, 1989.
449. **Van Gestel, J. F. E. and Van de Ven, M.-A. A. P.,** Observations on the antisporulant activity of imazalil (eniconazole), *Pestic. Sci.*, 15, 215, 1984.
450. **Abu-El Teen, K., Ghannum, M., and Stretton, R. J.,** Effects of sub-inhibitory concentrations of antifungal agents on adherence of *Candida* spp. to buccal epithelial cells in vitro, *Mycoses*, 32, 551, 1989.
451. **Smolka, S. and Wolf, G.,** Cytological studies on the mode of action of systemic fungicides on the host pathogen complex barley-powdery mildew (*Erysiphe graminis* f. sp. *hordei* Marchal), *Pestic. Sci.*, 17, 249, 1986.
452. **Heath, M. C.,** Effect of fungal death or inhibition induced by oxicarboxin or polyoxin D on the interaction between resistant or susceptible bean cultivars and the bean rust fungus, *Phytopathology*, 78, 1454, 1988.
453. **Marrichal, P., Gorrens, J., and Vanden Bossche, H.,** The action of itraconazole and ketoconazole on growth and sterol synthesis in *Aspergillus fumigatus* and *Aspergillus niger*, *J. Med. Vet. Mycol.*, 22, 13, 1984.
454. **Vanden Bossche, H., Marrichal, P., and Gorrens, J.,** Molecular basis for the activity of the azole derivatives ketokonazole and itraconazole. Model: *Candida lusitaniae*, in *In Vitro and In Vivo Evaluation of Antifungal Compounds*, Iwata, K. and Vanden Bossche, H., Eds., Elsevier, Amsterdam, 1986, 101.
455. **Vanden Bossche, H., Marichal, P., Gorrens, J., Geerts, H., and Janssen, P. A. J.,** Mode of action studies. Basis for the search of new antifungal drugs, *Ann. N.Y. Acad. Sci.*, 544, 191, 1988.
456. **Hitchcock, C. A., Barrett-Bee. K. J., and Russell, N. J.,** The lipid composition and permeability of the triazole antifungal antibiotic ICI 153066 of serum-grown mycelial cultures of *Candida albicans*, *J. Gen. Microbiol.*, 135, 1949, 1989.
457. **Köller, W. and Wubben, J. P.,** Variable resistance factors of fungicides acting as sterol demethylation inhibitors, *Pestic. Sci.*, 26, 133, 1989.
457a. **Nicholas, R. O. and Kerridge, D.,** Correlation of inhibition of sterol synthesis with growth-inhibitory action of imidazole antimycotics in *Candida albicans*, *J. Antimicrob. Chemother.*, 23, 7, 1989.
458. **Tamburini, P. P., White, R. E., and Schenkman, J. B.,** Chemical characterization of protein-protein interactions between cytochrome P450 and cytochrome b_5, *J. Biol. Chem.*, 260, 4007, 1985.
459. **Ohta, Y., Mitani, F., Ishimura, Y., Yanagibashi, K., Kawamura, M., and Kawato, S.,** Conversion of cholesterol to pregnenolone mobilizes cytochrome P-450 in the inner membrane of adrenocortical mitochondria: protein rotation study, *J. Biochem.*, 107, 97, 1990.
460. **Weete, J. D. and Wise, M. L.,** Effects of triazoles on fungi. V. Response by a naturally tolerant species, *Mucor rouxii*, *Exp. Mycol.*, 214, 1987.
461. **Köller, W. and Scheinpflug, H.,** Fungal resistance to sterol biosynthesis inhibitors: a new challenge, *Plant Dis.*, 71, 1066, 1987.
462. **Brent, K. J. and Hollomon, D. W.,** Risk of resistance against sterol biosynthesis inhibitors in plant protection, in *Sterol Biosynthesis Inhibitors — Pharmaceutical and Agrichemical Aspects*, Berg, D. and Plempel, M., Eds., Ellis Horwood, Chichester, 1988, 332.

463. **Köller, W.,** Fungicide resistance in plant pathogens, in *Handbook of Pest Management in Agriculture,* Vol. 3, 2nd ed., Pimentel, D., Ed., CRC Press, Boca Raton, FL, 1991, 679.
464. **Smith, F. D., Parker, D. M., and Köller, W.,** Sensitivity distribution of *Venturia inaequalis* to the sterol inhibitro flusilazole: baseline sensitivity and implications for resistance monitoring, *Phytopathology,* 81, 392, 1991.
465. **Lombardi, G., Gramegna, G., Cavanna, C., Poma, G., Marangoni, E., and Michelone, G.,** Itraconazole vs amphotericin B: *in vitro* comparative evaluation of the minimal inhibitory concentration (MIC) against clinically isolated yeasts, *Mycopathologia,* 106, 31, 1989.
466. **Van Cutsem, J., Van Gervan, F., and Janssen, P. A. J.,** The *in vitro* and *in vivo* antifungal activity of itraconazole, in *Recent Trends in the Discovery, Development and Evaluation of Antifungal Agents,* Fromtling, J. R., Ed., J. R. Prous Publishers, Barcelona, 1987, 177.
467. **Grant, S. M. and Clissold, S. P.,** Itraconazole — a review of its pharmacodynamic and pharmacokinetic properties, and therapeutic use in superficial and systemic mycoses, *Drugs,* 37, 310, 1989.
468. **Kerridge, D., Fasoli, M., and Wayman, F. J.,** Drug resistance in *Candida albicans* and *Candida glabrata, Ann. N.Y. Acad. Sci.,* 644, 245, 1988.
469. **Smith, K. J., Warnock, D. W., Kennedy, C. T. C., Johnson, E. M., Hopwood, V., Van Cutsem, J., and Vanden Bossche, H.,** Azole resistance in *Candida albicans, J. Med. Vet. Mycol.,* 24, 133, 1986.
470. **Kalamarakis, A. E., Demopoulos, V. P., Ziogas, B. N., and Georgopoulos, S. G.,** A highly mutable major gene for triadimenol resistance in *Nectria haematococca* var. *cucurbitae, Neth. J. Plant Pathol.,* 95(Suppl. 1), 109, 1989.
471. **Henry, M. J. and Trivellas, A. E.,** Laboratory-induced fungicide resistance to benzimidazole and azole fungicides in *Cercospora beticola, Pestic. Biochem. Physiol.,* 35, 89, 1989.
472. **Nunninger-Ney, Schwinn, F.-J., and Staub, T.,** *In vitro* selection of sterol-biosynthesis-inhibitor (SBI)-resistant mutants in *Monilinia fructicola* (Wint.) Honey, *Neth. J. Plant Pathol.,* 95(Suppl. 1), 137, 1989.
473. **De Waard, M. A. and Fuchs, A.,** Resistance to ergosterol biosynthesis inhibitors. II. Genetic and physiological aspects, in *Fungicide Resistance in Crop Protection,* Dekker, J. and Georgopoulos, S. G., Eds., Pudoc, Wageningen, 1982, 87.
474. **De Waard, M. A. and Fuchs, A.,** Resistance to ergosterol biosynthesis inhibiting fungicides, in *Systemische Fungizide und Antifungale Verbindungen,* Lyr, H. and Polter, C., Eds., Akademie Verlag, Berlin, 1983, 429.
475. **De Waard, M. A. and van Nistelrooy, J. G. M.,** Accumulation of SBI fungicides in wild type and fenarimol-resistant isolates of *Penicillium italicum, Pestic. Sci.,* 22, 371, 1988.
476. **De Waard, M. A. and van Nistelrooy, J. G. M.,** Inhibitors of energy-dependent efflux of the fungicide fenarimol by *Aspergillus nidulans, Exp. Mycol.,* 11, 1, 1987.
477. **Siegel, M. R. and Solel, Z.,** Effects of imazalil on a wild-type and fungicide-resistant strain of *Aspergillus nidulans, Pestic. Biochem. Physiol.,* 15, 222, 1981.
478. **De Waard, M. A. and van Nistelrooy, J. G. M.,** Differential accumulation of fenarimol by a wild-type isolate and fenarimol-resistant isolates of *Penicillium italicum, Neth. J. Plant Pathol.,* 90, 143, 1984.
479. **Siegel, M. R., Kerkenaar, A., and Sijpesteijn, A. K.,** Antifungal activity of the systemic fungicide imazalil, *Neth. J. Plant Pathol.,* 83(Suppl. 1), 121, 1977.
480. **Vanden Bossche, H.,** Biochemical effects of miconazole on fungi. I. Effects on the uptake and/or utilization of purines, pyrimidines, nucleosides, amino acids and glucose by *Candida albicans, Biochem. Pharmacol.,* 23, 887, 1974.
481. **Hitchcock, C. A., Barrett-Bee, K. J., and Russell, N. J.,** The lipid composition of azole-sensitive and azole-resistant strains of *Candida albicans, J. Gen. Microbiol.,* 132, 2421, 1986.
482. **Boiron, P., Drouhet, E., DuPont, B., and Improvisi, L.,** Entry of ketoconazole into *Candida albicans, Antimicrob. Agents Chemother.,* 31, 244, 1987.
483. **Hippe, S.,** Combined application of low temperature preparation and electron microscopic autoradiography for the localization of systemic fungicides, *Histochemistry,* 87, 309, 1987.
484. **Yoshida, Y., Aoyama, Y., Nishino, T., Katsuki, H., Maitra, U. S., Mohan, V. P., and Sprinson, D. B.,** Characterization of an altered cytochrome P-450 from a mutant yeast, in *Cytochrome P-450, Biochemistry, Biophysics, and Induction,* Vereczkey, L. and Magyar, K., Eds., Elsevier, Amsterdam, 1985, 439.
485. **Aoyama, Y., Yoshida, Y., Nishino, T., Katsuki, H., Maitra, U. S., Mohan, V. P., and Sprinson, D. B.,** Isolation and characterization of an altered cytochrome P-450 from a yeast defective in lanosterol 14α-demethylation, *J. Biol. Chem.,* 262, 14260, 1987.
486. **Leroux, P. and Gredt, M.,** Resistance to fungicides which inhibit ergosterol biosynthesis in laboratory strains of *Botrytis cinerea* and *Ustilago maydis, Pestic. Sci.,* 15, 85, 1984.
487. **Guan, J., Kerkenaar, A., and de Waard, M. A.,** Effects of imazalil on sterol composition of sensitive and DMI-resistant isolates of *Penicillium italicum, Neth. J. Plant Pathol.,* 95(Suppl. 1), 73, 1989.
488. **Carter, G. A., Kendall, S. J., Burden, R. S., James, C. S., and Clark, T.,** The lipid compositions of two isolates of *Cladosporium cucumerinum* do not explain their differences in sensitivity to fungicides which inhibit sterol biosynthesis, *Pestic. Sci.,* 26, 181, 1989.

489. **Kalb, V. F., Loper, J. C., Dey, C. R., Woods, C. W., and Sutter, T. R.,** Isolation of a cytochrome P-450 structural gene from *Saccharomyces cerevisiae, Gene,* 45, 237, 1986.

490. **Hitchcock, C. A., Barrett-Bee, K. J., and Russell, N. J.,** Inhibition of 14α-sterol demethylase activity in *Candida albicans* Darlington does not correlate with resistance to azole, *J. Med. Vet. Mycol.,* 25, 329, 1987.

491. **Köller, W.,** unpublished results.

492. **Vanden Bossche, H., Marichal, P., Gorrens, J., Bellens, D., Moereels, H., and Janssen, A. J. P.,** Mutation in cytochrome P-450-dependent 14α-demethylase results in decreased affinity for azole fungicides, *Biochem. Soc. Trans.,* 18, 56, 1990.

493. **Pommer, E. H.,** Chemical structure-fungicidal activity relationships in substituted morpholines, *Pestic. Sci.,* 15, 285, 1984.

494. **Isenring, H. P.,** Antifungal morpholine derivatives: chemistry and structure-activity relationships, in *Recent Trends in the Discovery, Development and Evaluation of Antifungal Agents,* Fromtling, R. A., Ed., J. R. Prous Science Publishers, Barcelona, 1987, 543.

495. **Polak, A. and Dixon, D. M.,** Antifungal activity of amorolfine (Ro 14-4767/002) *in vitro* and *in vivo,* in *Recent Trends in the Discovery, Development and Evaluation of Antifungal Agents,* Fromtling, R. A., Ed., J. R. Prous Science Publishers, Barcelona, 1987, 555.

496. **Polak, A.,** Morpholines in clinical use, in *Sterol Biosynthesis Inhibitors — Pharmaceutical and Agrochemical Aspects,* Berg, D. and Plempel, M., Eds., Ellis Horwood, Chichester, 1988, 431.

497. **Keerkenaar, A.,** Mechanism of action of morpholine fungicides, in *Modern Selective Fungicides — Properties, Applications and Mechanisms of Action,* Lyr, H., Ed., Gustav Fischer Verlag, Jena, 1987, 159.

498. **Keerkenaar, A.,** The mode of action of dimethylmorpholines, in *Recent Trends in the Discovery, Development and Evaluation of Antifungal Agents,* Fromtling, R. A., Ed., J. R. Prous, Science Publishers, Barcelona, 1987, 523.

499. **Mercer, E. I.,** The mode of action of morpholines, in *Sterol Biosynthesis Inhibitors — Pharmaceutical and Agrochemical Aspects,* Berg, D. and Plempel, M., Eds., Ellis Horwood, Chichester, 1988, 120.

500. **Kerkenaar, A.,** Inhibition of the sterol Δ^{14}-reductase and Δ^{8}-Δ^{7}-isomerase in fungi, *Biochem. Soc. Trans.,* 18, 59, 1990.

500a. **Akers, A., Ammermann, E., Buschmann, E., Götz, N., Himmele, W., Lorenz, G., Pommer, E.-H., Rentzea, C., Röhl, F., Siegel, H., Zipperer, B., Sauter, H., and Zipplies, M.,** Chemistry and biology of novel amine fungicides: Attempts to improve the antifungal activity of fenpropimorph, *Pestic. Sci.,* 31, 521, 1991.

501. **Kato, T., Shoami, M., and Kawase, Y.,** Comparison of tridemorph with buthiobate in antifungal mode of action, *J. Pestic. Sci.,* 5, 69, 1980.

502. **Kerkenaar, A., Uchiyama, M., and Versluis, G. G.,** Specific effects of tridemorph on sterol biosynthesis in *Ustilago maydis, Pestic. Biochem. Physiol.,* 16, 97, 1981.

503. **Berg, L. R., Patterson, G. W., and Lusby, W. R.,** Effects of triarimol and tridemorph on sterol biosynthesis of *Saprolegnia ferax, Lipids,* 18, 448, 1983.

504. **Leroux, P. and Gredt, M.,** Etudes sur les inhibiteurs de la biosynthese des stérols fongiques. I. Fongicides provoquant L'accumulation des desméthylstérols, *Agronomie,* 3, 123, 1983.

505. **Kerkenaar, A., Van Rossum, J. M., Versluis, G. G., and Marsman, J. W.,** Effect of fenpropimorph and imazalil on sterol biosynthesis in *Penicillium italicum, Pestic. Sci.,* 15, 177, 1984.

506. **Taton, M., Benveniste, P., and Rahier, A.,** Microsomal $\Delta^{8,14}$-sterol $\Delta^{8,14}$-reductase in higher plants. Characterization and inhibition by analogues of a presumptive carbocationic intermediate of the reduction reaction, *Eur. J. Biochem.,* 185, 605, 1989.

507. **Rahier, A., Schmitt, P., Huss, B., Benveniste, P., and Pommer, E. H.,** Chemical structure-activity relationships in the inhibition of sterol biosynthesis by N-substituted morpholines in higher plants, *Pestic. Biochem. Physiol.,* 25, 112, 1986.

508. **Polak, A.,** Mode of action of morpholine derivatives, *Ann. N.Y. Acad. Sci.,* 54, 221, 1988.

509. **Taton, M., Benveniste, P., and Rahier, A.,** Comparative study of the inhibition of sterol biosynthesis in *Rubus fructicosus* suspension cultures and *Zea mays* seedlings by *N*-(1,5,9-trimethyldecyl)-4α,10-dimethyl-8-aza-*trans*-decal-3β-ol and derivatives, *Phytochemistry,* 26, 385, 1987.

510. **Steel, C. C., Baloch, R. I., Mercer, E. I., and Baldwin, B. C.,** The intracellular location and physiological effects of abnormal sterols in fungi grown in the presence of morpholine and functionally related fungicides, *Pestic. Biochem. Physiol.,* 33, 101, 1989.

510a. **Marcireau, C., Guilloton, M., and Karst, F.,** In vivo effects of fenpropimorph on the yeast *Saccharomyces cerevisiae* and determination of the molecular basis of the antifungal property, *Antimicrob. Agents Chemother.,* 34, 989, 1990.

511. **Hays, P. R., Neal, W. D., and Parks, L. W.,** Physiological effects of an antimycotic azasterol on cultures of *Saccharomyces cerevisiae, Antimicrob. Agents Chemother.,* 12, 185, 1977.

512. **Taylor, F. R., Rodriguez, R. J., and Parks, L. W.,** Relationship between antifungal activity and inhibition of sterol biosynthesis in miconazole, clotrimazole, and 15-azasterol, *Antimicrob. Agents Chemother.,* 23, 515, 1983.

513. **Woloshuk, C. P., Sisler, H. D., and Dutky, S. R.,** Mode of action of the azasteroid antibiotic 15-aza-24-methylene-D-homocholesta-8,14-dien-3β-ol in *Ustilago maydis, Antimicrob. Agents Chemother.*, 16, 81, 1979.
514. **Bottema, C. K. and Parks, L. W.,** Δ^{14}-Sterol reductase in *Saccharomyces cerevisiae, Biochim. Biophys Acta,* 531, 301, 1978.
515. **Rodriguez, R. J. and Parks, L. W.,** Physiological response of *Saccharomyces cerevisiae* to 15-azasterol-mediated growth inhibition, *Antimicrob. Agents Chemother.*, 20, 184, 1981.
516. **Rahier, A., Taton, M., and Benveniste, P.,** Inhibition of sterol biosynthesis in higher plants by analogues of high energy carbocationic intermediates, in *Biochemistry of Cell Walls and Membranes in Fungi,* Kuhn, P. J., Trinci, A P. J., Jung, M. J., Goosey, M. W., and Copping, L. G., Eds., Springer-Verlag, Berlin, 1990, 205.
517. **Rahier, A., Taton, M., and Benveniste, P.,** Inhibition of sterol biosynthesis enzymes *in vitro* by analogues of high-energy carbocationic intermediates, *Biochem. Soc. Trans.*, 18, 48, 1990.
518. **Leroux, P., Gredt, M., and Boeda, P.,** Resistance to inhibitors of sterol biosynthesis in field isolates or laboratory strains of the eyespot pathogen *Pseudocercosporella herpotrichoides, Pestic, Sci.*, 23, 119, 1988.
519. **Campagnoni, C., Holmlund, C. E., and Whittaker, N.,** The effect of triparanol on the composition of free and esterified sterols of *Saccharomyces cerevisiae, Arch. Biochem. Biophys.*, 184, 555, 1977.
520. **Popják, G., Meenan, A., Parish, E. J., and Nes, W. D.,** Inhibition of cholesterol synthesis and cell growth by 24 (*R,S*), 25-iminolanosterol and triparanol in cultured rat hepatoma cells, *J. Biol. Chem.*, 264, 6230, 1989.
521. **Pierce, H. D., Pierce, A. M., Srinivasan, R., Unrau, A. M., and Oehlschlager, A. C.,** Azasterol inhibitors of yeast. Inhibition of 24-methylene sterol, Δ^{14}-reductase and Δ^{14}-sterol methyltransferase of *S. cerevisiae* by 23-azacholesterol, *Biochem. Biophys. Acta,* 529, 429, 1978.

Chapter 8

TARGET SITES OF STEROL BIOSYNTHESIS INHIBITORS IN PLANTS

P. Benveniste and A. Rahier

TABLE OF CONTENTS

I. INTRODUCTION

Sterols play important roles in most eukaryotic cells. The objective of the present article is to demonstrate that inhibitors of sterol biosynthesis have profound biological effects in higher plants and may contribute to unravel and to understand the different roles of plant sterols. In order to obtain conclusive results, an important requirement is to identify precisely the target enzymes for each inhibitor. Subsequent to the identification of interesting targets with regard to their physiological effects, the next goal would be to design chemical structures best suited to be recognized by a given enzyme and(or) to interfere with its catalytical mechanism. Consequently, the present article is devoted to concepts underlying the rational design of potent inhibitors of sterol biosynthesis. Prior to the treatise of these inhibitors, the current concepts of plant sterol functions will be summarized. Since sterols are present in almost all eukaryotic organisms, we will frequently refer to sterol functions in mammalian cells and in lower fungi, which have been more rigorously studied than those in higher plants. An excellent review on a similar subject has been recently published.[1]

II. PLANT STEROLS

A. FUNCTIONS OF PLANT STEROLS

Thus far, little is known about sterol functions in higher plants. There are several reasons to explain this lack of knowledge. A major one is that it is still not possible to obtain higher plant cells auxotrophic to sterols similar to yeast and some mammalian cell cultures. It is well established that yeast (*Saccharomyces cerevisiae*) in anaerobiosis can readily absorb exogenous sterols.[2] There is also a vast collection of auxotrophic mutants well suited for studies on the sterol requirements.[2-4] In addition, several Chinese hamster ovarian (CHO) cell lines dependent on an exogenous source of cholesterol for growth have been described.[5,6] Sterol structure activity relationships have been elucidated with these systems.[7] The results of these studies suggested that a distinction should be made between a bulk membrane function, requiring rather large amounts of sterols, and a "sparking" function requiring much smaller quantities. Whereas the structure specificity of the "bulk" fraction appears rather broad, that of the "sparking" function seems to be much narrower. Ergosterol would satisfy both bulk membrane and sparking functions in wild-type yeast, whereas lanosterol would satisfy only the bulk function.[7,8]

In plants, work with celery cell suspension cultures, employing paclobutrazol as a sterol biosynthesis inhibitor (SBI), has provided some clues regarding a better understanding of sterol function in this biological system.[9,10] When cell growth was inhibited by the azole, it could be partially restored by addition of cholesterol to the medium, although only stigmasterol was fully effective in this respect. Cholesterol (50 μM) and a trace amount of stigmasterol (0.05 μM) were as effective as stigmasterol (50 μM) alone in restoring growth, suggesting an essential regulatory role for stigmasterol and a bulk membrane function for cholesterol.

Work with model membranes[11,12] has shown that plant sterols (campesterol, stigmasterol, and sitosterol) are less effective than cholesterol in regulating membrane fluidity and permeability. This conclusion might debatable since most studies on plant sterols have been performed with artificial membranes (e.g., liposomes) composed of either egg yolk or synthetic phospholipids. It could be argued that a close fit must exist between the plant sterol mixture and some subtle features (e.g., highly unsaturated fatty acids) of polar lipids in higher plant membranes. Therefore, studies with artificial membranes should be ideally performed with liposomes containing phospholipids originating from purified plant membranes (e.g., plasma membranes). A satisfactory solution could also be to use soybean lecithins.

Studies directly performed with plant cells[13] have reached conclusions similar to those obtained with model membranes, i.e., typical plant sterols seem to be less efficient in regulating membrane permeability than cholesterol. However, it appears mandatory to be very careful with the interpretation of such *in vivo* studies, since the highly hydrophobic plant sterols may be less efficiently absorbed by whole tissues than cholesterol, which is more water-soluble.[14] Recent work from Arvinte and Steponkus[15] has convincingly shown that small unilamellar liposomes of defined phospholipid composition could readily fuse with protoplasts from barley leaves. Therefore, efficient incorporation of exogenous polar lipids in peri- and intracellular membranes from protoplasts could be demonstrated. Such a model could possibly be used to study the incorporation of exogenous sterols into plant membranes.

Finally, it has been shown that the effects of low temperature, short day-length, or high salinity on plants would lead to changes in sterol/phospholipid ratios[16] or in relative percentages of cholesterol, campesterol, stigmasterol, and sitosterol.[17-20] Whereas it is tempting to discuss these observations with regard to mechanisms by which plants adapt to their environment, definite conclusions still require further studies, in particular with suitable model systems. Within this context, recent studies performed with *Tetrahymenea pyriformis*, an organism that is able to incorporate exogenous sterols, have shown that addition of ergosterol to cells subjected to a growth temperature shift from 34 to 15°C led to a preferential increase in palmitoyl-CoA desaturase activity, resulting in a pronounced elevation of palmitoleate (16/1 Δ9).[21]

B. STEROL BIOSYNTHESIS IN HIGHER PLANTS

Since several reviews have been published on plant sterol biosynthesis,[22,23] we will only recollect the main features of this process. Figure 1 represents a comparative pathway of sterol synthesis in mammals, fungi, and higher plants. Upstream to 2(3)-oxidosqualene (OS) (VII) it is generally accepted that the three pathways are very similar. Identical intermediates have been found in all of these organisms. This does not implicate, however, that the enzymes catalyzing these steps are identical. Some structural differences may exist with regard to differences in enzyme kinetics or regulation.[24] In this context it has been recently demonstrated that the primary sequence of HMGCoA reductase of *Arabidopsis thaliana*, although sharing some homology with the primary sequence of the same enzyme in yeast, featured important departures.[25] Downstream to OS, profound differences exist between plants on one side, and yeast or mammalian sterol biosynthesis on the other. These differences may be summarized as follows: (1) cycloartenol (I) is the first cyclic product in photosynthetic eukaryotes, while lanosterol (II) is formed in nonphotosynthetic organisms;[26-28] (2) a cyclopropane cleaving enzyme, namely, cycloeucalenol (III)-obtusifoliol (IV) isomerase, has been found in higher plants only;[29,30] (3) the sequence of C-4 and C-14 methyl group removals is different;[119] (4) finally, the side chain methylations are restricted to plants and fungi, but they exhibit opposite stereochemistry at C-24 in these two latter organisms. The situation in filamentous fungi is more complex than in yeast since in the former the C-24 methylation has been shown to occur at an earlier stage of the biosynthetic pathway than in yeast, more precisely at the level of lanosterol, which is converted to 24-methylene-24-dihydro lanosterol (eburicol) (V). In addition to these four important differences, corresponding enzymes in the biosynthetic pathways may feature more subtle differences with regard to changes in substrate specificities or the susceptibility to a given inhibitor. As discussed below, these two distinctions are probably important for the understanding of the selectivity observed with some sterol biosynthesis inhibitors. For example, it can be presumed from the biosynthetic scheme (Figure 1) that lanosterol is indeed the best substrate for the 14-demethylase of mammals and yeast, whereas obtusifoliol is expected to be most suitable for the corresponding enzyme in higher plants. This latter assumption has been recently clarified in our

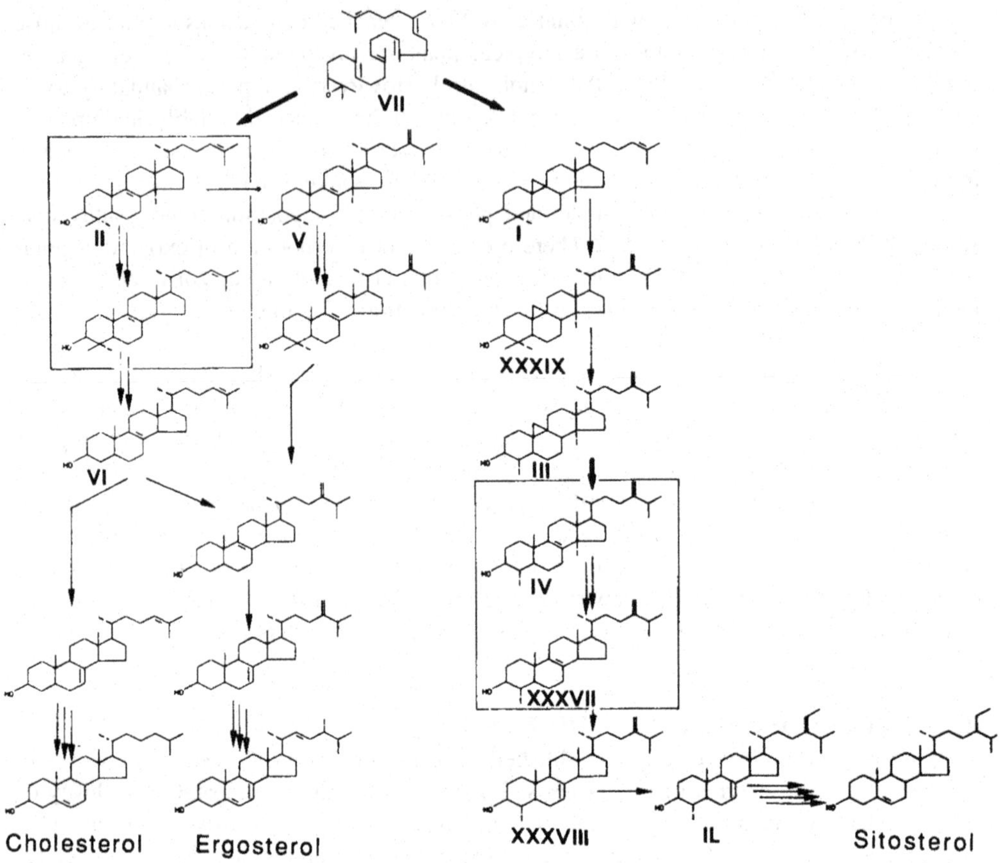

FIGURE 1. Comparative biosynthetic pathways yielding cholesterol in mammalian cells, ergosterol in *S. cerevisiae* and pathogenic fungi and sitosterol in higher plants. I: cycloartenol; II: lanosterol; IIII: cycloeucalenol; IV: obtusifoliol; V: eburicol(24-methylene dihydrolanosterol); VI: zymosterol; VII: 2(3)-oxidosqualene; XXXVII: 4α-methyl-ergosta-8,24(28)-dien-3β-ol; XXXVIII: 24-methylene lophenol; XXXIX: 24-methylene cycloartanol; IL: 24-ethylidene lophenol.

laboratory.[116] Another example of different substrate specificity has been found for the $\Delta^8 \rightarrow \Delta^7$-sterol isomerase. Whereas cholest-8-en-3β-ol and zymosterol (VI) were shown to be the best substrates of this enzyme in mammals and yeast,[31] we have recently demonstrated that 24-ethyl- or 24-ethylidene sterols such as 4α-methyl-5α-stigmasta-8,Z-24(28)-dien-3β-ol are better substrates in maize.[32] We suggest that such differences in substrate specificities between animal, fungal, and plant sterol biosynthetic enzymes are a common feature.

Only a few reports on the solubilization and purification of plant sterol biosynthetic enzymes operating downstream to squalene have been published in the past. However, 2(3)-oxidosqualene-, cycloartenol-, and β(α)-amyrin cyclases have been recently purified to homogeneity from pea seedlings[33] and from *Isodon japonicus* cell suspension cultures.[34] This is a far-reaching step in the understanding of functional relationships between these two cyclases. The extensive purification of other sterol biosynthetic enzymes from plants is undoubtedly a very important goal to be accomplished in the next few years.

Aiming at the rational design of inhibitors of sterol biosynthetic enzymes, it is important to emphasize that enzymes operating downstream to OS can be divided into at least two classes: oxydative enzymes (monooxygenases, desaturases) and enzymes that involve carbocationic intermediates during their course of catalysis (Table 1). Undoubtedly, the mechanism of enzyme action is profoundly different for these two types of enzymes. Therefore,

TABLE 1

Oxidative Enzymes	Enzymes involving carbocationic intermediates
Squalene-2(3)-epoxidase	Farnesyl-diphosphate-squalene synthetase
4,4-Dimethyl sterol-4α-methyl-demethylase	2(3)-Oxidosqualene-cycloartenol(lanosterol)cyclase
Lanosterol (obtusifoliol)-14-demethylase	*S*-Adenosylmethionine-sterol-C-24(C-28)-methyltransferases
4α-Methyl sterol-4α-methyl-demethylase	Cycloeucalenol-obtusifoliol isomerase
Δ^5-Desaturase	$\Delta^8 \rightarrow \Delta^7$-Sterol isomerase
Δ^{22}-Desaturase	$\Delta^{8,14}$-Sterol-Δ^{14}-reductase
	$\Delta^{5,7}$-Sterol-Δ^7-reductase

the strategy of inhibitor design concepts should also be different. Before describing details, the authors will first outline a few general principles in order to provide the theoretical foundation for the design of SBIs.

III. THE CONCEPT OF TRANSITION-STATE ANALOGUES

The concept of transition-state analogues is very useful for the design of potent enzyme inhibitors.[35,36] According to Bartlett and Marlowe and other authors,[37] the rationale for this approach is the recognition that "additional binding interactions can develop between an enzyme and its substrate when their complex reaches the transition state. These extra binding interactions which are not available in the ground state assist in lowering the activation energy of the reaction and are a mechanism for enzymatic rate enhancement. Transition-state analogue inhibitors seek to take advantage of these additional binding interactions in incorporating key structural elements of the unstable transition state in the stable molecule of the inhibitor. Consequently, transition-state analogue inhibitors may have a much higher affinity for the active site of the enzyme than traditional ground-state analogue inhibitors." How tight can an inhibitor be expected to bind? In a recent paper, Schloss[38] has discussed this issue. Within some limits it can be shown that the ratio between the rate of the enzymic reaction and the rate of the uncatalyzed reaction is equal to the ratio between the enzyme's dissociation constants for the substrate and the transition state. Therefore, a range of 10^{-11} to 10^{-20} *M* might be expected for dissociation constants of transition-state structures. A good transition-state analogue could then be expected to exhibit a dissociation constant in this range.[38]

Transition states are short lived and often lead to metastable high energy intermediates (HEI) interacting tightly with the enzyme. According to the Hammond principle, these HEIs are good approximations of the transition-state structure and can be considered as models for inhibitor design. However, it can be expected that HEIs should have less affinity for the active site of the enzyme than transition states. The conclusion reached by Schloss[38] is that "the more stable an intermediate of the enzymic reaction is, the less tightly bound by the enzyme should it, or its analogue, be".

A. DESIGN OF ANALOGUES OF CARBOCATIONIC HIGH ENERGY INTERMEDIATES

The results of mechanistic studies performed with sterol and terpenoid biosynthetic enzymes[39,40] have shown that several enzymatic reactions involve one or several carbocationic HEIs. This aspect has been discussed previously in detail, and the structures of the putative carbocationic HEIs have been described.[41,42] The enzymes listed in Table 1 belong to this class.

We have synthesized stable analogues distinguished by a positive charge at a position identical to that of the sp^2 carbon atom of the carbocationic HEI. A nitrogen atom proved to be useful in most cases. The tertiary amine with a pKa close to 10 is protonated at a physiological pH and generates a stable ammonium derivative displaying a positive charge at the expected position. We were indeed able to show that these analogues mimic efficiently these HEIs and display an affinity for their target enzyme three orders of magnitude higher than that of the best substrate.[41-44] We will later describe several examples for the design of potent inhibitors based on the preceding concepts.

B. INHIBITION OF ENZYMES UPSTREAM TO SQUALENE

Only a few examples are available for the inhibition of presqualene enzymes in higher plants. Whereas 3-hydroxy-3-methylglutaryl-CoA reductase is a key regulatory enzyme in mammalian cholesterol biosynthesis,[45] it does not seem to play an equivalent role in yeast[46] and in higher plant[47] sterol biosynthesis. Compactin and mevinolin are two compounds initially isolated from *Aspergillus terreus*. They act as extremely potent inhibitors of mammalian and yeast HMGCoA reductase (K_i = 0.25 n*M*).[48,49] It has been suggested[49] that the high affinity of these inhibitors for HMGCoA reductase is accounted for by the simultaneous interaction of these inhibitors with two separate binding sites at the enzyme. These compounds have also been shown to strongly inhibit the *Sinapis alba* HMGCoA reductase *in vivo* and *in vitro*.[50] The *in vivo* inhibition of this enzyme resulted in a strong decrease in sterol biosynthesis and root elongation.[51] In contrast, the biosynthesis of chloroplastic terpenoids was only slightly affected,[51] a result raising interesting questions with regard to the compartmentalization of terpenoid biosynthesis in higher plant cells. Exogenously applied gibberellins did not overcome growth inhibition produced by mevilonin.[51] Considering the vital functions of many terpenoids in higher plants, it could be expected that HMGCoA reductase might comprise an interesting target for the design of compounds with herbicidal properties. Two genes encoding HMGCoA reductase from *Arabidopsis thalianae* have been recently cloned.[25,47]

Isopentenyl diphosphate isomerase catalyzes the interconversion of isopentenyl diphosphate and dimethylallyl diphosphate. The purified enzyme from yeast has been shown to be irreversibly inactivated by 2-(dimethylamino)-ethyl-diphosphate (VIII) (Figure 2).[52,53] From the ratio of the rate constants of complex formation (k_{on}) and complex dissociation (k_{off}), a K_i value (10^{-10} to 10^{-11} *M*) has been calculated showing that the binding to the enzyme is extremely tight. These and other results indicate that VIII is probably a carbocationic HEI analogue inhibitor with the positively charged ammonium group as the analogue of a tertiary carbocationic species (IX) during the enzyme-catalyzed reaction. To our knowledge, VIII has never been tested with the corresponding plant enzyme.

Squalene synthetase catalyzes the 1′-1 condensation between two molecules of farnesyl diphosphate to yield squalene. This reaction is the result of two distinct transformations, the insertion of C(1) of one molecule of farnesyl diphosphate into the C(2)-C(3) double bond of the second molecule to generate presqualene diphosphate, followed by conversion of this molecule to squalene. Rilling et al.[54] proposed a three-step sequence involving the hypothetical tertiary cyclopropyl carbinyl intermediate (X). Sandifer et al.[55] have presented evidence that squalene synthetase binds an ammonium analogue (XI) that mimics the topological and electrostatic properties of X, and that the binding is dependent on pyrophosphate. The synergism found between XI and pyrophosphate suggests that the enzyme binds the carbocation-pyrophosphate ion pair more tightly than either partner. In order to give more strength to this conclusion, an analogue with a phosphonophosphate group linked to the amino group has been synthesized. This compound was shown to be a potent inhibitor of squalene synthesis in pyrophosphate-free buffers.[56] Diphosphate esters such as XI have never been tested *in vitro* with a squalene synthetase from higher plants. In addition, these molecules do not seem to be suitable for *in vivo* studies since they are unstable and probably not capable to cross cell membranes. However, squalene synthetase represents a challenging

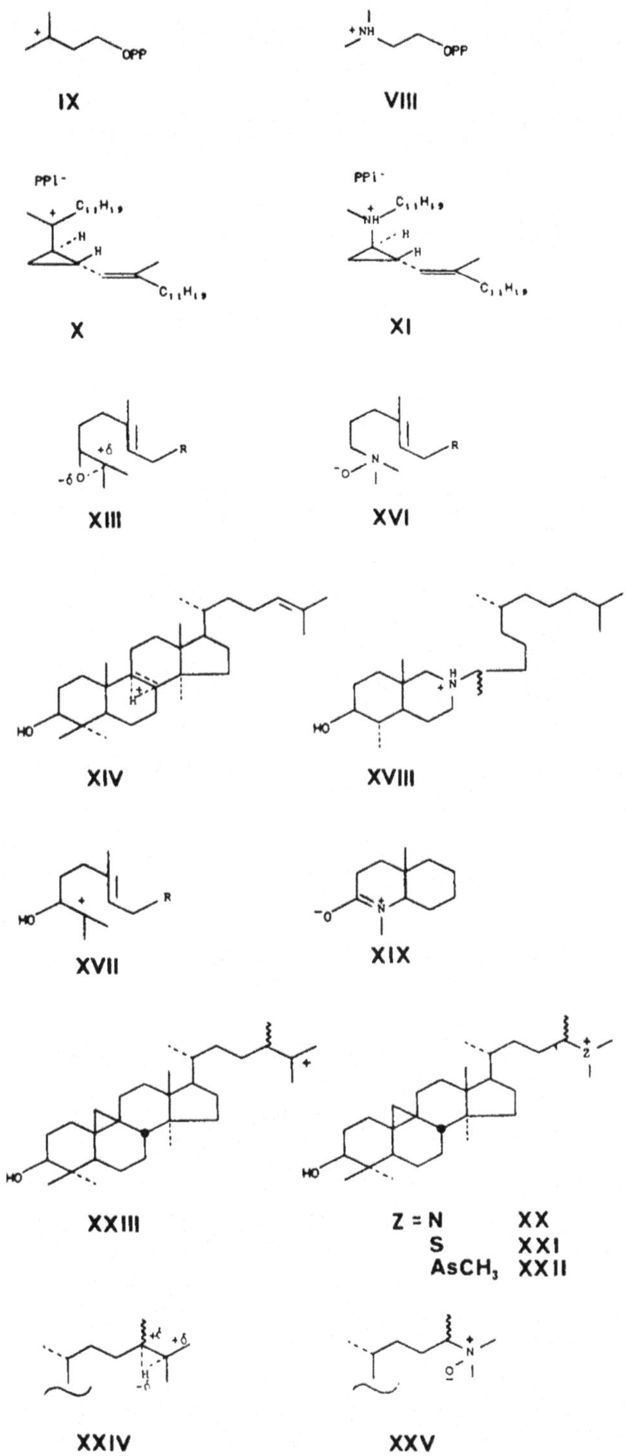

FIGURE 2. Chemical structures of compounds described in the text.

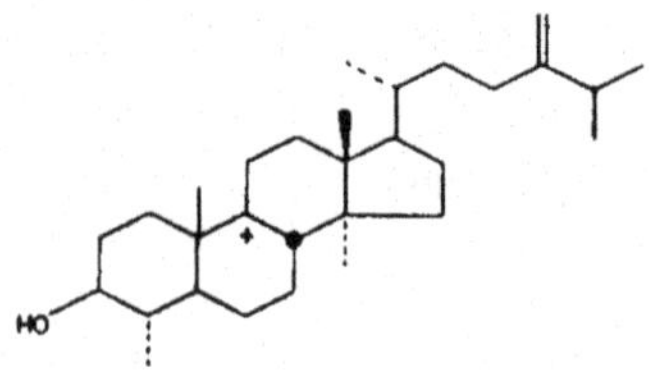

XXVI

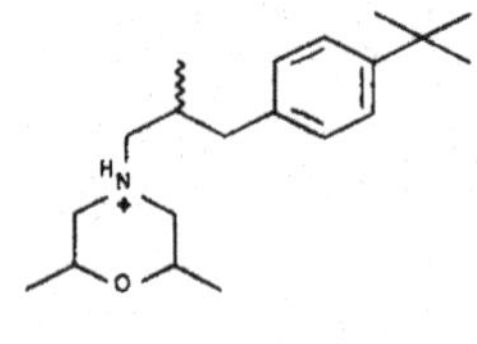

XXX

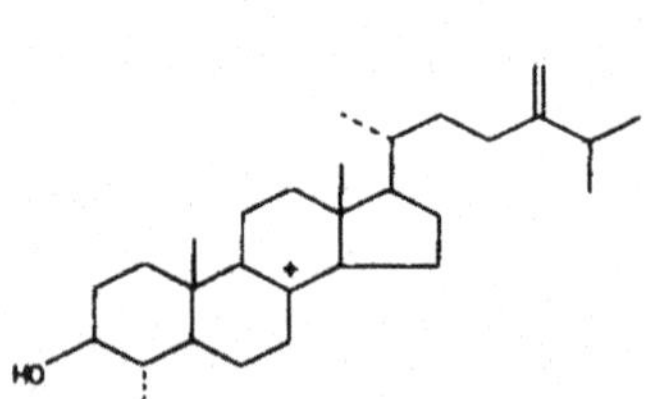

XXVIII

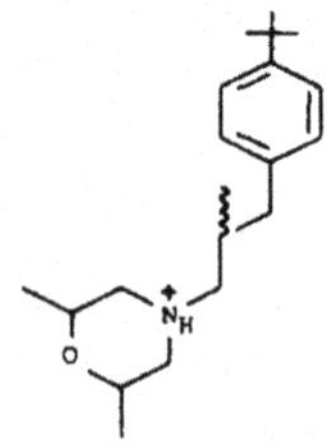

XXX

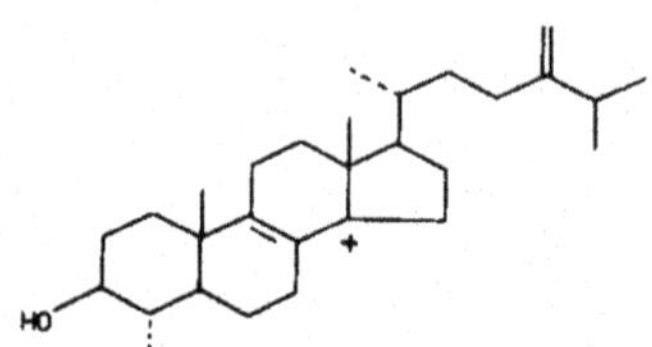

XXVII

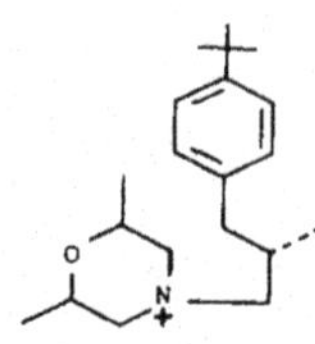

XXX

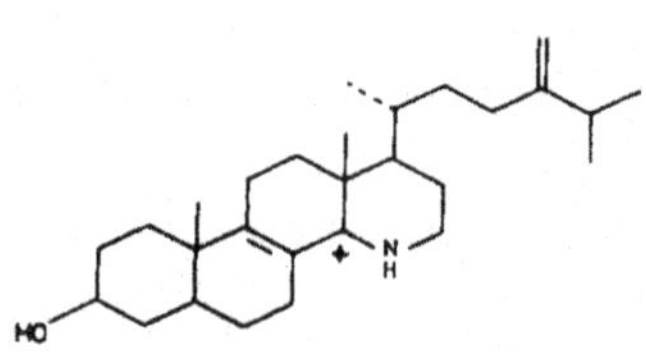

XXXII

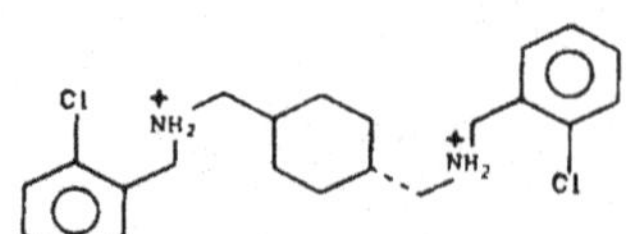

XXXIII

FIGURE 2 (continued).

target for inhibitor molecules. As discussed recently,[57] the enzyme comprises the link between the general isoprenoid metabolism and the specific biosynthetic pathway leading to sterols.

So far no potent inhibitors have been reported for the inhibition of mevalonate kinase, 5-phosphomevalonate kinase, 5-pyrophosphomevalonate decarboxylase, and the prenyltransferases.

C. 2(3)-OXIDOSQUALENE CYCLASE

The fascinating enzyme oxidosqualene (OS) cyclase catalyzes the cyclization of chair-boat-chair-boat folded (3S)-2(3)-OS, yielding lanosterol in nonphotosynthetic organisms and cycloartenol in photosynthetic organisms.[22,23,26] These authors and others have succeeded to design HEI or TS analogue inhibitors for two postulated steps of the reaction pathway: (1) the initial step, the polarization of the C-2 oxygen bond, leading to a charge deficiency at C-2 (XIII); and (2) the terminal step of the cyclization of OS to lanosterol and cycloartenol, i.e., the formation of the C-8(9) carbocationic HEI (XIV), and the subsequent proton elimination. For the first step we have synthesized 2-aza-2,3-dihydro-squalene, 2-aza-2,3-dihydro-squalene-*N*-oxide (XVI) (Figure 2), and derivatives.[58,59] These compounds were shown to be powerful inhibitors of OS-lanosterol cyclase from rat liver as well as OS-cycloartenol- and β(α)-amyrin cyclases from plant tissues. The results obtained with the tertiary amine-*N*-oxide derivatives such as XVI show that these compounds are more potent inhibitors than the parent amines.[59] This result is consistent with the fact that the *N*-oxide (XVI) is closer to the TS (XIII) while the ammonium derivative is more similar to HEI (XVII).[42,44] In this context it is worth noting that 2-aza-2,3-dihydro-squalene has been shown to inhibit lycopene cyclase.[60] In order to mimic the C-8(9) carbocationic HEI (XIV) involved in the last step of the reaction pathway, we decided to synthesize *N*-substituted-8-azadecalins such as the *N*-[(1,5,9)-trimethyl-decyl]-4α,10-dimethyl-8-aza-*trans*-decal-3β-ol (XVIII). The inhibition of the enzymatic activity by XVIII was measured *in vitro* in cell-free systems prepared from maize seedlings, pea cotyledons, and rat liver containing, respectively, OS-cycloartenol-, OS-β-amyrin-, and OS-lanosterol cyclases. It was shown that XVIII inhibited strongly the OS-cyclases from maize and from rat liver but, interestingly, did not affect the OS-β(α)-amyrin cyclases.[61] A distinctive feature of XVIII was its remarkable ability to block OS-lanosterol cyclase *in vivo*.[62] These results have been discussed recently with regard to our present knowledge of the mechanism of action of the OS-cyclases.[41,44] It is worth noting that Johnson et al.[63] have recently proposed a model for the enzymic conversion of 2(3)-OS to protolanosterol involving axial delivery of negative point-charge by the enzyme, which would stabilize the developing positive centers along the cyclization process. Such a model is consistent with the involvement of electrostatic interactions during the binding of ammonium derivatives such as XVIII to the active site of the OS cyclases. Recently, Ruhl et al.[64] synthesized a series of azadecalins such as XIX. These compounds, which structurally resemble the known cyclase inhibitor 4,4,10β-trimethyl-*trans*-decal-3β-ol (TMD), were selected because, under physiological conditions, they have at least partial positive character near the initial putative carbocationic position generated during the cyclization of OS, and thus were potential TS analogue inhibitors. Among these compounds XIX was shown to be most active but, disappointingly, not as active as TMD itself. Molecules such as XVI and XVIII have been shown to inhibit sterol biosynthesis in plant cell suspension cultures. Following treatment of bramble (*Rubus fruticosis*) cell suspension cultures with increasing amounts of XVI or similar derivatives, a correlation has been demonstrated between growth inhibition on one side, and OS accumulation and sterol biosynthesis inhibition on the other.[65]

D. INHIBITION OF *S*-ADENOSYL-L-METHIONINE-CYCLOARTENOL-C-24-METHYLTRANSFERASE (ADOMETCMT)

Since it had been observed previously that 24-dihydro-25-azazymosterol was a good inhibitor of the *in vivo* and *in vitro* C-24-methylation of zymosterol (VI) in yeast,[66,67] we have synthesized a series of compounds possessing a positively charged heteroatom (N,As,S) at C-25, such as XX, XXI, and XXII. These compounds were shown to inhibit strongly the AdoMetCMT.[68] These results could be explained by assuming that XX-XXII mimicked carbocationic HEIs such as XXIII. A thorough chemical structure-inhibitor activity relationship study[69] allowed to draw the following conclusions: (1) the presence of a positive charge at C-24 of the sterol side chain is of paramount importance; (2) a tetracyclic sterol skeleton is essential; (3) some molecular features that determine the affinity of the 25-aza-steroid are different from those affecting the recognition of the substrate in its ground state. We will discuss the latter conclusion in greater detail below. In addition to the involvement of a carbocationic HEI, it can be postulated that for the AdoMetCMT reaction a dipolar transition state (XXIV) exists during the C-24—C-25 hydride migration. A good mimic of this species is the stable analogue (XXV), featuring a tertiary amine *N*-oxide function. Indeed, XXV was shown to be even a more potent inhibitor than the parent amine (XX) (Figure 2).[41,69]

When the culture medium of suspension cultures of bramble (*R. fruticosus*) cells was supplemented with XX, the growth was severely inhibited at concentrations of XX close to 1 μM. The sterol profile of cells treated with XX showed a strong decrease in the sitosterol contents and a striking accumulation of sterols nonalkylated at C-24 such as cycloartenol (I), which became the major sterol of these cells. Also, 24-methylene cholesterol accumulated noticeably.[117] These results show that the two C-24 and C-28 methyltransferases catalyzing the insertion of the two extra carbon atoms present at position C-24 in sitosterol are inhibited *in vivo* in agreement with the *in vitro* experiments.[70] Therefore, C-25 heteroatom (N, As, and S)-substituted steroid derivatives are potent molecular tools for the manipulation of sterol alkylations in higher plants and fungi. One major question raised by these results is the relationships between C-24-methyltransferase inhibition and the severe growth inhibition observed upon treatment of bramble cells with XX.

E. CYCLOEUCALENOL-OBTUSIFOLIOL ISOMERASE (COI), Δ^{14}-STEROL REDUCTASE (SR), AND $\Delta^8 \rightarrow \Delta^7$-STEROL ISOMERASE (SI)

COI catalyzes the cleavage of the 9β,19-cyclopropane ring of cycloeucalenol (III) yielding obtusifoliol (IV). Its structural requirements for the transformation or binding of substrates as well as inhibitors have been recently studied.[71] From these data and other studies,[30] the catalytic pathway apparently involves the passage through a carbocationic HEI (XXVI). As discussed previously,[41] Δ^{14}-reductase and SI also entail HEIs, with positive charges localized at C-14 (e.g., XXVII) and C-8 (e.g., XXVIII), respectively. The protonated *N*-substituted 8-azadecalins (XVIII and XXIX) share electronic and structural similarities with the carbocationic HEIs (XXVI, XXVII, and XXVIII) and could be considered as potential inhibitors of COI, SR, and SI. Indeed, these low molecular weight analogues of HEIs were shown to strongly inhibit COI,[71,72] SI,[72,73] and SR[73] in cell-free extracts from maize seedlings. In particular, XVIII is a remarkably potent inhibitor of the COI *in vitro*[71] and *in vivo*.[92]

One important group of fungicides acting as inhibitors of sterol biosynthesis are morpholine derivatives such as tridemorph and fenpropimorph (XXX)[74] (see Chapter 7). In our laboratory, employing a microsomal preparation isolated from maize embryos, we have shown that XXX inhibited COI,[32,72,75] SI,[32] and more recently SR.[73] These results indicated that XXX and related compounds acted as analogues of the carbocationic HEIs (XXVI, and XXVII, and XXVIII). The structure flexibility of tridemorph or XXX could explain that the same morpholinium derivative is able to mimic three different HEIs with sp^2 carbon atoms in the C-9, C-14, and C-8 positions.[22,41,76] A tertiary amine *N*-oxide inhibited the COI with an IC_{50} similar to the parent amine.[75]

15-Aza-24-methylene-D-homocholestadiene-3β-ol (A25822B) (XXXII) has been shown to strongly inhibit the SR in yeast.[77] We have recently demonstrated that XXXII was also a potent inhibitor of the SR derived from maize embryos,[72] and we have explained this remarkable inhibition with the assumption that after protonation of XXXII, the iminium derivative would mimic the C-14 carbocationic HEI (XXVII) involved in the catalytic pathway. In contrast to morpholines, XXXII was shown to be highly specific toward the SR and did not at all inhibit the COI.[73] Finally we have shown that AY9944 (XXXIII) inhibited the SI but had little or no effect on the SR and the COI.[72] It is remarkable that AY 9944, A25822B, and the 8-azadecalin (XVIII) are specific inhibitors of the SI, SR, and COI. AY 9944, A25822B, and XVIII, when added to plant cell suspension cultures, were shown to lead to strong accumulation of Δ^8, $\Delta^{8,14}$-, and 9β,19-cyclopropyl sterols, respectively.[78,79,92] This result indicated that the SI, SR, and COI were also the main *in vivo* targets for the three inhibitors. Morpholine inhibitors such as fenpropimorph (XXX) were shown to be especially suitable for *in vivo* experiments since cyclopropyl sterols such as 24-methyl pollinastanol were shown to accumulate dramatically in several plant species treated with XXX. These precursors replaced almost completely the normally occurring Δ^5-sterols.[80-84,92] Cyclopropyl sterols have been found in membrane fractions isolated from maize seedlings treated with XXX.[85] Interestingly, endoplasmic reticulum membranes from XXX-treated plants were richer in sterols than the same membranes from control plants. Whereas the presence of cyclopropyl sterols in plasma membranes had no effects on the fatty acid profile of polar lipids, an increase of the desaturation index was found in endoplasmic reticulum-enriched membranes.[85]

To summarize the inhibition of enzymes that involved carbocationic intermediates during catalysis, results from our and other laboratories have shown that compounds containing a positively charged heteroatom at a position identical to that of the sp^2 carbon atom of the putative carbocationic HEI are in most cases strong enzyme inhibitors. However, this result is not sufficient to prove that the designed inhibitors are really TS or HEI analogue inhibitors. The following list of likely features of TS or HEI analogue inhibitors has been proposed by several authors[37,38,87] including ourselves:[69,71] (1) transition-state (high energy intermediate) resemblance; (2) tight binding;[86] (3) correlation between K_i values for inhibitors and K_m/k_{cat} values for the corresponding substrates;[37] (4) structural features essential for binding of the substrate affect the binding of TS or HEI analogue inhibitors to a much lesser extent;[69,71] (5) slow binding enzyme inhibition.[38] Only criterions 1, 2, and 4 were met in our own work.[69,71] Tests for criteria 3 and 5 would require kinetic studies which are difficult to accomplish with heterogenous enzymatic systems composed of membrane-bound enzymes and water-insoluble substrates and inhibitors. These tests might become feasible with solubilized and extensively purified enzymes.

Finally, the results described above shed light on the paramount importance of electrostatic interactions for the binding of HEI analogues to active sites. The nature of the negatively charged group present in the enzyme and interacting with these inhibitors is unknown, but it could be a delocalized carboxylate anion as shown in glycosidases,[87] or a thiolate as suggested for isopentenyl diphosphate isomerase.[52] It is worth mentioning that morpholine fungicides have been shown to inhibit bacterial cholesterol oxidase, an enzyme characterized by an aspartate residue essential for catalytical activity.[88]

IV. INHIBITION OF MONOOXYGENASES: OBTUSIFOLIOL-14-DEMETHYLASE

Among the enzymes listed in Table 1, only obtusifoliol-14α-demethylase and 24-methylene cycloartanol C-4 demethylase[119] have been studied in higher plants thus far. The squalene-2(3)-epoxidase has been investigated in yeast since this enzyme was shown to be the target for a family of new fungicidal inhibitors: the allylamines (see Chapter 7). To our knowledge, no data have been published on the plant squalene epoxidase and its inhibition by allylamines.

Both cholesterol biosynthesis in mammals and ergosterol biosynthesis in yeast involve a cytochrome P-450-catalyzed removal of the 14α-methyl group of lanosterol.[89,90] Until recently, a similar enzyme had not been characterized from higher plants. However, biosynthetic studies strongly suggested that the removal of the 14α-methyl group occurred subsequent to the elimination of the 4β-methyl group and cleavage of the 9β,19-cyclopropane ring.[22] This finding was in contrast to data obtained with animals and yeast systems, where the 14α-methyl group was shown to be removed at the 4,4-dimethyl stage.[89,90]

A microsomal system isolated from *Zea mays* embryos capable of demethylating efficiently the 14α-methyl group of obtusifoliol has been recently described.[91] Under interrupted conditions, i.e., in the presence of a potent inhibitor of the $\Delta^{8,14}$-sterol-Δ^{14}-reductase, the immediate demethylation product of obtusifoliol (IV) could be accumulated and thoroughly identified as 4α-methyl-5α-ergosta-8,14,24(28)-trien-3β-ol (XXXIV). Under these conditions no trace of $\Delta^{8(14)}$-sterol could be detected. These results, taken together with the accumulation of a $\Delta^{8(9)}$-metabolite in the presence of a specific inhibitor of the $\Delta^8 \rightarrow \Delta^7$-sterol isomerase, strongly support the following sequence of reactions in maize (Figure 3): (1) 14α-demethylation, (2) Δ^{14}-reduction of the $\Delta^{8,14}$-diene, (3) $\Delta^8 \rightarrow \Delta^7$-isomerization. This

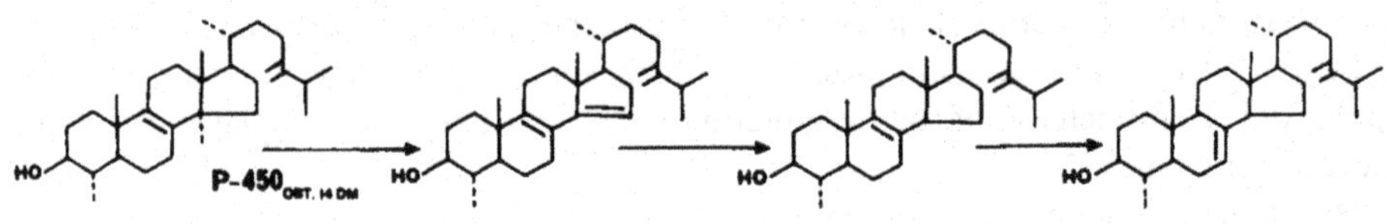

IV XXXIV XXXVII XXXVIII

FIGURE 3. Metabolism of obtusifoliol (IV) by microsomes from maize embryos in the presence of NADPH and O_2.

sequence is further supported by the recent characterization of a microsomal $\Delta^{8,14}$-sterol-Δ^{14}-reductase in the same material[73] and the accumulation of (XXXIV) as well as other $\Delta^{8,14}$-sterols in suspension cultures of *Rubus fruticosus* cells grown in the presence of Δ^{14}-reductase inhibitors.[79,92]

The maize microsomal 14α-demethylase activity was shown to be totally dependent on NADPH, whereas NADH proved to be an extremely poor reductant for the plant demethylase. In addition to its microsomal localization and NADPH requirement, the 14α-demethylase activity was fully dependent on molecular oxygen and showed a strong inhibition by carbon monoxide, this effect being partly reversed by light. Furthermore, compounds known to block electron transfer from NADPH cytochrome P-450 reductase (menadione, oxidized cytochrome c) to cytochrome P-450 potently inhibited the demethylation reaction. In addition, as discussed in detail, azole derivatives potently inhibited the demethylation reaction. These lines of evidence strongly support the involvement of cytochrome P-450 in the obtusifoliol-14α-demethylase reaction in agreement with the reported participation of cytochrome P-450 in the analogous reactions in mammals and yeast (Chapter 1).[89,90] Therefore this enzyme constituted a novel member of the increasing but still limited number of endobiotic cytochrome P-450s and associated monooxygenase activities identified in higher plants.[93]

Azoles, pyridines, and pyrimidines belong to a large group of fungicides used for the control of fungal diseases in agriculture and medicine (Chapter 7).[94,95] Structurally related compounds display potent plant growth regulator properties.[96,97] Recently a triazole derivative with herbicidal properties has been described.[98] The main inhibitory site of the fungicides appears to be the lanosterol-14α-methyl demethylase,[99,100] leading to the accumulation of lanosterol and other 14α-methyl sterols in fungi treated with these derivatives (Chapter 7). In some cases the direct inhibition of the isolated enzyme could be demonstrated.[101,102] Moreover, some experimental evidence supported the assumption that the molecular basis of 14α-demethylase inhibition is due to the strong binding of these chemicals to the ferric protoheme iron of cytochrome P-450.[93,102] In contrast to the extensive studies performed with fungi, much less attention has been paid to effects of such compounds in plants, since the early reports of the accumulation of 14α-methyl sterols in *Chlorella* grown in a culture medium containing triarimol[104] or suspension cultures of *R. fruticosus* cells treated with fenarimol.[105] However, in recent years some reports have described relevant experiments concerning the effects of azole compounds on sterol biosynthesis in higher plants. All these reports support the inhibition of the plant 14α-demethylase enzyme by these various fungicides. In most cases a strong accumulation of 14α-methyl sterols was observed.[84,106] For example, the watering of maize seedlings with a solution of LAB 170250F (a triazole fungicide synthesized at the B.A.S.F.) for 10 days led to a dramatic accumulation of 14α-methyl sterols in roots. These new sterols [essentially 24-dihydroobtusifoliol (XXXV) and 30 Nor-24-dihydroobtusifoliol (XXXVI)] replaced almost totally the initially present mixture of Δ^5-sterols (campesterol, stigmasterol, and sitosterol).[106] Similar results have been obtained in the case of barley plants treated with nuarimol, paclobutrazol, triapenthenol,[84] and a triazole herbicide.[98] Here again, the *in vivo* inhibition of the enzyme responsible for the removal of the C-14 methyl group led to the accumulation of 14α-methyl sterols at the expense of Δ^5-sterols.

Recently the *in vitro* inhibition of the cytochrome P-450-dependent obtusifoliol-14-demethylase (P-450-OBT.14DM) by heterocyclic derivatives containing a sp^2 nitrogen atom has been demonstrated.[106] The source of P-450-OBT.14DM for these assays was a microsomal preparation derived from maize embryos. After incubation with obtusifoliol in the presence of a range of fungicide or PGR concentrations, the resulting demethylation products, namely, 4α-methyl-5α-ergost-8,24-dien-3β-ol (XXXVII) and 24-methylene lophenol (XXXVIII), were quantified by GLC. A dose-response curve was then obtained and the I_{50} value determined. The results clearly indicated that many of these different azole derivatives are potent inhibitors of the P-450-OBT.14DM. In the case of fungicides, LAB 170250 F (I_{50} = 50 n*M*) and LAB 158241 F (I_{50} = 300 n*M*) are the most potent inhibitors, whereas propiconazole (I_{50} = 2 μ*M*) and triadimefon (I_{50} = 10 μ*M*) are less efficient. Most of the PGRs tested displayed low activity (I_{50} = 10 μ*M* for tetcyclacis and BAS 110).

Finally, using a microsomal preparation derived from maize embryos, azole derivatives (propiconazole and LAB 170250 F) and tetcyclacis have been found to yield type II binding spectra which are known to result from the displacement of the native sixth ligand of the heme iron by a sp^2 nitrogen atom of the inhibitor.[107] From the changes in absorbance differences between 391 and 428 nm in the difference spectra obtained with a range of inhibitor concentrations, Lineweaver-Burk double reciprocal plots were used to calculate K_d values. The values for the different compounds were in the same concentration range (20 to 30 μ*M*). Thus, the binding to the total cytochrome P-450 present in maize microsomes did not match the I_{50} values found for the same inhibitors. This result underlines the probable presence of multiple forms of cytochrome P-450 and associated monooxygenase activities in the microsomal preparation used.[93]

Several authors have provided evidence that fungicides could have side effects on the host plant. In the best case, a beneficial plant growth retardant (PGR) effect can be observed. In some cases undesirable phytotoxic effects can occur which may limit or even curtail the commercial development of a fungicide. What is the molecular basis of the PGR or the phytotoxic effects? The PGR effect is often associated with an inhibition of gibberellin biosynthesis at the level of ent-kaurene oxidation in cell-free systems.[96,97] In this case, reversions of plant growth retardation with exogenous gibberellic acids have been frequently observed. In a few cases the PGR or phytotoxic effects have been shown to parallel inhibition of plant sterol biosynthesis.[98,108,118] The nature and degree of the biological activity shown by inhibitors such as triadimenol (a fungicide) and paclobutrazol (a PGR) are considerably influenced by the absolute configuration of the two asymmetric carbon atoms present in the molecules (Chapter 7). The data obtained with paclobutrazol enantiomers indicated that the *R* configuration at the chiral carbon bearing the hydroxyl group is the prime determinant for fungitoxicity, whereas enantiomers with *S* configuration at this carbon are effective inhibitors of gibberellin biosynthesis.[109] Moreover, plant sterol C-14α demethylation was not always inhibited by enantiomers with high fungicidal activities, thus indicating that the structural features necessary for binding to the plant C-14α demethylase may differ from those suitable for the analogous fungal enzyme.[109,110] To be fully conclusive, these results should be validated by comparative *in vitro* assays employing the various enantiomers and on the isolated plant and fungal 14α-demethylases. Results obtained with triadimenol enantiomers on wheat seedlings suggest that the side effects of triadimenol on plant growth are best explained by the inhibition of plant sterol rather than gibberellin synthesis.[108] Here again, comparative *in vitro* inhibition of ent-kaurene hydroxylase and of 14α-demethylases would be required to strengthen such conclusions. These considerations emphasize the need for pure enantiomers of azole SBIs, especially when effects on distinct metabolic pathways are being studied.

Selectivity of azole inhibitors is indeed a very important property. For example, fungicides used in plant protection or in medicine should control the pathogenic fungus without affecting the plant or human host, respectively. Among other possibilities, which have been discussed in detail elsewhere,[93,111] this would imply that the azole inhibitor should bind more tightly to the fungal 14α-demethylase than to the host 14α-demethylases (Chapter 9).[112] Likewise, when azole inhibitors are being developed as potential herbicides, these molecules should control weeds without affecting humans and animals. This would imply that azole inhibitors should bind more tightly to the plant 14α-demethylase than to animal or fungal 14α-demethylases. Furthermore, it would be desirable that azole herbicides exert selectivity within the plant kingdom and control weeds without affecting crops. Such a property would imply that selective azole herbicides could discriminate between crop and weed 14α-demethylases. Most likely, the selectivity of azole inhibitors will originate in part from the substrate specificity of the three main 14α-demethylases found in nature, namely, the lanosterol-14α-demethylase in mammals and *Saccharomyces cerevisiae*, the eburicol-14α-demethylase in most filamentous fungi, and the obtusifoliol-14α-demethylase in higher plants. The designation of lanosterol (II), eburicol (V), and obtusifoliol (IV) as best substrates for the three demethylases, although reasonable with regard to the biosynthetic pathways, has still no direct experimental support except for higher plants. A thorough substrate structure-enzymatic activity study has been performed and has shown that among twenty assayed analogues, obtusifoliol was by far the best substrate of the plant 14α-demethylase. This result clearly demonstrated that the enzyme exhibits a considerable substrate specificity.[116]

The understanding of the molecular basis for substrate specificity and inhibitor selectivity of 14α-demethylases would require the determination of the primary sequence of substrate and inhibitor binding domains for each family of enzymes. Much work still needs to be done to reach this challenging goal. First of all it would be necessary to clone and to sequence the genes coding for the lanosterol-demethylases in rat liver and yeast, eburicol-demethylase in pathogenic fungi and obtusifoliol demethylase higher plants. So far, the lanosterol-14α-demethylase genes from *S. cerevisiae*[113,114] and from *Candida tropicalis* have been cloned and sequenced. The coding region of these genes contains sequences highly preserved in many other cytochrome P-450,[115] but in particular within the cytochrome P-450 that acts as the terminal monooxygenase in the bacterial d-camphor monooxygenase system.[115] Since the X-ray crystal structure has been elucidated in this case, key information will become available as soon as the primary sequences of the sterol-14DMs listed above will become known.

V. CONCLUSION

An impressive list of suitable *in vitro* and *in vivo* inhibitors to be applied as useful molecular tools in studies on enzymes of sterol biosynthesis in higher plants is currently available. The successful synthesis of inhibitors resembling and mimicking HEIs has provided insight into most of the enzymes with carbocationic intermediates during their path of catalysis (Table 1). Progress has been slower with the group of monooxygenases. Of particular interest are the two demethylations at the C-4 position of sterol precursors. Inhibition of this enzyme system would lead to the accumulation of 24-methylenecycloartenol (XXXXIX) or 24-ethylidene lophenol (IL), which may not be suitable to substitute for all of the functions inherent to Δ^5-sterols in higher plants. The Δ^{22}-desaturase comprises an additional and challenging target for chemical interference. Respective inhibitors would allow to manipulate the stigmasterol-to-sitosterol ratio in plant cells, and this manipulation would trigger the better understanding of functions necessitating the presence of a campesterol-sitosterol-stigmasterol "sterol cocktail" in membranes of higher plants.

REFERENCES

1. **Burden, R. S., Cooke, D. T., and Carter, G. A.,** Inhibition of sterol biosynthesis and growth in plants and fungi, *Phytochemistry,* 28, 1791, 1989.
2. **Servouse, M. and Karst, F.,** Regulation of early enzymes of ergosterol biosynthesis, in *Saccharomyces cerevisiae, Biochem. J.,* 240, 541, 1986.
3. **Gollub, E. G., Lies, K., Dayan, J., Adlesberg, M., and Sprinson, D. B.,** *J. Biol. Chem.,* 252, 2846, 1977.
4. **Buttke, T. and Bloch, K.,** Comparative responses of the yeast mutant strain GL7 to lanosterol, cycloartenol and cyclolaudenol, *Biochem. Biophys. Res. Commun.,* 92, 229, 1980.
5. **Chang, T. Y., Telakowski, C., Vanden Heuvel, W., Alberts, A. W., and Vagelos, P. R.,** Isolation and partial characterization of a cholesterol-requiring mutant of chinese hamster ovary cells, *Proc. Natl. Acad. Sci. U.S.A.,* 74, 832, 1977.
6. **Chen, H. W., Leonard, D. A., Fischer, R. T., and Trzaskos, J. M.,** A mammalian mutant cell lacking detectable lanosterol 14α-methyl demethylase activity, *J. Biol. Chem.,* 263, 1248, 1988.
7. **Rodriguez, R. J., Low, C., Bottema, C. D. K., and Parks, L. W.,** Multiple functions for sterols in *Saccharomyces cerevisiae, Biochim. Biophys. Acta,* 837, 336, 1985.
8. **Pinto, W. J., Lozano, R., Sekula, B. C., and Nes, W. R.,** Stereospecifically distinct roles for sterols in *Saccharomyces cerevisiae, Biochem. Biophys. Res. Commun.,* 112, 47, 1983.
9. **Haughan, P. A., Lenton, J. R., and Goad, L. J.,** Sterol requirements and paclobutrazol inhibition of a celery cell culture, *Phytochemistry,* 27, 2491, 1988.
10. **Haughan, P. A., Burden, R. S., Lenton, J. R., and Goad, L. J.,** Inhibition of celery cell growth and sterol biosynthesis by the enantiomers of paclobutrazol, *Phytochemistry,* 28, 781, 1989.
11. **Bloch, K. E.,** Sterol structure and membrane function, *CRC Crit. Rev. Biochem.,* 14, 47, 1983.
12. **Demel, R. A. and De Kruyff, B.,** The role of sterols in membranes, *Biochim. Biophys. Acta,* 457, 109, 1976.
13. **Grunwald, C.,** Effect of sterols on the permeability of alcohol-treated red beet tissue, *Plant Physiol.,* 43, 484, 1968.
14. **Windholz, M., Budavari, S., Stroumtsos, L. Y., and Fertig, M. V. Eds.,** *The Merck Index,* 9th ed., Merck & Co., Rahway, NJ, 1976, 283.
15. **Arvinte, T. and Steponkus, P. L.,** Characterization of the pH-induced fusion of liposomes with the plasma membrane of rye protoplasts, *Biochemistry,* 27, 5671, 1988.
16. **Horvath, I., Vigh, L., Woltjes, J., Farkas, T., van Hasselt, P., and Kuiper, P. J. C.,** Combined ESR, X-ray-diffraction studies on phospholipids vesicles obtained from cold-hardened wheats. II. The role of free sterols, *Planta,* 170, 20, 1987.
17. **Douglas, T. J. and Walker, R. R.,** 4-Desmethylsterol composition of citrus rootstocks of different salt exclusion capacity, *Physiol. Plant.,* 58, 69, 1983.
18. **Guye, M. G.,** Chilling and age related changes in the free sterol composition of *Phaseolus vulgaris* L. primary leaves, *Plant Sci.,* 53, 209, 1987.
19. **Kuiper, D. and Kuiper, P. J. C.,** Lipid composition of the roots of Plantago species: response to alteration of the level of mineral nutrition and ecological significance, *Physiol. Plant.,* 44, 81, 1978.
20. **Bae, M. and Mercer, E. I.,** The effect of long- and short-day photoperiods on the sterol levels in the leaves of *Solanum andigena, Phytochemistry,* 9, 63, 1970.
21. **Umeki, S. and Nozawa, Y.,** Effects of sterol manipulation on microsomal desaturase activities in *Tetrahymena pyriformis* with regard to thermal acclimatation, *Biochem. Biophys. Res. Commun.,* 113, 96, 1983.
22. **Benveniste, P.,** Sterol biosynthesis, *Annu. Rev. Plant Physiol.,* 37, 275, 1986.
23. **Goodwin, T. W.,** Biosynthesis of terpenoids, *Annu. Rev. Plant Physiol.,* 30, 369, 1979.
24. **Bach, T. J., Rogers, D. H., and Rudney, H.,** Detergent-solubilization, purification, and characterization of membrane-bound 3-hydroxy-3-methylglutaryl coenzyme A reductase from radish seedlings, *Eur. J. Biochem.,* 154, 103, 1986.
25. **Caelles, C., Ferrer, A., Balcells, L., Hegardt, F. G., and Boronat, A.,** Isolation and structural characterization of a cDNA encoding *Arabidopsis thaliana* 3-hydroxy-3-methylglutaryl coenzyme A reductase, *Plant Mol. Biol.,* 13, 627, 1989.
26. **Dean, P. D. G.,** The cyclases of triterpene and sterol biosynthesis, *Steroidologia,* 2, 143, 1971.
27. **Benveniste, P., Hirth, L., and Ourisson, G.,** La biosynthèse des stérols dans les tissus de tabac cultivés *in vitro.* I. Isolement de stérols et de triterpènes, *Phytochemistry,* 5, 31, 1966.
28. **Ardenne, M., Osske, G., Schreiber, K., Steinfelder, K., and Tümmler, R.,** Über die Sterine und Triterpenoide von *Solanum demissum* Lindl. und *Solanum polyadenium* Greenm., *Kulturpflanze,* 13, 115, 1965.
29. **Heintz, R. and Benveniste, P.,** Plant sterol metabolism. Enzymatic cleavage of the 9β,19-cyclopropane ring of cyclopropyl sterols in bramble tissue cultures, *J. Biol. Chem.,* 249, 4267, 1974.

30. **Rahier, A., Cattel, L., and Benveniste, P.,** Mechanism of the enzymatic cleavage of the 9β,19-cyclopropane ring of cycloeucalenol, *Phytochemistry,* 16, 1187, 1977.
31. **Gaylor, J. L., Delwiche, C. V., and Swindell, A. C.,** Enzymatic isomerization ($\Delta^8 \rightarrow \Delta^7$) of intermediates of sterol biosynthesis, *Steroids,* 8, 353, 1966.
32. **Taton, M., Benveniste, P., and Rahier, A.,** Mechanism of inhibition of sterol biosynthesis enzymes by N-substituted morpholins, *Pestic. Sci.,* 21, 269, 1987.
33. **Abe, I., Ebizuka, Y., and Sankawa, U.,** Purification of 2,3-oxido-squalene cycloartenol cyclase from pea seedlings, *Chem. Pharm. Bull.,* 36, 5031, 1988.
34. **Abe, I., Enizula, Y., Seo, S., and Sankawa, U.,** Purification of squalene-2,3-epoxide cyclases from cell suspension cultures of *Rabdosia japonica* Hara, *FEBS Lett.,* 249, 100, 1989.
35. **Wolfenden, R.,** Transition state analog inhibitors and enzyme catalysis, *Annu. Rev. Biophys. Bioeng.,* 5, 271, 1976.
36. **Pauling, L.,** Molecular architecture and biological reactions, *Chem. Eng. News,* 24, 1375, 1946.
37. **Bartlett, P. A. and Marlowe, C. K.,** Phosphonamidates as transition state analogue inhibitors of thermolysin, *Biochemistry,* 22, 4618, 1983.
38. **Schloss, J. V.,** Significance of slow-binding enzyme inhibition and its relationship to reaction-intermediate analogues, *Acc. Chem. Res.,* 21, 348, 1988.
39. **Arigoni, D.,** Stereochemical studies of enzymic C-methylations, *Ciba Found. Symp.,* 60, 243, 1978.
40. **Croteau, R.,** Biosynthesis and catabolism of monoterpenoids, *Chem. Rev.,* 87, 929, 1987.
41. **Rahier, A., Taton, M., and Benveniste, P.,** Inhibition of sterol biosynthesis in higher plants by analogues of high energy carbocationic intermediates, in *Biochemistry of Cell Walls and Membranes in Fungi,* Kuhn, P. J., Trinci, A. P. J., Jung, M. J., Goosey, M. W., and Copping, L. G., Eds., Springer-Verlag, Berlin, 1990, 206.
42. **Taton, M., Benveniste, P., and Rahier, A.,** Use of rationally designed inhibitors to study sterol and triterpenoid biosynthesis, *Pure Appl. Chem.,* 59, 287, 1987.
43. **Rahier, A., Taton, M., Bouvier-Nave, P., Schmitt, P., Benveniste, P., Schuber, F., Narula, A. S., Cattel, L., Anding, C., and Place, P.,** Design of high energy intermediate analogues to study sterol biosynthesis in higher plants, *Lipids,* 21, 52, 1986.
44. **Rahier, A., Taton, M., and Benveniste, P.,** *In vitro* inhibition of sterol biosynthesis enzymes by analogues of high energy carbocationic intermediates, *Biochem. Soc. Trans.,* 18, 48, 1990.
45. **Gil, G., Faust, J. R., Chin, D. J., Goldstein, J. L., and Brown, M. S.,** Membrane-bound domain of HMGCoA reductase is required for sterol-enhanced degradation of the enzyme, *Cell,* 41, 249, 1985.
46. **Basson, M. E., Thorsness, M., and Rine, J.,** *Saccharomyces cerevisiae* contains two functional genes encoding 3-hydroxy-3-methylglutaryl-coenzyme A reductase, *Proc. Natl. Acad. Sci. U.S.A.,* 83, 5563, 1986.
47. **Learned, R. M. and Fink, G. R.,** 3-Hydroxy-3-methylglutaryl-CoA reductase from *Arabidopsis thaliana* is structurally distinct from the yeast and animal enzymes, *Proc. Natl. Acad. Sci. U.S.A.,* 86, 2779, 1989.
48. **Alberts, A. W., Chen, J., Kuron, G., Hunt, V., Huff, J., Hoffman, C., Rothrock, J., Lopez, M., Joshua, H., Harris, E., Patchell, A., Monagan, R., Currie, S., Stapley, E., Albers-Schönberg, O., Hensens, O., Hirschfield, J., Hoogsteen, K., Liesch, J., and Springer, J.,** Mevinolin O a highly potent competitive inhibitor of HMGCoA reductase and a cholesterol-lowering agent, *Proc. Natl. Acad. Sci. U.S.A.,* 77, 3957, 1981.
49. **Nakamura, C. E. and Abeles, R. H.,** Mode of interaction of β-hydroxy-methylglutaryl Coenzyme A reductase with strong binding inhibitors: compactin and related compounds, *Biochemistry,* 24, 1364, 1985.
50. **Bach, T. J. and Lichtenthaler, H. K.,** Inhibition by mevinolin of plant growth, sterol formation and pigment accumulation, *Physiol. Plant.,* 59, 50, 1983.
51. **Bach, T. J. and Lichtenthaler, H. K.,** Mechanism of inhibition by mevinolin (MK 803) of microsome-bound radish and of partially purified yeast HMGCoA reductase, *Z. Naturforsch.,* 38c, 212, 1983.
52. **Rearden, J. E. and Abeles, R. H.,** Mechanism of action of isopentenyl pyrophosphate isomerase: evidence for a carbonium ion intermediate, *Biochemistry,* 25, 5609, 1986.
53. **Muehlbacher, M. and Dale Poulter, C.,** Isopentenyl-diphosphate isomerase: inactivation of the enzyme with active-site-directed irreversible inhibitors and transition state analogues, *Biochemistry,* 27, 7315, 1988.
54. **Rilling, H. C. and Poulter, C. D., Epstein, W. W., and Larson, B. R.,** *J. Am. Chem. Soc.,* 93, 1783, 1971.
55. **Sandifer, R. M., Thompson, M. D., Gaughan, R. G., and Poulter, C. D.,** Squalene synthetase. Inhibition by an ammonium analogue of a carbocationic intermediate in the conversion of presqualene pyrophosphate to squalene, *J. Am. Chem. Soc.,* 104, 7376, 1982.
56. **Poulter, C. D., Capson, T. L., Thompson, M. D., and Bard, R. S.,** Squalene synthetase. Inhibition by ammonium analogues of carbocationic intermediate in the conversion of presqualene diphosphate to squalene, *J. Am. Chem. Soc.,* 111, 3734, 1989.
57. **Vögeli, U. and Chappell, J.,** Induction of sesquiterpene cyclase and suppression of squalene synthetase activities in plant cell cultures treated with fungal elicitor, *Plant Physiol.,* 88, 1291, 1988.

58. **Duriatti, A., Bouvier-Navé, P., Benveniste, P., Schuber, F., Delprino, L., Balliano, G., and Cattel, L.,** *In vitro* inhibition of animal and higher plants 2,3-oxidosqualene-sterol cyclases by 2-aza-2,3-dihydrosqualene and derivatives, and by other ammonium-containing molecules, *Biochem. Pharmacol.*, 34, 2765, 1985.
59. **Cerutti, M., Delprino, L., Cattel, L., Bouvier-Navé, P., Duriatti, A., Schuber, F., and Benveniste, P.,** N-oxide as a potential function in the design of enzyme inhibitors. Application to 2,3-epoxysqualene-sterol cyclases, *J. Chem. Soc. Chem. Commun.*, p. 1054, 1985.
60. **Camara, B., Dogbo, O., D'harlingue, A., and Bardat, F.,** Inhibition lycopene cyclization by *Capsicum* chromoplasts membranes by 2-aza-2,3-dihydrosqualene, *Phytochemistry*, 24, 2751, 1985.
61. **Taton, M., Benveniste, P., and Rahier, A.,** N-[(1,5,9)-trimethyl-decyl]-4α,10-dimethyl-8-aza-decal-3β-ol, a novel potent inhibitor of 2,3-oxidosqualene cycloartenol and lanosterol cyclases, *Biochem. Biophys. Res. Commun.*, 138, 764, 1986.
62. **Gerst, N., Duriatti, A., Schuber, F., Taton, M., Benveniste, P., and Rahier, A.,** Potent inhibition of cholesterol biosynthesis in fibroblasts by N-[(1,5,9)-trimethyldecyl]-4α,10-dimethyl-8-aza-trans-decal-3β-ol, a new 2,3-oxidosqualene cyclase inhibitor, *Biochem. Pharmacol.*, 37, 1955, 1988.
63. **Johnson, W. S., Lindall, S. D., and Steele, J.,** Rate enhancement of biomimetic polyene cyclizations by a cation-stabilizing auxiliary, *J. Am. Chem. Soc.*, 109, 5852, 1987.
64. **Ruhl, K. K., Anzalone, L., Arguropoulos, E. D., Gayen, A. K., and Spencer, T. A.,** Azadecalin analogs of 4,4,10β-trimethyl-trans-decal-3β-ol: synthesis and assay as inhibitors of oxidosqualene cyclase, *Bioorg. Chem.*, 17, 108, 1989.
65. **Schmitt, P., Gonzales, R., Benveniste, P., Cerutti, M., and Cattel, L.,** Inhibition of sterol biosynthesis and accumulation of 2,3-oxido-squalene in bramble suspension cultures treated with 2-aza-2,3-dihydrosqualene and 2-aza-2,3-dihydrosqualene-N-oxide, *Phytochemistry*, 26, 2709, 1987.
66. **Oehlschlager, A. C., Angus, R. H., Pierce, A. M., Pierce, H. D., and Srinavasan, R.,** Azasterol inhibition of Δ^{24}-sterol-methyltransferase in *Saccharomyces cerevisiae*, *Biochemistry*, 29, 3582, 1984.
67. **Avruch, L., Fischer, S., Pierce, H., Jr., and Oehlschlager, A. C.,** The induced biosynthesis of 7-dehydrocholesterols in yeast: potential sources of new provitamin D3 analogs, *Can. J. Biochem.*, 54, 657, 1976.
68. **Narula, A. S., Rahier, A., Benveniste, P., and Schuber, F.,** 24-Methyl-25-azacycloartanol, an analogue of a carbonium ion high-energy intermediate, is a potent inhibitor of S-adenosyl-L-methionine-sterol C-24-methyltransferase in higher plant cells, *J. Am. Chem. Soc.*, 103, 2408, 1981.
69. **Rahier, A., Genot, J. C., Schuber, F., Benveniste, P., and Narula, A. S.,** Inhibition of S-adenosyl-L-methionine sterol C-24-methyltransferase by analogues of a carbocationic ion high-energy intermediate, *J. Biol. Chem.*, 259, 15215, 1984.
70. **Schmitt, P., Narula, A. S., Benveniste, P., and Rahier, A.,** Manipulation by 25-azacycloartanol of the relative percentage of C10, C9 and C8 side-chain sterols in suspension cultures of bramble cells, *Phytochemistry*, 20, 1097, 1981.
71. **Rahier, A., Taton, M., and Benveniste, P.,** Cycloeucalenol-obtusifoliol isomerase. Structural requirements for transformation or binding of substrates and inhibitors, *Eur. J. Biochem.*, 181, 615, 1989.
72. **Rahier, A., Taton, M., Schmitt, P., Benveniste, P., Place, P., and Anding, C.,** Inhibition of $\Delta^8 \rightarrow \Delta^7$-sterol isomerase and of cycloeucalenol-obtusifoliol isomerase by N-benzyl-8-aza-4α,10-dimethyl-trans-decal-3β-ol, an analogue of a carbocationic high energy intermediate, *Phytochemistry*, 24, 1223, 1985.
73. **Taton, M., Benveniste, P., and Rahier, A.,** Microsomal $\Delta^{8,14}$-sterol Δ^{14}-reductase in higher plants. Characterization and inhibition by analogues of a presumptive carbocationic intermediate of the reduction reaction, *Eur. J. Biochem.*, 185, 605, 1989.
74. **Pommer, E. H.,** Chemical structure-fungicidal activity relationships in substituted morpholins, *Pestic. Sci.*, 15, 285, 1984.
75. **Rahier, A., Schmitt, P., Huss, B., Benveniste, P., and Pommer, E. H.,** Chemical structure-activity relationships of the inhibition of sterol biosynthesis by N-substituted morpholins in higher plants, *Pestic. Biochem. Physiol.*, 25, 112, 1986.
76. **Mercer, E. I.,** The mode of action of morpholins, in *Sterol Biosynthesis Inhibitors*, Berg, D. and Plempel, M., Eds., Ellis Horwood, Chichester, 1988, 120.
77. **Bottema, C. K. and Parks, L. W.,** Δ^{14}-Sterol reductase in *Saccharomyces cerevisiae*, *Biochim. Biophys. Acta*, 531, 301, 1978.
78. **Schmitt, P. and Benveniste, P.,** Effect of AY 9944 on sterol biosynthesis in suspension cultures of bramble cells, *Phytochemistry*, 18, 445, 1979.
79. **Schmitt, P., Scheid, F., and Benveniste, P.,** Accumulation of $\Delta^{8,14}$-sterols in suspension cultures of bramble cells cultured with an azasterol antimycotic agent (A25822B), *Phytochemistry*, 19, 525, 1980.
80. **Schmitt, P., Benveniste, P., and Leroux, P.,** Accumulation of 9β,19-cyclopropyl sterols in suspension cultures of bramble cells cultured with tridemorph, *Phytochemistry*, 20, 2153, 1981.
81. **Costet-Corio, M. F. and Benveniste, P.,** Sterol metabolism in wheat treated by N-substituted morpholins, *Pestic. Sci.*, 22, 343, 1988.

82. **Bladocha, M. and Benveniste, P.**, Stereochemical aspects of the biosynthesis of the side-chain of 9β,19-cyclopropyl sterols in maize seedlings treated with Tridemorph, *Plant Physiol.*, 79, 1098, 1985.
83. **Hosakawa, G., Patterson, G. W., and Lusby, W. R.**, Effect of Triarimol, Tridemorph and Triparanol on sterol biosynthesis in carrot, tobacco, and soybean suspension cultures, *Lipids*, 19, 449, 1984.
84. **Burden, R. S., Clark, T., and Holloway, P. J.**, Effects of sterol biosynthesis-inhibiting fungicides and plant growth regulators on the sterol composition of barley plants, *Pestic. Biochem. Physiol.*, 27, 289, 1987.
85. **Grandmougin, A., Bouvier-Navé, P., Ullmann, P., Benveniste, P., and Hartmann, M. A.**, Cyclopropyl sterol and phospholipid composition of membrane fractions from maize roots treated with fenpropimorph, *Plant Physiol.*, 90, 591, 1989.
86. **Douglas, K. T.**, Transition-state analogues in drug design, *Chem. Ind.*, p. 311, 1983.
87. **Lalégerie, P., Legler, G., and Yon, J. M.**, The use of inhibitors in the study of glycosidases, *Biochimie*, 64, 977, 1982.
88. **Hesselink, P. G., Kerkenaar, A., and Witholt, B.**, Novel inhibitors of *Nocardia erythropolis* cholesterol oxidase: screening and characterization, *Pestic. Biochem. Physiol.*, 33, 69, 1989.
89. **Aoyama, Y., Yoshida, Y., and Sato, R.**, Yeast cytochrome P-450 catalyzing lanosterol 14α-demethylation. II. Lanosterol metabolism by purified P-450 14DM and by intact microsomes, *J. Biol. Chem.*, 259, 1661, 1984.
90. **Trzaskos, J. M., Bowen, W. D., and Shafiee, A., Fischer, R. T., and Gaylor, J. L.**, Cytochrome P-450-dependent oxidation of lanosterol in cholesterol biosynthesis. Microsomal electron transport and C-32 demethylation, *J. Biol. Chem.*, 259, 13402, 1984.
91. **Rahier, A. and Taton, M.**, The 14α-demethylation of obtusifoliol by a cytochrome P-450 monooxygenase from higher plant microsomes, *Biochem. Biophys. Res. Commun.*, 140, 1064, 1986.
92. **Taton, M., Benveniste, P., and Rahier, A.**, Comparative study of the inhibition of sterol biosynthesis in *Rubus fruticosus* suspension cultures and *Zea mays* seedlings by N-[(1,5,9)-trimethyldecyl)]-4α,10-dimethyl-8-azao-trans-decal-3β-ol and derivatives, *Phytochemistry*, 26, 385, 1987.
93. **Durst, F.**, Physiology and biochemistry of plant cytochrome P-450 enzymes, in *Frontiers in Biotransformation*, Vol. 6, RuckPaul, K., Ed., Academie Verlag, Berlin, in press.
94. **Vanden Bossche, H., Marichal, P., Gorrens, J., Bellens, D., Verhoeven, H., Coens, M. C., Lauwers, W., and Janssen, P. A.**, Interaction of azole derivatives with cytochrome P-450 isozymes in yeast, fungi, plants and mammalian cells, *Pestic. Sci.*, 21, 289, 1987.
95. **Schwinn, F. J.**, Ergosterol biosynthesis inhibitors. An overview of their history and contribution to medicine and agriculture, *Pestic. Sci.*, 15, 40, 1984.
96. **Coolbaugh, R. C., Swanson, D. J., and West, C. A.**, Comparative effects of ancymidol and its analogs on growth of peas and ent-kaurene oxidation in cell-free extracts of immature *Marah macrocarpus* endosperm, *Plant Physiol.*, 69, 707, 1982.
97. **Rademacher, W., Fritsch, H., Graebe, J. E., Sauter, H., and Jung, J.**, Tetcyclacis and triazole-type plant growth retardants: their influence in the biosynthesis of gibberellins and other metabolic processes, *Pestic. Sci.*, 21, 241, 1987.
98. **Burden, R. S., James, C. S., Cooke, D. T., and Anderson, N. H.**, C-14 demethylation in phytosterol biosynthesis. A new target site for herbicidal activity, *Proc. Br. Crop Prot. Conf. Weeds*, 1, 171, 1987.
99. **Mercer, E. I.**, The biosynthesis of ergosterol, *Pestic. Sci.*, 15, 133, 1984.
100. **Hancock, H. G. and Weete, J. D.**, Effects of triazoles on fungi. IV. Growth and and lipid composition of *Cercospora arahidicola* and *Cercosporidium personatum*, *Pestic. Biochem. Physiol.*, 24, 395, 1985.
101. **Gadher, P., Mercer, E. I., Baldwin, B. C., and Wiggins, T. E.**, A comparison of the potency of some fungicides as inhibitors of sterol 14-demethylation, *Pestic. Biochem. Physiol.*, 19, 1, 1983.
102. **Aoyama, Y., Yoshida, Y., Hata, S., Nishino, T., and Katsuki, H.**, Buthiobate: a potent inhibitor for yeast cytochrome-P-450 catalyzing 14α-demethylation of lanosterol, *Biochem. Biophys. Res. Commun.*, 115, 642, 1983.
103. **Yoshida, Y. and Aoyama, Y.**, Interaction of azole antifungal agents with cytochrome P-450 14DM purified from *Saccharomyces cerevisiae* microsomes, *Biochem. Pharmacol.*, 36, 229, 1987.
104. **Frasinel, C., Patterson, G. W., and Dutky, S. R.**, Effect of triarimol on sterol and fatty acid composition of three species of *Chlorella*, *Phytochemistry*, 17, 1567, 1978.
105. **Schmitt, P. and Benveniste, P.**, Effect of Fenarimol on sterol biosynthesis in suspension cultures of bramble cells, *Phytochemistry*, 18, 1659, 1979.
106. **Taton, M., Ullmann, P., Benveniste, P., and Rahier, A.**, Interaction of triazole fungicides and plant growth regulators with microsomal cytochrome P-450-dependent obtusifoliol 14α-methyl demethylase, *Pestic. Biochem. Physiol.*, 30, 178, 1988.
107. **Jefcoate, C. R.**, Measurement of substrate and inhibitor binding to microsomal cytochrome P-450 by optical-difference spectroscopy, *Methods Enzymol.*, 52, 258, 1978.
108. **Köller, W.**, Plant growth regulator activities of stereochemical isomers of triadimenol, *Physiol. Plant.*, 71, 309, 1987.

109. **Burden, R. S., Carter, G. A., Clark, T., Cooke, D. T., Croker, S. J., Deas, A. H. B., Hedden, P., James, C. S., and Lenton, J. R.,** Comparative activity of the enantiomers of Triadimenol and Paclobutrazol as inhibitors of fungal growth and plant sterol and gibberellin biosynthesis, *Pestic. Sci.*, 21, 253, 1987.
110. **Köller, W.,** Isomers of sterol synthesis inhibitors: fungicidal effects and plant growth regulator activities, *Pestic. Sci.*, 18, 129, 1987.
111. **Vanden Bossche, H.,** Mode of action of pyridine, pyrimidine and azole antifungals, in *Sterol Biosynthesis Inhibitors*, Berg, D. and Plempel, M., Eds., Ellis Horwood, Chichester, 1988, 79.
112. **Trzaskos, J. M. and Henry, M. J.,** Comparative effects of the azole-based fungicide Flusilazole on yeast and mammalian lanosterol 14α-methyl demethylase, *Antimicrob. Agents Chemother.*, 33, 1228, 1989.
113. **Chen, C., Kalb, V. F., Turi, T. G., and Loper, J. C.,** Primary structure of the cytochrome P-450 lanosterol 14α-demethylase gene from *Candida tropicalis*, *DNA*, 7, 617, 1988.
114. **Ishida, N., Aoyama, Y., Hatanaka, R., Oyama, Y., Imajo, S., Ishigun, M., Oshima, T., Nakazato, H., Nogushi, T., Maitra, U. S., Mohan, V. P., Sprinson, D. B., and Yoshida, Y.,** A single amino acid substitution converts cytochrome P-450 14DM to an inactive form, cytochrome P-450 SG1: complete primary structure deduced from cloned DNAs, *Biochem. Biophys. Res. Commun.*, 155, 317, 1988.
115. **Kalb, V. F. and Loper, J. C.,** Proteins, from eight eukaryotic cytochrome P-450 families share a segmented region of sequence similarity, *Proc. Natl. Acad. Sci. U.S.A.*, 85, 7221, 1988.
116. **Taton, M. and Rahier, A.,** Properties and structural requirements for substrate specificity of a cytochrome P-450-dependent obtusifoliol 14α-demethylase from maize (*Zea mays*) seedlings, *Biochem. J.*, 277, 483, 1991.
117. **Schmitt, P.,** unpublished results.
118. **Maillot-Vernier, P., Schaller, H., Benveniste, P., and Belliard, G.,** In vitro selection of calli resistant to a triazole cytochrome P-450-obtusifoliol-14-demethylase inhibitor from protoplasts of *Nicotiana tabacum* L. cv. Xanthii, *Plant Physiol.*, 93, 1190, 1990.
119. **Pascal, S., Taton, M., and Rahier, A.,** Oxidative demethylation of 24-methylene cycloartanol by a cyanide sensitive enzymatic system from higher plant microsomes, *Biochem, Biophys. Res. Commun.*, 172, 98, 1990.

Chapter 9

TARGET SITES OF STEROL BIOSYNTHESIS INHIBITORS: SECONDARY ACTIVITIES ON CYTOCHROME P-450-DEPENDENT REACTIONS

Hugo Vanden Bossche and Paul A. J. Janssen

TABLE OF CONTENTS

I. INTRODUCTION

During the last 20 years, an amazing number of antifungals have become available for the management of a wide variety of fungal diseases. Most of them belong to the class of ergosterol biosynthesis inhibitors (see Chapter 7). This is not surprising since ergosterol is the major sterol in most fungi pathogenic to plants and vertebrates, and its synthesis is somewhat different from that of cholesterol in mammalian cells and from phytosterols in plants.

A key step in the biosynthesis of the 24-alkylated sterol, ergosterol, is the cytochrome P-450-(P450) dependent 14α-demethylation of lanosterol (in *Saccharomyces cerevisiae, Torulopsis glabrta,* and some isolates of *Candida albicans*) or 24-methylenedihydrolanosterol in most pathogenic fungi. In both yeasts and fungi this cytochrome P-450 ($P450_{14DM}$, or according to the recommended nomenclature,[1] P450 LI) is the primary target for a number of pyrimidine, pyridine, imidazole, and triazole derivatives (for reviews see References 2 and 3 and Chapter 7).

Cytochrome P-450 (P450) isozymes are present in prokaryotes such as *Pseudomonas putida* and *Bacillus megaterium*, in protozoa, e.g., *Trypanosoma cruzi,* in yeast, fungi, plants, insects, invertebrates, fish, fowls, and mammals.

The name of this hemoprotein describes a carbon binding pigment which absorbs at 450 nm when reduced and complexed with carbon monoxide.[4]

These proteins containing iron protoporphyrin IX play a key function in the synthesis and metabolism of a huge number of endogenous compounds (endobiotics), and in the detoxication or activation of innumerable xenobiotics. P450 isozymes catalyze hydroxylations and epoxidations, *N*-, *O*-, and *S*-dealkylations, *S*-, *N*-, and P-oxidations, the desulfuration, and dehalogenation. For example, they oxidatively cleave C–N and C–O bonds in the metabolism of amines and ethers, and C–C bonds in the synthesis of steroids.[5,6] A P450 also activates the P=S group of organophosphorus pesticides to P=O, thus forming active inhibitors of the target enzyme acetylcholinesterease.[7] In prokaryotes the P450s are soluble proteins, whereas in all eukaryotes they are membrane bound. In insects and mammals P450 isozymes are localized in the endoplasmic reticulum and the inner mitochondrial membrane. Fungal P450s studied so far are all located in the endoplasmic reticulum. Examples of P450-dependent reactions in the endoplasmic reticulum or mitochondrial membranes are given in Tables 1 and 2, respectively.

The diversity of the reactions catalyzed and the broad substrate specificity are due to the presence of multiple P450s coded by a gene superfamily that contains more than 100 genes.[5] For example, some 20 forms are seen in the endoplasmic reticulum of rat liver, some of which are constitutive, whereas others are measurable only after the animal was challenged by external chemicals.

The presence of all these P450 isozymes in the host cells necessitates the study of interactions of an antifungal agent, e.g., ketoconazole, with P450s not only in the fungal target organism, but also in the host. It is the aim of this chapter to discuss the effects of a number of ergosterol biosynthesis inhibitors on P450-dependent reactions in mammalian cells. The main focus will be on those interactions that could contribute to the development of new possibilities in medical treatment.

II. THE P450 REACTION MECHANISM

To study the interactions of inhibitors with the P450 systems, some knowledge of the catalytic cycle of P450 is mandatory.

The cytochromes P-450 are classified as monooxygenases and catalyze the following reactions:

TABLE 1
Examples for P450-Dependent Reactions in the Endoplasmic Reticulum

Reaction	Protein name	Substrates	Tissues
21-Hydrozylation	XXIA	17α-OH-progesterone and progesterone	Adrenal
17α-Hydroxylation	XIIA1	Progesterone Pregnenolone	Adrenal, testis, and ovary
C-17,20-cleavage		17α-OH-progesterone 17α-OH-pregnenolone	Testis, adrenal, and ovary
Aromatase	XIXA1	Androgens	Ovary, placenta, adipose tissue brain (adrenal)
25-Hydroxylase[a]		Vitamin D	Liver
7α-Hydroxylase		Cholesterol	Liver
12α-Hydroxylase		3α, 7α-Dihydroxycholestane	Liver
25-Hydroxylase		3α, 7α, 12α-Trihydroxycholestane	Liver
14α-Demethylase:	LIA1	Lanosterol	Yeast liver and all sterol-forming cells

[a] The microsomal 25-hydroxylase was shown to be male specific and hardly detectable in adult female rat liver microsomes.[7a] Human liver microsomes are devoid of vitamin D_3 25-hydroxylase activity.[7b]

TABLE 2
Examples for Mitochondrial P450-Dependent Reactions

Reaction	Protein name	Substrate[a]	Tissues
20-Hydroxylation 22-Hydroxylation C-20,22-cleavage	XIA1	Cholesterol	Adrenal, testis, ovary, placenta
11β-Hydroxylation	XIB1	Deoxycortico. deoxycortisol	Adrenal
18-Hydroxylation		Corticosterone	Adrenal
Aldehyde synthetase		18-HO-cortico.	Adrenal
25-Hydroxylase		Vitamin D	Liver
1α-Hydroxylation		25-Hydroxyvit.D	Kidney, skin
26-Hydroxylation		Cholesterol	Liver, kidney, skin (?)

[a] Cortico. = corticosterone, vit. = vitamin.

$$RH + O_2 + NADPH + H^+ \rightarrow ROH + H_2O + NADP^+$$

The ability to activate oxygen for insertion into a substrate (RH) is the key property of most P450 enzymes.[6] It is based on the coordination of the heme iron with the thiolate group of a cysteine. A region close to the C terminal containing this cysteine residue is highly preserved in all P450 isozymes.[5] In other hemoproteins the fifth iron ligand is not a thiol sulfur but an imidazole. By quantum chemical calculations of the oxygen complex, considering thiol and imidazole, respectively, as the fifth ligand, Rein et al.[8] showed that the thiol sulfur induces a larger negative net charge of the oxygen molecule bound to P450. The distant oxygen atom of the iron is more negative than the proximate one (Figure 1). They assume that a proton can easily attach to the strongly negative oxygen. This is important for the activation process of oxygen by P450 described below.

The microsomal hydroxylation (insertion of an oxygen in the substance, RH) proceeds in a multistep reaction (Figure 2):

1. In the resting state the P450-Fe^{3+} is in a six-coordinated low-spin structure. This means that in addition to the porphyrin ring, the iron (in the ferric state) is attached to two

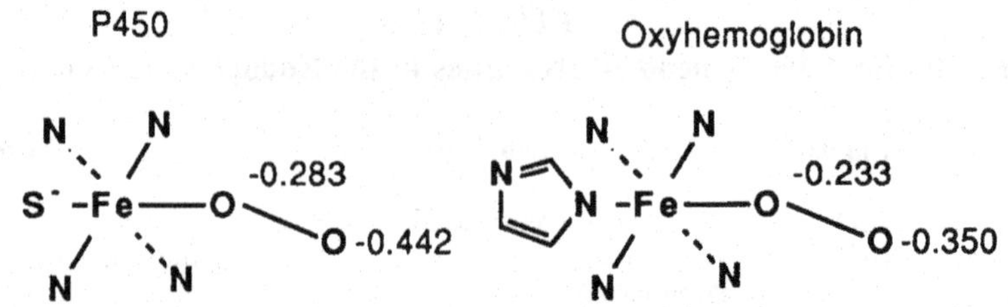

FIGURE 1. Models of P450 and oxyhemoglobin. S^- is the thiol sulfur of a cysteine residue; N represents the nitrogen atoms of the porphyrin; the imidazole group is part of the histidine residue. Values at the atoms represent the net charges. (Taken from Reference 8.)

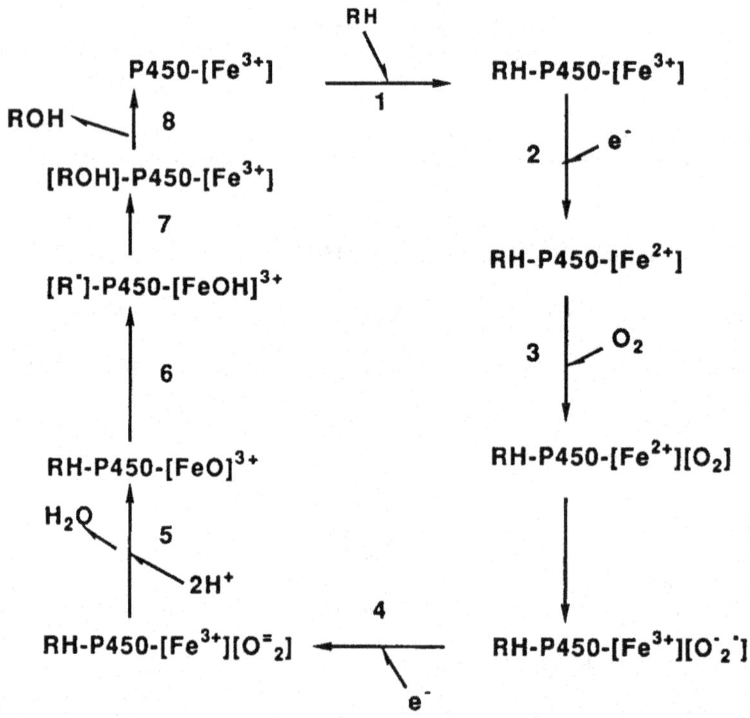

FIGURE 2. Proposed scheme for the P450 cycle.[21] $[Fe^{3+}]$ represents the heme iron, RH the substrate, and ROH the hydroxylated product. $[FeO]^{3+}$ is the activated oxygen intermediate, thought to be a ferryl species in which the oxygen is coordinated to a hypervalent state of the prosthetic heme iron.

endogenous axial ligands. One is the thiol sulfur of cysteine (the fifth ligand); the sixth ligand is still not identified. However, evidence is available that it might be a hydroxyl group from a seryl, threonyl, or tyrosyl residue.[9] Substrate binding represents the first step in the P450 reaction sequence. Three types of substrate complexes can be distinguished.

(a) Hydrophobic groups of the substrate interact with part(s) (active site) of the apoprotein moiety of the P450. Difference spectra show a maximum at 390 nm and a minimum at 420 nm. This spectral perturbation is called Type 1 spectrum. The spectral changes induced by substrate binding are paralled by a shift in the spin equilibrium from a low-spin *d*-electron configuration (S=1/2) toward the high-spin state (S=5/2). This perturbation is thought to originate from obstruction of the binding of the iron sixth ligand. With the release of the sixth ligand, the iron moves out of the plane of the porphyrin toward the fifth ligand.[10] The high-spin/low-spin equilibrium is important to the oxygen activation cycle because it makes the oxidation-reduction potential considerably more positive. For example, in the case of $P450_{CAM}$, the soluble P450 of the soil bacterium *Pseudomonas putida* catalyzing the hydroxylation of camphor, the redox potential in the absence of the substrate is near −340 mV, whereas in the presence of camphor the redox potential increases to −175 mV.[11] Thus, electron flow to the P450 iron and hence reduction is much more favored in the presence of bound substrate.

(b) A second class of substrates (the so-called Type 2 compounds) binds to the heme iron and to the protein moiety. The spectral alterations observed (minimum at 390 nm and maximum at 425 nm) indicate a spin shift toward the low-spin configuration.[12]

(c) Difference spectra obtained with the third group of substrates show maxima at 420 nm and minima at 390 nm. Thus the extremes correspond reversely to those of the Type 1 substrates. The binding mode of these substrates is thought to originate from competition with an endogenous substrate.[12]

2. The second step in the P450 cycle is the reduction of the P450-iron-substrate complex (RH-P450-Fe^{3+} $\rightarrow$ RH-P450-Fe^{2+}). In the endoplasmic reticulum this reduction is achieved by the transfer of an electron from NADPH, which is catalyzed by the NADPH-dependent cytochrome P-450 reductase. This membrane-bound flavin-containing enzyme contains per polypeptide two flavin molecules, FMN and FAD (1:1 ratio). Since, compared to FMN (−190 mV), FAD has a low redox potential (−328 mV), it has been postulated that FAD accepts an electron from NADPH, and that FMN is probably the flavin that directly reduces the P450 heme iron.[10]

In the adrenal cortex mitochondria, the transfer of electrons from NADPH to P450 is mediated by a NADPH-cytochrome P-450 reductase complex containing a FAD-flavoprotein, adrenodoxin reductase, and the iron-sulfur protein adrenodoxin. The latter is the actual heme iron reductant.[10]

In *P. putida*, a NADH-specific FAD-containing putidaredoxin reductase and the iron-sulfur protein putidaredoxin constitute the electron transfer system.[13] The associated electron transfer system of the 15β-steroid hydroxylase of *Bacillus megaterium* consists of an NADPH-specific FMN-containing megaredoxin reductase and an iron-sulfur protein, megaredoxin.[14,15]

3. P450 in the reduced state is capable of binding O_2, yielding a dioxygen complex (RH-P450-[Fe^{2+}] [O_2]). The oxygen is bound to the sixth coordination position of the heme iron. It is of interest that carbon monoxide can compete with oxygen for the binding site. CO-binding leads to inhibition of the P450-dependent reaction and to a typical absorption spectrum with a maximum at 450 nm. It should be noted that compounds, such as the N-1-substituted imidazole and triazole antifungals, also bind to the sixth coordination position and, thus, interfere with oxygen or CO-binding. Interference with CO-binding will result in a decreased absorption at 450 nm. Thus, this system can be used to measure the effects of, e.g., azole antifungals on P450 isozymes.

4. The fourth step in the P450 cycle consists of the second-electron transfer. This electron is delivered either by the P450 reductase or, in some P450 systems, by cytochrome b_5(b5). If the second electron is not delivered immediately, the oxygen complex may dissociate to ferric P450 and superoxide anion.[10] The presence of a covalent, functional P450-b5 complex has been found by Tamburini et al.[16] in $P450_{LM2}$. $P450_{LM2}$ is a phenobarbital-induced liver microsomal P450 of rabbits (a member of class IIB enzymes). P450 LM_2, also called $P450_2$, is the major enzyme induced by phenobarbital comprising as much as 50% of the total hepatic P450 after phenobarbital treatment.[17] Cytochrome b_5 causes a 13- and 7-fold increase

in the rate of P450 RLM_5- (a P450 isolated from hepatic microsomes from untreated male CD rats) and LM_2-dependent *p*-nitroanisole demethylation, respectively; it also causes a 7-fold increase in the binding affinity of LM_2 for benzphetamine.[18] These studies suggest that specific binding and electron transfer between b5 and P450 LM_2 and RLM_5 are physiologically important. It is of interest that phosphorylation of P450 isozymes from class II (P450 involved in drug metabolism) is inhibited by b5.[19] The site of cyclic AMP-dependent protein kinase phosphorylation of LM_2 has been reported to be serine 128. This residue is contained in the cyclic AMP-dependent protein kinase recognition sequence, Arg-Arg-Phe-Ser_{128}-Leu and is also found in a number of other P450s belonging to class II. This study by Epstein et al.[19] suggests that the b5 binding site may be located within or overlap this sequence.

5. The second electron, transferred to the ferrous dioxygen complex in step 4, initiates dioxygen cleavage with the formation of the hydroxylated substrate and water (step 5). The details of this dioxygen cleavage, the exact state of the oxygen, and the mechanism of the stereospecific insertion of a single atom of oxygen into a variety of substrates (steps 6 to 8) still continue to remain a black box and different opinions have been described.[8,10,20-24] One of the proposed schemes for the mechanism of P450 hydroxylations is shown in Figure 2.

III. INHIBITION OF P450-DEPENDENT ENZYMES

In the P450 cycle (Figure 2), a number of steps are vulnerable to inhibition. Examples are the binding of the substrate, the interaction between the P450 reductase and P450, the binding of molecular oxygen, and the transfer of an oxygen atom to the substrate. These inhibitors can be divided into three classes:[25]

1. Agents that bind reversibly, and in this way compete with a substrate for occupancy of the active site
2. Agents that form complexes with the heme iron and simultaneously bind to hydrophobic regions of the protein, at or near the substrate- or heme-binding domain
3. Agents that bind irreversibly to the apoprotein or the heme group

It should be noted that compounds, acting during or subsequent to the oxygen transfer step, are generally irreversible inhibitors and many fall into the category of suicide inhibitors. In contrast, agents that interfere in the catalytic cycle before the oxidative part of the cycle are reversible inhibitors.[25]

The most simple inhibitor that binds to the heme iron is carbon monoxide. It is bound to the Fe^{2+} form of P450, thus competing with oxygen binding and activation. Isocyanides also have a marked affinity for the reduced form of P450 (max absorption is at 455 nm, instead of 450 nm observed with CO). In contrast to CO, isocyanides bind also to the oxidized form. This gives a difference spectrum with an absorption peak at about 430 nm and a trough at about 390 nm,[26] similar to the spectra obtained with the second class of substrates (the so-called Type 2 compounds) described above. This type of spectrum is also obtained upon the addition of amines (e.g., aniline) or pyridines. Since aniline is capable of displacing CO from its binding site, it has been proposed that compounds causing Type 2 spectral changes also bind to the sixth coordination position of the heme iron.[27]

It is obvious that compounds that bind simultaneously to hydrophobic domains of the protein and to the heme iron are more active than those that bind either to the iron or to the apoprotein. This is best illustrated by the fact that imidazole (binding to the sixth coordination position of the heme iron only) is a weak inhibitor of the P450-dependent aldrin epoxidation in rat liver microsomes (IC_{50} value = 3.6 m*M*), whereas phenylimidazole is a much more potent inhibitor (IC_{50} value = 1.5 μ*M*).[28] Most of the compounds the authors will discuss in this chapter belong to the class of inhibitors that bind to both the heme iron and the apoprotein.

IV. EFFECTS OF ERGOSTEROL BIOSYNTHESIS INHIBITORS ON P450-DEPENDENT REACTIONS

A. THE 14α-DEMETHYLATION

Since Ragsdale and Sisler[29] found sterol biosynthesis in *Ustilago maydis* to be the primary site of interference for the pyrimidine fungicide triarimol, a great number of ergosterol biosynthesis inhibitors have been described (Chapter 7). Most of them inhibit the 14α-demethylation of lanosterol in yeast and 24-methylenedihydrolanosterol in fungi.[3,30,31] For example, Aoyama et al.[32] and Yoshida and Aoyama[33] showed that the pyridine-containing fungicide buthiobate specifically bound to $P450_{14DM}$ purified from *S. cerevisiae*, and inhibited lanosterol 14α-demethylation (IC_{50} value = 0.3 μM) in a reconstituted system consisting of NADPH P450 reductase and $P450_{14DM}$, both purified from *S. cerevisiae*. It should be noted that buthiobate binding is rather weak. Indeed, when it was added to the oxidized form of $P450_{14DM}$, then reduced, and the sample cuvet bubbled with CO, the buthiobate-P450 complex was replaced by a CO-P450 complex. This indicates that buthiobate has a lower affinity for the heme iron than CO.

Buthiobate also inhibits the 14α-demethylation of lanosterol in rat liver microsomes with an IC_{50} value = 0.4 μM.[33] This concentration is similar to the value reported for the yeast enzyme. However, buthiobate only partially inhibited the *O*-deethylation of 7-ethoxycoumarin (catalyzed by $P450_c$ or, according to the nomenclature of Nebert et al.,[1] IA1) by rat liver microsomes and neither the benzphetamine *N*-demethylation nor the *p*-nitrosoanisole *O*-demethylation (both catalyzed by $P450_b$, IIB1[1]) was inhibited.[33] It was also found that this pyridine derivative failed to inhibit the 22,23-desaturation of ergosta-5,7-dien-3β-ol, the last step in ergosterol synthesis.[34] Our studies (summarized in Reference 3) also indicate that the azole antifungals, such as ketoconazole and itraconazole, do not interfere with the 22,23-desaturase in *C. albicans*. However, Takano et al.[35] reported that in *U. maydis* the triazole antifungal, diniconazole, inhibited both the 14α-demethylase and the 22,23-desaturase.

As shown by Hata et al.,[36] this dehydrogenation requires NADPH and molecular oxygen, and is inhibited by CO and metyrapone. This indirect evidence suggests that this reaction is catalyzed by a P450. Further studies of the same group showed that the 22,23-desaturation is not blocked by rabbit antibodies against $P450_{14DM}$, and that in a *S. cerevisiae* mutant (N22), which is partly defective in 22,23-desaturase, the 14α-demethylation was not blocked.[34] These results indicate that $P450_{22\text{-}DS}$ is different from $P450_{14DM}$. The fact that P450 is able to function as a hydrocarbon dehydrogenase is not unique to *S. cerevisiae*. In a recent review, Ortiz de Montellano[24] presented evidence that P450 is able to function as a hydrocarbon dehydrogenase in the metabolism of the antiepileptic drug valproic acid. The results summarized in this review suggest that desaturation involves abstraction of a hydrogen, electron transfer from the radical to the ferryl species ($[FeO]^{3+}$), and deprotonation of the resulting carbocation.[24] It would be of interest to study the possible involvement of a P450 in the insertion of the Δ^5-double bond in the B-ring of ergosterol.

The studies summarized above already indicate that compounds that bind to particular P450s are not necessarily general inhibitors of P450-dependent reactions. However, the fact that buthiobate is an almost equipotent inhibitor of the 14α-demethylases in both yeast and rat liver microsomes exemplifies the difficulty to develop a P450 inhibitor that affects, for example, $P450_{14DM}$ in fungi but not in mammalian cells.

Comparison of the $P450_{14DM}$ present in yeast and rat liver microsomes supports this conclusion. $P450_{14DM}$, purified to homogeneity from rat liver[37] and *S. cerevisiae*[38] microsomes, catalyzes three oxidative steps:

1. Hydroxylation of the C-32 methyl (14α-methyl group) of lanosterol with the formation of an alcohol ($-CH_2OH$)
2. The oxidation of the C-32 alcohol to a C-32 aldehyde (–CHO)
3. The oxidative elimination of the aldehyde as formic acid with the formation of the 32-nor-14-desaturated metabolite, 4,4′-dimethyl-5α-cholesta-$\Delta^{8,14,24}$-trien-3β-ol.

Results obtained with intact rat liver microsomes suggested a cytochrome b_5 involvement for lanosterol C-32 demethylation under conditions of low NADPH concentrations, with NADH providing the necessary electrons for catalytic turnover of the enzyme.[39] However, further studies using purified components did not indicate a b5 involvement.[37]

The P450, purified from rat liver, appeared to be specific in its oxidation toward 14α-methyl sterols with very poor activity toward the xenobiotics aminopyrine, *p*-nitroanisole, or aniline.[37] A modest induction (1.6-fold) of the constitutive P450 isozyme of 14α-methyl sterol oxidase has been found in male Sprague-Dawley rats after 5 days of treatment (intraperitoneally) with 200 mg/kg imidazole.[39] Of interest is the finding that the rabbit $P450_{3a}$ (IIE1), ie., the alcohol inducible liver P450, is also induced by imidazole (about 4.5-fold).[40] In contrast, the ethanol-inducible rat P450 ($P450_j$) is not induced.[41] Thus the $P450_{14DM}$ from rat liver is different from $P450_j$. Pompon,[42] in a study on the functional expression in *S. cerevisiae* of β-naphthoflavone-induced rabbit liver $P450_{LM4}$ and $P450_{LM6}$, used imidazole to stabilize $P450_{LM4}$. He found that this nitrogen heterocycle induced an endogeneous yeast P450 that could be selectively complexed by ketoconazole. Although the 14α-demethylation was not investigated by direct means, $P450_{DM}$ is the predominant yeast microsomal P450.[31] Since the microsomes from imidazole-treated cells did not show an absorption peak at about 450 nm in the presence of 0.5 μM ketoconazole,[42] it is tempting to speculate that imidazole also induces the yeast $P450_{DM}$.

Trzaskos and Henry,[43] studying the effects of the triazole derivative flusilazole, found that the affinity of this silytriazole to a partly purified P450 from *S. cerevisiae* microsomes was similar to that observed with a partly purified P450 from rat liver microsomes. However, when the binding of flusilazole to P450(s) in microsomes was compared, the affinity for the yeast P450(s) was almost 24 times higher. The lanosterol 14α-demethylation in a hepatic microsomal preparation was found to be almost 100 times less sensitive than the enzyme in a yeast preparation.[43] Trzaskos and Henry concluded that the apparent increase in susceptibility of fungal preparations observed upon exposure to azole antifungals is due in part to a protective effect afforded by other susceptible P450 isozymes present in mammalian preparations. This is a conclusion that might be correct for a number of azoles interfering with liver P450 isozymes other than the $P450_{14DM}$. Most likely it cannot be applied to azole derivatives, such as itraconazole, which was found to be more selective than, for example, ketoconazole.[3,44-46] Furthermore, Trzaskos and Henry[43] used *S. cerevisiae* only. Although the $P450_{14DM}$ of both rat liver and *S. cerevisiae* might be quite similar it can not be excluded that $P450_{14DM}$ of other yeasts and fungi differ from the hepatic or *S. cerevisiae* counterparts. An alignment of the amino acid sequences of the two $P450_{14DM}$ should give more information, but the amino acid sequence of the rat liver P450 is still not available. Those of the $P450_{DM}$ from *C. albicans*,[47] *C. tropicalis*,[48] and *S. cerevisiae*[49-51] are known. Using a VGAP (alignment with a *v*ariable *gap* penalty) alignment program (Moereels, De Bie, Tollenaere, unpublished) the *C. albicans* and *C. tropicalis* $P450_{14DM}$ share 64.2 and 65.2% identical amino acids with the *S. cerevisiae* $P450_{14DM}$.[52] $P450_{14DM}$ from *C. albicans* and *C. tropicalis* share 83%, identical amino acids.[52] This might indicate that the $P450_{14DM}$ from *Candida* differs from that of *S. cerevisiae*. Results obtained with ketoconazole support this hypothesis. Indeed, when *S. cerevisiae* (ATCC 42407) and *C. albicans* (ATCC 28516) were grown for 24 h in a casein hydrolysate-yeast extract-glucose medium (CYG-medium)[53] in the presence of increasing concentrations of ketoconazole, 50% inhibition of ergosterol synthesis was found at 3.3×10^{-7} and 2.1×10^{-9} M, respectively (unpublished results).

The triazole antifungal itraconazole inhibits (IC_{50} value = 9.5×10^{-7} *M*) (cholesterol synthesis in mouse macrophages (J 774 G8).[54] When these macrophages were infected with *Leishmania mexicana mexicana* (MHOM/BZ/84/Bel 44) amastigotes cholesterol synthesis was blocked. This might originate from competition between the host cell and the parasite for the substrate [^{14}C]-acetate. Indeed, instead of cholesterol, ergosterol was found.[54] As in yeast and fungi, ergosterol synthesis was sensitive to ketoconazole and itraconazole. Fifty percent inhibition was found at 2.1×10^{-7} and 1.5×10^{-7} *M*. The inhibition of ergosterol synthesis not only coincided with the accumulation of 14-methylsterols, but also with an increased synthesis of cholesterol. The latter synthesis was inhibited at concentrations >1 μM. However, even at 5 μM, cholesterol synthesis was higher than under control conditions.[54] This study further proves that these azole antifungals are much more potent inhibitors of the $P450_{14DM}$ involved in ergosterol synthesis that of the $P450_{14DM}$ that catalyzes cholesterol synthesis in mammalian cells. Following are some other examples to substantiate this selectivity: to obtain 50% inhibition of the 14α-demethylation of lanosterol in phytohemagglutinin (PHA)-stimulated human peripheral lymphocytes, 10^{-7} *M* of ketoconazole is needed.[20] Furthermore, 50% inhibition of cholesterol synthesis in subcellular fractions of male rat liver is reached at 2×10^{-6} *M* ketoconazole, 6×10^{-6} *M* miconazole,[55] and 7×10^{-6} *M* itraconazole.[56] It should be noted that 50% inhibition of ergosterol synthesis by a subcellular fraction of *C. albicans* was reached at 7×10^{-8}, 9×10^{-8}, [55] and 8×10^{-8} *M*[56] of ketoconazole, miconazole, and itraconazole, respectively.

B. STEROID HORMONE SYNTHESIS

1. Androgen Synthesis

Ketoconazole is a well-tolerated oral antifungal imidazole with a broad spectrum of antifungal activity. The effective daily single dose of ketoconazole for most fungal infections is 200 mg. The incidence of side effects, mainly gastrointestinal, is low (3 to 10% of the patients).[57] However, during the widespread clinical use of ketoconazole two unexpected effects were observed. In 1980 a case of mild and reversible hepatitis was described.[58] Van Tyle estimated an incidence of about 1 in 12,000.[57] Gynecomastia was reported into patients during treatment with 200 mg ketoconazole daily[59] and in a few patients receiving high multiple doses (600 to 1200 mg/day) of ketoconazole.[60] The observation of De Felice et al.[59] triggered a multidisciplinary study of the endocrinological applications of ketoconazole and was at the onset of a series of in-depth studies on mammalian P450 isozymes. Indeed, this rare side effect suggested that in such patients ketoconazole affected the estrogen/androgen ratio.[61] However, no clear reduction in the plasma estradiol levels at 600 mg ketoconazole per day was observed.[61] This corresponds with the results obtained with human placental microsomes. An IC_{50} value was not reached up to 60 μM.[62] The 19- and 2-hydroxylations of androgens by pig testis microsomes are somewhat more sensitive. However, 7.5×10^{-6} *M* is still needed to reach 50% inhibition.[63] In rats, ketoconazole was five times less potent than aminoglutethimide, a nonsteroidal inhibitor of the P450-dependent aromatization ($P450_{AROM}$) of C-19 steroids (testosterone, androstenedione).[64] The aromatase will be discussed in greater detail below. Measuring the testosterone serum levels, a dose-related reversible decrease was observed.[65-70] After a 200-, 400-, or 600-mg dose, testosterone serum concentrations in male volunteers fell markedly, but returned toward normal levels 8 to 24 h later.[65] The exposure of rat Leydig cells to about 10 μM of ketoconazole reduced both basal and human chorionic gonadotropin-stimulated testosterone production.[65]

The finding of a marked but transient decrease in plasma testosterone levels initiated a series of studies in order to determine the site of action of ketoconazole in the androgen biosynthetic pathway (Figure 3). In a first attempt to elucidate the site of action, testosterone synthesis was studied in a rat testicular subcellular fraction (S10-fraction) containing the cytosol and microsomes.[71-72] Under the assay conditions employed, and with pregnenolone

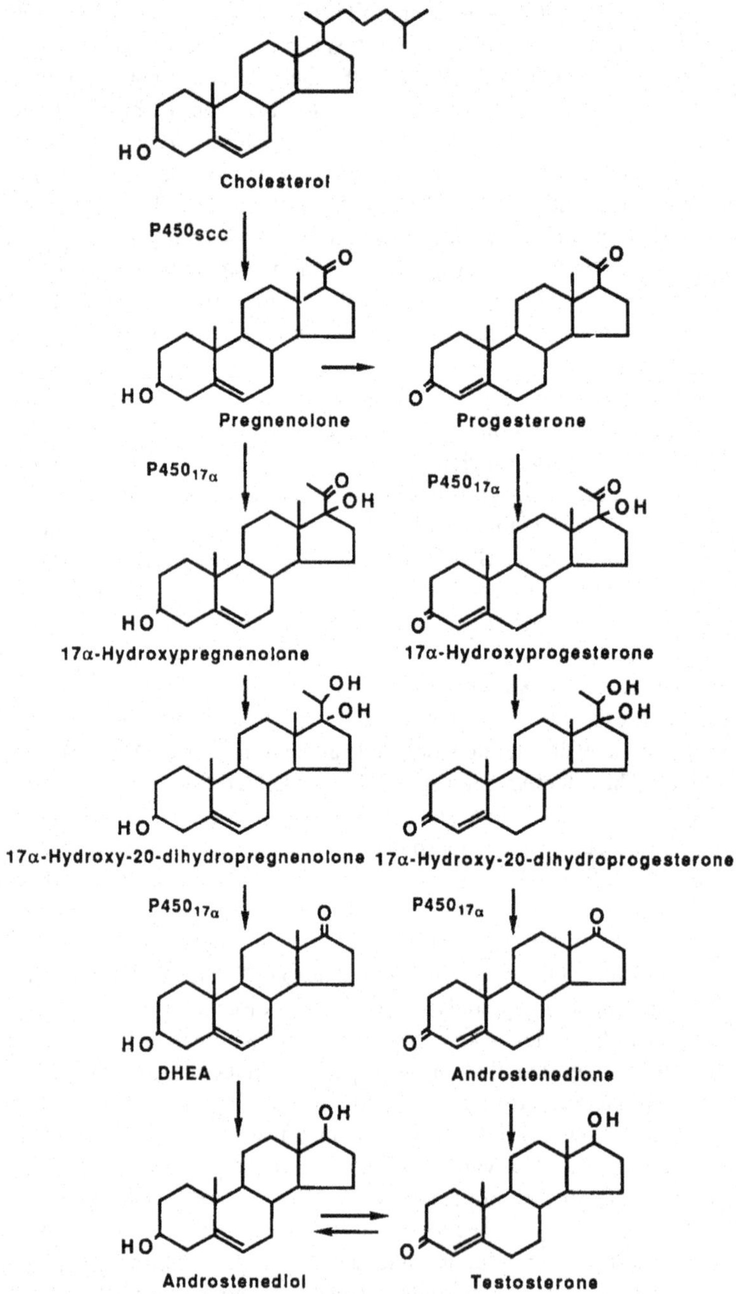

FIGURE 3. Synthesis of testosterone. $P450_{scc}$ = side-chain cleavage; $P450_{17\alpha}$ = 17 α-hydroxylase/17,20-lyase; DHEA = dehydroepiandrosterone.

as the substrate, a 50% inhibition of androstenedione and testosterone synthesis was achieved after a 3-h incubation period in the presence of 5 μM ketoconazole.[72] This inhibition coincided with an accumulation of 17α-hydroxy, 20-dihydropregnenolone. Maximum accumulation was reached at 0.5 μM. At concentrations >1 μM an accumulation of pregnenolone was observed with a concomitant decrease in 17α-hydroxy, 20-dihydropregnenolone.[72] Similar results were obtained when the conversion of progesterone into androgens was studied.[72-74] Further studies revealed that ketoconazole inhibited the conversion of 17α-hydroxy, 20-

dihydroprogesterone by rat testicular microsomes; 50% inhibition was achieved at 0.2 μM.[72] Studies by Rajfer et al.,[75] using minced decapsulated testes from Sprague-Dawley rats, indicated that ketoconazole did not inhibit the conversion of progesterone into 17α-hydroxyprogesterone. The inhibition of androgen synthesis originated from an inhibitory effect on the 17,20-lyase only.[75] Similar results were reported by Kan et al.[76] Higashi et al.[77] further proved that in rat testicular microsomes, ketoconazole is a more potent inhibitor of the 17,20-lyase than of the 17α-hydroxylase. The K_i values were 0.4 and 1.6 μM, respectively. The results indicate that, at least in rat testis, this azole is an inhibitor of the 17,20-lyase.

In a comparative study, the effects of ketoconazole on P450s in microsomal suspensions of *C. albicans, S. cerevisiae,* rat (basal and phenobarbital-induced), rabbit (3-methylcholanthrene- and phenobarbital-induced), male and female dog liver, dog and bovine adrenals, and dog and pig testes[56,72,73] were evaluated. It was found that the yeast microsomal P450 was the one that was most sensitive to ketoconazole (50% decrease in CO-binding at about 2×10^{-8} M). The second most sensitive microsomal P450 was that of piglet testes (50% decrease in CO-binding at about $4 \times 10^{-7}M$).[73] Ketoconazole added to the piglet microsomes caused a type II difference spectrum with maxima at 433 nm and a minimum at 412 nm with an isosbestic point at 423 nm.[73] This Type 2 spectrum indicated the binding of a ligand to a low spin hemoprotein. These results suggest that the observed decrease in testosterone levels might originate at least partly, from interaction with the heme iron of $P450_{17\alpha}$.

Cytochrome $P\text{-}450_{17\alpha}$ (encoded by gene[s]P450XVII) is found in all steroidogenic tissues, including adrenal cortex, ovarian theca cells, and testicular Leydig cells. Chung et al.[78] called the $P450_{17\alpha}$ enzyme a key branch point in human steroid hormone synthesis. Indeed, the 17α-hydroxylase activity distinguishes between the synthesis of mineralocorticoids (e.g., aldosterone) and glucocorticoids (cortisol), and the 17,20-lyase activity distinguishes between the synthesis of glucocorticoids and C-19 precursors of androgens. As with other P450 enzymes from the endosplasmic reticulum, the activity is dependent on electron transfer from the microsomal flavoprotein P450 reductase. Cytochrome b_5 may also be involved in electron transfer. $P450_{17\alpha}$ catalyzes the 17α-hydroxylation of progesterone and pregnenolone. These 17-hydroxylated steroids are converted to 17-hydroxy, 20-dihydroprogesterone and 17-hydroxy, 20-dihydropregnenolone, respectively, and may undergo scission of the c-17,20 carbon bond (17,20-lyase acivity) to yield androstenedione and dehydroepiandrosterone (DHEA). Studies of Nakajin et al.[79] have shown that, in the neonatal pig, the hydroxylase and 17,20-lyase activities of testicular smooth endoplasmic reticulum are attributed to a single P450. Similar results were obtained with a P450 purified from guinea pig,[80,81] bovine,[82,83] pig,[84] and human adrenal,[78] and rat,[84-86] pig,[84] and human[78,87] testis microsomes. Slight differences have been reported in porcine $P450_{17\alpha}$ isolated from adrenals and testes.[84] However, studies of Chung et al.[78] showed the same $P450_{17\alpha}$ mRNA in human adrenal gland and testis indicating that human testis contains a $P450_{17\alpha}$ identical to that in the adrenals.

Comparison of the 17α-hydroxylase and the 17,20-lyase activities in COS 1 cells transfected with the bovine, human, and rat forms of $P450_{17\alpha}$ showed interesting differences. The rat $P450_{17\alpha}$ is able to convert both 17α-hydroxypregnenolone and 17α-hydroxyprogesterone to the respective androgens, dehydroepiandrosterone and androstenedione, whereas the human and bovine enzymes catalyze only the conversion of 17α-hydroxypregnenolone to dehydroepiandrosterone.[86] Human $P450_{17\alpha}$ has also been shown to catalyze 16α-hydroxylation of progesterone, whereas the rat enzyme lacked this activity.[86] In addition, sequence analysis of $P450_{17\alpha}$ revealed 75 positions where the amino acids are common in the bovine and human enzyme but which differ in the rat enzyme.[86] These data illustrate that parallelism between an experimental model and man is often difficult to ascertain. Indeed, in a subcellular fraction of pig testis, ketoconazole induced accumulation of pregnenolone or progesterone only, indicating that in the pig testicular microsomes the primary target is the 17α-hydroxylase instead of the 17,20-lyase.[180] Our results are different from those obtained by Nagai et al.,[88] who found that in a reconstituted system consisting of the 17α-hydroxylase-17,20-lyase and its reductase purified from pig testis microsomes, ketoconazole inhibited the 17,20-lyase activity and not the 17α-hydroxylase activity. However, they used 11β-hydroxyprogesterone as substrate for the 17α-hydroxylase instead of progesterone. It is possible that progesterone and its 11α-hydroxy derivative have different affinities for the substrate binding site.

Higashi et al.[77] measured the effects of ketoconazole on human testicular microsome P450-dependent monooxygenases. using progesterone as substrate, the K_i values of ketoconazole were 9×10^{-8} *M* for the 17α-hydroxylase and 8×10^{-8} *M* for the 16α-hydroxylase. Inhibition of the 17,20-lyase, 17α-, and 16α-hydroxylase was also found when tissue slices of human testis were incubated with ketoconazole.[89] In both studies the 17,20-lyase was found to be the most sensitive.[77,89] Using 17α-hydroxyprogesterone as substrate, a K_i value of 4×10^{-8} *M* was found.[77] It should be noted that the K_i values are much lower than the above-mentioned values for rat testicular 17,20-lyase and 17α-hydroxylase. The results obtained by Higashi et al.[77] on the 17,20-lyase in human testis correspond with the results obtained *in vivo*.[71] In male volunteers, who received a single dose of 200 mg ketoconazole, total and free testosterone levels fell to levels 60% below basal within 4 to 8 h. The levels of androstenedione paralleled those of testosterone. This decrease coincided with increased 17α-hydroxyprogesterone plasma levels,[71] indicating an effect on the 17,20-lyase. De Coster et al.[90] determined the effects of a 2-week high-dose ketoconazole therapy (400 mg every 8 h) in 14 previously untreated patients with metastatic prostate cancer. Within 24 h, plasma testosterone levels decreased to about 2.5 nmol/l. Plasma androstenedione and dehydroepiandrosterone reached 0.64 and 2.82 nmol/l on day 14. In contrast, 17α-hydroxyprogesterone and progesterone levels increased two- and fivefold, respectively.

The ketoconazole-induced decrease in androgen levels originates not only from an effect on testicular $P450_{17\alpha}$. In the microsomes of the adrenal cortex, pregnenolone is converted into dehydroepiandrosterone, androstenedione, or 11-deoxycortisol and 11-deoxycorticosterone (Figure 4). An inhibition of the adrenocortical 17α-hydroxylase would result in a decreased synthesis of 11-deoxycortisol, and an effect on the 17,20-lyase would decrease the levels of the androgens, androstenedione and dehydroepiandrosterone. The availability of pregnenolone and/or progesterone might result in an enhanced synthesis of the mineralocorticoid, 11-deoxycorticosterone. Ketoconazole inhibited androgen synthesis when added to bovine adrenal microsomes, and 50% inhibition was achieved at 0.4 μ*M*. This inhibition coincided with an accumulation of pregnenolone and 11-deoxycorticosterone.[93] At higher concentrations, ketoconazole also inhibited the synthesis of 11-deoxycortisol coinciding with a further increase in 11-deoxycorticosterone. The accumulation of the latter mineralocorticoid indicates that ketoconazole does not interfere with the 11-hydroxylase. In rat adrenal microsomes a K_j value of 5 μ*M* was reported.[77] This is 12.5 times higher than the value found for the rat testicular 17,20-lyase.[77]

Ketoconazole also had little or no effect on 21-hydroxylase activity in human adrenal microsomes.[94] The low, if any, sensitivity of the 21-hydroxylase is of interest. Both the 17α- and 21-hydroxylase are using progesterone as the substrate, and the discrepancy suggests that ketoconazole is not binding to the steroid binding site. In fact, an alignment of the amino acid sequence of the human $P450_{17\alpha}$ with that of $P450_{C21}$ showed a well-conserved domain (from aspartic $acid_{345}$ up to $proline_{367}$ of the $P450_{17\alpha}$ sequence) (to be published elsewhere). This domain of the rat $P450_{17\alpha}$ has been proposed as the putative steroid binding site[85] and is also present in the human and bovine $P450_{17\alpha}$ (to be published elsewhere).

Ketoconazole also inhibited the 17α-hydroxylase (K_i value = 0.04 μ*M*) and 17,20-lyase (K_i value = 0.01 μ*M*) in human adrenal microsomes.[94] Ayub and Levell found that the human adrenal 17,20-lyase was about twice more sensitive to ketoconazole than the 17α-hydroxylase.[95] These investigators also compared the activity of a series of imidazole antifungals on steroidogenic enzymes of human adrenal microsomes. The order of decreasing

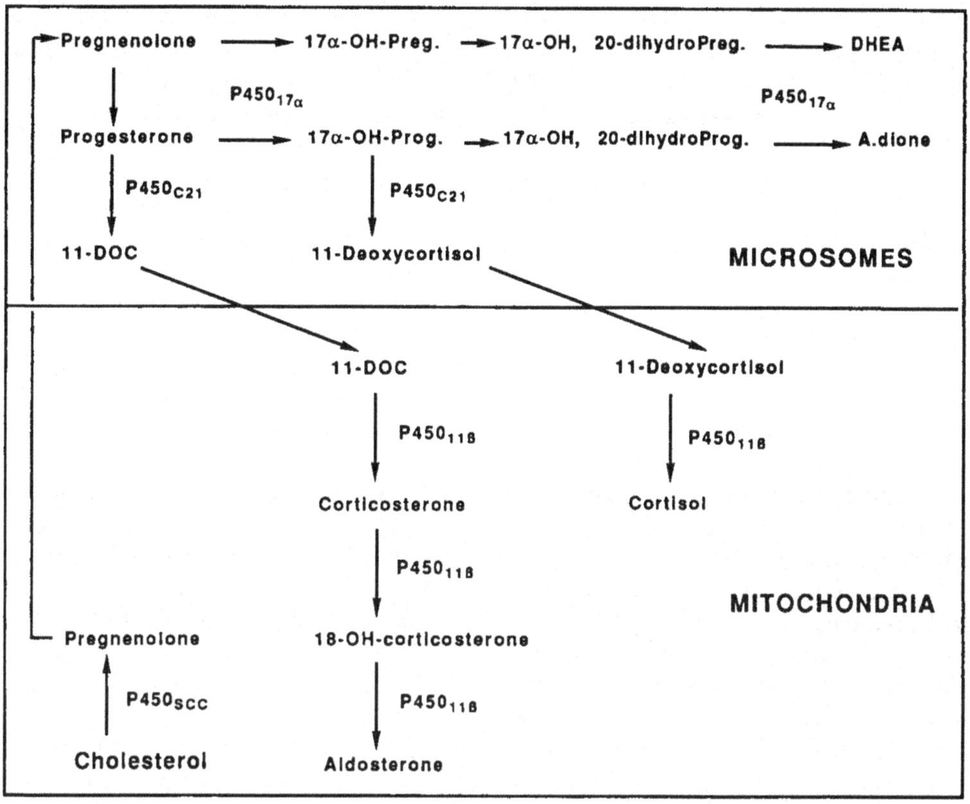

FIGURE 4. Scheme of the p450-dependent synthesis of androgens, cortisol, corticosterone, and aldosterone in adrenal cortex mitochondria and microsomes. Preg. = pregnenolone, Prog. = progesterone, DHEA = dehydroepiandrosterone, DOC = deoxycorticosterone, SCC = side-chain cleavage, $P450_{17\alpha}$ = P450 involved in the 17α-hydroxylase and 17,20-lyase activities, $P450_{C21}$ = 21-hydroxylase P450, $P450_{11\beta}$ = the P450 that in the adrenal cortex mediates not only the 11β-hydroxylation of 11-deoxycortisol and 11-deoxycorticosterone, but also the synthesis of aldosterone.[91,92]

inhibitory potency, as determined from values for both 17α-hydroxylase (IC_{50} values = 1.13 to 4.17 μM) and 17,20-lyase (IC_{50} values = 0.57 to 1.95 μM) activities, was bifonazole > clotrimazole > ketoconazole > miconazole > econazole > isoconazole > tioconazole.[95]

An effect of ketoconazole on the synthesis of adrenal steroids was also demonstrated *in vivo*.[90] Fourteen previously untreated patients with stage-D metastic prostate cancer received 400 mg ketoconazole every 8 h. In addition to a striking decrease of testosterone plasma levels, a decline in androstenedione and dehydroepiandrosterone, both mainly of adrenal origin, was noted.[90] As expected from the *in vitro* results, the decrease in androgen levels coincided with a rise in 17α-hydroxyprogesterone and progesterone plasma levels. Inhibition of human adrenal androgen secretion by ketoconazole (600 mg/day) was also noted in patients with elevated serum androgens.[96] The mean reduction in dehydroepiandrosterone sulfate serum levels was 32%, in dehydroepiandrosterone 54%, in androstenedione 52%, and in testosterone 43%. The reduction in androgen serum levels became significant 24 h after treatment and persisted as long as the drug was administered.[95]

The fact that ketoconazole interferes with testicular as well as adrenal androgen synthesis has made it a good candidate for the treatment of androgen-dependent prostate carcinoma.[60,93,97-104] With the frequent administration of high doses of ketoconazole (400 mg every 8 h) a rapid decline in serum testosterone, androstenedione, and DHEA has been observed. Relief of bone pain and sustained clinical remission has occurred in a great number of patients.[60,98,99,103,104] In various studies it has been noted that, parallel to testosterone, elevated levels of prostatic acid phosphatase fell dramatically within the first days of treatment.[60,98,104-106] Few side effects have occurred.[103] At this high dose the principal side effect is gastric discomfort with nausea,[60,104] which resolves with administration of antinauseants or after discontinuation of the drug.[107] From all these studies it can be deduced that ketoconàzole, at high doses, is effective both clinically and endocrinologically in the treatment of metastatic prostate carcinoma. Its use, however, is limited by gastric discomfort. Regardless of this limitation, the extensive investigations have triggered multidisciplinary efforts to open novel possibilities in medical treatments.

2. Mitochondrial P450 Isozymes

So far, our discussion was restricted to the effects of ketoconazole and similar compounds on microsomal P450 isozymes. As indicated in Figure 4, a number of mitochondrial P450 isozymes are involved in the synthesis of cortisol, corticosterone, and aldosterone. Furthermore, the cholesterol side-chain cleavage enzyme, catalyzing the conversion of cholesterol to pregnenolone, is the first rate-limiting step in the synthesis of all steroid hormones.

Cholesterol side-chain cleavage (scc) involves three consecutive monooxygenation steps and is catalyzed by $P450_{scc}$ (cholesterol, reduced-ferrodoxin/oxygen oxidoreductase).[92,108] The reaction proceeds via an hydroxylation at the 22-position of cholesterol to yield (22*R*)-hydroxycholesterol, which is then converted to a dihydroxy intermediate (20*R*,22*R*). The oxidative cleavage of the C20-C22 bond of the diol leads to pregnenolone and isocaproic aldehyde.[109] Protein purification data suggest the existence of a single immunologically identifiable species of $P450_{scc}$.[92] This P450 was found to be bound to the inner mitochondrial membrane as a multimer of 16 subunits totaling over 850,000 Da.[92] Reducing equivalents are provided to $P450_{scc}$ via an electron transfer system localized in the mitochondrial matrix. Electrons from NADPH are transferred to adrenodoxin reductase which then transfers electrons to adrenodoxin. This iron sulfur protein transfers electrons to $P450_{scc}$. It should be noted that adrenodoxin binding is essential to achieve the regioselective hydroxylation at the 22*R* position of the cholesterol side chain.[108] $P450_{scc}$ is a glycoprotein, and the sugar moiety may participate in the binding of adrenodoxin or the subsequent electron transfer.[110]

The $P450_{scc}$ system has been demonstrated in adrenal cortex, ovary, testes, and placenta. Chung et al.[111] cloned and sequenced full-length human $P450_{scc}$ cDNA and located the gene (termed P450XIA) on chromosome 15. The encoded human $preP450_{scc}$ contains 521 amino acids. The testis and adrenal amino acid sequences are identical, and are 72% homologous with the bovine $P450_{scc}$ sequence.[111] Chung et al. also found that the human 450_{scc} gene is expressed in the placenta in early and midgestation.

The importance of $P450_{scc}$ in the synthesis of all steroid hormones made it necessary to study its potential interaction with azole antifungals. Loose et al. found that ketoconazole decreased cholesterol side-chain cleavage activity in an enriched adrenal mitochondrial preparation from female Sprague-Dawley rats.[112] When 1.0 μM of ketoconazole was added to bovine or piglet mitochondrial suspensions, prior to the reduction of the P450 and treatment with carbon monoxide, approximately 50% decrease in the absorption at 450 nm was noted in both cases.[93] Using a reconstituted system consisting of $P450_{scc}$, adrenodoxin, and adrenodoxin reductase from bovine adrenal mitochondria, 50% inhibition of pregnenolone synthesis from cholesterol was obtained at 1.7 μM ketoconazole.[113] A similar inhibition was achieved with 3 μM ketoconazole when the conversion of cholesterol to pregnenolone was studied in a reconstituted system consisting of $P450_{scc}$, isolated from pig testis, and adrenodoxin and adrenodoxin reductase from bovine adrenals.[113] In both systems, 100% inhibition was not achieved at 10 μM, indicating that $P450_{scc}$ is less sensitive to ketoconazole than $P450_{17\alpha}$. However, when ketoconazole is used at high doses, this effect on pregnenolone synthesis might, together with the inhibition of the 17α-hydroxylase/17,20-lyase, contribute to the inhibition of androgen biosynthesis in both testis and adrenals.

The final enzyme in the synthesis of adrenal glucocorticoids and mineralocorticoids is the P450-dependent 11β-hydroxylase ($P450_{11\beta}$). This single enzyme (only present in the adrenal cortex) mediates not only the 11β-hydroxylation of 11-deoxycortisol, but also the synthesis of aldosterone (the most important mineralocorticoid) from 11-deoxycorticosterone or corticosterone.[91] This indicates that $P450_{11\beta}$ has 11β-hydroxylase, 18-hydroxylase, and aldehyde synthetase activities. In addition, the enzyme also has the capacity to catalyze the 19-hydroxylation of 11-deoxycorticosterone and androstenedione in gerbil adrenals.[114] A $P450_{11\beta}$-dependent conversion of 19-hydroxyandrostenedione to 19-oxoandrostenedione and further to 19-norandrostenedione (19-oxidase reaction) has also been described.[115,116] Thus, $P450_{11\beta}$ seems to possess 10-demethylase (C10-19-lyase) activity.[116] Suhara and colleagues[116] described a $P450_{11\beta}$-dependent conversion of androgen to estrogen indicating that this multifunctional enzyme catalyzes an aromatase reaction in the adrenal cortex. It should be noted that aldosterone is produced only in the zona glomerulosa, while $P450_{11\beta}$ exists and produces corticosterone and/or cortisol in all three zones of the adrenal cortex.

Recently, a P450, capable of producing aldosterone from 11-DOC, was purified from the mitochondria of the zona glomerulosa of adrenal cortex from sodium-depleted and potassium-repleted rats.[117] This P450 was distinguishable from the $P450_{11\beta}$ purified from the zonae fasciculata-reticularis mitochondria of the same rats. The former had a molecular weight of 49,500 and the latter 51,500 Da. The P450 from the zona glomerulosa catalyzed the hydroxylation of 11-DOC with the formation of 18-hydroxydeoxycorticosterone, 18-hydroxycorticosterone, and a significant amount of aldosterone. The enzyme from the zonae fasciculata-reticularis only catalyzed 11_{β}- and 18-hydroxylation reactions. Thus at least two forms of $P450_{11\beta}$ exist in rat adrenal cortex mitochondria, although aldosterone synthesis is catalyzed only by the form present in the zona glomerulosa.[117] In bovine adrenocortical mitochondria, again two distinct forms of $P450_{11\beta}$ could be identified.[118] However, both forms were able to catalyze aldosterone synthesis. As with rat and cattle, two genes encoding for $P450_{11\beta}$ have also been demonstrated in man.[119] Whether both encode for an active $P450_{11\beta}$ or one encodes for a nonfunctional pseudogene is not known yet. It is also unknown whether the human P450 resembles more closely the rat or the bovine $P450_{11\beta}$. The differences between rat and cattle $P450_{11\beta}$ represents another example of the difficulty to extrapolate data obtained from rats.

When bovine adrenal mitochondria were incubated in the presence of [^{3}H]-11-deoxycorticosterone or [^{3}H]-11-deoxycortisol, 50% inhibition was achieved at ketoconazole concentrations of 0.75 and 0.38 μM, respectively.[93,120] Nagai et al.[121] studied the inhibitory effect of ketoconazole on the 11β-hydroxylation of 11-deoxycorticosterone with a reconstituted system consisting of adrenodoxin and its reductase, and $P450_{11\beta}$ purified from bovine adrenal mitochondria. Fifty percent inhibition was achieved at about 1 μM. These investigators also found that ketoconazole inhibited the 18-hydroxylation of corticosterone (IC_{50} value = about 0.05 μM), whereas the deoxycorticosterone 18-hydroxylase was found to be ten times less sensitive. Using cells from rat adrenals, Loose et al.[112] reported a complete block of corticosterone synthesis from 11-deoxycorticosterone at about 0.9 μM ketoconazole.

The imidazole derivative imazalil inhibited the 11β-hydroxylase activity in gerbil adrenal mitochondria by 75% at 1 μM.[114] Miconazole was less active than either imazalil or methyrapone; a complete inhibition of 11β-hydroxylation was only achieved at 60 μM miconazole.[114] Ayub and Levell[95] compared the effects of ketoconazole, bifonazole, clotrimazole, miconazole, econazole, isoconazole, and tioconazole on the human adrenal mitochondrial 11β-hydroxylase activity. All imidazole derivatives caused a dose-related inhibition. Each of these drugs inhibited 100% of the activity at 5 μM, which was 25-fold higher than the substrate concentration (except tioconazole which inhibited 78% of the activity at this concentration). The IC_{50} value for ketoconazole was about 2.4 times higher than for the 17,20-lyase.[95] It should be noted that in patients receiving 400 mg at 8-h intervals, plasma cortisol and aldosterone levels remained unchanged during a 14-day treatment period, whereas 11-

deoxycorticosterone and 11-deoxycortisol levels rose 14- and 6.7-fold, respectively.[90] Trachtenberg[98] and Heyns et al.[100] also observed that the serum cortisol levels did not change from baseline values. Short- (200 mg orally every 5 h for 25 h) and long-term (200 mg orally every 8 h; 7 days to 12 months) administration of ketoconazole to patients with Cushing's syndrome indicated that ketoconazole is able to inhibit cortisol synthesis by blocking adrenal 11β-hydroxylase and 17α-hydroxylase.[122] However, this effect was partially compensated by an increased pituitary ACTH secretion in the majority of the patients with pituitary-dependent Cushing's disease. Therefore, Engelhardt et al. concluded that only patients with Cushing's syndrome due to an adrenal tumor or an ectopic ACTH syndrome can be treated effectively with ketoconazole.[122] Nevertheless, these results indicate that ketoconazole at a high dose might be of use in some steroid hormone-dependent diseases. Furthermore, the knowledge gathered with ketoconazole has provided us with a battery of P450 systems which have made it possible to predict *in vitro* activity and selectivity.

These P450 systems were of help in the selection of the potent and much more selective azole antifungals, itraconazole[3,44,45,52,56,123-127] and saperconazole.[128] For example, they were of help in the selection of the triazole antifungal itraconazole, which shares it broad spectrum of oral activity with ketoconazole but is much more active against, e.g., *Aspergillus fumigatus* and *Sporothrix schenckii,* and is highly active in meningeal cryptococcosis.[129] Itraconazole has little, if any, effect on cholesterol side-chain cleavage, 1-hydroxylation and 19-hydroxylation of testosterone.[44,45,72] Itraconazole is a poor inhibitor of the regio- and stereoselective P450-dependent hydroxylations of testosterone, progesterone, and estradiol by rat liver microsomes (unpublished results). Up to a dose of 40 mg/kg itraconazole did not prolongate methohexital hypnosis in female rats.[129a] With ketoconazole a median effective dose (ED_{50}) of 30 and 97 mg/kg was found with female and male rats, respectively.[129b] In contrast to itraconazole, fluconazole, when given as a single dose, prolonged hypnosis time (ED_{50} = 8.7 mg/kg).[129a] A decrease of the hypnosis time (due to enzyme induction) by more than 50% from day 1 to day 5 was marked with fluconazole (ED_{50} = 22.6 mg/kg) but not with itraconazole.[129a]

Itraconazole does not affect testosterone synthesis in subcellular fractions of rat testes.[72] It is of interest that fluconazole is also a poor inhibitor of testosterone synthesis and metabolism.[46-46b] Eleven percent inhibition of testosterone synthesis by intact rat Leydig cells was obtained at 16.3 μ*M* fluconazole.[46,461] So far no data are available on fluconazole's effects on both synthesis and metabolism of testosterone at the subcellular level.

High concentrations of itraconazole have an effect on testosterone synthesis by dispersed rat testicular cells. Without stimulation, 50% inhibition was found at 4.5 μ*M* itraconazole. In the presence of human chorionic gonadotropin (hCG), 50% inhibition was achieved at 1.6 μ*M*.[44] However, the production of the precursors 17α-hydroxyprogesterone and progesterone was affected in the same way as that of testosterone. This indicates that the effects of high concentrations of itraconazole are not related to interference with the 17α-hydroxylase and/or 17,20-lyase. The observed decrease might originate from an effect on cholesterol synthesis. Indeed 50% inhibition of cholesterol synthesis in Leydig cells was observed after 4 h of incubation in the presence of 1 μ*M* itraconazole (unpublished results). It should be noted that itraconazole at 0.005 μ*M* abolished ergosterol synthesis in *C. albicans*.[44] Itraconazole is devoid of any effect on the 11β-hydroxylase in adrenal cortex mitochondria and on the P450-dependent 1-hydroxylation of 25-hydroxyvitamin-D in kidney mitochondria.[44,45,56]

Itraconazole's selectivity has been confirmed *in vivo*. Indeed, itraconazole is not associated with changes in pituitary-testicular-adrenal function in human males, and in animals chronically treated with high doses of the drug.[130] Clinical pharmacology data also show no effect on 17β-estradiol levels,[131] indicating that *in vivo*, itraconazole has no effect on the aromatase. *In vitro*, using human placenta microsomes no effect was observed with itraconazole concentrations up to 10 μ*M*.[180]

The study of the different P450 systems also contributed to the finding of new P450 inhibitors that interfere with estrogen biosynthesis.[63]

3. Aromatase

The endoplasmic reticulum $P450_{AROM}$ converts androgens (androstenedione, testosterone, 16α-hydroxytestosterone, and 19-norandrostenedione) to estrogens by two hydroxylations at the C-19 methyl group and a third hydroxylation at C-2. This results in the loss of the C-19 methyl group and consequent aromatization of the A ring of the steroid.[92,132] The aromatase enzyme complex is present in the granulosa cells of the ovaries,[108,133,134] in adipose tissue,[135-137] in the Sertoli cells of the testis,[108] in several areas of the brain,[108,138] in placenta,[108,139] in rat[140] and human[141] prostate, in Dunning R3327H rat prostatic adenocarcinoma,[140] human breast tumor,[137,142,143] and neoplastic endometrium.[144]

Expression of cDNAs encoding, respectively, the chicken[145] and the human[146] $P450_{AROM}$ in nonsteroidogenic cells indicated that the several reactions involved in the aromatization of androgens are catalyzed by a single P450.[146] Analysis of the amino acid sequence of human placental aromatase and comparison with the sequences of other P450s by computer homology search revealed that this enzyme is a unique form of the P450 superfamily.[147]

As already mentioned above, ketoconazole is a poor inhibitor of the aromatase. However, the topically active imidazole antifungal, econazole, is a potent inhibitor of the enzyme in human placental microsomes,[148] and of the human placental $P450_{AROM}$ expressed in transfected Cos-1 cells.[146] In the latter cells an IC_{50} value of 0.04 μM was found.[146] Mason et al. also measured the inhibitory effects of imazalil, miconazole, prochloraz, clotrimazole, ketoconazole, and aminoglutethimide on the human placenta microsomal aromatase.[148] Fifty percent inhibition was achieved at 0.15, 0.6, 0.7, 1.8, 60, and 45 μM respectively. A similar study was done by Ayub and Levell.[149] They compared the activity of a number imidazole derivatives with the known aromatase inhibitors, 4-hydroxyandrostenedione (4-OHA) and aminoglutethimide (AG). The order of decreasing inhibitory effect was 4-OHA > tioconazole > econazole > bifonazole > clotrimazole > miconazole > isoconazole > ketoconazole > AG > nimorazole.

When miconazole was added to a purified human placental aromatase (ferri)P450, a low spin complex was formed.[150] The spectral characteristics indicate that the imidazole-nitrogen coordinates to the heme-iron.[150]

Inhibition of the aromatase activity by nonsteroidal compounds is not limited to imidazole derivatives. The triazole-substituted antifungal, fluconazole, caused at 10 μM 70% reduction of the estradiol production by isolated rat ovaries.[151] This concentration is needed to get 100% inhibition of ergosterol synthesis in *C. albicans*.[52]

Another ergosterol biosynthesis inhibitor (EBI), the pyrimidinylcarbinol agricultural fungicide fenarimol, caused a dose-related decrease in fertility in rats.[152] The infertility appeared to be a male-specific effect and was associated with a decrease in male sexual behavior. This effect was attributed to the inhibition of aromatase activity in the central nervous system.[152] Indeed, it was found that fenarimol localized in regions of the brain was associated with the development and expression of male sexual behavior.[153] Furthermore, fenarimol inhibited aromatase activity within the hypothalamic/preoptic area—amygdala regions of the brain.[153] In these regions the highest concentrations of aromatase are observed in adulthood.[138] In a rat ovarian microsome assay, 50% inhibition was found at 4.1 μM fenarimol.[152] In this system miconazole gave an IC_{50} value of 0.4 μM.[152]

According to Santen, effective and selective aromatase inhibitors should provide important therapeutic tools.[143] For example, first or second line treatment of postmenopausal women with breast or endometrial cancer with aromatase inhibitors is an expected application.[143] Aromatase inhibitors could also be of value in treating pubertal gynecomastia,[143] idiopathic oligospermia,[143,154] and perimenopausal uterine bleeding.[143]

As a result of the observation that a number of the ergosterol biosynthesis inhibitors also have inhibitory effects on the aromatase, derivatives have been synthesized. For example, Taylor et al.[152] replaced the hydroxyl group of fenarimol by a proton [α,α-bis(4-chlorophenyl)-5-methylpyridmidine; LY56110]. They found that the concentration needed to get 50% inhibition of the aromatase activity in rat ovarian microsomes was 0.055 μ*M* instead of the 4.1 μ*M* needed with the parent compound. LY56110 was orally active, blocking the testosterone-induced increase of uterine weight and ovarian estrogen biosynthesis in immature rats.[153] However, further development was stopped because of its effects on hepatic microsomal enzymes and a long half-life.[153] The research group at Eli Lilly started a number of structural modification and found that the indenopyrimidine, LY113174 [8-chloro-5-(4-chlorophenyl)-5H-indeno<1,2D>pyrimidine], is a potent inhibitor of the aromatase (IC_{50} value = 0.024 μ*M*) but is less potent than LY56110 with respect to induction of hepatic microsomal enzymes.[153]

The study of the different P450 systems also contributed to the selection of the new aromatase inhibitor, R 76713 (Janssen Research Foundation). This triazole derivative {(±)-6-[(4-chlorophenyl)(1*H*-1,2,4-triazol-1-yl)methyl]-1-methyl-1*H*-benzotriazole} has at 10 μ*M* almost no effect on microsomal P450 of *C. albicans* and, as a consequence, it does not affect ergosterol synthesis and growth.[63] It has high affinity for the human placenta microsomal P450. Fifty percent decrease in absorption at 453 nm is already achieved at 0.064 μ*M*. In contrast, R 76713 shows low, if any, affinity for other microsomal or mitochondrial P450s.[63] Measuring the effects on estrogen synthesis from androstenedione by human placenta microsomes, 50% inhibition is already obtained at 2.7 n*M*, i.e., at 1549 times lower concentration than that needed for ketoconazole.[63] At concentrations up to 10 μ*M*, this triazole derivative has no effect on cholesterol synthesis in rat liver subcellular fractions or PAH-stimulated human lymphocytes nor on cholesterol side-chain cleavage in bovine adrenal cortex mitochondria. At this high concentration a minor effect (30%) on the 11β-hydroxylase is observed. R 76713 is, in contrast to ketoconazole, a poor inhibition of androgen synthesis by subcellular fractions of Wistar rat testis; 50% inhibition is reached at 8 μ*M*.[63] R 76713 is also a very potent inhibitor of the aromatase activity in rat granulosa cells, showing an IC_{50} value of 3.0 n*M*.[155] *In vivo*, single oral doses of 0.05 mg/kg lowered plasma estradiol levels of pregnant mare serum gonadotropin (PMSG)-primed rats by more than 90%.[155]

Both LY113174 and R76713 are good examples to prove that P450 systems can be exploited in the construction of compounds that may become important tools in chemotherapeutic applications.

C. THROMBOXANE AND PROSTACYCLIN SYNTHESIS

The isomerization of prostaglandin-endoperoxide (9,11-endoperoxy, 15-hydroxy arachidonic acid; PGH_2) to prostacyclin (PGI_2) and to thromboxane (TXA_2) involved endoperoxide activation.[156] The spectral characteristics of PFI_2- and TXA_2-synthetase indicate that both enzymes are P450 proteins and therefore must contain a thiolate ligand at the heme iron. However, no monooxygenase activity is associated with their physiological function.[157] Therefore, Hecker et al.[157] proposed the name "heme-thiolate enzymes" instead of P450 enzymes. From optical and ESR studies on PGI_2- and TXA_2-synthetase Hecker and Ullrich[158,159] conclude that the ferric enzyme catalyzes the cleavage of the endoperoxide bond. These authors also proved that PGI_2-synthetase obtains its specificity through an interaction with the endoperoxide oxygen at C-11, whereas TXA_2-synthetase interacts with the C-9 endoperoxide oxygen.

The TXA_2-synthetase reacts with a number of imidazole derivatives[160] by nitrogen binding to the ferric heme iron, whereas not reaction occurs with prostacyclin synthetase.

Imidazole is a poor inhibitor of the thromboxane synthetase, solubilized from human platelet microsomes. An IC_{50} value was reached at 48 μ*M* only. However, dazoxiben, a *N*-substituted imidazole, showed an IC_{50} value of 0.011 μ*M*.[160] Using intact human platelets,

Tolman and Fuller found that clotrimazole was a potent inhibitor of thromboxane production.[161] Fifty percent inhibition was already achieved at 0.37 μ*M*. Miconazole and ketoconazole were about one and two orders of magnitude less potent, respectively.[160] The high concentration of ketoconazole needed to reduce thromboxane production has been confirmed by Beetens et al.[162]

Studies of Hecker et al.[160] proved that difference spectra provide a relatively simple means to establish structure-activity relationships. Their studies indicated that next to the liganding of the basic nitrogen of the imidazole or the pyridine moieties to the heme iron, an attachment of a hydrophobic carboxylic side chain to the active site at about 1 nm distance from the nitrogen is a prerequisite for high binding affinity and, thus, inhibitory activity.

A molecular that fits this model is ridogrel (R 68070; (E)-5-[[(3-pyridnyl)[3-(trifluoromethyl) methylen]amino]oxy]pentanoic acid; Janssen Research Foundation). This 3-pyridinyl derivative shows high affinity for P450 present inhuman platelet microsomes. Fifty percent inhibition of carbon monoxide binding is achieved at 19 n*M* (unpublished results). *In vitro* the compounds specifically inhibits the production of thromboxane A_2 from arachidonic acid by washed human platelets (IC_{50} = 8.2 n*M*) and by platelet microsomes (IC_{50} = 3.6 n*M*).[163] R 68070 has a dual mechanism of action, namely, thromboxane synthetase inhibition at low concentrations, plus, additionally, thromboxane/prostaglandin endoperoxide receptor blockade at higher concentrations.[163] This finding is of great interest since R 68070 will inhibit the binding of the prostaglandin endoperoxides, accumulating under the action of thromboxane synthetase inhibitors, tothethromboxane receptor, thus avoiding platelet aggregation.[163]

Due to its unique modulation of arachidonic metabolism and platelet function, ridogrel is indicated for the prevention of arterial thrombotic pathologies in man.

D. METABOLISM OF RETINOIC ACID AND P450-DEPENDENT REACTIONS IN THE SKIN

Vitamin A (retinol) must be converted in the target tissues to retinoic acid (RA) before carrying out is epithelial differentiation function.[164] However, unlike retinol, RA is rapidly converted into polar metabolites.[165] The oxidative metabolism of RA in liver and intestine of vitamin A-deficient hamsters requires NADPH and oxygen is inhibited by carbon monoxide and is associated with the microsomal fraction.[166-169] In rat liver microsomes, the involvement of P450 is indicated by the requirement of oxygen and NADPH, by the inhibitory action of prodifen-HCl (SKF-525A) and carbon monoxide, and by the five- to tenfold increase in the rate of retinol metabolism after administration of phenobarbital, an inductor of $P450_b$ (IIB1) and $P450_e$(IIB2.)[170,171] Interestingly, the P450-dependent metabolic activity in hamster liver microsomes is inducible by retinoids, but to a minor extent by phenobarbital or 3-methylcholanthrene. The major metabolites formed are 13-*cis* and all-*trans* isomers of 4-hydroxy- and 4-oxoretinoic acid.[166]

In vitro, ketoconazole inhibits, in an apparently competitive manner, the P450-mediated metabolism of RA to 4-hydroxy- and 4-keto-retinoic acids by hamster liver microsomes. The calculated K_i value is 0.75 μ*M*.[172] *In vivo*, ketoconazole suppresses the formation of polar RA metabolites by Wistar rats. After oral treatment with ketoconazole, administered 1 h before the [^{3}H]RA injection, a significant decrease in the formation of polar RA metabolites is observed at a dose of 2.5 mg/kg.[172]

Epidermal microsomes, prepared from neonatal (4-day-old) Wistar or Sprague-Dawley rats, showed low levels of retinoic acid metabolism.[173] However, 24 h after topical application of RA, a 3.3- and 3.9-fold increase in epidermis microsomal RA metabolism is found. A 4.6- to 8.1-fold increase is observed 24 h after topical application of 3-methylcholanthrene. Phenobarbital has a much smaller inducing effect.[173] The RA metabolism in these microsomes shows the characteristics of a P450-dependent enzyme system, i.e., a requirement for NADPH and oxygen and inhibition by carbon monoxide and SKF-525A.[173] So far the chemical structure of the polar metabolites has not been identified. The R values of two major compounds correspond with those of 4-hydroxy- and 4-ketoretinoic acid, formed after incubation of hamster liver microsomes in the presence of [15-^{14}C]-RA.[172,173]

Ketoconazole and miconazole inhibit RA metabolism *in vitro* in rat epidermal microsomes.[173] Fifty percent inhibition is achieved at 6.5×10^{-7} *M* and $\geq 10^{-5}$ *M*, respectively. The triazole antifungal agent, itraconazole, has no effect at concentrations up to 10^{-5} *M*. Topical application of ketoconazole (1, 5, or 10 mg/kg) on 4-day-old Wistar rats, 1 h before the application of RA (1 mg/rat), results in a dose-dependent inhibition of RA metabolism by epidermal microsomes, prepared 24 h later.[173] These results suggest that ketoconazole may prove to be effective in maintaining biologically active levels of RA in epidermal cells.

The skin is not only capable of modulating compounds such as RA. Indeed, Bickers and colleagues proved that the skin actively processes not only drugs,[174] but also testosterone by hydroxylations at the 6β-, 7α-, and 16α-position.[175] Their studies suggest that multiple isozymes of P450 are present and prove that skin P450s are involved in the metabolism of both exogenous and endogenous substrates. It has been shown recently that human foreskin keratinocytes are able to hydroxylate 25-hydroxyvitamin D_3.[176] 1,25-Dihydroxy vitamin D_3 appears to play a role in epidermopoiesis and melanin pigmentation.[176]

In an interesting study Merk et al.[177] show that exposure of volunteers to a coal tar containing shampoo, a preparation rich in polycyclic aromatic hydrocarbons, resulted in an increase in the P450-dependent aryl hydrocarbon hydroxylase (AHH) activity in human hair follicles. The AHH ($P450_c$-dependent, protein name: IA1[1]) is catalyzing steps in the conversion of the polycyclic aromatic hydrocarbon, benzo[a]pyrene, into the carcinogenic, (+)-7β-8α-dihydroxyl-9α-10α-epoxy-7,8,9,10-tetrahydrobenzo[a]pyrene. Topical application of a crude coal tar solution to the skin of neonatal rats is shown to induced not only epidermal, but also hepatic P450-dependent monooxygenase activities.[178] Using monoclonal antibodies, Khan et al.[178] prove that topical application of this crude coal tar solution results in induction of $P450_c$ in epidermis and $P450_b$ (protein name: IIB1[1]) and $P450_c$ in liver. Induction of $P450_c$ in both epidermis and skin by treatment with coal tar-containing preparations suggests that prolonged use may lead to adverse reaction not only in this skin, but also in other tissues. Clotrimazole is found to inhibit AHH in human hair follicles, and oral administration of ketoconazole (200 mg daily for 5 days) resulted in 90% inhibition of hair follicle AHH activity.[177] These results suggest that some imidazole antifungals may prove effective in protecting against, e.g., skin cancers caused by polycyclic aromatic hydrocarbons.

V. CONCLUSION

The examples discussed in this overview demonstrate the importance of cytochrome P-450-dependent enzyme activities in the synthesis and metabolism of many endogenous substances in different organisms and organs. They also demonstrate that the in-depth research on the effect of ergosterol biosynthesis inhibitors on different P450 isozymes not only helps in the selection of new antifungals with improved selectivity toward fungal P450, but also opens new possibilities for therapeutic applications of azole antifungals and other P450 inhibitors.

REFERENCES

1. **Nebert, D. W., Nelson, D. R., Adesnik, M., Coon, M. J., Estabrook, R. W., Gonzalez, F. J., Guengerich, F. P., Gunsalus, I. C., Johnson, E. F., Kemper, B., Levin, W., Phillips, I. R., Sato, R., and Waterman, M. R.,** The P450 superfamily: updating listing of all genes and recommended nomenclature for the chromosomal loci, *DNA*, 8, 1,1989.
2. **Kato, T.,** Sterol biosynthesis in fungi, a target for broad spectrum fungicides, in *Chemistry of Plant Protection. Vol. 1 Sterol biosynthesis Inhibitors and Anti-Feeding Compounds*, Haug, G. and Hoffman, H., Eds., Springer-Verlag, Berlin, 1986, 1.
3. **Vanden Bossche, H.,** Mode of action of pyridine, pyrimidine and azole antifungals, in *Sterol Biosynthesis Inhibitors, Pharmaceutical and Agrochemical Aspects*, Berg, D. and Plempel, M., Eds., Ellis Horwood, Chichester, England, 1988, 79.
4. **Omura, T. and Sato, R.,** The carbon monoxide-binding pigment of liver microsomes. I. Evidence for its hemoprotein nature, *J. Biol. Chem.*, 239, 2370, 1964.
5. **Nebert, D. W. and Gonzalez, F. J.,** P450 genes: structure, evolution, and regulation, *Annu. Rev. Biochem.*, 56, 945, 1987.
6. **Dawson, J. H.,** Probing structure-function relation in heme-containing oxygenases and peroxidases, *Science*, 240, 433, 1988.
7. **Hodgson, E.,** Bioactivation of pesticides by cytochrome P-450, presented at Nato Advanced Study Institute on Molecular Aspects of Monooxygenases and Bioactivation of Toxic Compounds, Çesme, Turkey, August 27 to September 7, 1989, 36.

7a. **Anderson, S. and Jörnvall, H.,** Sex differences in cytochrome P-450-dependent 25-hydroxylation of C27-steroids in rat liver microsomes, *J. Biol. Chem.*, 261, 16932, 1986.

7b. **Saarem, K. and Pedersen, J. I.,** 25-Hydroxylation of 1α-hydroxyvitamin D-3 in rat and human liver, *Biochim. Biophys. Acta*, 840, 117, 1985.

8. **Rein, H., Jung, C. H., Ristau, O., and Friedrich, J.,** Biophysical properties of cytochrome P-450, analysis of the reaction mechanism—thermodynamic aspects, in *Cytochrome P-450*, Ruckpaul, K. and Rein, H., Eds., Akademie-Verlag, Berlin, 1984, 163.
9. **Ruckpaul, K. and Bernhardt, R.,** Biochemical aspects of the monooxygenase system in the endoplasmic reticulum of mammalian liver, in *Cytochrome P-450*, Ruckpaul, K. and Rein, H., Eds., Akademie-Verlag, Berlin, 1984, 9.
10. **White, R. E. and Coon, M. J.,** Oxygen activation by cytochrome P-40, *Annu. Rev. Biochem.*, 49, 315, 1980.
11. **Sligar, S. G. and Gunsalus, I. C.,** A thermodynamic model of regulation: modulation of redox equilibria in camphor monooxygenase, *Proc. Natl. Acad. Sci. U.S.A.*, 73, 1078, 1976.
12. **Blanck, J., Smettan, G., and Greschner, S.,** The cytochrome P-450 reaction mechanism—kinetic aspects, in *Cytochrome P-450*, Ruckpaul, Kl. and Rein, H., Eds., Akademie-Verlag, Berlin, 1984, 111.
13. **Sligar, S. G. and Murray, R. I.,** Cytochrome P-450$_{CAM}$ and other bacterial P-450 enzymes, in *Cytochrome P-450 Structure, Mechanism, and Biochemistry*, Ortiz de Montellano, P. R., Eds., Plenum Press, New York, 1986, 429.
14. **Berg, A., Carlström, K., Gustafsson, J.-Å., and Ingelman-Sunberg, M.,** Demonstration of a cytochrome P-450-dependent steroid 15β hydroxylase in *Bacillus megaterium*, *Biochem. Biophys. Res. Commun.*, 66, 1414, 1975.
15. **Berg, A., Gustafsson, J.-Å., Ingelman-Sunberg, M., and Carlström, K.,** Characterisation of a cytochrome P-450-dependent steroid hydroxylase system present in *Bacillus megaterium*, *J. Biol. Chem.*, 251, 2831, 1976.
16. **Tamburini, P. P. and Schenkman, J. B.,** Purification to homogeneity and enzymological characterization of a functional covalent complex composed of cytochrome P-450 isozyme 2 and *b5* from rabbit liver, *Proc. Natl. Acad. Sci. U.S.A.*, 84, 11, 1987.
17. **Schwab, G. E. and Johnson, E. F.,** Enzymology of rabbit cytochrome P-450, in *Mammalian Cytochromes P-450*, Vol. 1. Guengerich, F. P., Ed., CRC Press, Boca Raton, FL, 1987, 55.
18. **Tamburini, P. P., White, R. E., and Schenkman, J. B.,** Chemical characterisation of protein-protein interactions between cytochrome P-450 and cytochrome b5, *J. Biol. Chem.*, 260, 4007, 1985.
19. **Epstein, P. M., Curti, M., Jansson, I., Huang, C.-K., and Schenkman, J. B.,** Phosphorylation of cytochrome P450: regulation by cytochrome b_5, *Arch. Biochem. Biophys.*, 271, 424, 1989.
20. **Janssen, P. A. J. and Vanden Bossche, H.,** Mode of action of cytochrome P-450 monooxygenase inhibitors. Focus on azole derivatives, *Arch. Pharm. Chem.*, 15, 23, 1987.
21. **Coon, M. J. and Usanov, S. A.,** Cytochrome P-450 structure and function: summary and discussion, in *Cytochrome P-450: Biochemistry and Biophysics*, Schuster, I., Ed., Taylor & Francis, London, 1989, 235.
22. **Dawson, J. H.,** Probing structure-function relations in heme-containing oxygenases and peroxidases, *Science*, 240, 433, 1988.

23. **Champion, P. M.,** Elementary electronic excitations and the mechanism of cytochrome P450, *J. Am. Chem. Soc.*, 111, 3433, 1989.
24. **Ortiz de Montellano, P. R.,** Cytochrome P-450 catalysis: radical intermediates and dehydrogenation reactions, *Trends Pharmacol. Sci.*, 10, 354, 1989.
25. **Ortiz de Montellano, P. R. and Reich, N. O.,** Inhibition of cytochrome P-450 enzymes, in *Cytochrome P-450 Structure, Mechanism, and Biochemistry,* Ortiz de Montellano, P. R., Ed., Plenum Press, New York, 1986, 273.
26. **Testa, B. and Jenner, P.,** Inhibitors of cytochrome P-450s and their mechanism of action, *Drug Metab. Rev.*, 12, 1, 1981.
27. **Schenkman, J. B., Remmer, H., and Estabrook, R. W.,** Spectral studies of drug interaction with hepatic microsomal cytochrome, *Mol. Pharmacol.*, 3, 113, 1967.
28. **Rogerson, T. D., Wilkinson, C. F., and Hetarski, K.,** Steric factors in the inhibitory interaction of imidazoles with microsomal enzymes, *Biochem. Pharmacol.*, 26, 1039, 1977.
29. **Ragsdale, N. N. and Sisler, H. D.,** Mode of action of triarimol in *Ustilago maydis, Pestic. Biochem.*, 3, 20, 1973.
30. **Vanden Bossche, H.,** Biochemical targets for antifungal azole derivatives: hypothesis on the mode of action, in *Current Topics in Medical Mycology,* Vol. 1, McGinnis, M. R., Ed., Springer-Verlag, New York, 1985, 313.
31. **Yoshida, Y.,** Cytochrome P450 of fungi: primary target for azole antifungal agents, in *Current Topics in Medical Mycology,* Vol. 2, McGinnis, M. R., Ed., Springer-Verlag, New York, 1988, 388.
32. **Aoyama, Y., Yoshida, Y., Hata, S., Nishino, T., and Katsuki, H.,** Buthiobate: a potent inhibitor for yeast cytochrome P-450 catalysing 14α-demethylation of lanosterol, *Biochem. Biophys. Res. Commun.*, 115, 642, 1983.
33. **Yoshida, Y. and Aoyama, Y.,** Effects of buthiobates, a fungicide of cytochrome P-450 of rat liver microsomes, *J. Pharmacobiol. Dyn.*, 8, 432, 1985.
34. **Hata, S., Nishino, T., Katsuki, H., Aoyama, Y., and Yoshida, Y.,** Two species of cytochrome P-450 in ergosterol biosynthesis of yeast, *Biochem. Biophys. Res. Commun.*, 116, 162, 1983.
35. **Takano, H., Oguri, Y., and Kato, T.,** Mode of action of (E)-1-(2,4-dichlorophenyl)-4,4-dimethyl-2-(1,2,4-triazol-1-yl)-1-penten-3-ol (S-3308) in *Ustilago maydis, J. Pestic. Sci.*, 8, 575, 1983.
36. **Hata, S., Nishino, T., Komori, M., and Katsuki, H.,** Involvement of cytochrome P-450 in Δ^{22}-desaturation in ergosterol biosynthesis in yeast, *Biochem. Biophys. Res. Commun.*, 103, 272, 1981.
37. **Trzaskos, J., Kawata, S., and Gaylor, J. L.,** Microsomal enzymes of cholesterol biosynthesis, purification of lanosterol 14α-methyl demethylase cytochrome P-450 from hepatic microsomes, *J. Biol. Chem.*, 261, 14651, 1986.
38. **Aoyama, Y., Yoshida, Y., Sonoda, Y., and Sato, Y.,** Metabolism of 32-hydroxy-24,25-dihydrolanosterol by purified cytochrome $P\text{-}450_{14DM}$ from yeast. Evidence for contribution of the cytochrome to the whole process of lanosterol 14α-demethylation, *J. Biol. Chem.*, 262, 1239, 1987.
39. **Trzaskos, J. M., Bowen, W. D., Shafiee, A., Fischer, R. T., and Gaylor, J. L.,** Cytochrome P-450-dependent oxidation of lanosterol in cholesterol biosynthesis. Microsomal electron transport and C-32 demethylation, *J. Biol. Chem.*, 259, 13402, 1984.
40. **Coon, M. J. and Koop, D. R.,** Alcohol-inducible cytochrome P-450 ($P\text{-}450_{ALC}$), *Arch. Toxicol.*, 60, 16, 1987.
41. **Song, B.-J., Gelboin, H. V., Park, S.-S., Yang, C. S., and Gonzalez, F. J.,** Complementary DNA and protein sequences of ethanol-inducible rat and human cytochrome P-450s, *J. Biol. Chem.*, 261, 16689, 1986.
42. **Pompon, D.,** cDNA cloning and functional expression in yeast *Saccharomyces cerevisiae* of β-naphthoflavone-induced rabbit liver P-450LM4 and LM6, *Eur. J. Biochem.*, 177, 285, 1988.
43. **Trzaskos, J. M. and Henry, M. J.,** Comparative effects of the azole-based fungicide flusilazole on yeast and mammalian lanosterol 14α-methyl demethylase, *Antimicrob. Agents Chemother.*, 33, 1228, 1989.
44. **Vanden Bossche, H., Bellens, D., Cools, W., Gorrens, J., Marichal, P., Verhoeven, H., Willemsens, G., De Coster, R., Beerens, D., Haelterman, C., Coene, M.-C., Lauwers, W., and Le Jeune, L.,** Cytochrome P-450: target for itraconazole, *Drug Dev. Res.*, 8, 287, 1986.
45. **Vanden Bossche, H.,** Itraconazole: a selective inhibitor of the cytochrome P-450 dependent ergosterol biosynthesis, in *Recent Trends in the Discovery, Development and Evaluation of Antifungal Agents*, Fromtling, R. A., Eds., J. R. Prous Science Publishers, Barcelona, 1987, 207.
46. **Shaw, J. T. B., Tarbit, M. H., and Troke, P. F.,** Cytochrome P-450 mediated sterol synthesis and metabolism: differences in sensitivity to fluconazole and other azoles, in *Recent Trends in the Discovery, Development and Evaluation of Antifungal Agents,* Fromtling, R. A., Eds., J. R. Prous Science Publishers, Barcelona, 1987, 125.

46a. **Hanger, D. P., Jevons, S., and Shaw, J. T. B.,** Fluconazole and testosterone: in vivo and in vitro studies, *Antimicrob. Agents Chemother.*, 32, 646, 1988.

46b. **Houston, J. B., Humphrey, M. J., Matthew, D. E., and Tarbit, M. H.,** Comparison of two azole antifungal drugs, ketoconazole and fluconazole, as modifiers of rat hepatic monooxygenase activity, *Biochem. Pharmacol.*, 37, 401, 1988.

47. **Lai, M. H. and Kirsch, D. R.,** Nucleotide sequence of cytochrome P450LIA1 (lanosterol 14α-demethylase) from *Candida albicans, Nucleic Acids Res.*, 17, 804, 1989.

48. **Chen, C., Kalb, V. F., Turi, T. G., and Loper, J. C.,** Primary structure of the cytochrome P-450 lanosterol 14α-demethylase gene from *Candida tropicalis DNA*, 7, 617, 1988.

49. **Kalb, V. F., Loper, J. C., Dey, C. R., Woods, C. W., and Sutter, T. R.,** Isolation of a cytochrome P-450 structural gene from *Saccharomyces cerevisiae, Gene*, 45, 237, 1986.

50. **Kalb, V. F., Woods, C. W., Turi, T. G., Dey, C. R., Sutter, T. R., and Loper, L. C.,** Primary structure of the P450 lanosterol demethylase ene from *Saccharomyces cerevisiae, DNA*, 6, 529, 1988.

51. **Ishida, N. Y., Aoyama, Y., Hatanaka, R., Oyama, Y., Imajo, S., Ishiguro, M., Oshima, T., Nakazato, H., Noguchi, T., Maitra, U. S., Mohan, V. P., Sprinson, D. B., and Yoshida, Y.,** A single amino acid substitution converts cytochrome $P450_{14\,DM}$ to an inactive form cytochrome $P450_{SG1}$: complete primary structures deduced from cloned DNAs, *Biochem. Biophys. Res. Commun.*, 155, 317, 1988.

52. **Vanden Bossche, H., Marichal, P., Gorrens, J., Coene, M.-C., Willemsens, G., Bellens, D., Roels, I., Moereels, H., and Janssen, P. A. J.,** Biochemical approaches to selective antifungal activity. Focus on azole antifungals, *Mycoses*, 32(Suppl. 1), 35, 1989.

53. **Vanden Bossche, H., Willemsens, G., Cools, W., Lauwers, W. F. J., and Le Jeune, L.,** Biochemical effects of miconazole on fungi. II. Inhibition of ergosterol biosynthesis in *Candida albicans, Chem. Biol. Interact.* 21, 59, 1978.

54. **Hart, D. T., Lauwers, W. J., Willemsens, G., Vanden Bossche, H., and Opperdoes, F. R.,** Perturbation of sterol biosynthesis by itraconazole and ketoconazole in *Leishmania mexicana mexicana* infected macrophages, *Mol. Biochem. Parasitol.*, 33, 123, 1989.

55. **Willemsens, G., Cools, W., and Vanden Bossche, H.,** Effects of miconazole and ketoconazole on sterol synthesis in a subcellular fraction of yeast and mammalian cells, in *The Host Invader Interplay*, Vanden Bossche, H., Ed., Elsevier/North-Holland, Amsterdam, 1980, 691.

56. **Vanden Bossche, H., Willemsens, G., Marichal, P., Cools, W., and Lauwers, W.,** The molecular basis for the antifungal activities of N-substituted azole derivatives. Focus on R 51 211, in *Mode of Action of Antifungal Agents*, Trinci, A. P. J. and Ryley, J. F., Eds., Cambridge University Press, Cambridge, 1984, 321.

57. **Van Tyle, J. H.,** Ketoconazole mechanism of action, spectrum of activity, pharmacokinetics, drug interactions, adverse reactions and therapeutic use, *Pharmacotherapy*, 4, 343, 1984.

58. **Petersen, E. A., Alling, D. W., and Kirkpatrick, C. H.,** Treatment of chronic mucocutaneous candidiasis with ketoconazole: a controlled clinical study, *Ann. Intern. Med.*, 93, 791, 1980.

59. **De Felice, R., Johnson, D. G., and Galgiani, J. N.,** Gynecomastia with ketoconazole, *Antimicrob. Agents. Chermother.*, 19, 1073, 1981.

60. **Amery, W. K., De Coster, R., and Caers, Y.,** Ketoconazole: from an antimycotic to a drug for prostate cancer, *Drug Dev. Res.*, 8, 299, 1986.

61. **Pont, A., Goldman, E. S., Sugar, A. M., Siiteri, P. K., and Stevens, D. A.,** Ketoconazole-induced increase in estradiol-testosterone ratio, *Arch. Intern. Med.*, 145, 1429, 1985.

62. **Mason, J. I., Murry, B. A., Olcott, M., and Sheets, J. J.,** Imidazole antimycotics: inhibitors of steroid aromatase, *Biochem. Pharmacol.*, 34, 1087, 1985.

63. **Vanden Bossche, H., Willemsens, G., Bellens, D., Roels, I., and Janssen, P. A. J.,** From 14α-demethylase inhibitors in fungal cells to androgen and estrogen biosynthesis inhibitors in mammalian cells, *Biochem. Soc. Trans.*, 18, 10, 1990.

64. **Wouters, W., De Coster, R., Goeminne, N., Beerens, D., and van Dun, J.,** Aromatase inhibition by the antifungal ketoconazole, *J. Steroid Biochem.*, 30, 387, 1988.

65. **Pont, A., Williams, P. L., Azhar, S., and Reitz, R. E.,** Ketoconazole blocks testosterone synthesis, *Arch. Intern. Med.*, 142, 2137, 1982.

66. **De Coster, R., Caers, I., Haelterman, C., and Debroye, M.,** Effect of a single administration of ketoconazole on total and physiologically free plasma testosterone and 17β-oestradiol levels in healthy male volunteers, *Eur. J. Clin. Pharmacol.*, 29, 489, 1985.

67. **Feldman, D.,** Ketoconazole and other imidazole derivatives as inhibitors of steroidogenesis, *Endocr. Rev.*, 7, 409, 1986.

68. **Pont, A., Williams, P. I., Loose, D. S., Feldman, D., Bochra, C., Retiz, R. E., and Stevens, D. A.,** Ketoconazole inhibits adrenal steroid synthesis, *Clin. Res.*, 30, 99A, 1982.

69. **Pont, A., Williams, P. I., Loose, D. S., Feldman, D., Retiz, R. E., Bochra, C., and Stevens, D. A.,** Ketoconazole blocks adrenal steroid synthesis, *Ann. Int. Med.*, 97, 370, 1982.

70. **Schürmeyer, Th. and Nieschiag, E.,** Ketoconazole-induced drop in serum and saliva testosterone, *Lancet*, p. 1098, 1982.

71. **Santen, R. J., Vanden Bossche, H., Symoens, J., Brugmans, J., and De Coster, R.,** Site of action of low dose ketoconazole on androgen biosynthesis in men, *J. Clin. Endocrinol. Metab.*, 57, 732, 1983.
72. **Vanden Bossche, H., Lauwers, W., Willemsens, G., and Cools, W.,** The cytochrome P-450 dependent C17,20-lyase in subcellular fractions of the rat testis: differences in sensitivity to ketoconazole and itraconazole, in *Microsomes and Drug Oxidations*, Boobis, A. R., Caldwell, J., de Matteis, F., and Elcombe, C. R., Eds., Taylor & Francis, London, 1985, 63.
73. **Vanden Bossche, H., Lauwers, W., Willemsens, G., and Cools, W.,** Ketoconazole an inhibitor of the cytochrome P-450 dependent testosterone biosynthesis, in *Therapeutic Principles in Metastatic Prostatic Cancer. Progress in Clinical and Biological Research*, Vol. 185A, Schroeder, F. H. and Richards, B., Eds., Alan R. Liss, New York, 1985, 187.
74. **Lauwers, W., Le Jeune, L., Vanden Bossche, H., and Willemsens, G.,** Identification of 17α,20α-dihydroxyprogesterone in testicular extracts after incubation with ketoconazole, *Biomed. Mass Spectrom.*, 12, 296, 1985.
75. **Rajfer, J., Sikka, S. C., Xie, H. W., and Swerdloff, R. S.,** Effect of *in vitro* ketoconazole on steroid production in rat testis, *Steroids*, 46, 867, 1985.
76. **Kan, P., Hirst, M. A., and Feldman, D.,** Inhibition of steroidogenic cytochrome P-450 enzymes in rat testis by ketoconazole and related imidazole anti-fungal drugs, *J. Steroid Biochem.*, 23, 1023, 1985.
77. **Higashi, Y., Omura, M., Suzuki, K., Inano, H., and Oshima, H.,** Ketoconazole as a possible universal inhibitor of cytochrome P-450 dependent enzymes: its mode of inhibition, *Endocrinol. Jpn.*, 34, 105, 1987.
78. **Chung, B.-C., Picado-Leonard, J., Haniu, M., Bienkowski, M., Hall, P. F., Shively, J. E., and Miller, W. L.,** Cytochrome P450C17 (steroid 17α-hydroxylase/17,20 lyase): cloning of human adrenal and testis cDNAs indicates the same gene is expressed in both tissues, *Proc. Natl. Acad. Sci. U.S.A.*, 84, 407, 1987.
79. **Nakajin, S., Shively, J. E., Yuan, P.-M., and Hall, F.,** Microsomal cytochrome P-450 from neonatal pig testis: two enzymatic activities (17α-hydroxylase and C17,20-lyase) associated with one protein, *Biochemistry*, 20, 4037, 1981.
80. **Kominami, S., Shinzawa, K., and Takemori, S.,** Purification and some properties of cytochrome P-450 specific for steroid 17α-hydroxylation and C17-C20 bond cleavage from guinea pig adrenal microsomes, *Biochem. Biophys. Res. Commun.*, 109, 916, 1982.
81. **Kominami, S., Higuchi, A., and Takemori, S.,** Interaction of steroids with adrenal cytochrome P-450 (P-$450_{17\alpha,\text{lyase}}$) in liposome membranes, *Biochem. Biophys. Acta*, 937, 177, 1988.
82. **Zuber, M. X., John, M. E., Okamura, T., Simpson, E. R., and Waterman, M. R.,** Bovine adrenocortical cytochrome P-$450_{17\alpha}$, regulation of gene expression by ACTH and elucidation of primary sequence, *J. Biol. Chem.*, 261, 2475, 1986.
83. **Zuber, M. X., Simpson, E. R., and Waterman, M. R.,** Expression of bovine 17α-hydroxylase cytochrome P-450 cDNA in non-steroidogenic (COS1) cells, *Science*, 234, 1258, 1986.
84. **Nakajin, S., Shinoda, M., Haniu, M., Shively, J. E., and Hall, P. F.,** C21 steroid side chain cleavage enzyme from porcine adrenal microsomes. Purification and characterization of the 17α-hydroxylase/C17,20-lyase cytochrome P-450, *J. Biol. Chem.*, 259, 3971, 1984.
85. **Namiki, M., Kitamura, M., Buczko, E., and Dufau, M. L.,** Rat testis P-$450_{17\alpha}$ cDNA: the deduced amino acid sequence, expression and secondary structural configuration, *Biochem. Biophys. Res. Commun.*, 157, 705, 1988.
86. **Fevold, H. R., Lorence, M. C., McCarthy, J. L., Trant, J. M., Kagimoto, M., Waterman, M. R., and Mason, J. I.,** Rat $P450_{17\alpha}$ from testis: characterization of a full-length cDNA encoding a unique steroid hydroxylase capable of catalyzing both Δ^4- and Δ^5-steroid-17,20-lyase reactions, *Mol. Endocrinol.*, 3, 968, 1989.
87. **Kagimoto, M., Winter, J. S. D., Kagimoto, K., Simpson, E. R., and Waterman, M. R.,** Structural characterization of normal and mutant steroid 17α-hydroxylase genes: molecular basis of one example of combined 17α-hydroxylase/17,20 lyase deficiency, *Mol. Endocrinol.*, 2, 564, 1988.
88. **Nagai, K., Miyamori, I., Takeda, R., Suhara, K., and Katagiri, M.,** Effect of ketoconazole, etomidate and other inhibitors of steroidogenesis on cytochrome P450sccII-catalyzed reactions, *J. Steroid Biochem.*, 28, 333, 1987.
89. **Zeller, M. and Engelhardt, D.,** Inhibition of $C_{17\text{-}20}$ desmolase, 17α-hydroxylase- and 16α-hydroxylase activity in human testis by ketoconazole, *Acta Endocrinol.*, 267, 99, 1985.
90. **De Coster, R., Caers, I., Coene, M.-C., Amery, W., Beerens, D., and Haelterman, C.,** Effects of high dose ketoconazole therapy on the main plasma testicular and adrenal steroids in previously untreated prostatic cancer patients, *Clin. Endocrinol.*, 24, 657, 1986.
91. **Yanagibashi, K., Haniu, M., Shively, J. E., Shen, W. H., and Hall, P.,** The synthesis of aldosterone by the adrenal cortex. Two zones (fasciculata and glomerulosa) possess one enzyme for 11β-, 18-hydroxylation and aldehyde synthesis, *J. Biol. Chem.*, 261, 3556, 1986.
92. **Miller, W. L.,** Molecular biology of steroid hormone synthesis, *Endocr. Rev.*, 9, 295, 1988.

93. **Vanden Bossche, H., De Coster, R., and Amery, W.**, Pharmacology and clinical uses of ketoconazole, in *Pharmacology and Clinical Uses of Inhibitors of Hormone Secretion and Action*, Furr, B. J. A. and Wakeling, A. E., Eds., Balliere Tindall, London, 1987, 288.
94. **Couch, R. M., Muller, J., Perry, Y. S., and Winter, J. S. D.**, Kinetic analysis of inhibition of human adrenal steroidogenesis by ketoconazole, *J. Clin. Endocrinol. Metab.*, 65, 551, 1987.
95. **Ayub, M. and Levell, M. J.**, Inhibition of human adrenal steroidogenic enzymes *in vitro* by imidazole drugs including ketoconazole, *J. Steroid Biochem.*, 32, 515, 1989.
96. **Weber, M. M., Iuppa, P., and Engelhardt, D.**, Inhibition of human adrenal androgen secretion by ketoconazole, *Klin. Wochenschr.*, 67, 707, 1989.
97. **Trachtenberg, J., Halpern, N., and Pont, A.**, Ketoconazole: a novel and rapid treatment for advanced prostatic cancer, *J. Urol.*, 130, 152, 1983.
98. **Trachtenberg, J.**, Ketoconazole therapy in advanced prostatic cancer, *J. Urol.*, 132, 61, 1984.
99. **Trachtenberg, J. and Pont, A.**, Ketoconazole therapy for advanced prostate cancer, *Lancet*, August 25, 433, 1984.
100. **Heyns, W., Drochmans, A., vander Schueren, E., and Verhoeven, F.**, Endocrine effects of high dose ketoconazole therapy in advanced prostatic cancer, *Acta Endocrinol.*, 110, 276, 1985.
101. **Trachtenberg, J.**, Ketoconazole therapy in advanced prostatic cancer, *J. Urol.*, 137, 959, 1987.
102. **Sonino, N.**, Current trends in the endocrine use of ketoconazole, *J. Endocrinol. Invest.*, 11, 741, 1988.
103. **Trachtenberg, J. and Zadra, J.**, Steroid synthesis inhibition by ketoconazole: sites of action, *Clin. Invest. Med.*, 11, 1, 1988.
104. **Witjes, J. A., Debruyne, F. M. J., Fernandez del Moral, P., Geboers, A. D. H., Baert, L., Casselman, J., Stragier, M., and Debruyne, F. H. J.**, Ketoconazole high dose in the treatment of metastatic prostate carcinoma, *Akt. Urol.*, 20, 29, 1989.
105. **Tapazoglou, E., Subramanian, M. G., Al-Sarraf, M., Kresge, C., and Decker, D. A.**, Hygh-dose ketoconazole therapy in patients with metastatic prostate cancer, *Am. J. Clin. Oncol.*, 9, 369, 1986.
106. **Vanuytsel, L., Ang, K. K., Vantongelen, K., Drochmans, A., Baert, L., and van der Schueren, E.**, Ketoconazole therapy for advanced prostatic cancer: feasibility and treatment results, *J. Urol.*, 137, 905, 1987.
107. **Eichenberger, T., Trachtenberg, J., Toor, P., and Keating, A.**, Ketoconazole: a possible direct cytotoxic effect on prostate carcinoma cells, *J. Urol.*, 141, 190, 1989.
108. **Waterman, M. R., John, M. E., and Simpson, E. R.**, Regulation of synthesis and activity of cytochrome P-450 enzymes in physiological pathways, in *Cytochrome P-450. Structure, Mechanism, and Function*, Ortiz de Montellano, Ed., Plenum Press, New York, 1986, 345.
109. **Tsubaki, M., Hiwatashi, A., Ichikawa, Y., Fujimoto, Y., Ikekawa, N., and Hori, H.**, Electron paramagnetic resonance study of ferrous cytochrome P-450_{scc}-nitric oxide complexes: effects of 20(R), 22(R)-dihydroxycholesterol and reduced adrenodoxin, *Biochemistry*, 27, 4856, 1988.
110. **Ichikawa, Y. and Hiwatashi, A.**, The role of the sugar regions of components of the P-450-linked mixed-function oxidase (monooxygenase) system of bovine adrenocortical mitochondria, *Biochim. Biophys. Acta*, 705, 82, 1982.
111. **Chung, B.-C., Matteson, K. J., Voutilanen, R., Mohandas, T. K., and Miller, W. L.**, Human cholesterol side-chain cleavage enzyme, P450_{scc}: cDNA cloning, assignment of the gene to chromosome 15, expression in the placenta, *Proc. Natl. Acad. Sci. U.S.A.*, 83, 8962, 1986.
112. **Loose, D. S., Kan, P. B., Hirst, M. A., Marcus, R. A., and Feldman, D.**, Ketoconazole blocks adrenal steroidogenesis by inhibiting cytochrome P-450-dependent enzymes, *J. Clin. Invest.*, 71, 1495, 1983.
113. **Willemsens, G. and Vanden Bossche, H.**, Effects of ketoconazole on the testicular and adrenal cholesterol side-chain cleavage, in *Cytochrome P-450-, Biochemistry, Biophysics and Induction*, Vereczkey, L. and Magyar, K., Eds., Akadémiai Kiadó, Budapest, 1985, 203.
114. **Drummond, T. D., Mason, J. I., and McCarthy, J. L.**, Gerbil adrenal 11β- and 19-hydroxylating activities respond similarly to inhibitory or stimulating agents: two activities of a single enzyme, *J. Steroid Biochem.*, 29, 641, 1988.
115. **Suhara, K., Takeda, K., and Katagiri, M.**, P-$450_{11\beta}$-dependent conversion of cortisol to cortisone, and 19-hydroxyandrostenedione to 19-oxo-androstenedione, *Biochem. Biophys. Res. Commun.*, 136, 369, 1986.
116. **Suhara, K., Ohashi, K., Takahashi, K., and Katagiri, M.**, Aromatase and nonaromatizing 10-demethylase activity of adrenal cortex mitochondrial P-$450_{11\beta}$, *Arch. Biochem. Biophys.*, 267, 31, 1988.
117. **Ogishima, T., Mitani, F., and Ishimura, Y.**, Isolation of aldosterone synthase cytochrome P-450 from zona glomerulosa mitochondria of rat adrenal cortex, *J. Biol. Chem.*, 264, 10935, 1989.
118. **Ogishima, T., Mitani, F., and Ishimura, Y.**, Isolation of two distinct cytochromes P-$450_{11\beta}$ with aldosterone synthase activity from bovine adrenocortical mitochondria, *J. Biochem.*, 105, 497, 1989.
119. **Globerman, H., Rösler, A., Theodor, R., New, M. I., and White, P. C.**, An inherited defect in aldosterone biosynthesis caused by a mutation in or near the gene for steroid 11-hydroxylase, *N. Engl. J. Med.*, 319, 1193, 1988.

120. **Vanden Bossche, H., Coene, M.-C., Cools, W., and Willemsens, G.,** Cytochrome P-450 dependent steroid biosynthesis—targets for *N*-substituted imidazoles and triazoles, *J. Steroid Biochem. Suppl.*, 25, 57S, 1986.
121. **Nagai, K., Miyamori, I., Takeda, R., Itagaki, E., and Katagiri, M.,** The effect of ketoconazole and some other inhibitors of steroidogenesis on P-$450_{11\beta}$-catalyzed reactions, *Folia Endocrinol.*, 61, 90, 1985.
122. **Engelhardt, D., Jacob, K., and Doerr, H. G.,** Different therapeutic efficacy of ketoconazole in patients with Cushing's syndrome, *Klin. Wochenshr.*, 67, 241, 1989.
123. **Vanden Bossche, H., Marichal, P., Geerts, H., and Janssen, P. A. J.,** The molecular basis for itraconazole's activity against *Aspergillus fumigatus*, in *Aspergillus and Aspergillosis*, Vanden Bossche, H., Mackenzie, D. W. R., and Cauwenbergh, G., Eds., Plenum Press, New York, 1988, 171.
124. **Marichal, P., Gorrens, J., and Vanden Bossche, H.,** The action of itraconazole and ketoconazole on growth and sterol synthesis in *Aspergillus fumigatus and Aspergillus niger, Sabouraudia, J. Med. Vet. Mycol.*, 23, 13, 1985.
125. **Vanden Bossche, H., Willemsens, G., and Marichal, P.,** Anti-*Candida* drugs—the biochemical basis for their activity, *CRC Crit. Rev. Microbiol.*, 15, 57, 1987.
126. **Marichal, P., Gorrens, J., Van Cutsem, J., Van Gerven, F., and Vanden Bossche, H.,** Effects of ketoconazole and itraconazole on growth and sterol synthesis in *Pityrosporum ovale, J. Med. Vet. Mycol.*, 24, 487, 1986.
127. **Vanden Bossche, H., Marichal, P., Gorrens, J., Geerts, H., and Janssen, P. A. J.,** Mode of action studies. Basis for the search of new antifungal agents, *Ann. N.Y. Acad. Sci.*, 544, 191, 1988.
128. **Vanden Bossche, H.,** Anti-candida drugs—mechanisms of action, in *Candida and Candidamycoses*, Tumbay, E., Seeliger, H. P. R., and Ang, O., Eds., Plenum Press, New York, 1991, 83.
129. **Van Cutsem, J., Van Gerven, F., and Janssen, P. A. J.,** The *in vitro* and *in vivo* antifungal activity of itraconazole, in *Recent Trends in the Discovery, Development and Evaluation of Antifungal Agents*, Fromtling, R. A., Ed., J. R. Prous Publishers, Barcelona, 1987, 177.

129a. **Awouters, F., Niemegeers, C. J. E., and Janssen, P. A. J.,** Interference of antimycotics and other drugs with methohexital hypnosis in rats, *Drug Dev. Res.*, 16, 85, 1989.

129b. **Niemegeers, C. J. E., Levron, J. Cl., Awouters, F., and Janssen, P. A. J.,** Inhibition and induction of microsomal enzymes in the rat. A comparative study of four antimycotics: miconazole, econazole, clotrimazole and ketoconazole, *Arch. Int. Pharmacol. Ther.*, 251, 26, 1981.

130. **De Coster, R., Beerens, D., Haelterman, C., and Doolaege, R.,** Effect of itraconazole on the pituitary-testicular-adrenal axis: an overview of preclinical and clinical studies, in *Recent Trends in the Discovery, Development and Evaluation of Antifungal Agents*, Fromtling, R. A., Ed., J. R. Prous Publishers, Barcelona, 1987, 251.
131. **Vander Pas, H.,** International clinical experience with itraconazole in the treatment of vaginal candidosis, *Int. J. Exp. Clin. Chemother.*, 2(Suppl. 1), 43, 1989.
132. **Kellis, J. T. and Vickery, L. E.,** The active site of aromatase cytochrome P-450, *J. Biol. Chem.*, 262, 8840, 1987.
133. **Hickey, G. J., Chen, S., Besman, M. J., Shively, J. E., Hall, P. F., Gaddy-Kurten, D., and Richards, J. S.,** Hormonal regulation, tissue distribution, and content of aromatase cytochrome P450 messenger ribonucleic acid and enzyme in rat ovarian follicles and corpora lutea: relationship to estradiol biosynthesis, *Endocrinology*, 122, 1426, 1988.
134. **Newton, C. J., Samuel, D. I., and James, V. H. T.,** Aromatase activity and concentrations of cortisol, progesterone and testosterone in breast and abdominal adipose tissue, *J. Steroid Biochem.*, 24, 1033, 1986.
135. **Fujimoto, M., Berkovitz, G. D., Brown, T. R., and Migeon, C. J.,** Time-dependent biphasic response of aromatase to dexamethasone in cultured human skin fibroblasts, *J. Clin. Endocrinol. Metab.*, 63, 468, 1986.
136. **Mendelson, C. R., Merrill, J. C., Steinkampf, M. P., and Simpson, E. R.,** Regulation of the synthesis of aromatase cytochrome P-450 in human adipose stromal and ovarian granulosa cells, *Steroids*, 50, 51, 1987.
137. **Killinger, D. W., Pewrel, E., Danillescu, D., Kharlip, L., and Blackstein, M. E.,** Aromatase activity in the breast and other peripheral tissues and its therapeutic regulation, *Steroids*, 50, 523, 1987.
138. **MacLusky, N. J., Clark, A. S., Naftolin, F., and Goldman-Rakic, P. S.,** Estrogen formation in the mammalian brain: possible role of aromatase in sexual differentation of the hippocampus and neocortex, *Steroids*, 50, 459, 1987.
139. **Hagerman, D. D.,** Human placenta estrogen synthetase (aromatase) purified by affinity chromatography, *J. Biol. Chem.*, 262, 2398, 1987.
140. **Marts, S. A., Padila, G. M., and Petrow, V.,** Aromatase activity in microsomes from rat ventral prostate and Dunning R3327H rat prostatic adenocarcinoma, *J. Steroid Biochem.*, 26, 25, 1987.
141. **Kaburachi, Y., Marino, M. B., Kirdani, R. Y., Greco, J. P., Karr, J. P., and Sandberg, A. A.,** The possibility of aromatization of androgen in human prostate, *J. Steroid Biochem.*, 26, 739, 1987.

142. **James, V. H. T., McNeill, J. M., Lai, L. C., Newton, C. J., Ghilchik, M. W., and Reed, M. J.,** Aromatase activity in normal breast tumor tissues: *in vivo* and *in vitro* studies, *Steroids,* 50, 270, 1987.
143. **Santen, R. J.,** Potential clinical role of new aromatase inhibitors, *Steroids,* 50, 575, 1987.
144. **Tseng, L., Mazella, J., Funt, M. I., Mann, W. J., and Stone, M. L.,** Preliminary studies of aromatase in human neoplastic endometrium, *Obstet. Gynecol.,* 63, 150, 1984.
145. **McPhaul, M. J., Noble, J. F., Simpson, E. R., Mendelson, C. R., and Wilson, J. D.,** The expression of a functional cDNA encoding the chicken cytochrome $P450_{AROM}$ (aromatase) that catalyses the formation of estrogen from androgen, *J. Biol. Chem.,* 263, 16358, 1988.
146. **Corbin, C. J., Graham-Lorence, S., McPhaul, M., Mason, J. I., Mendelson, C. R., and Simpson, E. R.,** Isolation of a full-length cDNA insert encoding human aromatase system cytochrome P-450 and its expression in nonsteroidogenic cells, *Proc. Natl. Acad. Sci. U.S.A.,* 85, 8948, 1988.
147. **Harada, N.,** Cloning of a complete cDNA encoding human aromatase: immunochemical identification and sequence analysis, *Biochem. Biophys. Res. Commun.,* 156, 725, 1988.
148. **Mason, J. I., Carr, B. R., and Murry, B. A.,** Imidazole antimycotics: selective inhibitors of steroid aromatization and progesterone hydroxylation, *Steroids,* 50, 179, 1987.
149. **Ayug, M. and Levell, M. J.,** Structure-activity relationships of the inhibition of human placental aromatase by imidazole drugs including ketoconazole, *J. Steroid Biochem.,* 31, 65, 1988.
150. **Vickery, L. E., Kellis, J. T.,** Aromatase cytochrome P-450. Purification and characterization of the enzyme from human placenta, *Steroids,* 50, 29, 1987.
151. **Latrille, F., Charuel, C., and Lodola, A.,** A comparative study of the effects of ketoconazole and fluconazole on 17-β estradiol production by rat ovaries *in vitro, Res. Commun. Chem. Pathol. Pharmacol.,* 64, 173, 1989.
152. **Taylor, H. M., Jones, C. D., Davenport, J. D., Hirsch, K. S., Kress, T. J., and Weaver, D.,** Aromatase inhibition by 5-substituted pyrimidines and dihydropyrimidines, *J. Med. Chem.,* 30, 1359, 1987.
153. **Hirsch, K. S., Jones, C. D., Lindstrom, T. D., Stamm, N. B., Sutton, G. P., Taylor, H. M., and Weaver, D. E.,** Discovery and development of a novel class of nonsteroidal aromatase inhibitors, *Steroids,* 50, 201, 1987.
154. **Brodie, A. M. H.,** Aromatase inhibition and its pharmacological implications, *Biochem. Pharmacol.,* 34, 3213, 1985.
155. **Wouters, W., De Coster, R., Krekels, M., van Dun, J., Beerens, D., Haelterman, C., Raeymaekers, A., Freyne, E., Van Gelder, J., Vewnet, M., and Janssen, P. A. J.,** R 76713, a new specific non-steroidal aromatase inhibitor, *J. Steroid Biochem.,* 32, 781, 1989.
156. **Ullrich, V. and Graff, H.,** Prostacyclin and thromboxane synthase as P-450 enzymes, *Trends Pharmacol. Sci.,* 5, 352, 1984.
157. **Hecker, M., Baader, W. J., Weber, P., and Ullrich, V.,** Thromboxane synthase catalyses hydroxylations of prostaglandin H2 analogs in the presence of iodosylbenzene, *Eur. J. Biochem.,* 169, 563, 1987.
158. **Hecker, M. and Ullrich, V.,** On the mechanism of prostacyclin and thromboxane A_2 biosynthesis, *J. Biol. Chem.,* 264, 141, 1988.
159. **Ullrich, V., Hauand, M., and Hecker, M.,** The significance of the heme-thiolate structure of prostacyclin and thromboxane synthase, in *Cytochrome P-450: Biochemistry and Biophysics,* Schuster, I., Ed., Taylor & Francis, London, 1989, 41.
160. **Hecker, M., Haurand, M., Ullrich, V., and Terao, S.,** Spectral studies on structure-activity relationships of thromboxane synthase inhibitors, *Eur. J. Biochem.,* 157, 217, 1986.
161. **Tolman, E. L. and Fuller, B. L.,** Inhibition of thromboxane synthesis in guinea pig lung and human platelets by clotrimazole and other imidazole antifungals, *Biochem. Pharmacol.,* 32, 3488, 1983.
162. **Beetens, J. R., Loots, W., Somer, Y., Coene, M.-C., and De Clerck, F.,** Ketoconazole inhibits the biosynthesis of leukotrienes *in vitro* and *in vivo, Biochem. Pharmacol.,* 35, 883, 1986.
163. **De Clerck, F., Beetens, J., de Chaffoy de Courcelles, D., Freyne, E., and Janssen, P. A. J.,** R 68070: thromboxane A_2 synthetase inhibition and thromboxane A_2/prostaglandin endoperoxide receptor blockade combined in one molecule. I. Biochemical profile *in vitro, Thromb. Haemostasis,* 61, 35, 1989.
164. **Deluca, H. F.,** Peripheral metabolism of retinoids, in *Retinoids: New Trends in Research and Therapy,* Saurat, J. H., Eds., Karger, Basel, 1985, 12.
165. **Roberts, A. B. and DeLuca, H. F.,** Pathways of retinol and retinoic acid metabolism in the rat, *Biochem. J.,* 102, 600, 1967.
166. **Roberts, A. B.,** Microsomal oxidation of retinoic acid in hamster liver, intestine, and testis, *Ann. N.Y. Acad. Sci.,* 359, 45, 1981.
167. **Roberts, A. B., Nichols, M. D., Newton, D. L., and Sporn, M. B.,** *In vitro* metabolism of retinoic acid in hamster intestine and liver, *J. Biol. Chem.,* 254, 6296, 1979.
168. **Roberts, A. B., Frolik, C. A., Nichols, M. D., and Sporn, M. B.,** Retinoid-dependent induction of the *in vivo* and *in vitro* metabolism of retinoic acid in tissues of the vitamin A-deficient hamster, *J. Biol. Chem.,* 254, 6303, 1979.
169. **Frolik, C. A.,** *In vitro* and *in vivo* metabolism of all-*trans* and 13-*cis*-retinoic acid in the hamster, *Ann. N.Y. Acad. Sci.,* 359, 37, 1981.

170. **Leo, M. A. and Lieber, C. S.,** New pathway for retinol metabolism in liver microsomes, *J. Biol. Chem.*, 260, 5228, 1985.
171. **Leo, M. A., Iida, S., and Lieber, C. S.,** Retinoic acid metabolism by a system reconstituted with cytochrome P-450, *Arch. Biochem. Biophys.*, 234, 305, 1984.
172. **Van Wauwe, J. P., Coene, M.-C., Goossens, J., Van Nijen, G., Cools, W., and Lauwers, W.,** Ketoconazole inhibits the *in vitro* and *in vivo* metabolism of all-*trans*-retinoic acid, *J. Pharmacol. Exp. Ther.*, 245, 718, 1988.
173. **Vanden Bossche, H., Willemsens, G., and Janssen, P. A. J.,** Cytochrome P-450-dependent metabolism of retinoic acid in rat skin microsomes: inhibition by ketoconazole, *Skin Pharmacol.*, 1, 176, 1988.
174. **Bickers, D. R., Dutta-Choudhury, T., and Mukhtar, H.,** Epidermis: a site of drug metabolism in neonatal rat skin. Studies on cytochrome P-450 content and mixed-function oxidase and epoxide hydrolase activity, *Mol. Pharmacol.*, 21, 239, 1982.
175. **Mukhtar, H., Athar, M., and Bickers, D. R.,** Cytochrome P-450 dependent metabolism of testosterone in rat skin, *Biochem. Biophys. Res. Commun.*, 145, 749, 1987.
176. **Pillai, S., Bikie, D. D., and Elias, P. M.,** Vitamin D and epidermal differentiation: evidence for a role of endogenously produced vitamin D metabolites in keratinocyte differentiation, *Skin Pharmacol.*, 1, 149, 1988.
177. **Merk, H. F., Mukhtar, H., Kaufmann, I., Das, M., and Bickers, D. R.,** Human hair follicle benzo[a]pyrene and benzo[a]pyrene-7,9-diol metabolism: effect of exposure to a coal tar-containing shampoo, *J. Invest. Dermatol.*, 88, 71, 1987.
178. **Khan, W. A., Park, S. S., Gelboin, H. V., Bickers, D. R., and Mukhtar, H.,** Monoclonal antibodies directed characterization of epidermal and hepatic cytochrome P-450 isozymes induced by skin application of therapeutic crude coal tar, *J. Invest. Dermatol.*, 93, 40, 1989.

Chapter 10

TARGET RESEARCH IN THE DISCOVERY AND DEVELOPMENT OF ANTIFUNGAL INHIBITORS

Wolfram Köller

TABLE OF CONTENTS

I. INTRODUCTION

The first class of organic fungicides with captan or the ethylene(bisdithio)-carbamates as typical representatives is classified as "multisite inhibitors". This term described their nonspecific desactivation of several and thus rarely specified enzymes, most often via the chemical modification of essential SH groups.[1,2] The advantage of this particular mode of action is the strong inhibition of spore germination mediated by the inhibition of enzymes involved in the mobilization of storage products and respiration. The disadvantage of the multisite mode of action are toxic side effects if antifungal compounds would enter the tissue of the respective host. Systemic movement within plant tissue and, thus, a curative activity is not feasible. The treatment of deep-seated and systemic mycotic diseases with multisite inhibitors would lead to similar toxic side effects.

Several classes of antifungal fungicides and drugs with specific modes of action have been developed over the last two decades. The benzimidazoles (Chapter 3), the phenylamides (Chapter 5), and the large group of sterol biosynthesis inhibitors (Chapter 7) have reached a high level of acceptance due to their broad spectrum of important diseases. In particular, the latter group has become an example for the feasibility to develop a very similar class of inhibitors for both the control of plant diseases and for the treatment of human mycoses. The first representatives of most of our current antifungal groups with site-specific properties were introduced between 1965 and 1975. Since then, many derivatives with identical modes of action have been developed, but progress with regard to inhibitors with novel target sites has been slow.

For the farmer the lack of "new chemistries" has led to a sometimes alarming situation with respect to the development of fungicide resistance.[3] Regardless of the availability of excellent fungicides for most of our plant diseases, the development of resistance has led to a revival of traditional contact fungicides, used either alone as alternatives or in mixtures with the systemic compounds.[3] However, some of the older and nonspecific fungicides are, in many countries under toxicological scrutiny,[4] and the development of new fungicides with different modes of action will become important in the future, either as alternatives to fungicides with resistance problems, or as components of antiresistance strategies. The demand for novel site-specific antifungal inhibitors has, at the same time, dramatically increased in the field of medicine.[5] Immunocompromised patients (AIDS, cancer therapy, organ transplants) are prone to deep-seated and systemic mycoses, which are sometimes hard to treat with the currently available arsenal of antimycotic drugs.[6-12] Future antimycotic drugs must be fungal specific to warrant a long-term treatment of internal mycoses without side effects on the patient. Scientists engaged in the discovery of new antifungal agents might be advised to cross disciplines more frequently in the future.

Most of our modern fungicides and antifungal drugs have been found via random synthesis, biological screening, and empirical optimization of lead structures. Some have been derived from natural product screening, an approach not fundamentally different from random synthesis. The modes of action of particular antifungal groups were only identified in the late stages of development. Most often, this initial biochemical work was done by scientists at academic institutions and, thus, outside corporate research. Undoubtedly, the "screening approach" has been remarkably successful in the past, and the role and importance of molecular tools in the discovery of new antifungal inhibitors is not immediately apparent. It is certainly justified that the "traditional" way of fungicide discovery and the eminent importance of appropriate and reliable screening systems and methods have been emphasized in recent reviews.[13,14] Regardless of the undisputed merits of random synthesis and screening, however, continuous biochemical studies have yielded valuable contributions to fungicide discovery programs, in particular in the group of sterol demethylation inhibitors (Chapter 7) with a mode of action identified in 1974 and under continuous development for more than two decades.[15-18]

Although a large number of antifungal agents have been introduced in the 1980s, most of these new compounds were derivatives of the groups discovered and biochemically characterized in the 1970s, and progress with regard to new major antifungal groups with novel modes of action has slowed down. This lack of success is surprising, because efforts in fungicide discovery programs did not diminish. Considering an average number of 5000 to 10,000 compounds screened for fungicidal activity per year at a major company, and 10 to 15 companies engaged in the development of fungicides and antifungal agents, approximately 1 million compounds have been synthesized during the last decade. Triggered by the slower pace of success, the value of molecular information received increasing recognition, and the integration of biochemistry into fungicide discovery programs was frequently proposed as a more rational tool in the search for novel inhibitor lead structures.[19-29] The advent of fungal molecular biology has added, without any doubt, a new dimension to these molecular programs.

Unfortunately, the term "rational" approach is frequently used to distinguish between a biochemically guided discovery program and the traditional screening approach. This appears to imply that the screening approach lacked rationale, which is not the case. The organic chemist engaged in the synthesis of new inhibitor structures applies and contributes to the progress made in the field of (economical) chemical synthesis, and the screening biologist is continuously challenged by the improvement and adjustment of screening systems. Furthermore, the application of QSAR studies has a long history not only in drug but also in fungicide research.[30,31] In broad terms, QSAR analysis is the correlation of biological activity with those physicochemical parameters of the inhibitors that are easy to determine. Although QSAR studies can be applied to the analysis of particular enzymes,[32] knowledge of the mode of action of an antifungal compound is not essential for these studies. Rather, QSAR studies reflect the rationale of the "random" discovery approach.

Unlike the clear goals of biological screening of compounds derived from random synthesis, the role of biochemistry remains less focused and has been defined in various ways. In principle, two distinctly different strategies are feasible. One approach would call for the identification of the mode of action soon after biological activities of a randomly synthesized inhibitor have been identified. Within the framework of this approach, molecular studies would be initiated as soon as a "new chemistry" has emerged from random synthesis and screening. The clear goal would be the contribution of biochemical information to inhibitor optimization. A second strategy would call for the rational design of antifungal inhibitors. This strategy would involve, as a prerequisite, the definition and characterization of a suitable target first, followed by the design of respective inhibitors. Goals such as high inhibitor specificity could guide the initial search for appropriate target sites prior to the initiation of synthesis programs.

The two general strategies are not exclusive. Important for future success is not a theoretical search for the "right" or "wrong" way of conducting research. More important will be a high degree of flexibility within an interdisciplinary environment. Jolles[33] stated in his opening remark to a round-table discussion on drug design: "Everything has its share of reality and dream. In the best champagne, we must distinguish the foam, which we see, from the wine, which we drink, and what is expected from a good drug research scientist is that he acts intelligent not that he looks intelligent." The aim of this chapter is to highlight some "act intelligent" opportunities of molecular research in fungicide discovery.

II. INHIBITOR LEAD STRUCTURES DERIVED FROM RANDOM SCREENING

The overall antifungal activity evaluated in biological screening tests is the most integral and almost perfect biological parameter of a newly synthesized compound. The antifungal activity describes not only the intrinsic inhibitory potency, but also additional and essential parameters such as chemical stability, penetration into the fungal cell, and the (systemic) distribution within the host. Considering this wealth of information gained from the overall parameter "biological performance", it is not immediately apparent why the mode of action and the availability of *in vitro* enzyme tests should contribute to success. This concern is justified, and the role of biochemistry within the framework of random synthesis can neither be a substitution of biological screening, nor the mere identification of a fungicidal mode of action. It is the application and integration of biochemical information as tool for optimization of lead structures, once fungicidal activity of "new chemistries" has been discovered.

A. BIOCHEMISTRY AND ONGOING PROGRAMS

Molecular studies integrated into the random screening of synthetic chemicals or natural products are, by definition, not directly involved in the discovery of novel inhibitor. It would be of restricted value to establish *in vitro* tests for several selected enzymes, and then to search for particular enzyme inhibitors among compounds derived from random synthesis. The response of a whole fungal organism to a putative inhibitor constitutes an elegant, simple, and simultaneous "assay" of hundreds of potential targets, and the statistical chances to find an inhibitor of one specified enzyme by chance are magnitudes lower than those experienced with *in vivo* tests. The goal of biochemical work at this early stage should be the identification of respective target site(s) in order to provide biochemical information for the optimization of lead structures. The advantage of this approach is that the antifungal potency of a lead structure has already been validated. The disadvantage of the approach is that the time required for the precise identification of an inhibitor target site is unpredictable. It has to be emphasized that biochemical information employed in optimization efforts must be precise. Any synthetic work guided by the catalytic mechanism of a particular enzyme identified as the target site would become counteractive if the proposed mode of action is incorrect.

In order to avoid misperceptions, the inhibitory site of "new chemistries" should be narrowed down in a systematic step-by-step procedure. A scheme for the early steps in mode of action evaluation is shown in Figure 1; appropriate experimental procedures have been described in numerous publications, e.g., References 34 to 37. The first step is the careful choice of the fungal test organism according to the antifungal spectrum of the compound. Mode of action studies done with relatively insensitive fungi might lead to results reflecting secondary rather than primary target sites. Once a suitable organism has been chosen, it will be required to determine a suitable time period of inhibitor pretreatment prior to biochemical tests. In order to correlate the inhibition of certain metabolic and anabolic pathways with *in vivo* activities on growth parameters, the inhibitor should be fully penetrated into the cell, but inhibitor action should not be allowed to proceed for long periods. Prolonged inhibitor treatment might lead to metabolic effects reflecting the consequence rather than the cause of inhibitor action.

The first test should be the inhibitor action on respiration. Any target site in catabolic routes, but also the inhibition of nutrient uptake will be reflected in a decreased oxygen consumption of whole cells. Pronounced inhibition of respiration implies that ATP synthesis is decreased or abolished, and all catabolic sequences would be inhibited as a consequence of inhibitor action. Inhibitors that test positive should be tested with host cell tissue, for agricultural fungicides most appropriately with plant cell cultures. This would provide first indications for the suitability of catabolic inhibitors to be optimized toward systemic movement within the host. The lack of effects on respiration are a first indication for a specific inhibitor site in anabolic pathways. The identity of this site can then be narrowed down by incorporation of early radioactive precursors of different classes of end products (Figure 1). The incorporation of radioactivity into end products can be easily measured; it requires an

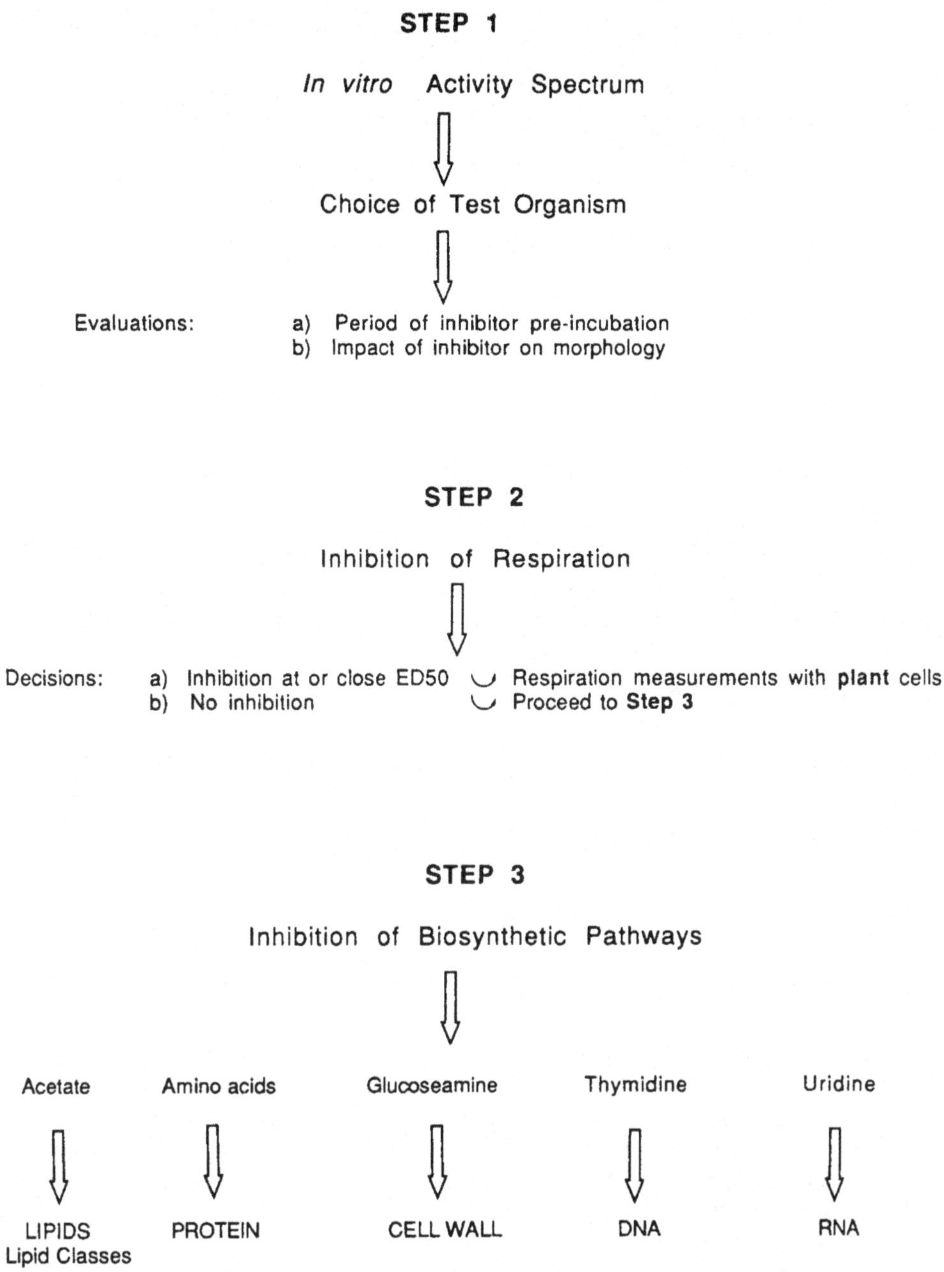

FIGURE 1. Investigative steps of initial mode of action studies.

additional analytical step for acetate incorporated into lipids. Here, a particular precursor can accumulate to high levels without effect on the incorporation of acetate into the total lipid fraction.

The initial steps of evaluation described above are relatively easy to perform, not overly time-consuming, and well suited for a routine setting. It has to be pointed out that the set of data derived from this preliminary evaluation only provides guidance for subsequent studies. However, the informative value should not be underestimated. The data will be too preliminary for the guidance of structural optimization efforts, but they are a valuable contribution to research decisions. For example, extensive efforts to modify a promising

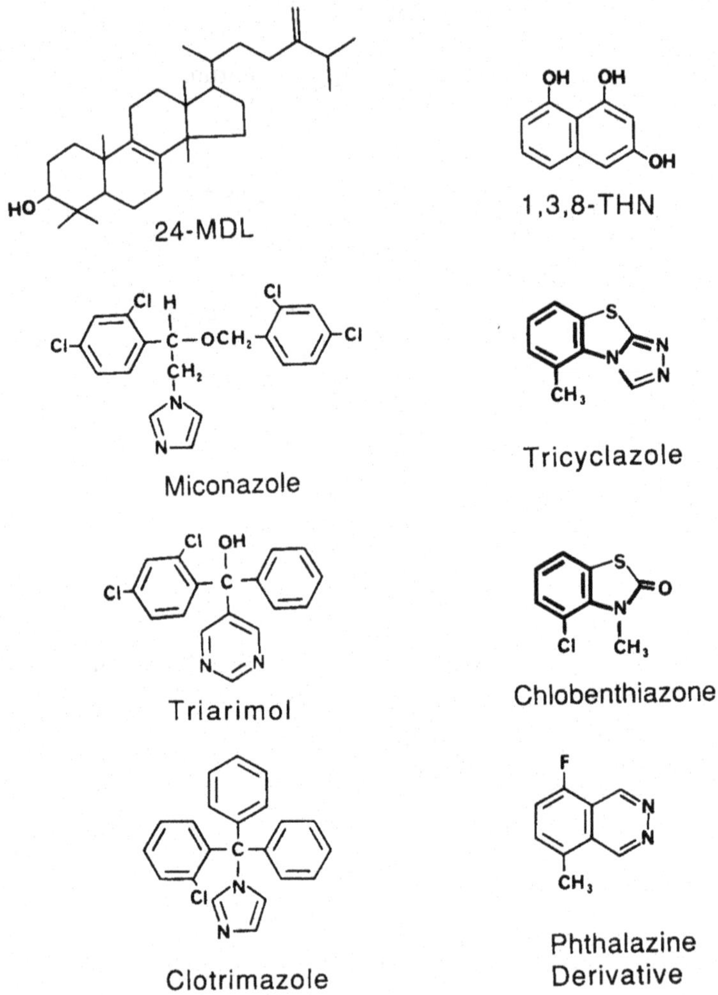

FIGURE 2. Comparison of enzyme substrates with inhibitor structures.

antifungal group toward systemic movement in plants would have little chances for success if the first representatives of the group were identified as equally potent inhibitors of respiration in both the fungal and the plant system. In this case, efforts should concentrate on the question whether the inhibitors represent a multisite or site-specific mode of action (the carboxamides described in Chapter 1 are an example for site-specific respiration inhibitors), and whether selectivity could be improved toward fungal respiration once a specific mode of action has been identified. To the contrary, a clear inhibitory action in a biosynthetic pathway not yet reported for one of the established antifungal agents is good indication for a novel and site-specific mode of action and might make a new class of compounds considerably more attractive.

It might be argued that a new chemistry combined with antifungal and probably even systemic activity in screening systems is sufficient indication for a novel mode of action. Many examples from the past have shown that this is not the case. For instance, the common mode of action of the first representatives of sterol demethylation inhibitors (DMIs; Chapter 7), triarimol, clotrimazole, and miconazole, is not immediately apparent from their structures (Figure 2). In addition, even the trained eye of an experienced biochemist would not have recognized the potential of any of the three inhibitors to compete with 24-methylenedihydrolanosterol for binding to the respective enzyme, the 14α-sterol demethylase (Chapter 7). The common mode of action of the three structurally different DMIs, however, would have been recognized in an initial mode of action evaluation according to the scheme outlined above. This would have opened the opportunity to broaden the chemistry of DMIs much earlier than it eventually happened (Chapter 7).

Even if the data derived from a systematic and early biochemical evaluation comprise a contribution to research and development decisions, the question remains whether an in-depth study aimed at the precise characterization of the respective target site would add more information and opportunities to the development of a particular group. In contrast to the initial evaluation, these studies may consume considerable time. Furthermore, the dicarboximides (Chapter 4) and benzimidazoles (Chapter 3) are good examples for inhibitors, where the schematic initial evaluation would have provided very little information with regard to the mode of action. Even if these concerns are justified at times, the identification of the target site has its clear intrinsic values and might open opportunities for efforts to reinitiate research on a group of already developed inhibitors. The melanin biosynthesis inhibitors described by Yamaguchi and Kubo in Chapter 6 are an excellent example to demonstrate this opportunity. The chemical structures of melanin biosynthesis inhibitors (MBIs) are diverse and are not indicative for a common mode of action. Once the mode of action had been clarified for tricyclazole (Figure 2), it became apparent that compounds such as pyroquilon and others were identical with regard to their modes of action. Employing the antimelanization activity as criterion, two approaches for the synthesis of new MBI structures have been described. In their search for structural features common to all MBIs, Inoue et al.[38] identified a benzo-bicyclic ring system containing a substituted nitrogen in one ring in α-position to the benzene ring, and two additional substitutions at specific sites (Figure 2) was common structural features; new MBIs previously not reported were synthesized. Although this structure-activity study was guided by the known mode of action of structurally different compounds, the approach to utilize this information was basically chemical in nature. A more biochemically oriented approach was reported by Omata et al.[39] According to the published mode of MBI action (Chapter 6), the authors assumed a competitive inhibition of the 1,3,8-trihydroxynaphthalene (Figure 2) and compared the known MBIs with the structure of the respective enzyme substrate. Based on this structural information, a series of phthalazine derivatives was synthesized, and several representatives of these designed inhibitors were indeed active as MBIs.

Lead structures of novel antifungal agents are not only derived from organic synthesis programs, but also from natural products. The screening of microbial products or plant extracts is, most often, random and not principally different from synthetic chemicals. However, the selection of microorganisms or plants can be based on existing information. For example, an antifungal component present in garlic extracts, a component of "folklore" medicine, has been found to inhibit lipid biosynthesis in *Candida albicans*.[40] Although the structures of numerous antifungal inhibitors produced by microorganisms or extracted from plants have been described in the past,[41-43] respective modes of action are rarely known. This lack of information, however, does not exclude that interesting lead structures can be derived from natural sources. An example of biochemical considerations in the characterization of a plant product, guided by a comparison to known fungicides with similar biological activities, is a study on the rice blast activity of the diterpene pisiferic acid (Figure 3), a preformed antifungal plant product isolated from Cupressaceae.[44] Computer graphic analysis of pisiferic analysis revealed similarities with the rice blast fungicide probendazole, a saccharin derivative (Figure 3). Indeed, the two compounds shared similarities when the effects in the host-pathogen interaction were compared, indicating a similar mode of action. Although not all details have been clarified yet, the mode of action of probendazole, as discussed in greater detail below, has been shown to be indirect via the activation of defense reactions within the host plant rice.[45-47]

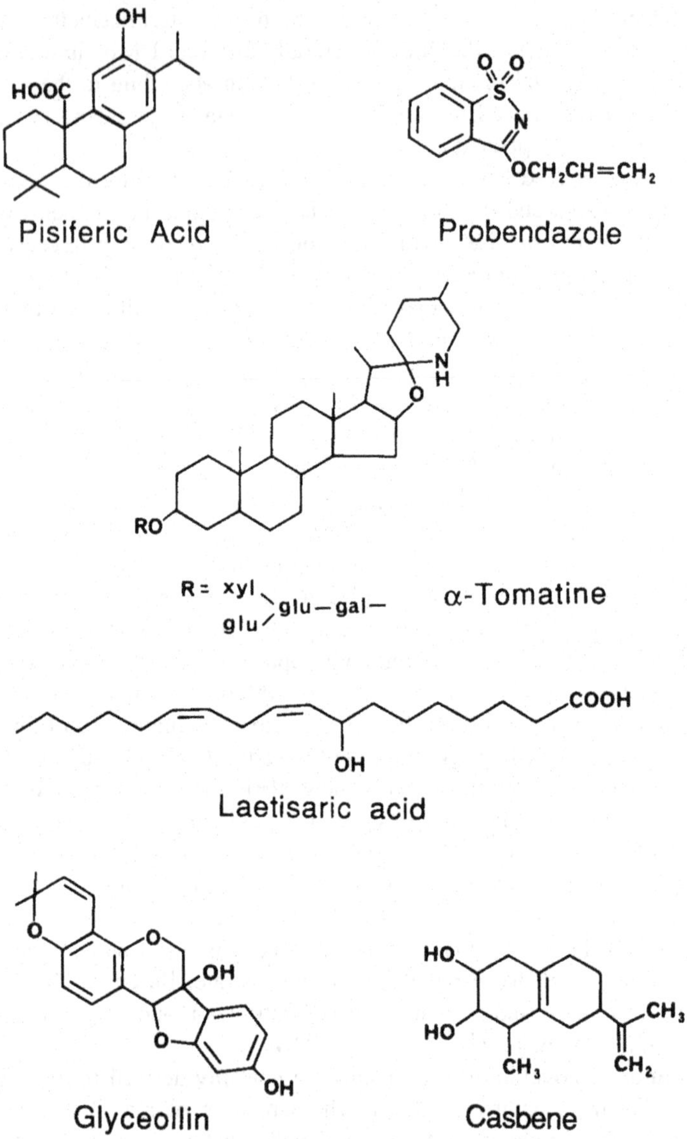

FIGURE 3. Natural products with antifungal activities.

A second class of preformed antifungal plant metabolites are two groups of plant steroids, the saponines and glycoalkaloids of the *Solanum* type.[48,49] These compounds are characterized by a steroidal aglycone attached to a carbohydrate moiety; the structure of α-tomatine as a typical representative of this class of compound with a relatively broad spectrum of antifungal activities is shown in Figure 3. Although the question exists of whether the glycosides or the aglycons formed by glycosidase action are the more active compounds, little doubt exists that the saponines act by forming 1:1 complexes with sterols in fungal membranes.[48] The mechanism of membrane disturbance is very similar to the action of the polyene macrolide antibiotic described in Chapter 7. The expression of fungal resistance to saponins has been related to a decreased ergosterol content in fungal membranes, again very similar to the macrolides (Chapter 7).[50,51] The more indirect and not highly selective action on biological membranes implies that saponines have a restricted value in providing lead structures for novel modes of antifungal action. A nonspecific mode of action might also apply to laetisaric acid (Fig.3), a midchain hydroxylated linoleic acid produced as a allelopathic agent by the soil fungus *Laetisaria* arvalis a potential biocontrol agent.[52] Laetisaric acid caused rapid lysis of *Phytium* membranes.[52] Similar fatty acid derivatives isolated from plants have been reported to act as antifungal agents.[43]

A second class of antifungal plant metabolites are the phytoalexins, which are synthesized *de novo* as a defensive response to fungal attack, but also under abiotic stress conditions.[53,54] Since the development of the phytoalexin concept in 1940 and the first structural identification of the first phytoalexin pisatin in 1960, research in this area has been a major subject of molecular plant pathology. The most extensively studied phytoalexins belong to the class of isoflavanoid-derived compounds such as glyceollin, and to terpenoids such as rishitin.[53,54] Considering the wealth of information, it is surprising that studies on the mode of action are relatively rare. VanEtten and Bateman[55] first demonstrated that phaseollin, an isoflavonoid-derived phytoalexin, drastically affected the functions and integrity of plasma membranes of *Rhizoctonia solani*. A nonspecific membrane effect was indicated by damages to plant cells as well as erythrocytes. These results were confirmed and substantiated by Yoshikawa et al., working with *Phytophthora capsii*.[56] The fact that artificial liposomes became permeable after treatment with phaseollin, and that the presence of sterols had no effect on the membrane-disturbing action indicated that the phytoalexin acts as a general membrane antagonist. Fluorescent depolarization studies showed that glyceollin increased membrane fluidity. Moreover, ultrastructural studies supported the view that fungal plasma membranes are the most sensitive site of action.[56] A nonspecific membrane action of glyceollin is supported by the inhibition of electron transport in isolated soybean mitochondria,[57] and the plasma membrane ATPase isolated from red beet.[58] A similar membrane-disturbing effect has been reported for other isoflavonoid-derived phytoalexins, but also for the structurally different terpenoid phytoalexins such as rishitin (Figure 3).[56,59] Overall, the phytoalexins appear to act, similar to most of the investigated preformed antifungal plant metabolites, as multisite toxicants with biocidal effects on membranes rather than as site-specific compounds with distinct target sites.

Considering the widespread occurrence of secondary plant metabolites with antifungal activities, be it preformed or induced under stress conditions, it is astounding that no site-specific inhibitor has been identified thus far. Consequently, the chemical structures of known antifungal plant products are not ideal chemical leads for the synthesis of specific inhibitors. As discussed by McLaren,[60] phytoalexins as such have little chances to be considered as chemical control agents. The vast majority is less active than chemical fungicides. ED_{50} values are typically in the range of 20 to 50 μg/ml, and chemical derivatives are normally not dramatically more active.[61,62] The current lack of specific antifungal inhibitors of plant origin, however, does not necessarily indicate that plant metabolites of this quality do not exist, and screening programs combined with mode of action studies might be successful. The question of whether these programs would have advantages over conventional synthetic programs is not easy to assess. It should be mentioned, however, that both screening strategies are mainly based on chance, and that the toxicological risk of man-made vs. natural chemicals appears to be more or less equal.[63-65]

The future potentials of most antifungal natural products such as phytoalexins might be in the genetic engineering of disease-resistant plants rather than in their usefulness as leads for novel synthetic fungicides. Although phytoalexins have been the subject of extensive studies over the last three decades, the question of to what degree these antifungal metabolites are involved in the expression of resistance is still under debate.[53,54] Perhaps the most convincing evidence for a decisive role of phytoalexins has emerged from work with *Fusarium solanum* f.sp. *pisi* and its host pea, a plant that responds to fungal attack and other stress by the synthesis of pisatin. Van Etten and co-workers have shown that pathogenicity of *F. solani* was dependent on the detoxification of pisatin by an inducible cytochrome P-

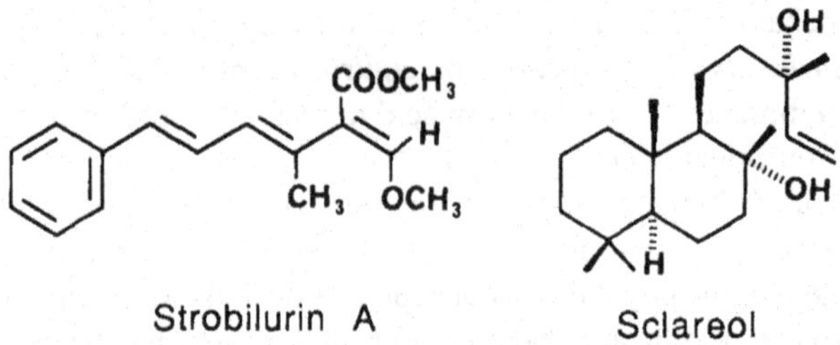

FIGURE 4. Differences in chemical complexity of natural inhibitors. There has been confusion over the correct geometry of strobilurin A. The (*E,E,E*)-configuration as shown above has been corrected to (*E,Z,E*).

450-monooxygenase (pisatin demethylase).[66] Furthermore, the transformation of the corn pathogen *Cochliobolus heterostrophus* with the gene for pisatin demethylation rendered pathogenicity to this pathogen on the nonhost pea.[67] As critically discussed by VanEtten,[66] the alteration of phytoalexin synthesis in a given plant would implicate that a given pathogen would not be able to tolerate this "foreign" antifungal metabolite. The success of this and similar strategies would rely on several premises. For example, the new phytoalexin would have to be under the control of an inducible promoter (constitutively expressed phytoalexins would by phytotoxic), and even the tissue-specific expression might be required (potential mammalian toxicity of phytoalexin-containing plant tissues). However, rapid progress triggered by the novel techniques of molecular genetics is being made in this field, and it can be expected that the merits of these strategies will be tested in the nearer future.

A disadvantage inherent to may natural products is their complex chemical structures. For example, diterpenes such as sclareol (Figure 4) isolated from *Nicotiana* spp. have been shown to control rusts and powdery mildews.[68,69] The stereochemcial complexity of these secondary plant products would prohibit the chemical variation of the basic structures. Even the chemical modification of these compounds, a strategy frequently applied to natural products produced by microorganisms, would not be applicable. The yields obtained from plant extracts are too low to be of economical interest. The only approach could be the identification of the mode of action, followed by attempts to simplify the inhibitors toward chemically accessible structures. In some cases, however, the structures of antifungal natural products are simple enough to be utilized in synthesis programs. Methoxyacrylates such as strobilurin A (Figure 4) are an example of recent interest.[70] These structures, which have been identified as respiration inhibitors with a complex III target site in beef heart mitochondria,[71] are of simple nature and have served as chemical leads in synthetic programs.[70a] A relatively simple structure has also been identified for harzianopyridone, an antifungal metabolite produced by *Trichoderma harzianum*.[72] Antifungal activities of a small number of chemical derivatives were not largely improved.[72]

B. LEAD STRUCTURES FROM KNOWN INHIBITORS

The examples described above illustrate the opportunities of early and systematically initiated mode of action studies. It has to be emphasized, however, that the inherent value of information on the mode of action is not entirely dependent on timely progress. In particular with the earlier representatives of modern antifungal inhibitors, the modes of actions were clarified long after the respective fungicides and antifungal agents had been introduced. Consequently, active research on many of the chemical classes had been discontinued already when the biochemical information became available. This lag between random synthesis, inhibitors optimization, and development on one side, and the identification of the respective modes of action on the other offers ample opportunities for renewed considerations. In contrast to programs dedicated to the search for completely new inhibitors, research decisions

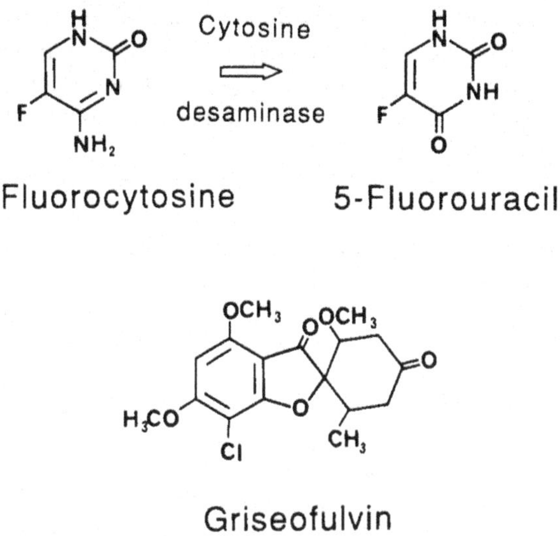

FIGURE 5. Antifungal inhibitors of nucleic acid metabolism and nuclear function.

can be guided by considerations derived from the former experience with particular classes of established fungicides. A major concern, for example, might be the experienced risk of resistance, which proved to be high for some of the inhibitors described below.

The modes of action of most of the major classes of antifungal compounds have been treated in the preceding chapters of this book and thus will only be discussed with regard to interesting chances for improvement based on biochemical information. However, several of the known antifungal inhibitors did not justify a separate chapter and will be described in greater detail. Several comprehensive reviews on the biochemistry of current groups of fungicides[73,75] and antifungal agents,[76] and a book on pesticide modes of action[77] are additional sources of comprehensive information.

1. Inhibition of Nucleic Acid Metabolism and Nuclear Division

One of the classes of fungicides that interferes with nucleic acid metabolism are the **hydroxypyrimidines** described in Chapter 2. The mode of action is the specific inhibition of adenosine deaminase (ADAases), one of the enzymes active in the purine salvage pathway. Striking is the strict confinement of antifungal action to powdery mildews, a selectivity that has been explained by the unique sensitivity of the powdery mildew enzyme. ADAases from other fungal sources were not inhibited. The same applies to the mammalian enzyme, whereas in plants, ADAase appears to be absent. The widespread occurrence of ADAase in a broader spectrum of fungal organisms implies that the antifungal spectrum could be extended to pathogens other than powdery mildews. This would require the molecular understanding of the precise mechanisms responsible for the selective inhibition of the powder mildew enzyme. Furthermore, the vital importance of the respective enzyme in other fungal organisms and, thus, the broader suitability of ADAse to serve as a more general antifungal target would have to be established, before design efforts aimed at the broadening of ADAase inhibitor activities could be initiated.

The pyrimidine base analogue **5-fluorocytosine** (5-FC), originally designed to act as an anticancer drug, is active against *Candida albicans* and is in use as an orally applied antifungal agent (Figure 5).[76,78] The mode of action clarified during the 1970s has been summarized in recent reviews.[76,78,69] 5-FC is actively transported into the fungal cell, but it is as such metabolically inactive and not incorporated into nucleic acids. The first key step toward antifungal activity is the deamination to 5-fluorouracil (5-FU), an enzymatic step confined to sensitive fungi. This unique activation step is thought to explain the low mammalian toxicity of the drug. A second key enzyme necessary for activation is the conversion of 5-fluorouracil to 5-fluorouridine monophosphate (5-FUMP) catalyzed by uridine monophosphate pyrophosphorylase. 5-FUMP is than converted to 5-FUTP and incorporated into RNA, leading to substantial impairment of RNA functions. Probably more important with regard to the antifungal action is the conversion of 5-UMP to 5-fluorodesoxyuridine monophosphate (5-dUMP). The position normally methylated in the conversion of dUMP to desoxythymidine monophosphate as the precursor for DNA synthesis is blocked by the fluoro substituent, and thymidylate synthase is inhibited by 5-FdUMP. Consequently, DNA synthesis is blocked. Since the antifungal activity of 5-FC is dependent on a series of events, the active transport into the cell, the conversion to 5-FU (5-FU is not taken up by the cell), and finally the conversion of 5-FU to 5-FUMP and 5-FdUMP, a potential for inhibitor improvement is not immediately apparent. The example shows, however, that alternative and unique pathways in fungal anabolism, in this case the desamination of cytosine, can be utilized in the design of specific antifungal agents.

The group of highly Oomycete-specific **phenylamides** has been identified as specific inhibitors of RNA polymerase I active in the synthesis of ribosomal RNA (Chapter 5). The dependency of *in vitro* inhibitory activity on isolated nuclei and the lack of inhibition observed with partially purified RNA polymerase I suggest that the inhibitors are not active site-directed. They most likely interact with a yet unidentified protein involved in the transcription of the respective genes. Ribosomal RNA synthesis is clearly an ubiquitous process in all eukaryotes, and structural differences of the Oomycete target rather than the lack of the target site will most likely be identified as an explanation for the high selectivity and specificity of the phenylamides. The identification and characterization of the inhibitor binding site, and the comparison of the Oomycete target site with the respective target operative in different fungal genera but also in mammals and plants appear to comprise an attractive strategy to evaluate the potentials for the design of similar inhibitors targeted against a broader spectrum of fungal pathogens.

The mode of action **benzimidazoles** is described in Chapter 3. As inhibitors of tubulin polymerization these fungicides are not directly involved in nucleotide metabolism, but they inhibit spindle formation and, thus, nuclear division. The fact that the binding of benzimidazoles to tubulin is specific to sensitive fungi opens the opportunity to broaden the antifungal spectrum of this class of inhibitors to benzimidazole-insensitive fungi such as the group of oomycetes. This would require the three-dimensional analysis of benzimidazole-sensitive tubulin binding sites, and the comparison of these sites with the structure of respective sites in insensitive tubulin subunits. A second specific inhibitor of fungal mitosis is **griseofulvin** (Figure 5), an antibiotic isolated from *Penicillium* spp. The mode of action, however, is not related to microtubule polymerization and thus different from the benzimidazole action.[73-75] Although first described in 1939, the precise mode of griseovulvin action has not been identified. Antifungal activities of several semisynthetic derivatives were not improved over the parent compound.[80] Considering the low toxicity to nontarget organisms in combination with a relatively simple structure, intensified mode of action work might well lead to a valuable new target site of antifungal action. Rhizoxin, an inhibitor of microtubule formation with a binding site different from benzimidazoles and a broader spectrum of biological activities, has been described as a third member of tubulin-directed inhibitors (Chapter 3).

2. Inhibition of Lipid Synthesis and Metabolism

As described in Chapter 7, the most important but also most diversified class of antifungal inhibitors are targeted against enzymes active in fungal sterol biosynthesis, making this biosynthetic pathway the perhaps best investigated target for antifungal drugs and fungicides.

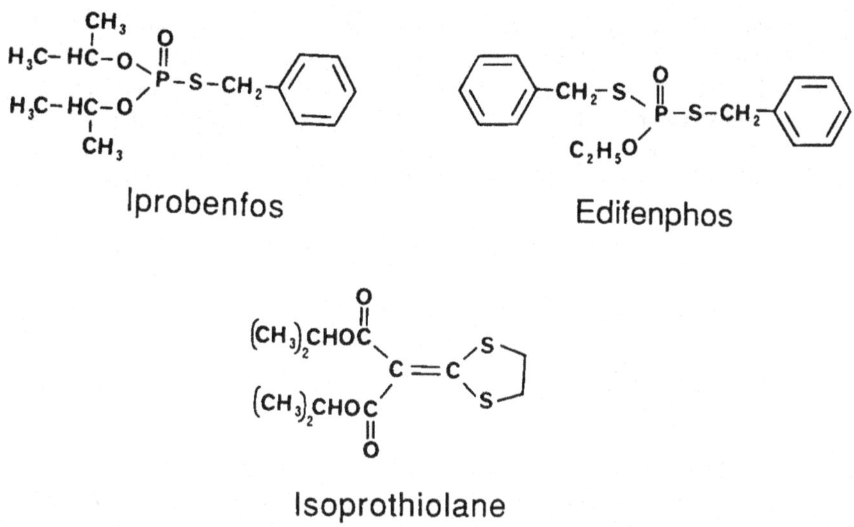

FIGURE 6. Antifungal inhibitors of lipid metabolism.

Antifungal inhibitors of five enzymes involved in the formation of lanosterol, the first sterol structure in sterol biosynthesis, and five additional groups of inhibitors active toward enzymes of sterol modification steps have been described, and opportunities for inhibitor improvements are discussed in detail in Chapter 7. Among the various sterol biosynthesis inhibitors, three groups have been developed to commercial products.

The **allylamines** as antifungal drugs have been identified as specific inhibitors of fungal squalene epoxidases. The goal for the improvement of this group of inhibitors is the broadening of the antifungal spectrum. This task would require the clarification of causes responsible for their selective antifungal spectrum. The most diversified group of sterol biosynthesis inhibitors are the DMIs as inhibitors of lanosterol or 24-methylenedihydrolanosterol demethylation at the C-14 position. This group of antifungal drugs and fungicides is perhaps the best example for inhibitor improvements guided by biochemical considerations. As described by Vanden Bossche and Janssen in Chapter 9, the identification of the target site and the subsequent refinement of molecular models of inhibitor action have had considerable impact on the design and development of new inhibitors with improved properties. The third group of commercial sterol biosynthesis inhibitors are the **morpholines** as inhibitors of both the sterol Δ^8-Δ^7-isomerization and Δ^{14}-reduction. Similar to the allylamines, the relatively restricted antifungal spectrum leaves room for future progress and several lead structures for both enzymes (Chapter 7), but also biochemical research on the causes for the selective action should be a valuable contribution to this task.

A second class of fungicides with a target site in lipid biosynthesis are the two phosphororganic compounds **edifenphos** and **iprobenfos** (IBP) (Figure 6). Both are phosphororganic compounds developed in the 1960s for the control of rice blast.[73-75] Based on the inhibition of glucoseamine incorporation into the cell walls of *Pyricularia oryzae*, the mode of action was initially related to the inhibition of chitin synthesis. Phosphatidyl choline synthesis was later shown to be considerably more sensitive to both edifenphos and iprobenfos,[81-83] and this target site is now accepted as the primary mode of action. In eukaryotes, phosphatidyl choline is synthesized via two alternative pathways, the CDP choline and the phosphatidyl ethanolamine methylation pathway. In fungi, the successive methylation of phosphatidylethanolamine with *S*-adenosylmethionine as the methyl donor predominates as the major route of synthesis.[84-90] In *Saccharomyces*, two transferases are involved in the process. One is exclusively active in the first methylation step; a second catalyzes all three methylations with strong preference for the second and third methyl transfer.[88] The phosphatidylethanolamine methyltransferases are the target sites of the two inhibitors,[81-83] but the differential inhibition of the two transferases involved has not been investigated. Most likely, a third and structurally different rice blast compound, isoprothiolane (Figure 6), can be included in this list of phospholipid biosynthesis inhibitors.[73-75] For two reasons, edifenphos, iprobenfos, and perhaps isoprothiolane are interesting antifungal inhibitors. The chemical structures in combination with the biochemical information gathered since the compounds were synthesized and introduced appear to provide opportunities for new structural evaluations. With antifungal activities around concentrations of 10 μg/ml, all three inhibitors are relatively weak fungicides compared to more recently developed inhibitors, and the improvement of activity, which might explain the restriction to rice blast control, might be feasible. The second interesting property resides in their reported inhibitory activity toward chitin synthesis. This aspect will be discussed later.

As described by Edlich and Lyr in Chapter 4, the primary effects of the **dicarboximides** and **aromatic hydrocarbons** are the artificial peroxidation of polyunsaturated fatty acids in phospholipids and, thus, a disturbance of lipid function rather than biosynthesis. Thus far, the molecular details of inhibitor action have not been entirely clarified, and more studies need to be completed before the potentials for inhibitor improvement can be evaluated. The dicarboximide example demonstrates, however, that interference with active-oxygen-controlled lipid metabolism[91] comprises a potential antifungal target.

3. Inhibition of Cell Wall Synthesis

Fungal cell walls consist primarily of polysaccharides. The skeletal and water-insoluble elements are a homopolymer of glucose, (1-3)-β-D-glucans with (1-6)-β-linkages as branching point, and with the exception of all Oomycetes, chitin composed of (1-4)-β-N-acetyl-D-glucosamine.[92,93] Cellulose comprises the polysaccharide in addition to (1-3)-β-D-glucans in cell walls of Oomycetes.[92-94]

Since chitin is only found in fungi and intervertebrates but not in mammals and plants, the inhibition of chitin and thus cell wall biosynthesis appears to be an excellent target for highly selective fungicides. The fungal chitin synthase is by far the best-characterized enzyme active in fungal cell wall biosynthesis.[95] The enzyme is membrane bound and synthesized as a inactive precursor activated by proteolysis. Although the localization of chitin synthase in actively growing cells is still a matter of conflicting views (chitosomes, plasma membranes, or both),[95-97] the enzymatic properties are well established. The polymer chitin is synthesized from the sugar nucleotide UDP-GlcNAc (Figure 7), with K_m values around 1 mM:[95]

$$2n \text{ UDP-GlcNac} \rightarrow 2n \text{ UDP} + \text{obGlcNAc-}\beta\text{-(1-4)GlcNac]}_n$$

Relevant is the finding that two chitin synthases coded by two different genes have been identified in *S. cerevisiae*.[98] Chitin synthase 1 (Chs1) coded by the gene *CHS1* comprises the bulk activity present in yeast cells. Surprisingly, disruption of this gene yielded viable transformants. This result indicated that Chs1 was not essential for chitin synthesis *in vivo*. The disruption of a second gene, *CHS2*, provided evidence for the essential character of Chs2. This enzyme is vital for septum formation and thus cell division and cannot be substituted by Chs1.[98] This latter enzyme might have a function in the repair of cell wall damages caused by chitinase.[99] Although the two genes share only 36% identical nucleotides, a higher degree of similarities in the carboxyl two thirds of the protein is probably related to the common catalytic function.[100] First evidence for the presence of similar isoenzymes has been presented for *C. albicans*, and the enzymes were present in both the yeast and mycelial stages.[101] A functional diversity of chitin synthetases has yet not been reported for filamentous fungi, but the availability of specific DNA-probes will trigger progress in this interesting field of cell wall biosynthesis.

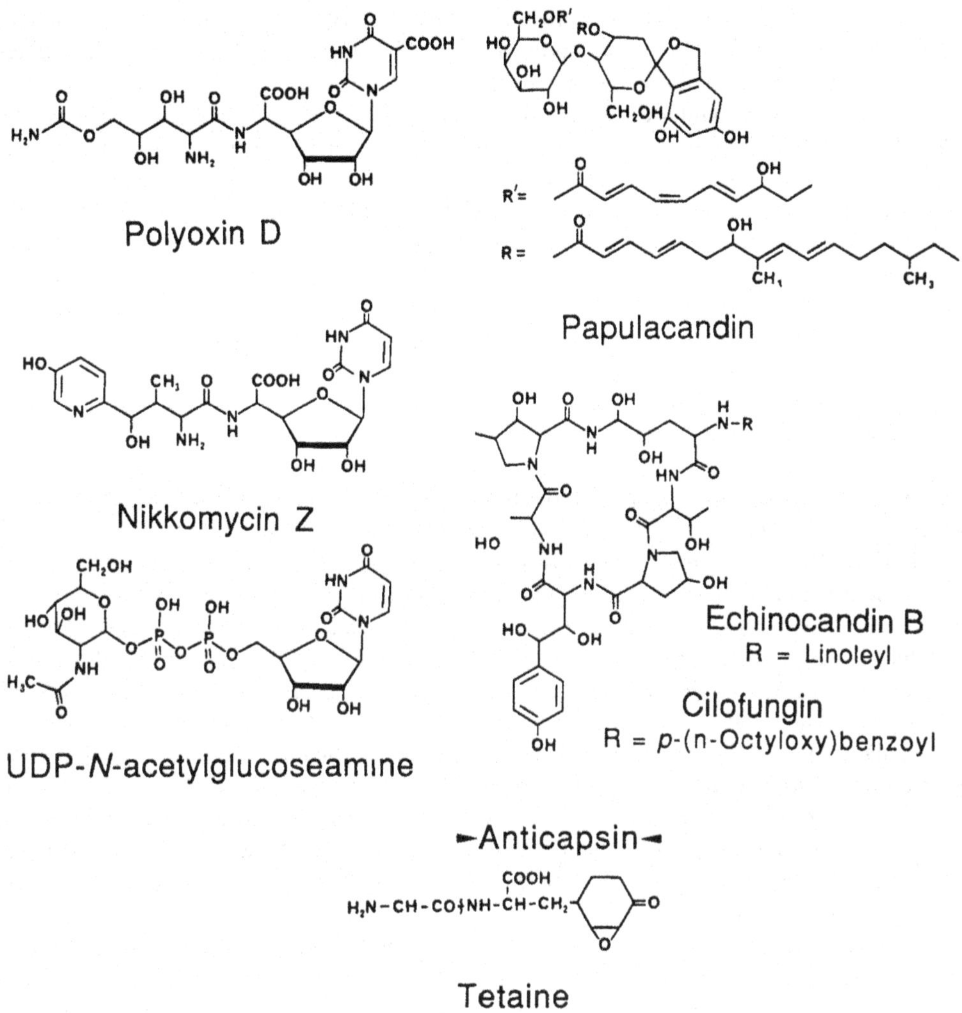

FIGURE 7. Antifungal inhibitors of cell wall biosynthesis.

The oldest group of potent chitin synthase inhibitors is the antibiotics polyoxins and nikkomycins (Figure 7), produced by *Streptomyces* spp. Polyoxin has been developed for the control of several plant diseases.[73-75] Nikkomycins continue to be candidates of clinical testing as drugs against mycotic diseases.[102] Both inhibitors are nucleoside di- and tripeptides with close structural similarities to the enzyme substrate UDP-GlcNAc (Figure 7). Consequently, polyoxins and nikkomycins have been identified as competitive inhibitors of chitin synthase, with K_i values around 1 μM and thus several hundred-fold lower than the K_m of the substrate.[103] Both isoenzymes present in yeast are equally inhibited by polyoxin D.[103] The high inhibitory potency *in vitro* is not always reflected by the antifungal activity achieved *in vivo*. This discrepancy is related to both the active uptake via peptide transport systems and the desactivation of the inhibitors by peptidases.[103] Attempts to substantially improve these properties by chemical modifications have not been successful thus far.

The antibiotic anticapsin (Figure 7) has been identified as an inhibitor of glutamine-fructose-6-P-aminotransferase derived from *C. albicans*, an enzyme catalyzing the first step in UDP-GlcNac synthesis.[104] Anticapsin is a structural analogue of glutamate, derived by peptidase action from the peptide antibiotic tetaine. Other synthetic glutamate analogues have been shown to act as equally potent inhibitors,[105] and this early enzyme in chitin synthesis appears to represent an interesting antifungal target. It has not been determined, however, to what extend other aminotransferases with glutamate as one of the substrates are inhibited, and the inhibitor specificity has yet to be determined.

Fungal β-glucan synthases, which are active in the synthesis of the second skeletal element of cell walls, are similar to chitin synthase membrane-bound enzymes and utilize UDG-glucose as substrate.[94,106] The known inhibitors of fungal β-glucan synthesis are, like chitin synthase inhibitors, mostly natural products produced by microorganisms. The antifungal activity of echinocandins, all cyclic lipopeptides produced by *Aspergillus* spp., was discovered in the early 1970s.[107] Cilofungin (LY121019), a semisynthetic derivative of echinocandin B (Figure 7), has a narrow fungicidal spectrum against *Candida* spp.[108,109] (but also *Neurospora crassa*) and is undergoing extensive preclinical studies.[102] The antibiotic has been shown to act as a specific inhibitor of fungal (1-3)-β-D-glucan synthases.[110,111] In contrast to the polyoxins and nikkomycins, cilofungin acts as a noncompetitive inhibitor with K_i values of 16 μ*M* with the *N. crassa*[110] and 2.5 μ*M* with the *C. albicans*[111] enzyme. *C. albicans* chitin synthase was not affected,[111] indicating a specific effect on β-glucan synthesis in cell wall formation. Although the lipophilic side chain has been identified as an essential structural feature,[112] the precise molecular mechanism of inhibition remains to be defined. However, direct interactions with membranes as indicated by the hemolytic potential of natural echinocandins is a likely side effect of these antibiotics.[107] A substantial improvement with regard to this side effect could be achieved by modifications of the lipoidic residue, a semisynthetic approach that has yielded cilofungin.[107] Several structurally similar lipopeptides such as molundocandin and sporiofungin have been described more recently.[107]

A second antibiotic active as an inhibitor of β-glucan synthesis but not chitin synthase is papulacandin B (Figure 7). The activity of these antibiotics, originally isolated from *Papularia sphaerosperma* and more recently described as semisynthetic derivatives,[113] is largely restricted to *C. albicans*. The antibiotic has been identified as inhibitor of glucan synthesis in several fungi with a mode of action similar to cilofungin.[114-116] A third potent inhibitor of β-glucan synthesis is aculeacin A, like papulacandin and cilofungin an antibiotic of amphophilic nature.[117]

It is striking that all of the reported antifungal inhibitors with target sites in cell wall biosynthesis are antibiotics produced by microorganisms. Some of the antibiotics described above are complex in nature and thus provide only restricted leads for chemical synthesis. One exception, however, exists. As discussed above, edifenphos and iprobenfos (Figure 6) have been identified as inhibitors of phosphatidylethanolamin methylation in phosphatidylcholine synthesis. Kodama et al.[82] showed that the inhibition of chitin synthase preparation derived from *Pyricularia oryzae* was approximately 100 times less sensitive to edifenphos than the *in vivo* activity would suggest. The inhibition of glucosamine incorporation into cell walls had been discussed earlier as the mode of action of edifenphos. Based on the low *in vitro* activity of edifenphos as chitin synthase inhibitor, it was assumed that the effect on cell wall biosynthesis was indirect through the alteration of membrane properties.[73-75] The concentration of iprobenfos required for 50% inhibition of a *Pyricularia* preparation of chitin synthase was high, but some lower values could be achieved with some iprobenfos analogues.[118] The inhibition type of iprobenfos, when tested with the *Pyricularia* enzyme, was reported to be noncompetitive in character. This noncompetitive inhibition was contrasted by results obtained with chitin synthase derived from *Aspergillus nidulans*.[119] Although high concentration of iprobenfos were required, inhibition was clearly competitive, indicating a binding of the inhibitor into the active site of the *Aspergillus* chitin synthase.

It is interesting to compare the structure of iprobenfos and iprobenfos analogues with the substrates of the two inhibited enzymes, UDP-*N*-acetylglucosamine for chitin synthethase and *S*-adenosylmethionine for the phosphatidylethanolamine methylases (Figure 8). Misato and Kakiki[118] have reported a structure-activity relationship of iprobenfos analogues with

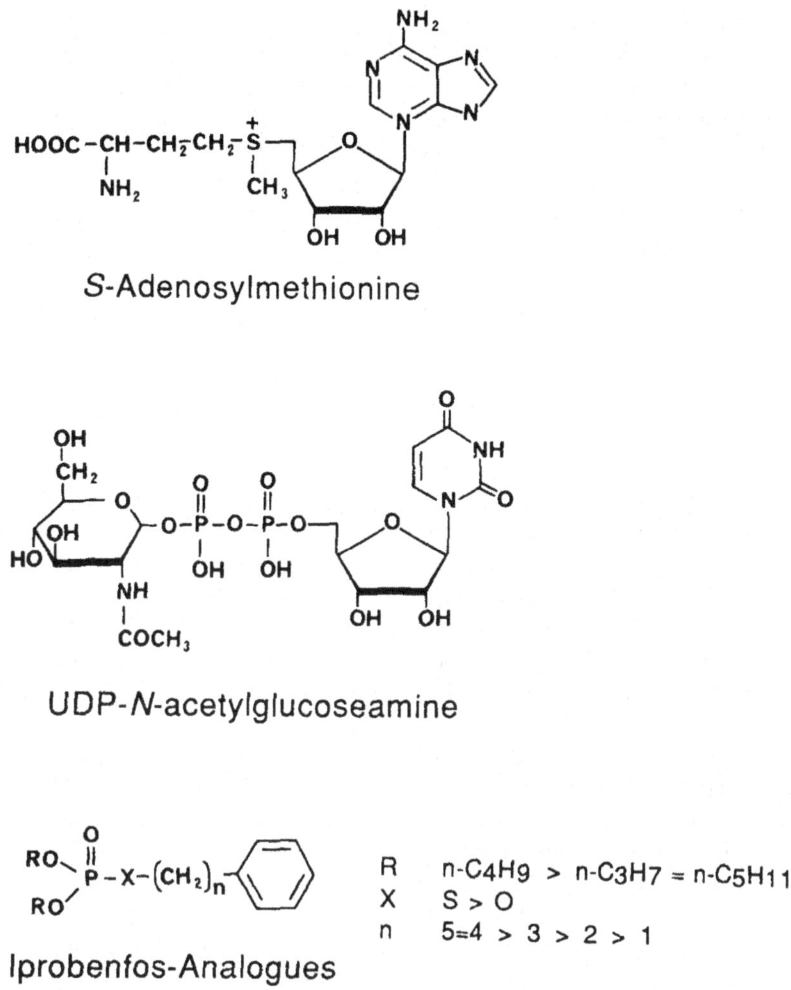

FIGURE 8. Comparison of iprobenfos (analogues) with substrates of proposed target sites.

regard to the inhibition of chitin synthase prepared from *P. oryzae* (Figure 8). Unfortunately, comparative data for the inhibition of phosphatidylethanolamine methylation are not available. A structural substrate-inhibitor analogy is not immediately apparent for either of the enzymes reported to be inhibited. Interesting, however, is the fact that both substrates contain a nucleotide portion, indicating partially related binding sites. A computer-aided analysis of the structural similarities, including the inhibitor data of iprobenfos analogues toward chitin syntethase, might reveal similarities not yet recognized.

Of the various matrix components of fungal cell walls, only the melanins comprise a target for plant disease control. The pigments are though to contribute to the survival of fungal propagules, although this role cannot be regarded as the target for the melanin biosynthesis inhibitors used in the control of rice blast disease (Chapter 6). Here, the failure of heavily melanized appressoria of *P. oryzae* and *Colletrotrichum* spp. to produce melanin is the cause for a failure to penetrate the plant cuticle. As discussed above, biochemical considerations have already been taken into account in the design of structurally different inhibitors with a target site identical to the enzyme blocked by the initially developed inhibitors. Yamaguchi and Kubo (Chapter 6) discuss, furthermore, that alternative and fungal-specific targets in melanin biosynthesis do exist.

4. Inhibition of Respiration

The only commercial group of fungicides with a target site in fungal respiration are the carboxamindes described by White and Georgopoulos in Chapter 1. These compounds inhibit the mitochondrial succinate dehydrogenase complex by binding to a component of the electron transport chain rather than by acting as a substrate-related inhibitor. All of the commercial carboxamides are highly specific toward Basidiomycetes, and the *in vivo* activity against Ascomycetes is extremely poor. Evidence exists, however, that the specificity is related to uptake parameters rather than target insensitivities (Chapter 1). With isolated mitochondria derived from Ascomycetes, carboxamides have been shown to be active, and some alternative structures are available that exhibit a broader *in vivo* spectrum. Thus it appears feasible to direct synthetic efforts to the improvement of uptake characteristics rather than target affinities. The carboxamides are an excellent example of new opportunities guided by results of continuous biochemical research.

III. DESIGN OF ANTIFUNGAL INHIBITORS

The strategies of molecular research in the discovery of new antifungal inhibitors discussed above rely on the biochemical information gathered with inhibitor structures derived from random synthesis and biological screening programs. Consequently, the target sites blocked or influenced by the respective compounds are not predefined and are identified only after biological activity has been found. A radically different approach would be the design of inhibitors directed toward enzymes with potentials to serve as an antifungal target. This approach would take advantage of the progress being made in the field of enzyme inhibitors and antagonists of receptors. The disadvantage would be that fungal enzymology (with the exception of *Saccharomyces cerevisiae*) has remained a "Sleeping Beauty" of biochemical research. Furthermore, excellent enzyme inhibitors are not necessarily excellent fungicides and antifungal drugs. Unrestricted penetration through fungal cell walls and membranes, adequate chemical stability, and optimal distribution within the host tissue are equally important parameters. Although these additional parameters are increasingly understood, the rational integration of all important features in inhibitor design programs will be extremely demanding at the present stage of knowledge.

Similar to drug design efforts,[120] the design of antifungal inhibitors would depend on the availability of enzymatically well-characterized target enzymes or receptors and the evaluation of the most appropriate inhibitor type for a particular target. It will be a delicate but crucial task to decide at what stage of physiological and biochemical knowledge a putative fungal target is adequately characterized to initiate a biochemically guided synthesis of inhibitors. Consequently, it will also be a delicate task to identify adequate and promising targets. With this limitation in mind, an attempt will be made to provide an overview on principle considerations, and to describe several "case studies."

A. GENERAL STRATEGIES OF INHIBITOR DESIGN

Although the current arsenal of modern antifungal agents is not restricted to inhibitors of enzymes, design efforts might be mostly restricted to this particular class of antifungal compounds. The understanding of enzyme catalytic mechanisms has rapidly advanced in the past, and the success rate of the inhibitor design efforts based on this knowledge has been high.[120] Several classes of inhibitors have been identified and characterized in the past, depending on their particular mechanism of inhibition. An in-depth treatise of the theoretical foundations, in particular with regard to sometimes sophisticated and complex kinetic models, is beyond the scope of this chapter. Only a brief description of different principles of enzyme inhibition will be provided. An excellent overview on this subject applied to pesticide design strategies has been compiled by Schloss,[121] and more focused review articles are available for particular classes of inhibitors.

1. Types of Inhibitors

The oldest class of enzyme inhibitors are **reactive** and **group-specific reagents** that bind to or modify essential amino acids involved in enzyme catalysis. The most simple fungicide within this group is the Bordaux mixture, with copper ions binding to sulfhyryl groups. It is interesting to note that even this simple fungicide is intellectually charming. The general toxicity of copper ions is circumvented by the deposition of an insoluble copper salt on the leaf surface. By the acidification of a restricted area surrounding the germinating fungal spores, copper ions are dissolved from the deposit, and the desactivation of fungal enzymes can take place.[122] It is remarkable that this simple but ''rational'' principle of ''fixed coppers'' has withstood the 100 years of competition with increasingly sophisticated fungicides developed for use in agriculture. The evolution of group-specific fungicides proceeded from the development of organomercuries and other organometallic compounds[123,124] to organic sulfur compounds such as the EBDCs[125] and the ''R-$SCCl_3$'' group such as captan.[126] The integration of reactive functional groups into larger molecules would make it possible to direct the reactive group into the active site of a particular enzyme. The site-specificity of these active-site-directed inhibitors (affinity labels) would be greatly improved,[127] but in many cases it remains doubtful that the reactivity could be entirely restricted to a specific target site. This restriction should apply also to light-activated reactive compounds.[128] For the development of new antifungal drugs, this lack of specificity and thus potential for general toxicity might be prohibitive. However, group-specific inhibitors are still an important class of protective fungicides, and new developments might be desirable in view of the current problems with fungal resistance to site-specific inhibitors.[3] It should be emphasized that many representatives of this older class of fungicides are under toxicological scrutiny,[4] and some of these problems could be inherent to their nonselective reactivity.

Three different types of specific enzyme inhibitors have been described. The oldest class of reversible inhibitors are the **substrate** or **ground-state inhibitors**. These inhibitors are substrate analogues competing with the natural substrates for the formation of the enzyme-substrate complex. Reversible inhibitors have one major limitation with the regard to design strategies. Within a normal chain of enzyme reactions, a ground-state inhibitor will initially inhibit the conversion of the respective substrate. Consequently, the intracellular substrate concentration will rise, and the equilibrium of substrate-inhibitor binding will be shifted toward substrate binding. Eventually, sufficient amounts of substrate will be converted to restore the initial flow of metabolites through the respective pathway, and the inhibitory action would be diminished. Exceptions, however, exist, and the design of ground-state inhibitors can be a viable approach under certain conditions. For example, the initial increase of enzyme substrates might cause the down-regulation of the entire pathway at allosterically regulated enzymes. A valid example for the combination of a competitive ground-state inhibitor with feedback inhibition are the polyoxins and nikkomycins described above. These chitin synthase inhibitors should lead to a substantial increase of UDP-*N*-acetylglucoseamine. This substrate, however, is known to down-regulate glutamine-fructose-6-P-aminotransferase, a key enzyme involved in the synthesis of UDP-*N*-acetylglucoseamine,[129] and the intracellular concentration of the chitin synthase substrate is not dependent on the consumption of the substrate.[130] No principle limitations to reversible ground-state inhibitors are also apparent whenever enzyme substrates are not subject of fluctuating concentrations. The mobilization of storage products in fungal spores is a good example for this type of situation. A third scenario for useful ground-state inhibitors would be a strong toxic effect of accumulating precursors. The concentration to which substrates of inhibited enzymes would increase depends on the ratio between the K_i of the inhibitor and the K_m of the substrate. High ratios and, thus, tight inhibitor binding would require high substrate concentrations before a reaction pathway would proceed, regardless of the presence of an inhibitor.

Since enzymes are known to bind high-energy intermediates of enzymatic reactions more tightly than the substrates, analogues of these high energy intermediates are a group of inhibitors with great promise in the design of potent antifungal agents. These **analogues of transition states** and **high-energy intermediates** have received increasing attention as a group of powerful enzyme inhibitors.[131,132] For example, the substitution of hydrogen, which is frequently involved as H^+ in enzyme catalysis, by fluorine has been a successful and more general design approach.[133] It has been noticed that the binding of many representatives of this inhibitor class is time dependent and slow when compared to ground-state inhibitors.[134] With regard to antifungal inhibitors, the morpholines as one class of the sterol biosynthesis inhibitors have been shown to act as analogues of high-energy carbocationic intermediates, and many more inhibitors of enzymes involved in the sterol pathway have been designed based on this concept (Chapters 7 and 8). Whenever the intermediate of an enzyme reaction is known or can be deduced from analogies, the concept of high energy intermediate analogues is probably the most fruitful and most flexible inhibitor design strategy.[121,131,132]

A second group of highly potent inhibitors is the class of **mechanism-based inhibitors or suicide substrates.**[135-137] These inhibitors with structural resemblance to particular enzyme substrates contain a latent reactive entity that is activated by the catalytic action of a particular enzyme. The liberated active group reacts with a functional group within the active site, and the inhibitor is bound irreversibly. The active enzyme is slowly "titrated out", and provided each activated group reacts before the inhibitor is released from the enzyme, a 1:1 inhibitor/enzyme ratio would be sufficient to desactivate a given enzyme. None the existing antifungal inhibitors acts as a pure suicide substrate, but this type of irreversible inhibitor is a promising strategy for the design of new antifungal inhibitors. Latent functional groups activated by the catalytic pathway of an enzyme have been described for numerous classes of enzyme reactions. The design of specific inhibitors would depend on the adaptation of inhibitor principles to the structural requirements of a particular substrate binding site, and the presence of a reactive amino acid in vicinity to the reactive group generated within the active site.

Not all of our current antifungal agents fall into one of the inhibitor groups in its purest definition. The azole fungicides and antifungal drugs as inhibitors of C-14 lanosterol demethylation described in Chapter 7 are an example of a more complex inhibitors mechanism. This is not surprising, because the respective enzyme is a cytochrome P-450 monooxygenase and, thus, exhibits a more complex mechanism of enzyme action.[138] Furthermore, not all enzyme inhibitors have to be bound to the substrate binding site. For example, the allylamines as inhibitors of squalene epoxidases (Chapter 7) have been described as noncompetitive inhibitors of the fungal enzyme. Surprisingly, the inhibition of the mammalian enzyme appeared to be competitive.[139] This example, however, might also demonstrate a problem inherent to studies on the kinetics of inhibitor action. As discussed by Balliano et al.[140] for a series of inhibitors of fungal 2,3-oxidosqualene cyclase, not all enzymes are accessible to normal assay procedures, and the differentiation between competitive and noncompetitive inhibitors based on classical data evaluation is not always easily made. Naturally, for the design of inhibitors this information would be of utmost importance. Critical scepticism with regard to the interpretation of apparently sound kinetic data might be advisable in some cases, dependent on the particular nature of the enzyme-inhibitor under question.

The ideal antifungal inhibitor should be highly enzyme specific, it should easily reach the target site, and it should react in a 1:1 inhibitor/enzyme ratio. The latter demand implies that enzymes present in low concentrations would require less quantities of inhibitors and, thus, would be superior targets. Considering the minimal enzyme concentration (one single enzyme molecule per fungal cell) the theoretical minimal inhibitor concentration would be around 10^{-12} *M*. A thousand enzyme molecules per cell would imply theoretical MIC values of 10^{-9} *M*; some of the azole fungicides are approaching these limits under certain conditions

(Chapter 7). Unfortunately, the axiom that low abundancy enzymes make better antifungal targets implies that the most valuable target enzymes are also most difficult to purify in sufficient quantities for enzymatic characterization. As described below, the opportunities to express or overexpress genes in transformed organisms provide future opportunities for elegant solutions.

2. Design Strategies

In theory, the design of inhibitors can be guided and even accomplished with the aid of increasingly available computer modeling procedures, provided detailed structural information is available for an enzyme chosen as a target.[141-145] Most valuable are still high resolution crystal structures. In this case it would even be possible to extend the specific binding of inhibitors to surface areas close to but outside the substrate binding site. Unfortunately, not many structures of fungal enzymes have been elucidated by X-ray analysis. In particular for "low abundancy" and membrane-bound enzymes, the sophisticated structural analysis cannot be expected to provide a rapidly growing set of data. It might become possible to compute the precise three-dimensional structure of a protein based on the amino acid sequence, which is readily available from cDNA sequences. This ultimate goal has not been accomplished, and substantial refinement of the current models in addition to supercomputers will be required to model the precise three-dimensional structure of catalytic sites.[146,147] Similar size restrictions apply to protein structure determinations by NMR.[148] A promising emerging approach for a facilitated evaluation of three-dimensional structures might be based on the capacity of antibodies to recognize and thus picture a three-dimensional structural feature.[149]

Regardless of the current restrictions imposed on a rational design of antifungal inhibitors, it would be unwise to select antifungal targets based exclusively on the wealth of structural and enzymatic information. The legitimate question is sometimes raised whether any of the most successful target sites of existing fungicides would have been chosen as a target of inhibitor design programs. Considering a purely theoretical design approach, the answer would probably be no. More flexible strategies, however, might have been successful. The azole fungicides (Chapter 7) are a good example for a hypothetical evaluation of this question, assuming that the current information would be available but the inhibitors not yet synthesized. A first hint that fungal lanosterol demethylase could serve as an antifungal target might have been the report that a nystatin-resistant mutant of *Ustilago maydis* accumulated 14-methylsterol precursors, leading to substantial impairment of growth and development.[150] This strong accumulation of a precursor would have indicated a poorly feedback-regulated pathway and would have placed emphasis on very tight-binding inhibitors. However, inhibitors of the enzyme could have been expected to be active toward a broad spectrum of ergosterol-producing fungi, with the exception of Oomycetes (Chapter 7). Consultation of the literature would have revealed that the enzyme responsible for the C-14 demethylation is a typical cytochrome P-450 monooxygenase.[151] Because the mechanism of this class of enzymes is complex, *in vitro* enzyme assays are difficult to perform, all P-450 systems are membrane-bound, and data on the active site geometry are not available, the target would have been most likely dismissed in theoretically oriented inhibitor design programs. Furthermore, the respective lanosterol demethylase present in mammalian liver could have been expected to be very similar to the fungal counterpart, and "rational" considerations with regard to toxicological properties of sterol demethylation inhibitors would not have been in favor of this target. Based on both shortcomings, the target would have been dismissed.

A more flexible and less theoretically oriented program might have proceeded, and it would have been recognized that one class of known P450 inhibitors is based on the concept of blocking the binding of oxygen to the porphyrin iron by heteroatoms such as nitrogen.[138] Based on the structure of the enzyme substrate lanosterol, an initial design approach could have been the synthesis of lanosterol analogues with respective groups in vicinity to the C-14 methyl group, and most likely essential groups under consideration would have included a basic imidazole or pyrimidine nitrogen. Derivatives of this type have indeed been shown to be active as demethylation inhibitors.[152] The next logical step would have been the substitution of the complex and synthetically too demanding sterol ring structure by aromatic substituents as well as cyclic and bulky aliphatic groups, a general structure common to all of the commercial compounds (Chapter 7).

The admittedly theoretical and hypothetical scenario describes a reversion of actual events. It demonstrates, however, that the design of sterol demethylation inhibitors could have been possible, but only if relatively unfavorable properties and characteristics of a putative target site would have been accepted. This implies that design programs will have to rely on semi-empirical approaches whenever a target enzyme is not entirely understood and difficult to handle. The evaluation of available information, the initiation of basic but focused research if it become necessary, and the integration of modern tools as they fit will have considerable impact on the success of such efforts.

B. CHOICE OF ANTIFUNGAL TARGETS

The most crucial question in the biorational design of inhibitors will be the selection of appropriate target sites. Considering the goal of designing highly fungal-specific inhibitors with minimal side effects on nontarget organisms, any enzyme or receptor unique to fungal organisms, and in particular to fungal pathogens, would have a clear preference. The evaluation of target sites identified for established fungicides and antifungal drugs, however, does not necessarily support the stringent validity of this premise. In many cases, functionally identical enzymes present in different classes of organisms have evolved substantial structural differences, and opportunities for highly fungal-specific inhibitors exist even for ubiquitous enzymes. The degree of inhibitor specificity is sometimes dramatic, and these examples might suggest that almost all fungal enzymes could be considered as potential antifungal targets. Consequently, "lead structures" for inhibitors of antifungal targets would be almost countless. In his *Handbook of Enzyme Inhibitors,* Zollner listed more than 5000 described inhibitors for about 1000 enzymes,[153] and although only a fraction of these inhibitors would be accessible to synthetic variation, a principal restriction of target sites does not appear to impose a major limitation.

The actual number and nature of enzymes and receptors affected by existing site-specific antifungal agents is in surprising contrast to this assumption. All of the "established" inhibitors with proven records in fungal control have evolved from random screening programs and, thus, were selected and developed according to their biological performance rather than their respective target sites. The modes of action of these fungicides and antifungal drugs are listed in Table 1 and provide a valuable, although largely circumstantial, account for the intrinsic quality of antifungal target sites. The vast majority of our antifungal inhibitors acts either on lipid metabolism, or they interfere with the synthesis of polymers such as nucleic acids and cell wall components. A common feature of this relatively short list of targets is the fact that most of the enzyme products are insoluble in water. Even the exceptions from this rule are relatively unique. The hydroxypyrimidines inhibit the synthesis of a water-soluble product, but as discussed by Hollomon in Chapter 1, they are highly powdery mildew specific and inhibit, most likely, a crucial step in the complex and highly regulated pattern of nucleotide metabolism. Inhibitors within this framework of nucleotide synthesis and interconversion[154] could well be suited to become attractive candidates for antifungal agents. The carboxamides as inhibitors of the mitochondrial succinate dehydrogenase complex are not active site-directed and inhibit the oxidation of succinate indirectly by disrupting the electron transport (Chapter 2). Furthermore, the benzimidazoles and griseofulvine are not active as enzyme inhibitors but interfere with cytoskeleton functions.

TABLE 1
Antifungal Modes of Action

Inhibitor class	Target site(s)
Nucleic acid metabolism and nuclear division	
Hydroxypyrimidine	Adenosine deaminase
5-Florocytosine	DNA synthesis
Benzimidazoles	Tubulin polymerization
Griseofulvin	Mitosis (?)
Phenylamides	Ribosomal RNA synthesis
Lipid biosynthesis and metabolism	
Pyridine, pyrimidines, azoles	C-14 sterol demethylation
Morpholines	Sterol Δ^8-Δ^7-isomerization and Δ^{14}-reduction
Allylamines	Squalene epoxidase
Ediphenphos and iprobenfos	Phosphatidyl ethanolamine *N*-methyltransferase
Dicarboximides and aromatic hydrocarbons	Lipid peroxidation
Cell wall biosynthesis	
Polyoxins, nikkomycins	Chitin synthase
Anticapsin	UDP-GlcNAc synthesis
Cilofungin, papulacandin, aculeacin A	β-Glucan synthesis
Melanin biosynthesis inhibitors	Cell wall melanin synthesis
Inhibition of respiration	
Carboxamides	Succinate dehydrogenase
Methoxyacrylates	Complex III

It is highly unlikely that specific inhibitors of enzymes with soluble products have never been synthesized in the decades of random synthesis programs. It is more likely that these inhibitors did not meet the standards set by commercial products and, consequently, that the respective target sites were of limited practical value and thus never identified. For example, inhibitors of amino acid biosynthesis comprise a successful class of modern herbicides,[155,156] and in theory several enzymes involved in fungal amino acid biosynthesis should be well suited to serve as specific targets for antifungal inhibitors. Indeed, glyphosate as an inhibitor of bacterial and plant 5-enolpyrovylshikimate-3-phosphate synthase was identified as a potent inhibitor of the respective fungal enzyme in *Candida maltosa*.[157] Likewise, sulfonylurea herbicides as inhibitors of branched-chain amino acids inhibit acetolactate synthase in plants but also in *S. cerevisiae,* and sulfonylurea-resistant yeast strains with a mutated gene coding for the enzyme have been described.[158] The antifungal amino acid analogue 2-amino-4-oxo-5-hydroxypentanoic acid (RI-331) was shown to inhibit yeast homoserine dehydrogenase and, thus, the synthesis of methionine, threonine, and isoleucin.[159,160] A design program aimed at the optimization of inhibitor structures toward the fungal enzymes should be feasible and attractive, in particular in case of inhibitors of dietary essential amino acids.

Would these inhibitors of amino acid synthesis become successful fungicides and antimycotics? This question is probably best answered with a second question. Why are the target sites of current antifungal inhibitors restricted to enzymes involved in the synthesis of water-insoluble products. A likely though circumstantial explanation is the fact that during their parasitic disease cycle, fungal pathogens are in intimate contact with their respective hosts. The inhibition of metabolic steps yielding soluble products could be circumvented by uptake of suppressed metabolites from the host. Under these conditions, the action of the inhibitor would either be nullified, or inhibition of growth would be incomplete. Indeed, the inhibition of growth of *C. maltosa* by glyphosate and RI-331 could be completely

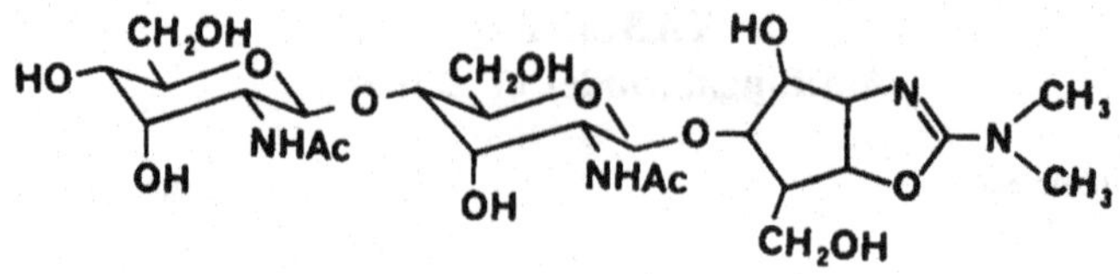

Allosamidine

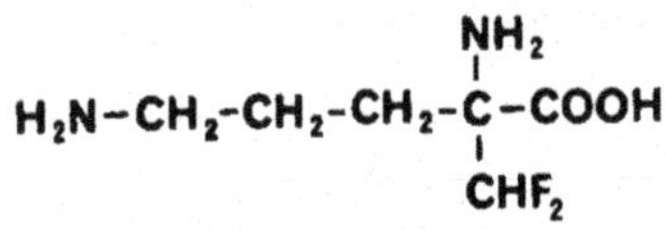

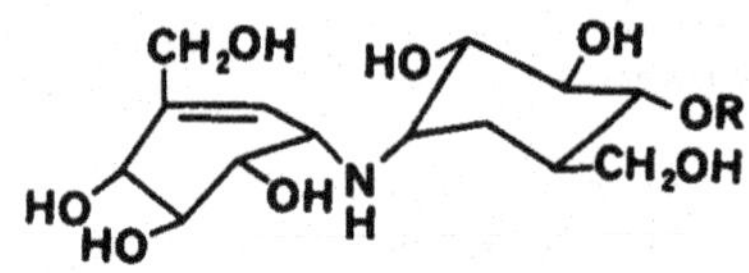

DFMO; Eflornithine

Validamycin A R = β-D-glucose
Validoxylamine A R = H

FIGURE 9. Lead structures for antifungal inhibitors.

compensated with amino acids representing the end products of the inhibited biosynthetic pathways.[157,160] The large number of described fungal auxotrophic mutants with defined enzyme deletions are an additional support of the hypothesis that inhibitor sites can be circumvented by an exogenous supply of biosynthetic products from the host. Moreover, obligate fungal plant pathogens have apparent metabolic deficiencies and thus can rarely be grown on artificial media. However, they are viable and destructive pathogens because they acquire essential components from their hosts. The circumvention of inhibitory blocks by water-soluble products provided by the host is certainly not an absolute necessity. It depends on the presence and concentration of the suppressed fungal metabolites in host tissues, and on the uptake of these compounds by fungal cells. For the majority of soluble products of putative target enzymes, however, the question of inhibitor antagonism by soluble host metabolites would have to be investigated prior to research decisions.

Regardless of these limitations, the number of novel and promising target enzymes is too large to be treated in an overview. The following discussion will thus be restricted to enzymes that have been under discussion as targets for fungal growth inhibition, and to targets with crucial functions in host-pathogen interactions rather than fungal metabolism. The latter type of inhibitors would be targeted against highly specialized events in host-pathogen interactions, and they would act toward the establishment of a disease without direct interference with fungal growth. Considering this highest possible degree of specificity it should be expected that the impact of these antidisease agents on nontarget organisms should be extremely small.

1. Inhibitors with Direct Targets in Fungal Metabolism

As discussed above, biosynthetic steps in **fungal cell wall synthesis** can be expected to comprise excellent targets for antifungal inhibitors. Not only biosynthetic, but also lytic enzymes are involved in the formation of cell walls in growing hyphal tips,[95,161,162] and some of these enzymes should be considered as target sites for interference. For example, allosamidine (Figure 9), an antibiotic produced by *Streptomycete* sp., has been identified as a potent inhibitor of various **chitinases**.[103] The degree of inhibitor specificity is dramatic. Reported I_{50} values range from 0.2 n*M* for a chitinase isolated from a nematode to over 500 n*M* for the enzyme isolated from *Neurospora crassa* to no effect at 100,000 n*M* for a plant chitinase.[103] This wide range of enzyme specificities is promising with regard to opportunities for the design of lead structures that would act as specific inhibitors of fungal chitinases.

Inhibitors of **polyamine biosynthesis** with their broad biological activities have received considerable attention in the past. The polyamines putrecine, spermidine, and spermine are ubiquitous constituents of both prokaryotes and eukaryotes.[163] Based on the polycationic nature under physiological conditions, polyamines bind to and interact with DNA and other helical nucleic acid structures and contribute to the regulation of cell growth and differentiation.[163] Inhibition of polyamine synthesis leads to growth inhibition of a variety of organisms. The synthesis of putrescine as the direct precursor of spermidine and spermine proceeds via decarboxylation of ornithin, or it originates from arginine.[164] Most fungi and protozoa are thought to lack the arginine pathway, and it is not surprising that specific inhibitors of ornithin decarboxylase have received preference. Inhibition of this enzyme would deplete the polyamine pool only in organisms that are entirely dependent on the ornithin pathway and opens the opportunity for specific inhibitors.

A great number of 2-substituted ornithin derivatives have been identified as inhibitors of **ornithin decarboxylase**.[165] One of these derivatives, 2-difluomethylornithin (DFMO; eflornithine) (Figure 9), is of considerable chemotherapeutic interest in the treatment of African sleeping sickness caused by parasitic protozoa and AIDS-related pneumonia caused by *Pneumocystis carinii*.[166] The classification of this organism as protozoae has been recently challenged. rRNA analysis revealed that the organism should be classified as a fungus.[167] DFMO has indeed been shown to inhibit a broad spectrum of fungal organisms, including *Candida* spp., but also a considerable number of plant pathogens.[163,168-173] The perspectives for plant disease control by inhibitors of ornithin decarboxylase are attractive. DFMO is taken up and translocated by plants, and it lacks phytotoxicity.[173] The latter beneficial aspect is most likely explained by a pronounced biosynthetic difference in plants and fungi. Plant putrescine biosynthesis can proceed via both ornithin and arginine, whereas fungi are normally restricted to ornithin decarboxylation.[168-170] Although this distinction might not be absolute,[174] ornithin decarboxylase appears more attractive as an antifungal target than alternative enzymes involved in polyamine biosynthesis. Most studies on the activity of ornithin decarboxylase have been done with DFMO, and data on optimization attempts toward the increase of antifungal activities are rare.[175] Although in theory less fungal-specific, the design of inhibitors of alternative targets might be feasible. Enzymatic differences between fungal and other organisms with regard to alternative enzymes could be pronounced enough to allow the synthesis of fungal-specific inhibitors. Inhibitors of SAM decarboxylase and spermidine synthase have indeed been shown to act antifungal.[171]

Trehalose is a nonreducing disaccharide isolated from different classes of prokaryotes, plants, animals, and fungi.[176] In fungi, trehalose is thought to comprise a storage carbohydrate in mycelia and, in particular, in spores.[176,177] Evidence exists that trehalose is also synthesized in response to stress.[178] Rapid catabolism of trehalose by **trehalase** is a characteristic feature of several developmental steps in fungi such as spore germination.[177] Validamycin A (Figure 9), an antibiotic isolated from *Streptomyces hygroscopicus* and in use for the control of rice sheath blight and some other sclerotial diseases in Japan,[179] has been recently identified as competitive inhibitor of trehalase isolated from *Rhizoctonia solani*.[180,181] The inhibitor activity of this antibiotic, however, is not restricted to fungi. Validamycin A was shown to inhibit trehalases in all organisms thus far examined.[182-185] I_{50} values are in the range of 10^{-7} *M* for trehalases of porcine, plant, and insect origin[183-185] and 10^{-5} *M* for fungal trehalases derived from *Rhizoctonia* and *Sclerotium* spp.[181] The inhibitory potency of validoxylamine A, a hydrolysis product of validamycin A (Figure 9), was increased by a factor of 10 to 20 when tested with the pig kidney and insect enzyme.[183,185] Castanospermine, an aminocyclitol and thus simpler in structure than validamycins, has been identified as an inhibitor of porcine trehalase;[183,185] simple amino sugars were inhibitors of an insect enzyme.[186] Both inhibitors, however, are also powerful inhibitors of glycohydrolases other than trehalase.

Although the information available for fungal trehalases remains very limited, the inhibitors described above comprise lead structures for more specific fungal trehalase inhibitors with antifungal activities. A distinct difference within the group of fungal trehalases has to be considered. The fungal enzymes are divided into two types: a nonregulatory "acid" trehalase apparently not directly involved in the regulation of trehalose hydrolysis, and a regulatory "neutral" trehalose activated by a typical cAMP-dependent phosphorylation.[187] A clear functional distinction has not yet been assigned. Indications exist, however, that the "regulatory" enzyme might be involved in cell cycle control of yeast.[187] The evaluation of trehalase as an antifungal target appears to require more information on the relative functions of the two enzyme types and should involve *in vitro* systems employing both trehalases. Fungal organisms that contain both types such as *S. cerevisiae* and *Candida utilis*[187] might be a valuable source for such programs.

Fungal pathogens in their parasitic phase are dependent on nutrients taken up from the host. The inhibition of **nutrient uptake** would thus comprise a valuable and attractive antifungal target, leading to the starvation of the pathogen. The fungal plasma membrane **H^+-ATPase** (proton pump) is by far the best-studied key component of transport systems. The enzyme is not directly involved in transport of compounds such as sugars, amino acids, and peptides. Acting as an energy-consuming proton pump, it is responsible for the generation of an electrochemical gradient across the plasma membrane, the driving force for membrane transprot across a concentration gradient. Much recent progress has been made in the understanding of H^+-ATPases,[188-191] and many vital functions other than the involvement in solute transport have been assigned to the enzyme.[191] There are two current drawbacks to inhibitor design efforts. Although various H^+-ATPase inhibitors have been described,[189-191] most lack specificity and exert side effects. Furthermore, fungal H^+-ATPases are closely related to the plant enzyme and to other cation-pumping ATPases. Thus, the magnitude of phytotoxic and toxic side effects of inhibitors is relatively hard to predict. However, the availability of cloned genes opens opportunities for rapid progress in the understanding of enzymatic properties, catalytic mechanisms, and differences among the enzymes of different origin.

2. Inhibitors with Targets in Host-Pathogen Interactions

The **melanin biosynthesis inhibitors** are good examples for a group of existing plant disease control agents that lack a direct inhibitory effect on the growth of fungal organisms (Chapter 6). They rather inhibit the formation of melanin in cell walls and act as antipenetrants. Antipenetrant activity, however, is restricted to the group of plant pathogens that form heavily melanized appressoria such as *Pyricularia oryzae* and *Colletotrichum* spp. There is increasing evidence that intact melanin biosynthesis is also involved as a virulence factor of human fungal pathogens,[192,193] and respective inhibitors might have chances to become useful as antimycotic drugs.

A more generally important target for antipenetrants could be the highly specialized fungal enzyme **cutinase**. These enzymes are produced by fungal pathogens during the penetration of the plant cuticles, the first barrier to be overcome by a pathogen. The majority of fungal pathogens gains entry through the intact leaf surface and, thus, has to breach the cuticle layer.[194-196] Plant cuticles consist of a complex mixture of impregnating material collectively called waxes and the polymer cutin which is best described as a fatty acid polyester.[197-199] The precise mechanism of cuticle penetration by invading plant pathogens has been long debated. The cuticle might be disrupted by the mere mechanical force of the growing infection structure, or the enzymatic cleavage of the polymeric cutin layer might be an inevitable prerequisite for a successful attempt to penetrate the layer.[200] Considerable progress has been made in the understanding of this initial step in host-pathogen interactions. The evidence in support of a crucial role for cutinase has been summarized in several reviews.[201-204]

The basic strategy to demonstrate the unequivocal existence of cutinolytic enzymes produced by plant pathogenic fungi was to grow a fungal pathogen on highly purified cutin as the sole carbon source. The success of this approach was first reported for *Fusarium solani* f. sp. *pisi*, and the corresponding cutinase was purified shortly thereafter.[201-204] Most of its properties have been revealed, including the enzyme mechanism of ester bond cleavage,[205] the primary protein structure derived from the cDNA sequence,[206] and the characterization of the cutinase gene including its promoter region.[207,208] However, the mere presence of cutinase in culture fluids of fungi **grown** saprophytically on cutin as the sole carbon source cannot be rated as conclusive proof for a crucial role in the penetration of plant cuticles by invading pathogenic fungi. This role has been investigated by more direct means with *F. solani* and its host *Pisum sativum*. Conclusive proof has been derived from several lines of evidence: (1) the presence of cutinase at the penetration site was demonstrated by electron microscopy combined with immunocytology;[209] (2) antibodies and chemical inhibitors against cutinase effectively prevented the infection of pea stems, presumably by inactivation of the enzyme after its excretion from the penetrating infection structure;[210-212] (3) a cutinase-deficient mutant of *Fusarium solani* was shown to express virulence exclusively on wounded but not on intact pea stem surfaces.[211] The initially excreted small amounts of enzyme were shown to act as a trigger, releasing small amounts of fatty acids from the cutin. These cutin monomers serve, in a second step, as specific inducers of cutinase biosynthesis.[213,214] The virulence of the cutinase-deficient mutant on intact plant surfaces could be reestablished by application of exogenous cutinase and a set of cell wall-degrading enzymes to the infection droplet,[211] substituting for the artificial would normally necessary for a successful infection with the cutinase-deficient mutant; (4) the application of molecular biology techniques has yielded additional and conclusive proof that cutinase produced during the early steps of plant infection is crucial. The introduction of the cutinase gene derived from *F. solani* into the wound parasite *Mycospherella* spp. rendered transformants pathogenic on intact cuticle-sealed surfaces of papaya fruits,[215] and the addition of the same gene into a cutinase-less mutant of *F. solani* restored pathogenicity on pea stems.[216]

The basic lines of evidence for the involvement of cutinase in plant infection have been presented for a second plant pathogen, *Colletotrichum gloeosporioides,* infecting immature papaya fruits.[217] Similar evidence for cutinase involvement has also been provided for *Alternaria alternata* infecting pear,[218,219] and *Venturia inaequalis* infecting apple leaves.[220] Many more cutinases have been purified or identified in culture fluids of pathogens grown in the presence of cutin.[204] The evidence for the role of cutinases as a penetration enzyme indicates that they are a crucial factor of virulence in a large number of important plant pathogens, although a mechanical rather than enzymatic penetration of plant cuticles remains as an alternative route of entrance. The mechanical mode of penetration, however, appears restricted to a relatively small group of pathogens that form appressoria with heavily melanized cell walls such as *Pyricularia oryzae*.[221-223] In summary, the role of cutinase as a crucial penetration enzyme is undisputed in most cases. For the group of pathogens producing melanized appressoria the role of the enzyme in pathogenicity remains less clear.

Although the functions of fungal cutinases as penetration enzymes appear to be generally similar, there is a remarkable molecular heterogeneity among cutinases.[204] One example for this diversity is the optimal pH for hydrolysis of the natural cutinase substrate, the polymer cutin. Cutinases purified from *Fusarium* spp. were found to be most active at pH 10.[202,204] In contrast, a cutinase preparation derived from *Botrytis cinerea*,[224] and a cutinase purified from *Venturia inaequalis*,[225] was found to exhibit optimal cutinolytic activity at the slightly acidic pH of 6.5. Interestingly, the cutinases with alkaline pH optima had been purified from stem-infecting pathogens, whereas the corresponding enzymes with slightly acidic optima were derived from pathogens primarily infecting leaves. The potential role of cutinase in the determination of tissue susceptibility and resistance has been substantiated in a recent study by Trail and Köller.[226] The impact of this more recent identification of two types of cutinases on the design of inhibitors is only poorly understood and will be discussed below.

The studies on the role of fungal cutinases are a typical example for the evaluation of putative targets for disease control agents. It has to be proven first that a target enzyme is indeed a crucial component of pathogenicity. The results with regard to cutinase indicate that a great number of plant pathogens depend on the presence of these enzymes. Only pathogens with heavily melanized appressoria appear to present potential exceptions.[204] Furthermore, the precisely defined role of cutinases allows one to predict the characteristics of cutinase inhibitors applied as plant disease control agents. They would inhibit the penetration of a pathogen into the host plant and, thus, would act as strict protectants rather than curative compounds. In contrast to conventional protective fungicides with their multisite mode of action, cutinase inhibitors would be expected to act specific, and they could move systemically impregnating the cuticle from the inside, very similar to the melanin biosynthesis inhibitors that also act as antipenetrants (Chapter 6).

The second step in the evaluation of cutinases as a target was the indentification of enzyme mechanism and, based on this, the search for enzyme inhibitors. The cutinase from *F. solani* was shown to act as a typical serine hydrolase,[205] a feature shared with the enzyme purified from *V. inaequalis*,[225] regardless of the different enzymatic parameters described above. Since the organophophorus insecticides are known mechanism-based inhibitors of acetylcholinesterase,[77] similar to cutinase a serine esterase, a series of these insecticides was investigated with regard to their effect on fungal cutinases from *F. solani* and *C. gloeosporioides*.[212,227] The general results were positive. The tested compounds were powerful cutinase inhibitors and inhibited penetration of the two pathogens in to their respective hosts. Considering the fatty acid nature of the natural substrate cutin, it was not surprising that a chain elongation of substituents had a drastic effect on the inhibitory potency. For example, the transition from methyl- to ethyl- in the 3,5,6-trichloro-2-pyridylphosphate shown in Figure 10 increased the cutinase inhibitor activity by a factor of 400 for the *Fusarium enzyme*,[221] and by a factor of 10,000 for the *Colletotrichum* cutinase.[227] This drastic difference in the response to structural alterations of inhibitors might be related to the type of cutinase. A strikingly different response has been observed for ebelactone A and B (Figure 10), the first "natural" cutinase inhibitor identified, when cutinases form *V. inaeuqlais* (acidic pH-optimim) and *R. solani* (alkaline pH-optimum) were assayed.[228] The slight variation from a methyl to ethyl substituent had a dramatic effect on inhibitor potency, but structural preference was reversed for the two enzymes. This differential response of the two types of cutinases can be rated as first evidence for differences in optimal inhibitor fits.

The strategy of cutinase inhibitor designs can be based on the mechanism of enzyme action (serine hydrolase) and the structure of the substrate cutin. Considering the fatty acid nature of the polyester to be hydrolyzed, the strategy would call for the elongation of aliphatic inhibitor substituents in order to mimic fatty acids. In case of symmetrically substituted phosphoorganic esters, however, the presence of two long-chain residues might be too bulky to be bound into the active site, and the design of asymmetric "short-long" derivatives should be of advantage. The validity of this approach has been demonstrated with the 3,5,6-trichloro-2-pyridylphosphate derivatives shown in Figure 10. A methyl *n*-hexyl substitution pattern increased the inhibitor potency in *in vitro* tests with *Fusarium* cutinase, and the derivative had good protective activity against apple scab in greenhouse studies.[229] Cutinase from the respective pathogen, *V. inaequalis*, was inhibited at a inhibitor/enzyme ratio approximating 1:1.[220] At a concentration of 0.5 μg/ml, the penetration of the pathogen into apple leaves was completely inhibited, whereas spore germination and the formation of appressoria as prepenetration events were not affected. These results indicate that the general perspectives of cutinase inhibitors is promising.

Are phosphororganic esters as powerful inhibitors of serine esterases an ideal lead

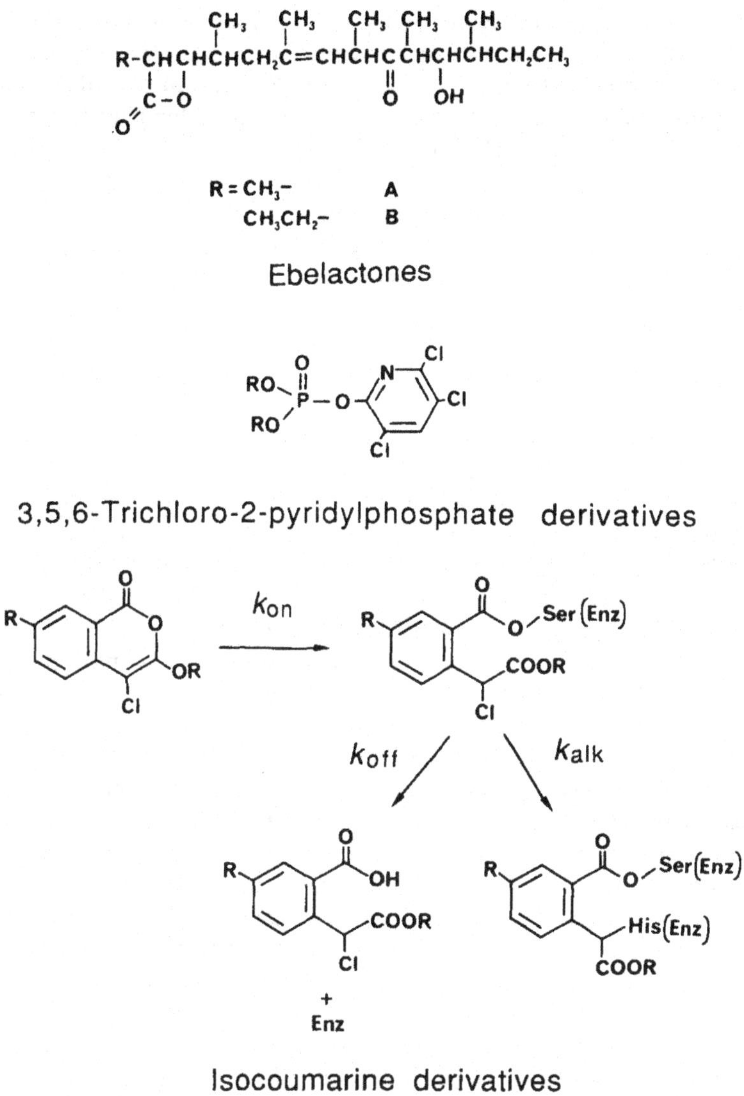

FIGURE 10. Lead structures for cutinase inhibitors.

structure for cutinase inhibitors? Probably not. They are known to act as inhibitors of mammalian esterases, and although the adaptation to the structure of the substrate cutin can be expected to improve inhibitor specificity and, thus, toxicological properties, the degree of specificity ultimately achieved remains uncertain. A second drawback of these compounds is a sometimes low hydrolytic stability. In contrast to a contact insecticide, a substantially increased chemical stability would be required for a cutinase inhibitor acting as a plant protectant. It should be mentioned, however, that a tremendous wealth of information is available on all details of phosphororganic ester action,[77] and the stringent application of this knowledge to the design of cutinase inhibitors might be successful.

Phosphoorganic esters are only one out of numerous chemical groups that have been described as suicide substrates or transition-state analogues for both serine esterases and serine proteases. The principles of enzyme catalysis are identical for both classes of enzymes,[205] and inhibitors of both types of serine hydrolases can serve as a lead for cutinase inhibitors. None of these inhibitor classes has been evaluated according to their potencies to act as cutinase inhibitors. This applies to the second group of insecticidal serine esterase inhibitors, the carbamates,[77] but also to numerous other lead structures that have been described as inhibitors of serine esterases and serine proteases. Most of these inhibitors act as suicide substrates or analogues of high energy intermediates. Inhibitory structures include haloenol lactones,[232,233] isatoic anhydrates,[234] haloketone transition-state analogues,[235-240] isocoumraine derivatives,[241-243] oxazine dioners,[244] isobenzofuranones,[245] benzoxazinones,[246] and β-lactam derivatives.[247] Many of these structures are typical mechanism-based inhibitors activated through the acetylation of the active serine; the course of inhibition is outlined for a halogenated isocoumarine derivative as a representative example for a mechanism-based serine esterase inhibitor in Figure 10.

Not all of these inhibitor classes will be suitable candidates for the design of cutinase inhibitors. The abundancy of different lead structures, however, appears promising. It will have to be determined whether a high specificity of putative cutinase inhibitors without effects on other esterases can be achieved. An advantage will be that toxicologically important esterases are well characterized,[230] and evaluative *in vitro* tests employing acetylcholine and choline esterase, liver esterase, or cholesterol esterase can be implemented as integrative parts of design efforts. In summary, fungal cutinases are one of the best-characterized targets for specific plant disease control agents. Only the initiation of inhibitor design efforts based on the accumulated knowledge will answer the question of whether inhibitors will be vital under practical conditions. It might even be possible to combine a cutinase inhibitor with an inhibitor of metabolic reactions and, thus, two mode of action principles within one molecule. Benomyl serves as a model case. This fungicide is spontaneously converted to methyl-2-benzimidazole, the actual inhibitor of tubulin polymerization (Chapter 3), and butylisocyanate, an inhibitor of acetylcholinesterase. The latter breakdown product was shown to exert cutinase inhibitor and, under laboratory conditions, antipenetrant activity.[248] The development of resistance to benomyl indicates that, under practical conditions, this secondary mode of antipenetrant action is of minor importance. The example demonstrates, however, that the design of "combination molecules" are a potential alternative.

Not only plant but also human pathogens are challenged by a passive barrier to be overcome in early steps of host invasion. Increasing evidence exists that human pathogens actively breach barriers of epithelial tissue, possibly associated with proteolytic activity as a factor of virulence.[249] For example, a mutant of *Candida albicans* deficient in an extracellular **aspertate proteinase** was 1000-fold less virulent than a wild-type and proteinase-secreting strain.[250] Consequently, it might become possible to design inhibitors with antipenetrant activities that would protect from the invasion of human tissue.

Many specialized steps in the initiation and development of fungal diseases are not sufficiently understood to guide the synthesis of chemical antagonists with indirect modes of action. Progress, however, is rapid, and future developments can be expected to render opportunities for the design of such antidisease compounds. For example, penetration of host tissue is preceded by the attachment to and recognition of an appropriate host.[251-254] Some antimycotic drugs such as amphotericin A, ketoconazole, and itraconazole (Chapter 7) have been reported to interfere, at subinhibitory doses, with initial adherence of *C. albicans* to host tissue.[255] Unfortunately, the molecular mechanism of this effect remains unknown.

It is a common feature of human pathogenic fungi that the morphology found in infected tissue is different from that of the disease-initiating propagule. This dimorphism is believed to comprise a factor of virulence, although experimental evidence is not without contradictions.[256] The application of molecular genetic tools will allow one to study the genetic control of this dimorphism, and to generate highly specific mutants by gene disruption for studies aimed at the importance of dimorphism and other traits as determinants of virulence.[257-259] Similar strategies are being applied to identify the nature and regulation of specific genes involved in the initiation and propagation of plant diseases.[260-262] For example, specific genes expressed during the stage of appressoria formation, a prerequisite for host invasion, have been identified.[263,264] Similarly, genes specifically involved in the formation of fungal conidia have been cloned and characterized.[265,266] Circumstantial evidence indicates that sterols in their regulatory mode are involved in both the induction of appressoria formation and conidiation.[267,268] The combination of molecular genetic tools with biochemistry will undoubtedly elucidate many of the regulatory mechanisms involved in the expression of developmental genes relevant for the initiation, establishment, and propagation of fungal diseases of both plants and human. This knowledge will identify future targets for chemical interference. The inhibition of gene expression rather than the inhibition of gene products might become a realistic avenue in the development of specific antifungal agents.

IV. INDIRECT MODES OF ACTION

The class of compounds targeted against highly specialized steps in disease initiation and development are often referred to as inhibitors with an indirect mode of action. This term might be misleading. Corresponding inhibitors are by definition nontoxic to the saprophytic growth of a fungus, but they do interfere directly with fungal gene products. A term such as antidisease inhibitors might be more appropriate. The term indirect mode of action should imply that the development of a fungal pathogen is halted by indirect means, most often by the strengthening or the restoration of the host's resistance response. For example, any compound that would stimulate or restore the immunoresponse of a patient would thus have an indirect mode of antifungal action.[269]

Although distinctly different from a mammalian immunosystem, plants have developed a highly effective arsenal of defense reactions, which is only overcome as an exception in a compatible host-pathogen interaction. Most of the plant defense reactions are induced by fungal attack, and they comprise the synthesis of phytoalexins, the synthesis of pathogenesis-related proteins such as chitinases, and the erection of physical barriers such as papillae and lignin.[270-275] These defense responses of plants to fungal attack are most often local events in vicinity to the cells attacked, but it is long known that resistance in remote tissues of a plant can be induced by fungal infections at other sites.[276]

The stimulation of plant host resistance responses by plant disease control agents comprises an attractive concept with a relatively long history of attention.[277] These compounds would not exert inherent toxic properties; they rather would trigger natural defense reactions of plants that are overcome by the pathogen in a compatible interaction. Indirect effects via host resistance responses have been reported for several classes of compounds. For example, many of our modern systemic fungicides such as the benzimidazoles,[278] sterol demethylation inhibitors (Chapter 7),[279] phenylamides (Chapter 5),[280] carboxamides (Chapter 2),[281] polyoxin D (Figure 7),[281] and fosetyl-Al (Chapter 5)[282] have been discussed as activators of host defense reactions. In most cases it has not been clarified, however, whether these host responses are triggered directly by the fungicides, or indirectly through the damaging effect on the pathogen, which in turn could lead to the release of fungal elicitors of plant defense reactions. The fact that resistance developed to most of the fungicides mentioned above,[3] and the results described by Heath[281] appear to be in favor of the latter explanation.

Although many herbicides have been shown to increase the incidence of disease or to exert direct effects on (soilborne) pathogens,[283] some herbicides have been described to protect plants from disease via the induction of host resistance responses.[284] For example, the dinitroaniline herbicide trifluralin was reported to induce resistance of tomato against *Fusarium oxysporum*, and to induce the synthesis of fungitoxic compounds different from typical solaceous phytoalexins in inoculated plants.[285] Unfortunately, exploratory follow-up studies to these reports are extremely rare, and the molecular mechanisms of induced responses remain unknown.

Two compounds, both active in the control of rice blast disease and described in detail in previous review articles,[284,286] are likely to act as inducers of defense reactions in susceptible rice plants. Probendazole, a saccharin derivative (Figure 3), is in use as a prophylactic rice blast fungicide with an indirect mode of action. The precise site of action within the cascade of defense reactions, however, has not been clarified.[45-47] The second antiblast compound with an indirect mode of action,2,2-dichloro-3,3-dimethylcyclopropane carboxylic acid (WL 28325), has not been developed to a market product. Although the effects triggered by this compound, lignification of infected tissue and enhancement of phytoalexin production, has been described,[287] the molecular mechanisms of defense gene activation have not been elucidated.

Only the precise understanding of events leading to triggered defense reactions can provide leads for the rational design of chemicals with indirect modes of action. Recent progress in the identification of plant defense genes and the understanding of regulatory mechanisms involved in plant defense reactions has been rapid.[270,275,288] This emerging knowledge has great chance to be utilized in the genetic engineering of disease-resistant plants, but it will also provide rational means for concepts in the design of future chemicals with indirect modes of actions. In particular attractive would be the identification of signal molecules involved in the phenomenon of induced resistance,[276] a task not fully accomplished. However, first evidence was presented that salicyclic acid could act as an endogenous signal in the resistance of tobacco to viral infection,[289] and of cucumber to infection by *Colletotrichum lagenarium*.[290]

V. TARGET RESEARCH AND FUNGAL RESISTANCE

The introduction of many of our modern and systemic fungicides used in plant disease control was followed by the development of resistance.[3] Likewise, some of the antimycotic drugs have been affected by identical drawbacks, with little current impact on the polyene antimycotics and, thus far, the azole compounds, but severely affecting 5-fluorocytosine.[291,292] Although the problems with resistance of fungi to antifungal inhibitors appear to be very similar, there has been little cooperation between the two disciplines of applied mycology. This applies in particular to the terminology in use and the understanding of the selection of resistant subpopulations of fungal pathogens.

In the field of plant disease control, resistance is defined as a stable and inheritable adjustment of a fungus to a fungicide resulting in a significantly less than normal sensitivity.[3] The origin of resistant genotypes is by spontaneous mutation, and a small number of these genotypes will reside in a fungal population prior to the exposure with a given antifungal inhibitor. Under the selective advantage in the presence of the inhibitor, the resistant genotypes will increase in frequency. Two types of population responses have been identified and defined for plant pathogens. The original population can consist of a large, sensitive population and a small but highly resistant subpopulation. Since the inhibitor is virtually inactive for the resistant subpopulation, the entire population changes in a qualitative manner, and selection might proceed fast. A second population response is quantitative rather qualitative in nature. In this case, the original sensitivity distribution is unimodal and continuous in character ranging from highly sensitive to resistant. Exposure of the population to the selection force of a fungicide will lead to a shift of the sensitivity distribution toward higher frequencies of resistant genotypes, but it will not change the principle character of the continuous distribution. Almost classical examples for a qualitative change of population responses are the benzimidazoles and phenylamides; the best documented case for a quantitative change are the sterol demethylation inhibitors.[3,293]

In the field of antimycotic drugs, the two terms "primary" and "secondary" resistance are in use.[291] Primary resistance describes a case where a high frequency of resistant phenotypes (>1%) can be identified prior to any contact to the drug. The term secondary resistance is used when resistance develops as the result of exposure to the drug, in particular during chemotherapy. In the light of two distinctly different sensitivity distributions found in plant pathogen populations and the two different pathways leading to the selection of resistant phenotypes, these terms should be critically discussed and eventually revised. Secondary resistance (no resistant strains are discovered prior to the exposure of a pathogen population to the drug) might coincide with a highly resistant and separated subpopulation. Here, the frequency of resistant genotypes in wild-type populations is minute and most often beyond detection when finite number of isolates are tested.[3,293] In the case of a continuous sensitivity distribution within a wild-type population, the presence of resistant phenotypes is more frequent and could correspond to the term primary resistance. For example, the sensitivity distribution within a population of 300-wild-type isolates of *Venturia inaequalis*, the causal agent of apple scab, to the sterol demethylation inhibitor flusilazole was continuous in character, ranging from highly sensitive to resistant phenotypes separated by a factor of 300.[294] The distribution found in a population of *Candida albicans* to itraconazole was very similar, and MIC values were separated by an even greater margin.[295] It should be emphasized that under conditions of continuous sensitivity distributions, the definition of the term resistance is not without problems.[294] It should also be emphasized that the speed of resistance development is not only determined by the risk inherent to the molecular mechanism of a given inhibitor, but also by other factors such as the biology and epidemiology of a given pathogen.[294] In particular, the epidemiology of human pathogens in relation to the development of resistance to antimycotic drugs deserves increased attention. The frequency of resistant genotypes residing in wild-type populations is largely dependent on the size of the population.[296] The population sizes that infect field crops can be expected to be much larger than those infecting individual human patients, and the risk of resistance development might be substantially influenced by these differences.

Biochemical research aimed at the phenomenon of fungal resistance to antifungal inhibitors has two different goals: (1) predicting and minimizing the risk of resistance in inhibitor design programs, and (2) the exploration of specific "resistance breakers" designed to counteract resistance once it became established in fungal pathogen populations. Based on past experiences with fungicide resistance, three risk categories have been proposed.[3] A very low risk of resistance has been assigned to biocidal compounds with multisite modes of action, a moderate risk became apparent for several classes of site-specific fungicides such as the sterol biosynthesis inhibitors, whereas other representatives of site-specific inhibitors such as the phenylamides or the benzimidazoles are classified as high risk compounds. Evidence exists that the differentiation of site-specific fungicides into moderate and high risk fungicides is related to the qualitative and quantitative pattern of resistance development. The two different population responses are, on the other hand, related to the genetics and, ultimately, the molecular mechanisms of resistance.

In the following section, the question will be addressed how this information can be utilized in the design of novel and site-specific inhibitors with low to moderate risks of resistance, and how the mechanism of resistance can guide the synthesis of antiresistance compounds. The former question is closely related to the fitness penalty imposed on fungal genotypes resistant to an antifungal inhibitor. If the cost for a high level of resistance is substantial, the risk of resistance should be small. Opportunities for the design of resistance breakers are dependent on the precise identification of resistance mechanisms,[297] and on the question of how many optional resistance mechanisms exist within a given population of a fungal pathogen. Since putative resistance breakers will have to be used in mixture with the original inhibitor, this option would be restricted by economical constraints if several mechanisms with equal impact can be identified. In his comprehensive review article on fungal resistance to fungicides used in agriculture, Dekker[298] listed several different mechanisms of resistance, which can be partly combined into (1) mutational change of the target site leading to decreased inhibitor affinity, (2) decreased intracellular inhibitor concentration, (3) compensation of inhibitor by an increased level of target sites, and (4) circumvention of inhibitor action.

A. TARGET SITE MUTATIONS

The **benzimidazoles** as inhibitors of tubulin polymerization comprise the first class of broad-spectrum fungicides used in agriculture with severe problems of resistance. As described by Ishii in Chapter 3, a point mutation in β-tubulin, the target protein, has been identified as the mutational change responsible for a drastically decreased binding affinity. Phenylcarbamates have been found to bind selectively to the mutated target, and antifungal activity was restored. Unfortunately, the phenylcarbamates only bind to one out of several classes of mutated tubulin subunits and, thus, are inhibitory to most but not all of the resistant genotypes. The resistance risk of benzimidazoles is considered high.[3]

A second class of compounds characterized by a mutational change of the target site in combination with a high risk of resistance are the **phenylamides** described in detail by Griffith et al. in Chapter 5. The phenylamides are inhibitors of ribosomal RNA synthesis. They bind to the intact polymerase-DNA complex and, thus, are not active site-directed inhibitors of RNA polymerase. Although the phenylamid binding site within the RNA polymerase complex has not been identified thus far, a mutational alteration of that site is with all likelihood responsible for resistance. The combination of a target mutation with an enzyme inhibitor (succinate dehydrogenase) that is not active site-directed has also been reported for the **carboxamides** (White and Georgopoulos in Chapter 1). The relatively positive experience with regard to the speed of resistance development to carboximides is most likely related to the biology of the spectrum of controlled pathogens rather than the inherent risk of the compounds.[3]

A target mutation has also been identified as the mechanism of resistance of *Pyricularia oryzae* to **kasugamycin**, an antibiotic used in the control of rice blast. The mode of action is the inhibition of protein synthesis by binding to ribosomes.[73-75] Binding affinity of kasugamycin to ribosomes isolated from a resistant mutant *P. oryzae* was reduced,[298,299] indicating a mutational change within the ribosome binding site. It was observed that kasugamycin-resistant field populations were less fit than the sensitive counterparts. This fitness penalty might explain why the development of widespread practical resistance was relatively slow.[298,299]

All examples of target mutations listed above represent antifungal inhibitors that are not targeted at the substrate binding site of an enzyme. This restriction might be different for the allylamine **terbinafine** (Chapter 7). Orth et al.[300] investigated a *Ustilago maydis* mutant with a high level of terbinafine resistance and provided strong evidence for a decreased affinity of the inhibitor to the target enzyme squalene epoxidase. Resistance was not accompanied by any signs of a fitness penalty.[301] As discussed in Chapter 7, it has not been entirely clarified whether allylamines act as competitive inhibitors of the fungal squalene epoxidase or bind to sites other than the active site.

Only very limited evidence exists that a target mutation is involved in the mechanism of fungal resistance to DMIs (Chapter 7). Mutants of *Saccharomyces cerevisiae* with an altered P450 lanosterol demethylase have been reported.[302,303] The binding of DMIs and also the binding of lanosterol as the enzyme substrate were drastically reduced. This mechanism of resistance is unlikely to occur under natural conditions, because the viability of the deficient mutants is greatly impaired. Interesting with regard to altered DMI affinities is the report that the disruption of the NADPH-cytochrome P-450 reductase gene in *S. cerevisiae* generated a transformant without an expected defect in lanosterol demethylation.[304] However, the transformant was 200 times more sensitive to ketoconazole. Apparently, an alternative redox partner for the P450 lanosterol demethylase exists, and the interaction with this alternative component might have impact on the binding affinity of DMIs. It is not known whether a similar mechanism is in operation for DMI-resistant isolates of *C. albicans* with reduced affinities of DMIs to the target site.[305]

B. DECREASED INHIBITOR CONCENTRATION

Metabolic detoxification has long been identified as the prime mechanism of insecticide resistance.[306] For most of the site-specific fungicides, detoxification appears to be of minor importance. However, detoxification in a broader sense has been reported for **copper compounds** and for **captan**, both multisite inhibitors without serious problems with practical resistance, regardless of their use for many decades.[3] Copper resistance of *S. cerevisiae* has been explained by an amplification of a copper-induced gene coding for a cystein-rich metallothionine that binds and thus ''detoxifies'' intracellular copper ions.[307] Evidence exists that resistance of *Botrytis cinerea* to captan is correlated with an increased level of reduced glutathion, creating a higher level of nonvital and captan-desactivation thiol groups.[308] It is intriguing to speculate that the necessity to maintain a high and continuous level of detoxifying scavengers requires so much energy that a fitness penalty tolerable for survival under outside conditions is reached at relatively low levels of resistance.

The synthesis of detoxifying enzymes could be expected to be less energy demanding and, thus, a more risky resistance mechanism. This type of detoxification had been identified as the mechanism of resistance of *Fusarium solani* to the phytoalexin **pisatin** present in peas. Pisatin induces the synthesis of a specific pisatin-demethylating cytochrome P-450 monooxygenase, and the presence of this detoxifying enzyme has been related to the virulence of this pathogen to peas.[66,309] The fact that the induction of the detoxification enzyme is a prerequisite for virulence indicates the lack of a severe fitness penalty imposed by pisatin resistance. The presence of a relatively substrate-specific P450 phytoalexin detoxification system has been reported for other pathogens and apparently evolved over a relatively long period of time.[66] This requirement for coevolution and selection might explain why P450-catalyzed detoxifications are rare mechanisms of resistance to synthetic antifungal agents. Enzymatic detoxification has been reported for the rice blast fungicide iprobenfos (IBP) (Figure 6). In resistant phenotypes most frequently collected from field populations, the oxidation of the phosphorus-sulfur bond, which also takes place in sensitive isolates, is accelerated in resistant phenotypes.[299,310] The metabolic function of the corresponding enzyme remains unknown.

Inhibitor detoxification has also been reported for some experimental inhibitors with close structural resemblance to respective enzyme substrates. Inhibition of growth of *Gibberella fujikuroi* by oxidosqualene cyclase inhibitor 2,3-epiminosqualene (Chapter 7) was only transient. The inhibitor was enzymatically converted to 24,25-epiminolanosterol, 25-aminolanosterol, and eventually lanosterol.[311,312] This example demonstrates a potential shortcoming of inhibitors designed as very close analogues of enzyme substrates. They might not only act as inhibitors, but also could undergo slow enzymatic conversion and thus detoxification.

An alternative way for a fungal cell to decrease intracellular inhibitor concentrations is to reduce the uptake into the cell. **Polyoxins** and **nikkomycins**, both inhibitors of fungal chitin synthetase, are one of the rare examples of antifungal inhibitors that depend on active transport into fungal cells. These inhibitors are taken up via a peptide transport system, and effective transport was impaired in resistant isolates of several fungi.[103,313] The relative expression of multiple peptide transporters appears to be involved in the expression of resistance. For example, a peptide transport-deficient and nikomycin-resistant mutant of *C. albicans* was shown to have lost the dipeptide transport system with a simultaneous increase in tripeptide transport.[314] Synthetic modifications of the peptide moieties of the inhibitors could restore uptake; however, the modified compounds were either less active as chitin synthetase inhibitors, or more prone to peptidase degradation.[103,313] Some indications exist that polyoxin-resistant field isolated of *Alternaria kikuchiana* were less fit than their sensitive counterparts.[298] The modification of the peptide transport system toward decreased inhibitor uptake was apparently accompanied by a fitness penalty.

The initial model of DMI resistance proposed by de Waard and originating from studies with *Aspergillus nidulans, Pennicillium italicum,* and *C. albicans*[315-318] is related to a mechanism of decreased intracellular inhibitor concentrations. The model implies an energy-dependent efflux of the inhibitors, induced by the inhibitor in sensitive strains, and constitutive in resistant mutants. The immediate efflux from resistant cells prevents an initial burst of intracellular inhibitor levels and, thus, the saturation of the target site. A second mechanism of DMI removal, a segregation of triadimenol into the vacuoles, has been reported for a DMI-resistant mutant of *U. avenae*.[319]

C. COMPENSATION

Compensation of inhibitor action by an increased synthesis of the respective target is relatively rare among the reported mechanisms of resistance. Evidence for the suitability of such gene-dosage effects[320] comes from *S. cerevisiae* transformed with the cytochrome P-450 lanosterol demethylase gene coding for the target site of DMIs (Chapter 7). A higher copy number of the gene led to ketoconazole resistance of the transformants.[321,322] A gene dosage effect was also identified as the cause for a relatively low level of resistance of *U. maydis* to the squalene epoxidase inhibitor terbinafine.[300]

5-Fluorocytosine (5-FC), as outlined above, is a good example for an antifungal prodrug that is actively taken up and requires a series of metabolic conversions for full activity. As discussed by Kerridge et al.,[292] the most widespread mechanism of resistance among clinical isolates of *Candida* spp. is the down-regulation of the enzyme involved in the conversion of 5-FU to 5-FUMP and 5-FdUMP. It has not been clarified why other essential components involved in the transport and conversion of the drug are less prone to changes leading to resistance. A likely explanation is that drastically decreased levels of essential enzymes are normally accompanied by substantial fitness penalties. The example demonstrates, however, that a manipulation of enzymes to lower levels, in opposition to a typical compensation mechanism, can serve as means of resistance development whenever a prodrug requires enzymatic activation.

D. CIRCUMVENTION

The circumvention of inhibitor action has been reported as one of the major mechanisms of fungal resistance to polyene macrolides such as amphotericine B (Chapter 7). These antibiotics interact with ergosterol present in plasma membranes of sensitive isolates. The decrease of sterol contents, or the structural alteration of sterol end products both lead to lowered affinities of the membranes for the antibiotic and, consequently, to reduced sensitivities.[291,292] The levels of resistance achieved with this mechanism are relatively low. Widespread problems with practical resistance have not been reported, probably because the fitness penalties for drastic alterations of membrane compositions and high levels of resistance are substantial.[291] A similar mechanism of resistance, a decreased content of sterols, was reported for tomatine-resistant mutants of *F. solani*.[50,51]

An interesting type of inhibitor circumvention has been reported for the effects of the hypocholestereolemic drugs AY-9944 and trifluperidol on yeast, a fungal organism relatively insensitive to the inhibitor of the Δ^8-Δ^7-isomerase (Chapter 7). Although Δ^8-sterol precursors accumulated, the precursors were predominantly found in the pool of sterols esters and, thus, less critical for growth than free sterols incorporated into membranes.[323,324] However, a segregation of sterol precursors into sterol esters is not a general escape from the action of sterol biosynthesis inhibitors. The proportion of sterol precursors accumulating under the action of triadimenol in a resistant mutant of *U. avenae* was not increased in sterol esters, and the segregation of sterol precursors into sterol esters was not involved in the expression of resistance of *U. avenae* to DMIs.[325]

Several other options of circumvention, however, have been reported for a number of fungal mutants resistant to DMIs (Chapter 7). A resistant mutant of *S. cerevisiae* was shown to lack the enzymatic system for the hydroxylation of accumulating 14-methyl sterols at the 6-position. These dihydroxysterols are thought to exert more serious membrane-damaging effects.[326] Elevated membrane-damaging effects have also been reported for 14-methyl sterols containing additional C-4 methyl groups. Consequently, a circumvention of effects inherent to lanosterol accumulating under the action of a DMI would be the removal of the C-4 methyl groups from the precursor. Indeed, a mutant of *U. maydis* that lacked the system for the removal of both C-4 methyl groups from the precursor was supersensitive to DMIs.[301] A change of sterol structures has also been reported for DMI-resistant mutants of *Cercospora beticola*. All but one mutant produced Δ^{22}-sterols; this side-chain desaturation was absent in sterols isolated from the sensitive parent strain.[327] In summary, several examples indicate that structural changes of sterol structures and membrane properties can circumvent the inhibitory effects of accumulating 14-methyl sterols. A similar mechanism has been reported for *Mucor rouxii*, a fungus relatively insensitive to DMIs. Here, 14-methyl sterols accumulated to high levels without effect on fungal growth.[328]

E. CONCLUSION

The arsenal of escape options outlined above indicates that fungal organisms inherited diverse mechanisms to resist inhibitor action. With the exception of detoxifications, which appear to be surprisingly rare, almost all other possible mechanisms have been reported. However, the escape mechanisms are by and large highly specific, and it is surprising that pleiotropic drug resistance, a form of resistance that mediates insensitivity of mammalian tumors (but also of *S. cerevisiae*) to a broad spectrum of structurally diverse drugs,[329] has not been reported to play a major role. In almost all case investigated, fungal resistance was restricted to groups of antifungal inhibitors with identical modes of action, and cross-resistance between unrelated compounds is rare.

Based on experience mainly derived from the development of resistance to agricultural fungicides, the multisite inhibitors were classified as low risk compounds. To the contrary, resistance to one of the first groups of systemic and site-specific fungicides, the benzimidazoles, developed rapidly, and it was proposed that the risk inherent to single-site inhibitors was generally high.[3] Subsequent groups of specific inhibitors have shown, however, that this generalization is not always valid. The development of resistance was relatively slow for the DMIs; practical resistance has not been reported for a second group of sterol biosynthesis inhibitors, the morpholines (Chapter 7), and no resistance has been observed for the melanin biosynthesis inhibitors (Chapter 6), regardless of their specific mode of action.[3] This relatively promising experience has led to increased efforts in the development of investigative systems for the assessment and prediction of risks inherent to new fungicides before they are used on a wide scale.[330,331] Knowledge about the mechanism of resistance comprises one of the key elements of these prediction schemes.

The involvement of molecular considerations regarding resistance goes beyond its value in predictive studies. In particular in the design of inhibitors of predefined target sites, it might become possible to incorporate this question into target considerations. The current information on the speed and thus the risk of resistance development indicates that differences between site-specific inhibitors exist. Although circumstantial in character, evidence appears to indicate that active site-directed inhibitors of enzymes (e.g., DMIs, morpholines, melanin biosynthesis inhibitors) are less prone to the development of resistance. Rapid development of resistance has been reported for inhibitors that bind to nonenzyme targets (benzimidazoles, phenylamides). Here, a mutational change leading to a drastically reduced inhibitor affinity can occur without adverse effects on the functionality of the respective target. A similar mutation without any impairment of the target site's function and, thus, impact on the relative fitness of the mutant is less likely to occur for the preserved structure of catalytic sites of enzymes. Consequently, key enzymes and active site-directed inhibitors might be the better choice in design programs.

It has to be emphasized that the preference of enzyme inhibitors can only be rated as guidance for target considerations. Decreased affinities of enzyme inhibitors as mechanisms of resistance have been reported for the insecticidal inhibitors of acetylcholine esterase,[306] and for several herbicides.[332] Furthermore, resistance to enzyme inhibitors can develop by other means than target mutations. It follows that a risk assessment should be conducted whenever biological activities of designed inhibitors have been identified. Traditional approaches are the generation of resistant mutants, and assessments of these mutants regarding their fitness in relation to the levels of resistance achieved.[330,331] Recombinant DNA techniques will be an additional and valuable asset of these efforts. For example, it has become possible to transform a sensitive fungus with genomic DNA fragments derived from a resistant counterpart and to identify (and sequence) the gene that renders resistance.[333] This approach has been introduced and applied to the experimental fungicide LY214352.[334] The "resistance genes", once identified, can be compared with the known target site genes and thus would rapidly identify the presence of a point mutation as the mechanism of resistance. They can also be used to address the question of whether all resistant mutants, but also resistant phenotypes of other species, contain this gene and, thus, share a common mechanism of resistance. Target considerations prior to the initiation of synthesis programs in combination with resistance risk assessments including recombinant DNA techniques will be able to minimize the threat of rapid resistance development for future fungicides and antifungal drugs. Some of our newer specific inhibitors appear to indicate that this goal can be reached.

As mentioned above, a second aim of molecular studies in the field of fungicide resistance would be the synthesis of "resistance breakers" for antifungal inhibitors already hampered by resistance. Fundamental for such programs is a profound knowledge of the mechanism of resistance. For example inhibitor detoxification would call for an inhibitor of the detoxifying enzyme, decreased uptake for compounds that synergize uptake, and target site modifications for a structural readjustment of the inhibitor into the mutated target site. The latter approach has been described for the phenylcarbamates with activity against benzimidazole-resistant mutants (Chapter 3). There seems to be little chance for chemical interference with a compensation or circumvention mechanism. Since fungal populations will never become entirely resistant, potential resistance breakers have to be used in mixtures with the original fungicide. Consequently, it will be of utmost importance to determine the number and relative importance of resistance mechanism operative in field and clinical populations of fungal pathogens. The experience with the phenylcarbamates has shown that most but not all of the tubulin mutations leading to resistance were sensitive to these compounds (Chapter 3). Thus, even mixtures of benzimidazoles and phenylcarbamates were inactive toward a small portion of phenotypes,[3] and the risk of resistance remained high even for the mixture.[3]

Both questions, the mechanism of resistance and the presence and relative importance of alternative mechanisms, have not been clearly answered for many of our antifungal inhibitors. In addition, the resistance mechanisms reported in the literature are sometimes diverse, and it is rarely known whether all alternatives are equally relevant under practical and competitive conditions. The group of DMIs is a typical example for this mechanistical diversity. As described above, with the exception of inhibitor detoxification, all other options of resistance mechanisms have been suggested. Nevertheless, the conclusion that this variety of options would make any efforts to design resistance breakers obsolete appears premature.

First, many of the proposed models are in an infant stage of experimental refinement and require more work for full confirmation. Second, the levels of resistance and the fitness penalties inherent to the proposed resistance mechanisms are rarely known. Indications exist that genes responsible for low levels of DMI resistance can act additive, but more importantly, even synergistically.[335,336] It thus remains possible that minor mechanisms of resistance are superimposd by a major one, and chemical interference with the most important mechanism might be sufficient to render restored DMI performance. Interesting with regard to the mechanism of DMI resistance is the recent finding that expression of DMI resistance in *U. avenae* was an induced response rather than a constitutive trait.[325] It will thus become possible to clone the genes transcribed during the recovery phase of resistant cells, and it will open the opportunity to search for antiresistance compounds that interfere with the induction process rather thant the expressed resistance trait.

Even if promising antiresistance compounds could be found, they would have to be used in combination with DMIs, and developmental costs would be comparable with those required for a new fungicide. In many cases, economical considerations might thus prohibit this option. An alternative to resistance breakers might be the search for synergists among fungicides already available.[337] For the DMIs, a third alternative might exist. It has become evident that all DMIs are cross-resistant to each other. However, the levels of resistance are greatly variable and reach from high to very low.[335,338] Extrapolation of this trait would indicate that it might be possible to synthesize new DMIs without cross-resistance to the previous representatives. A molecular explanation for the phenomenon of variable resistance factors dependent on the DMI structure would certainly assist the evaluation of this option. Overall, the DMIs exemplify that opportunities for engagement of biochemistry and molecular biology in the field of fungicide resistance are real.

VI. BIOCHEMISTRY OF FUNGICIDE ACTION AND MOLECULAR GENETICS

The analysis and manipulation of genes through recombinant DNA techniques have become routine during the last decade and have been increasingly applied to fungal organisms, and some of the opportunities, such as the convenience and ease of protein sequence determination, have been emphasized already throughout the chapter. However, it can be expected that the impact of these techniques on fungicide research will be substantial in the near future. It is clearly beyond the scope of this article to treat the field of fungal molecular biology in detail. Rather, an attempt will be made to outline general principles and to highlight some potential fields of application in the field of antifungal target research. Excellent review articles on basic principles of recombinant DNA technologies applied to fungi have been published,[260-262] and numerous manuals such as *Molecular Cloning — A Laboratory Manual*[339] are available as sources for techniques and approaches.

The key operations in studies employing molecular genetics are the cloning and isolation of cDNAs (*in vitro* prepared DNA copies of mRNAs) and genes coding for a protein of interest. In brief, cDNA libraries prepared from the total of mRNAs isolated from a fungal organism are screened for the bacterial clone containing the cDNA under question. Screening methods rely either on antibodies against the enzyme or protein or on nucleic acid probes that specifically hybridize with the cDNA under question. Various nucleic acid probes can be used for this purpose. Most convenient is the availability of probes, either genes or cDNAs, that contain the coding region for a functionally identical protein expressed by a different fungus or even a different organism. These heterologous probes are versatile tools whenever the sequence of a protein was preserved through evolution. A good example for the suitability of heterologous nucleic acids as specific probes is HMG-CoA reductase, a key enzyme in sterol biosynthesis (Chapter 7). A high degree of sequence conservation was discovered between the two isozymes present in yeast and the single enzyme expressed in mammalian cells.[340] The plant enzyme appeared structurally distinct, but it was possible to screen for the respective plant cDNA by using a DNA fragment containing the most conserved sequence in the yeast and mammalian HMG-CoA genes.[341] Other examples for heterologous probes are the isolation of the lanosterol 14α-demethylase gene from *Candida tropicalis* accomplished with a DNA probe containing coding regions of the respective yeast gene.[322] The opposite scenario, the availability of a *C. tropicalis* gene fragment as a probe, was successful in the isolation of the yeast gene coding for NADPH-cytochrome P-450 reductase.[304] As an alternative to cloned DNA fragments, nucleotide probes can also be synthesized to match the coding sequence of a gene if the amino acid sequence of a highly conserved protein region is known, or if portions of a protein under investigation have been sequenced.[342] The usefulness of heterologous nucleic acid probes can be restricted whenever the codon preferences are too different from organism to organism and not known for the fungus under investigation.[343] In these cases, the identification of a specific cDNA clone via antibody screening and the application of these homologous probes for the isolation of the gene might remain the method of choice.

In general, the development of techniques and tools for the isolation of fungal cDNAs and genes has been highly successful during the past decade, and the number of characterized fungal genes is rapidly increasing. A second important breakthrough in fungal molecular genetics has been the development of procedures to manipulate the expression and character of defined genes present in the fungal genome. Initially developed for *Saccharomyces cerevisiae*, fungal transformation procedures were soon adopted for mycelial fungi, including plant pathogens.[260-262,344,345] Leong and Holden[262] list 17 transformed plant pathogens from nine genera. More recent examples are *Alternaria alternata*,[346] *Botryotinia squamosa*,[347] *Pyricularia oryzae*,[348] and *Fusarium oxysporum*.[349] The transformation of filamentous fungi depends on the incorporation of exogenously added DNA into the fungal genome. Consequently, the steps of genetic engineering involve the construction of transformation vectors, the transfer of these vectors into fungal cells, the genomic incorporation of the foreign DNA, and the selection of transformants from nontransformed cells. Most often, transformation vectors consist of a bacterial plasmid containing a fungal promoter and the gene under question. The promoters can be isolated from the fungus to be transformed,[350] or from a different fungus. For example, *B. squamosa* and *P. oryzae* have been transformed with regulatory sequences derived from *Aspergillus nidulans*.[347,348] The transformation of *Ustilago violacea* and *Colletotrichum lindemuthianum* was accomplished with promoters derived from *Cochliobolus heterostrophus*.[351,352] The transfer of the transformation vectors into fungal cells is most frequently accomplished by the preparation of protoplasts and fusion of protoplasts in the presence of polyethylene glycol (PEG).[345] The plasmids are taken up during fusion. Slight variations of these procedures are often required to increase transformation efficacy.[349] Alternatives to the protoplast-PEG method are the preparation of competent cells with lithium-acetate treatment,[352] electroporation,[353] and biolistic procedures.[354] Selectable marker genes most frequently incorporated into the transformation vector are the *hygB* gene of *Escherichia coli* encoding hygromycin B phosphotransferase and permitting growth on hygromycin,[262] the *N. crassa* gene encoding the β-tublin gene which confers resistance to benomyl,[355] and the *amdS* (acetamidase) gene from *A. nidulans*, which permits growth on acetamide as the sole nitrogen source.[345]

Modern techniques of fungal molecular genetics applied to molecular aspects of antifungal inhibitor action and design are of great potential, although the examples from the literature remain few at present. One important field of application will be the opportunity to **overexpress low abundancy enzymes**. As discussed above, a low cellular level of potential targets is one of the ideal requirements for novel inhibitors, but it imposes, at the same time, difficulties in the purification, characterization, and *in vitro* testing of these enzymes. This obstacle can be elegantly overcome by the isolation of the respective enzymes from genetically altered expression vectors. For example, a fungal acetolactate synthase as a low abundancy enzyme was recently purified after expression in transformed *E. coli.*[356] A prokaryote expression vector, although most convenient, might not always be the adequate choice for the production of eukaryote proteins, and the expression and overproduction of heterologous enzymes would have to be accomplished in transformed yeast or filamentous fungi. The state of the art of heterologous gene expression in fungi is well advanced and sufficiently diversified to warrant a strategy most adequate for the enzyme under question.[357-359] Even overexpession of homologous enzymes has become possible by genetic engineering of fungal organisms.[360] Moreover, transformed fungi expressing heterologous genes can be constructed to serve as elegant screening systems. For example, *S. cerevisiae* expressing β-galactosidase as an enzyme that can be easily tested has been described as a suitable test system for the identification of cell wall biosynthesis inhibitors.[361] Overall, the availability of fungal target sites for biochemical analysis and testing of inhibitors, which has been a sometimes severe limitation in the past, imposes a clearly less pronounced handicap in the future.

A second opportunity of molecular genetic tools is the application of **site-directed mutagenesis** to studies on **enzyme mechanisms.**[362,363] The strategy is to produce an artifical gene coding for an enzyme with specifically altered amino acids, to express this gene in a transformed microorganism, and to study the impact of this alteration on enzymatic parameters such as substrate binding, substrate specificity, and inhibitor binding. Identification of these features is required for the design of transition-state analogues and suicide substrates described above, and effects of amino acid alterations can provide valuable information on the structural configuration of substrate binding sites.

A third but equally attractive field of application is the disruption and thus specific desactivation of a fungal gene once it has been isolated.[260-262] Most frequently, the gene under question is disrupted within the coding region and fused to a selectable marker gene, for example, the gene conferring resistance to hygromycin B or benomyl. The desactivated gene is incorporated into the genome by homologous recombination, and the originating transformants are null-mutants that lack the specified gene product. This technique allows one to study the effect of the desactivation of a particular enzyme on growth and reproduction of fungal organisms and, thus, comprises an ideal tool for the **evaluation of antifungal targets**. Three examples, all with *S. cerevisiae*, have already been discussed above. The disruption of two gene both coding for chitin synthetase[98-101] and HMG-CoA reductase isozymes[340,341] were necessary to abolish growth. Although the presence of similar conditions has not been reported for other fungi thus far, the identification of these sometimes unexpected effects are mandatory for strategies of inhibitor design programs. The disruption of particular genes coding for isoenzymes would also allow one to design test organisms for screening of inhibitors almost by will. The strategy chosen for the characterization of the two yeast methyltransferases involved in the biosynthesis of phosphatidylcholine represents an excellent example.[88] The disruption of either one of the genes clearly indicated which one of the isoenzymes was more essential for growth and thus the more appropriate target for inhibitors. Respective transformants would also comprise a very suitable source for the discriminatory testing of inhibitors, both *in vivo* and *in vitro*.

The desactivation of specified genes in yeast opens, as a second opportunity, the option to construct elegant systems for tests of **target specificities**. Again, the two isoenzymes of yeast HMG-CoA reductase can serve as an excellent example for this opportunity.[340,341] The disruption of either one of the genes had little effect on the viability of the transformants, and the desactivation of both genes was required to generate greatly impaired mutants. Viability of these mutants, however, could be restored by the addition of genes encoding the mammalian and the plant HMG-CoA reductase. The selective effect of inhibitors on the mammalian and plant HMG-CoA reductase could be easily tested using the respective transformants as a test system. The functional replacement of yeast genes with respective genes isolated from mammals and plants, but also from other fungal species, will comprise a more general future approach to simple and convenient test systems for evaluations aimed at inhibitor selectivity and specificity.

Last but not least, the advent of fungal molecular genetics might lead to the **discovery of novel target sites** for disease control agents that interfere with regulatory events in disease initiation and the establishment of a pathogen in its host. As described above, the availability of cutinase cDNA has made it possible to study the precise mechanisms of cutinase induction with isolated nuclei.[214] The induction of cutinase transcription is dependent on the cutin monomers as specific inducers and a cytoplasmic protein. The identification of the mechanisms of cutin monomer binding to this protein and the initiation of gene transcription might open the possibility to chemically antagonize the induction process at hormonal concentrations. These antagonists would prevent the synthesis of cutinase by an attacking fungus and, thus, would act as plant protectants. Even if the function of gene products remains unknown, the technique of differential or cascade hybridization allows one to identify mRNAs specifically induced and synthesized during differential steps in disease development,[364] and respective cDNAs will be valuable tools for studies on the regulation of these events. The understanding of the molecular mechanism involved in the regulation of developmental genes can be expected to identify interesting targets for chemical interference. This research is underway, and substantial progress can be expected in the future.

VII. FUTURE PERSPECTIVES

In the past, chemists, biologists, and toxicologists have played the most important roles in the discovery and development of new fungicides and antifungal drugs. Will biochemistry and molecular biology play a similarly important role in the future, and will this discipline be able to innovate the process? The examples and opportunities described and discussed above appear to justify an optimistic view. Progress in fungal biochemistry and physiology, the increasing understanding of molecular events involved in the development of diseases, and the establishment of recombinant DNA technologies will have impact in this field of applied mycology. The innovative benefit of antifungal target research, however, will have to be carefully defined to warrant future success. It cannot be expected that biochemistry will become a substitution for already existing programs in synthetic chemistry and biological screening. The goal of biochemical programs should rather be the interaction with chemists and biologists in order to evaluate inhibitor targets and lead structures, and to facilitate and innovate biological screening.

Biochemical target research is frequently justified with the assumption that the escalation of costs as a result of decreasing success rates in the discovery of new inhibitors could be halted by an increased emphasis on molecular target research. However, increasing costs of pesticide and drug development are largely caused by the increasing demands on toxicological and environmental tests, and the cost increases due to the synthesis and screening of increasing numbers of compounds are not nearly as extensive. Most likely, the future benefit of molecular target research will indeed not be of a direct economical nature. The inherent opportunities will rather be on the side of innovation, the chance to discover and develop disease control agents for an increasingly demanding market.

A second requirement for success will be to improve the bridges between research at academic institutions largely funded by the public sector and research programs at private corporation. The magnitude of gaps between these two disciplines can be a substantial hurdle. A company biochemist will have to decide how long it would take and how expensive it would be to transfer progress being made with *Saccharomyces cerevisiae* to a plant or human pathogen, whereas research directions of university scientists are frequently restricted by the goal as of public funding agencies. It is surprising that, from a molecular point of view, *Fusarium solani* f. sp. *pisi, Cochliobolus heterostrophus, Ustilago maydis*, and the bean rust fungus are perhaps the best understood plant pathogens. None of these fungi causes a disease of outstanding economical or clinical importance. One possible solution is the establishment of basic research programs within companies. This basic research is time-consuming and requires a substantial long-term commitment. An alternative would be the cooperation between university and company biochemists on a case-by-case basis, be it through the exchange of ideas and research tools or through joint projects. In particular, in the field of molecular genetics this approach appears to be highly attractive and promising.

The character of biochemical and molecular research is distinct from synthetic chemistry and biological screening. The base of knowledge is more diversified, and biochemical programs are less suited to be managed in a schematic way. It follows that the infrastructure of research integrating fungicide biochemistry and target research should reflect this character. The ideal surrounding might be a highly interactive team integrating chemists, biologists, and biochemists, which is flexible enough to act and react in a timely manner. It will be an interesting challenge for a fungicide biochemist to find the most efficient compromise between realistic opportunities and too ambitious goals. On one side is should not be underestimated that in-depth research requires time. The time frame for intensive projects might not be compatible with the goals and interests of a discovery team. On the other side, the groundwork on which biochemical guidance is based has to be solid. Superficial results will have little impact and might even be misleading. Each problem will be different and will require a separate strategic evaluation and approach. "What is expected from a good drug scientist is that he acts intelligent not that he looks intelligent." Jolles' view stated in the introduction can certainly be extended to the field of fungicide biochemistry.

REFERENCES

1. **Sijpestein, A. K.,** Mode of action of some traditional fungicides, in *Mode of Action of Antifungal Agents,* Trinci, A. P. J. and Ryley, J. F., Eds., Cambridge University Press, Cambridge, 1984, 135.
2. **Owen, R. G.,** Chemistry and physiology of fungicidal action, *Annu. Rev. Phytopathol.*, 1, 77, 1963.
3. **Köller, W.,** Fungicide resistance in plant pathogens, in *Handbook of Pest Management,* Vol. 3, 2nd ed., Pimentel, D., Ed., CRC Press, Boca Raton, FL, 1991, 679.
4. **Norman, C.,** EPA sets new policy on pesticides cancer risks, *Science,* 242, 366, 1988.
5. **Fromtling, R. A.,** *Recent Trends in the Discovery, Development and Evaluation of Antifungal Agents,* J. R. Prous Science, Barcelona, 1987.
6. **Spencer, P. M. and Jackson, G. G.,** Fungal and mycobacterial infections in patients infected with the human immunodeficiency virus, *J. Antimicrob. Chemother.*, 23(Suppl. A), 107, 1989.
7. **Bodey, G. P.,** Fungal infections in cancer patients, *Ann. N.Y. Acad. Sci.*, 544, 431, 1988.
8. **Armstrong, D.,** Life-threatening opportunistic fungal infection in patients with the acquired immunodeficiency syndrome, *Ann. N.Y. Acad. Sci.*, 544, 443, 1988.
9. **Simberkoff, M. S.,** Cryptococcosis in patients with acquired immunodeficiency syndrome, in *Opportunistic Infections in Patients with the Acquired Immunodeficient Syndrome,* Leoung, G. and Mills, J., Eds., Marcel Dekker, New York, 1989, 267.
10. **Trelkeld, M. G. and Dismukes, W. E.,** Endemic mycoses, in *Opportunistic Infections in Patients with the Acquired Immunodeficiency Syndrome,* Leoung, G. and Mills, J. Eds., Marcel Dekker, New York, 1989, 285.
11. **Ghamnoum, M. A.,** Mechanisms potentiating *Candida* infections, *Mycoses,* 31, 543, 1988.
12. **Richardson, S. E., Bannatyne, R. M., Summerbell, R. C., Milliken, J., Gold, R., and Weitzman, S. S.,** Disseminated fusarial infection in the immunocompromised host, *Rev. Infect. Dis.*, 10, 1171, 1988.
13. **Ryley, J. F. and Rathmell, W. G.,** Discovery of antifungal agents: *in vitro* and *in vivo* testing, in *Mode of Action of Antifungal Agents,* Trinci, A. P. J. and Ryley, J. F., Eds., Cambridge University Press, Cambridge, 1984, 63.
14. **Shephard, M. C.,** Screening for fungicides, *Annu. Rev. Phytopathol.*, 25, 189, 1987.

15. **Janssen, P. A. J.,** New antifungal agents: a result of target-oriented drug design, in *Frontiers in Microbiology,* de Clerk, E., Ed., Martinus Nijhoff, Dordrecht, 1987, 29.
16. **Vanden Bossche, H., Marichal, P., Gorrens, J., Geerts, H., and Janssen, P. A. J.,** Mode of action studies—basis for the search of new antifungal compounds, *Ann. N.Y. Acad. Sci.,* 544, 191, 1988.
17. **Vanden Bossche, H., Marichal, P., Gorrens, J., Coene, M.-C., Willemsens, G., Bellens, D., Roels, I., Moereels, H., and Janssen, P. A. J.,** Biochemical approaches to selective antifungal activity. Focus on azole antifungals, *Mycoses,* 32(Suppl. 1), 35, 1989.
18. **Berg, D., Büchel, K. H., Krämer, W., Plempel, M., and Scheinpflug, H.,** Mechanistic studies as a tool for the development of new compounds, in *Sterol Biosynthesis Inhibitors—Pharmaceutical and Agricultural Aspects,* Berg, D. and Plempel, M., Eds., Ellis Horwood, Chichester, 1988.
19. **Langcake, P.,** Alternative chemical agents for controlling plant diseases, *Philos. Trans. R. Soc. London, Ser. B,* 295, 83, 1981.
20. **Geissbuhler, H., Muller, U., Pachlatko, J. P., and Waespe, H.-R.,** Biorational design of chemicals, in *Chemistry and World Food Supplies: The New Frontiers,* Shemilt, L. W., Ed., Pergamon Press, Oxford, 1983, 643.
21. **Brent, K. J.,** Biochemical plant pathology and plant disease control, in *Biochemical Plant Pathology,* Callow, J. A., Ed., John Wiley & Sons, Chichester, 1983, 435.
22. **Rathmell, W. G.,** The discovery of new methods of chemical disease control: current development, future prospects and the role of biochemical and physiological research, *Adv. Plant Pathol.,* 2, 259, 1984.
23. **Baldwin, B. C.,** Potential targets for the selective inhibition of fungal growth, in *Mode of Action of Antifungal Agents,* Trinci, A. P. J. and Ryley, J. F., Eds., Cambridge University Press, Cambridge, 1984, 43.
24. **Alder, E. F., Wright, W. L., and Brown, I. F.,** Future trends in controlling weeds and plant pathogens, in *Agricultural Chemicals of the Future,* Hilton, J. L., Ed., Rowman & Allanheld, Totowa, 1985, 147.
25. **Ragsdale, N. N. and Siegle, M. R.,** New approaches to chemical control of plant pathogens, in *Bioregulators for Pest Control,* Hedin, P. A., Ed., American Chemical Society, Washington, D.C., 1985, 34.
26. **Schwinn, F. and Geissbühler, H.,** Toward a more rational approach to fungicide design, *Crop Prot.,* 5, 33, 1986.
27. **Marchington, A. F. and Lambros, S.,** Computer design of fungicides, in *Modern Selective Fungicides—Properties, Application and Mechanism of Action,* Lyr, H., Ed., Gustav Fischer Verlag, Berlin, 1987, 325.
28. **Baldwin, B. C. and Rathmell, W. G.,** Evolution of concepts for chemical control of plant disease, *Annu. Rev. Phytopathol.,* 26, 265, 1988.
29. **Kuhn, P. J.,** The discovery and development of fungicides—does biochemistry have a role, *Pestic. Sci.,* 25, 123, 1989.
30. **Iwamura, H. and Fujita, T.,** QSAR studies in pesticide research in Japan, *J. Pestic. Sci.,* 7, 289, 1982.
31. **Fujita, T. and Iwamura, H.,** Applications of various steric constants to quantitative analysis of structure-activity relationships, in *Steric Effects in Drug Design,* Charton, M. and Motoc, I., Eds., Springer-Verlag, Berlin, 1983, 119.
32. **Dearden, J. C.,** Quantitative structure-activity studies of enzyme inhibition, in *Design of Enzyme Inhibitors as Drugs,* Sandler, M. and Smith, H. J., Eds., Oxford University Press, Oxford, 1989, 120.
33. **Jolles, G.,** *Drug Design: Fact or Fantasy?,* Academic Press, London, 1984.
34. **Davidse, L. C., Hofman, A. E., and Velthuis, G. C. M.,** Specific interference of metalaxyl with endogenous RNA polymerase activity in isolated nuclei from *Phytophthora megasperma* f. sp. *medicaginis, Exp. Mycol.,* 7, 344, 1983.
35. **Ziogas, B. N. and Davidse, L. C.,** Studies on the mechanism of action of cymoxanil in *Phytophthora infestans, Pestic. Biochem. Physiol.,* 29, 89, 1987.
36. **Clemons, G. P. and Sisler, H. D.,** Localization of the site of action of a fungitoxic benomyl derivative, *Pestic. Biochem. Physiol.,* 1, 32, 1971.
37. **Nakanishi, T. and Sisler, H. D.,** Mode of action of hymexazole in *Phythium aphanidermatum, J. Pestic. Sci.,* 8, 173, 1983.
38. **Inoue, S., Uematsu, T., Kato, T., and Ueda, K.,** New melanin biosynthesis inhibitors and their structural similarities, *Pestic. Sci.,* 16, 589, 1985.
39. **Omata, K., Tomita, H., Nakajima, T., and Natsume, B.,** Design of new melanin biosynthesis inhibitors, *Pestic. Sci.,* 26, 271, 1989.
40. **Adetumbi, M., Javor, G. T., and Lau, B. H. S.,** *Allium sativum* (Garlic) inhibits lipid biosynthesis by *Candida albicans, Antimicrob. Agents Chemother.,* 30, 499, 1986.
41. **Marston, A. and Hostettman, K.,** Antifungal, molluscicidal and cytotoxic compounds from plants used in traditional medicine, in *Biologically Active Natural Products,* Hostettman, K. and Lea, P. J., Eds., Clarendon Press, Oxford, 1987, 65.
42. **Fawcett, C. H. and Spencer, D. M.,** Natural antifungal compounds, in *Fungicides—An Advanced Treatise,* Vol. 2, Torgeson, D. C., Ed., Academic Press, New York, 1969, 637.

43. **Harborne, J. B.,** Natural fungitoxins, in *Biologically Active Natural Products,* Hostettman, K. and Lea, P. J., Eds., Clarendon Press, Oxford, 1987, 195.
44. **Kobayashi, K., Nishino, C., Tomita, H., and Fukushima, M.,** Antifungal activity of pisiferic acid derivatives against the rice blast fungus, *Phytochemistry,* 26, 3175, 1987.
45. **Watanabe, T., Sekizawa, Y., Shimura, M., Susuki, Y., Matsumoto, K., Iwata, M., and Mase, S.,** Effects of probendazole (Oryzemate) on rice plants with reference to controlling rice blast, *J. Pestic. Sci.,* 4, 53, 1979.
46. **Sekizawa, Y. and Mase, S.,** Mode of controlling action of probendazole against rice blast disease with reference to the induced resistance mechanism in rice plant, *J. Pestic. Sci.,* 6, 91, 1981.
47. **Sekizawa, Y., Haga, M., Iwata, M., Hamamoto, A., Chihara, C., and Takino, Y.,** Probenazole and burst of respiration in rice leaf tissue infected with blast fungus, *J. Pestic. Sci.,* 10, 225, 1985.
48. **Roddick, J. G.,** Antifungal activity of plant steroids, in *Ecology and Metabolism of Plant Lipids,* Fuller, G. and Nes, W. D., Eds., American Chemical Society, Washington, D.C., 1987, 286.
49. **Schlösser, E.,** Preformed chemical barriers in host-parasite incompatibility, in *Experimental and Conceptual Plant Pathology,* Vol. 3, Singh, R. S., Singh, U. S., Hess, W. M., and Weber, D. J., Eds., Gordon and Breach, New York, 1988, 465.
50. **Défago, G. and Kern, H.,** Induction of *Fusarium solani* mutants insensitive to tomatine, their pathogenicity and aggressiveness to tomato fruits and pea plants, *Physiol. Plant Pathol.,* 22, 29, 1983.
51. **Défago, G., Kern, H., and Sedlar, L.,** Genetic analysis of tomatine insensitivity, sterol content and pathogenicity for green tomato fruit in mutants of *Fusarium solani, Physiol. Plant Pathol.,* 22, 39, 1983.
52. **Bowers, W. S., Hoch, H. C., Evans, P. H., and Katayama, M.,** Thallophytic allelopathy: isolation and identification of laetisaric acid, *Science,* 232, 105, 1986.
53. **Baily, J. A. and Mansfield, J. W., Eds.,** *Phytoalexins,* Blackie, Glasgow, 1982.
54. **Harborne, J. B.,** The role of phytoalexins in natural plant resistance, in *Natural Resistance of Plants to Pests,* Green, M. B. and Hedin, P. A., Eds., American Chemical Society, Washington, D.C., 1986, 22.
55. **VanEtten, H. D. and Bateman, D. F.,** Studies on the mode of action of the phytoalexin phaseollin, *Phytopathology,* 61, 1363, 1971.
56. **Yoshikawa, M., Masago, H., Onoe, T., and Matsuda, K.,** Mode of biochemical action of phytoalexins, in *Molecular Determinants of Plant Diseases,* Nishimura, S., Vance, C. P., and Doke, N., Eds., Springer-Verlag, Berlin, 1987, 253.
57. **Boydston, R., Paxton, J. D., and Koeppe, D. E.,** Glyceollin: a site-specific inhibitor of electron transport in isolated soybean mitochondria, *Plant Physiol.,* 72, 151, 1983.
58. **Giannini, J. L., Briskin, D. P., Holt, J. S., and Paxton, J. D.,** Inhibition of plasma membrane and H^+-transporting ATPases by glyceollin, *Phytopathology,* 78, 1000, 1988.
59. **Smith, D. A.,** Toxicity of phytoalexins, in *Phytoalexins,* Bailey, J. A. and Mansfield, J. W., Eds., Blackie, Glasgow, 1982, 218.
60. **McLaren, J. S.,** Biologically active substances from higher plants: status and future potential, *Pestic. Sci.,* 17, 559, 1986.
61. **Schultz, T. P., Hubbard, T. F., Jin, L., Fisher, T. H., and Nicholas, D. D.,** Role of stilbenes in the natural durability of wood: fungicidal structure-activity relationships, *Phytochemistry,* 29, 1501, 1990.
62. **Laks, P. E. and Pruner, M. S.,** Flavonoid biocides: structure/activity relations of flavonoid phytoalexin analogues, *Phytochemistry,* 28, 87, 1990.
63. **Ames, B. N. and Gold, L. S.,** Chemical carcinogenesis: too many rodent carcinogens, *Proc. Natl. Acad. Sci. U.S.A.,* 87, 7772, 1990.
64. **Ames, B. N., Profet, M., and Gold, L. S.,** Dietary pesticides (99.99% all natural), *Proc. Natl. Acad. Sci. U.S.A.,* 87, 7777, 1990.
65. **Ames, B. N., Profet, M., and Gold, L. S.,** Nature's chemicals and cancer risk, *Proc. Natl. Acad. Sci. U.S.A.,* 87, 7782, 1990.
66. **VanEtten, H. D., Matthews, D. E., and Matthews, P. S.,** Phytoalexin detoxification: importance for pathogenicity and practical applications, *Annu. Rev. Phytopathol.,* 27, 143, 1989.
67. **Schäfer, W., Straney, D., Ciufetti, L., VanEtten, H. D., and Yoder, O. C.,** One enzyme makes a fungal pathogen, but not a saprophyte, virulent on a new host, *Science,* 246, 247, 1989.
68. **Severson, R. F., Cutler, H. G., Cole, P. D., Jackson, D. M., Sisson, V. A., Johnson, A. M., Herzog, G. A., and Stephenson,** Biologically active components in the cuticula of the *Nicotiana* species, *Proc. PGR Soc. Am.,* p. 175, 1985.
69. **Banthorpe, D. V., Brown, J. T., and Morris, G. S.,** Accumulation of the antifungal diterpene sclareol by cell cultures of *Salvia sclarea* and *Nicotiana glutinosa, Phytochemistry,* 29, 2145, 1990.
70. **Anke, T., Besl, H., Mocek, U., and Steglich, W.,** Antibiotics from basidiomycetes. XVIII. Strobilurin C and Oudemansin B, two new antifungal metabolites from *Xerula* species (Agaricales), *J. Antibiot.,* 36, 661, 1983.

70a. **Beautement, K., Clough, J. M., deFraine, P. J., and Godfrey, C. R. A.,** Fungicidal β-methoxyacrylates: from natural products to novel synthetic agricultural fungicides, *Pestic. Sci.,* 31, 499, 1991.

71. **Becker, W. F., von Jagow, G., Anke, T., and Steglich, W.,** Oudemansin, strobilurin A, strobilurin B and myxothiazol: new inhibitors of the *bc*1 segment of the respiratory chain with an E-β-methoxyacrylate system as common structural element, *FEBS Lett.*, 132, 329, 1981.
72. **Rodgers, P. B.,** Potential of biological control organisms as a source of antifungal compounds for agrochemical and pharmaceutical product development, *Pestic. Sci.*, 27, 155, 1989.
73. **Langkake, P., Kuhn, P. J., and Wade, M.,** The mode of action of systemic fungicides, *Prog. Pestic. Biochem. Toxicol.*, 3, 1, 1983.
74. **Davidse, L. and de Waard, M. A.,** Systemic fungicides, *Adv. Plant Pathol.*, 2, 191, 1984.
75. **Buchenauer, H.,** Physiological reactions in the inhibition of plant pathogenic fungi, in *Chemistry of Plant Protection*, Vol. 6, Haug, G. and Hoffmann, H., Eds., Springer-Verlag, Berlin, 1990, 217.
76. **Kerridge, D.,** Mode of action of clinically important antifungal drugs, *Adv. Microbiol. Physiol.*, 27, 1, 1986.
77. **Corbett, J. R., Wright, K., and Baillie, A. C.,** *The Biochemical Mode of Action of Pesticides*, Academic Press, New York, 1984.
78. **Vanden Bossche, H., Willemsens, G., and Marichal, P.,** Anti-*Candida* drugs—the biochemical basis for their activity, *CRC Crit. Rev. Microbiol.*, 15, 57, 1987.
79. **Polak, A.,** Mode of action of 5-fluorocytosine in *Aspergillus fumigatus*, in *Aspergillus and Aspergillosis*, Vanden Bossche, H., Mackenzie, D. W. R., and Cauwenbergh, G., Eds., Plenum Press, New York, 1987, 165.
80. **Ko, B.-S., Oritani, T., and Yamashita, K.,** Synthesis and biological activities of griseofulvin analogs, *Agric. Biol. Chem.*, 54, 2199, 1990.
81. **Kodama, O., Yamada, H., and Akatsuka, T.,** Kitazin P, inhibitor of phosphatidyl choline biosynthesis in *Pyricularia oryzae*, *Agric. Biol. Chem.*, 43, 1719, 1979.
82. **Kodama, O., Yamashita, H., and Akatsuka, T.,** Edifenphos, inhibitor of phosphatidyl choline biosynthesis in *Pyricularia oryaze*, *Agric. Biol. Chem.*, 44, 1015, 1980.
83. **Yoshida, M., Moriya, S., and Uesugi, Y.,** Observation of transmethylation from methionine into choline in the intact mycelia of *Pyricularia oryzae* by ^{13}C NMR under the influence of fungicides, *J. Pestic. Sci.*, 9, 703, 1984.
84. **Pieringer, R. A.,** Biosynthesis of non-terpenoid lipids, in *Microbial Lipids*, Vol. 2, Ratledge, C. and Wilkinson, S. G., Eds., Academic Press, New York, 1989, 51.
85. **Carman, G. M. and Henry, S. A.,** Phospholipid biosynthesis in yeast, *Annu. Rev. Biochem.*, 58, 635, 1989.
86. **Robson, G. D., Wiebe, M., Kuhn, P. J., and Trinci, A. P. J.,** Inhibitors of phospholipid biosynthesis, in *Biochemistry of Cell Walls and Membranes in Fungi*, Kuhn, P. J., Trinci, A. P. J., Jung, M. J., Goosey, M. W., and Copping, L. G., Eds., Springer-Verlag, Berlin, 1990, 261.
87. **Chin, J. and Bloch, K.,** Phosphatidylcholine synthesis in yeast, *J. Lipid Res.*, 29, 9, 1988.
88. **Kodaki, T. and Yamashita, S.,** Characterization of the methyltransferases in the yeast phosphatidylethanolamine methylation pathway by selective gene disruption, *Eur. J. Biochem.*, 185, 243, 1989.
89. **Mago, N. and Khuller, G. K.,** Biosynthesis of major phospholipids in *Candida albicans*, *Curr. Microbiol.*, 20, 369, 1990.
90. **Gaynor, P. M. and Carman, G. M.,** Phosphatidylethanolamine methyltransferase and phospholipid methyltransferase activities from *Saccharomyces cerevisiae*, *Biochim. Biophys. Acta*, 1045, 156, 1990.
91. **Gutteridge, J. M. C. and Halliwell, B.,** The measurement and mechanism of lipid peroxidation in biological systems, *Trends Biochem. Sci.*, 15, 129, 1990.
92. **Wessels, J. G. H. and Sietsma, J. H.,** Fungal cell walls: a survey, in *Plant Carbohydrates II—Extracellular Carbohydrates*, Tanner, W. and Loewus, F. A., Eds., Springer-Verlag, Berlin, 1981, 352.
93. **Perberdy, J. F.,** Fungal cell walls—a review, in *Biochemistry of Cell Walls and Membranes in Fungi*, Kuhn, P. J., Trinci, A. P. J., Jung, M. J., Goosey, M. W., and Copping, L. G., Eds., Springer-Verlag, Berlin 1990, 5.
94. **Févre, M., Girard, V., and Nodet, P.,** Cellulose and β-glucan synthesis in *Saprolegnia*, in *Biochemistry of Cell Walls and Membranes in Fungi*, Kuhn, P. J., Trinci, A. P. J., Jung, M. J., Goosey, M. W., and Copping, L. G., Eds., Springer-Verlag, Berlin, 1990, 97.
95. **Cabib, E.,** The synthesis and degradation of chitin, *Adv. Enzymol.*, 59, 59, 1987.
96. **Gozalbo, D., Dubon, F., Schwencke, J., and Sentandreu, R.,** Characterization of chitosomes in *Candida albicans* protoplasts, *Exp. Mycol.*, 11, 331, 1987.
97. **Martinez, A. and Schwencke, J.,** Chitin synthetase is bound to chitosomes and to the plasma membrane in protoplasts of *Saccharomyces cerevisiae*, *Biochim. Biophys. Acta*, 946, 328, 1988.
98. **Cabib, E., Silverman, S. J., Sburlati, A., and Slater, M. L.,** Chitin synthesis in yeast (*Saccharomyces cerevisiae*), in *Biochemistry of Cell Walls and Membranes in Fungi*, Kuhn, P. J., Trinci, A. P. J., Jung, M. J., Goosey, M. W., and Copping, L. G., Eds., Springer-Verlag, Berlin, 1990, 31.
99. **Cabib, E., Sburlati, A., Bowers, B., and Silverman, S. J.,** Chitin synthase 1, an auxiliary enzyme for chitin synthesis in *Saccharomyces cerevisiae*, *J. Cell. Biol.*, 108, 1665, 1989.

100. **Silverman, S. J.,** Similar and differnt domains of chitin synthases 1 and 2 of *S. cerevisiae*: tow isozymes with distinct functions, *Yeast*, 5, 459, 1989.
101. **Au-Young, J. and Robbins, P. W.,** Isolation of chitin synthase gene (*CHSI*) from *Candida albicans* by expression in *Saccharomyces cerevisiae*, *Mol. Microbiol.*, 4, 197, 1990.
102. **Saksena, A. K., Girijavallabhan, V. M., Cooper, A. B., and Loebenberg, D.,** Recent advances in antifungal agents, *Annu. Rep. Med. Chem.*, 24, 111, 1989.
103. **Gooday, G. W.,** Inhibition of chitin metabolism, in *Biochemistry of Cell Walls and Membranes in Fungi*, Kuhn, P. J., Trinci, A. P. J., Jung, M. J., Goosey, M. W., and Copping, L. G., Eds., Springer-Verlag, Berlin, 1990, 61.
104. **Milewski, S., Chmara, H., and Borowski, E.,** Anticapsin: an active site directed inhibitor of glucoseamine 6-phosphate synthetase from *Candida albicans*, *Drugs Exp. Clin. Res.*, 12, 577, 1986.
105. **Milewski, S., Chamara, H., Andruszliewicz, R., and Borowski, E.,** Synthetic derivatives of 3*N*-fumaroyl-L-2,3-diaminopropanoic acid inactivate glucoseamine synthetase from *Candida albicans*, *Biochim. Biophys. Acta*, 828, 247, 1985.
106. **Févre, M., Girard, V., and Nodet, P.,** Cellulose and β-glucan synthesis in *Saprolegnia*, in *Biochemistry of Cell Walls and Membranes in Fungi*, Kuhn, P. J., Trinci, A. P. J., Jung, M. J., Goosey, M. W., and Copping, L. G., Eds., Springer-Verlag, Berlin, 1990, 97.
107. **Debono, M., Abbott, B. J., Turner, J. R., Howard, L. C., Gordee, A. S., Hunt, A. S., Barnhart, M., Molloy, R. M., Willard, K. E., Fukuda, D., Butler, T. F., and Zeckner, D. J.,** Synthesis and evaluation of LY121019, a member of a series of semisynthetic analogues of the antifungal lipopeptide echinocandin B, *Ann. N.Y. Acad. Sci.*, 544, 152, 1988.
108. **Hall, G. S., Myles, C., Pratt, K. J., and Washington, J. A.,** Cilofungin (LY 12109), and antifungal agent with specific activity against *Candida albicans* and *Candida tropicalis*, *Antimicrob. Agents Chemother.*, 32, 1331, 1988.
109. **Rennie, R. P. and Hellman, L.,** Comparative activity of cilofungin (LY 121019) with other agents used for treatment of deep-seated *Candida* infections, *Mycoses*, 32, 145, 1988.
110. **Taft, C. S. and Selitrennikoff, C. P.,** LY121019 inhibits *Neurospora crassa* growth and (1-3)-β-D-glucan synthase, *J. Antibiot.*, 41, 697, 1988.
111. **Taft, C. S., Stark, T., and Selitrennikoff, C. P.,** Cilofungin (LY121019) inhibits *Candida albicans* (1-3)-D-glucan synthase activity, *Antimicrob. Agents Chemother.*, 32, 1901, 1988.
112. **Taft, C. S. and Selitrennikoff, C. P.,** Cilofungin inhibition of (1-3)-β-D-glucan synthase: the lipophilic side chain is essential for inhibition of enzyme activity, *J. Antibiot.*, 43, 433, 1990.
113. **Traxler, P., Tosch, W., and Zak, O.,** Papulacandins—synthesis and biological activity of papulacandin B derivatives, *J. Antibiot.*, 40, 1146, 1987.
114. **Baguley, B. C., Roemmele, G., Gruner, J., and Wehrli, W.,** Papulacandin B: an inhibitor of glucan synthesis in yeast spheroplasts, *Eur. J. Biochem.*, 97, 345, 1979.
115. **Pérez, P., García-Acha, I., and Durán, A.,** Effect of papulacandin B on the cell wall and growth of *Geotrichum lactis*, *J. Gen. Microbiol.*, 129, 245, 1983.
116. **Varona, R., Pérez, P., and Durán, A.,** Effect of papulacandin B on β-glucan synthesis in *Schizosaccharomyces pombe*, *FEMS Microbiol. Lett.*, 200, 243, 1983.
117. **Yamaguchi, H., Hiratani, T., and Baba, M.,** Effect of aculeacin A on reverting protoplasts of *Candida albicans*, *Microbiol. Immunol.*, 31, 625, 1987.
118. **Misato, T. and Kakiki, K.,** Inhibition of fungal cell wall synthesis and cell membrane function, in *Antifungal Compounds*, Vol. 2, Siegel, M. R. and Sisler, H. D., Eds., Marcel Dekker, New York, 1977, 277.
119. **Craig, G. D. and Peberdy, J. F.,** The mode of action of *S*-benzyl *O,O*-di-isopropyl phosphorotioate and of dicloran on *Aspergillus nidulans*, *Pestic. Sci.*, 14, 17, 1983.
120. **Sandler, M. and Smith, H. J., Eds.,** *Design of Enzyme Inhibitors as Drugs*, Oxford University Press, Oxford, 1989.
121. **Schloss, J. V.,** Modern aspects of enzyme inhibition with particular emphasis on reaction-intermediate analogs and other potent, reversible inhibitors, in *Target Sites of Herbicide Action*, Böger, P. and Sandmann, G., Eds., CRC Press, Boca Raton, FL 1989, 165.
122. **Martin, H.,** Inorganics, in *Fungicides—An Advanced Treatise*, Vol. 2, Torgeson, D. C., Ed., Academic Press, New York, 1969, 101.
123. **Ulfvarson, U.,** Organic mercuries, in *Fungicides—An Advanced Treatise*, Vol. 2, Torgeson, D. C., Ed., Academic Press, New York, 1969, 303.
124. **Sijpesteijn, A. K., Luijten, J. G. A., and Van der Kerk, G. J. M.,** Organometallic fungicides, in *Fungicides—An Advanced Treatise*, Vol. 2, Torgeson, D. C., Ed., Academic Press, New York, 1969, 331.
125. **Owens, R. G.,** Organic sulfur compounds, in *Fungicides—An Advanced Treatise*, Vol. 2, Torgeson, D. C., Ed., Academic Press, New York, 1969, 147.

126. **Lukens, R. J.,** Heterocyclic nitrogen compounds, in *Fungicides—An Advanced Treatise,* Vol. 2, Torgeson, D. C., Ed., Academic Press, New York, 1969, 396.
127. **Shaw, E.,** Active-site-directed irreversible inhibitors, in *Design of Enzyme Inhibitors as Drugs,* Sandler, M. and Smith, H. J., Eds., Oxford University Press, Oxford, 1989, 49.
128. **Neil, G. H. and Champagne, D. E.,** Fungicidal activity of naturally occurring photosensitizers, in *Light-Activated Pesticides,* Hetiz, J. R. and Downum, K. R., Eds., American Chemical Society, Washington, D.C., 1987, 231.
129. **Endo, A., Kakiki, K., and Misato, T.,** Feedback inhibition of L glutamine D fructose-6-phosphate amino transferase by UDP-*N*-acetylglucoseamine in *Neurospora crassa, J. Bacteriol.,* 103, 588, 1970.
130. **Martinez, J. P., Gimenez, G., and Batnicki-Garcia, S.,** Intracellular localization of UDP-*N*-acetylglucoseamine in *Neurospora crassa* wild-type and slime mutant strains, *Exp. Mycol.,* 11, 278, 1987.
131. **Wolfenden, R. and Frick, L.,** Transition state affinity and the design of enzyme inhibitors, in *Enzyme Mechanisms,* Page, M. I. and Williams, A., Eds., Royal Society of Chemistry, London, 1987, 97.
132. **Frick, L. and Wolfenden, R.,** Substrate and transition state analogue inhibitors, in *Design of Enzyme Inhibitors as Drugs,* Sandler, M. and Smith, H. J., Eds., Oxford University, Oxford, 1989, 19.
133. **Abeles, R. H. and Alston, T. A.,** Enzyme inhibition by fluoro compounds, *J. Biol. Chem.,* 265, 16705, 1990.
134. **Morrison, J. F. and Walsh, C. T.,** The behaviour and significance of slow-binding enzyme inhibitors, *Adv. Enzymol.,* 61, 201, 1988.
135. **Tipton, K. F.,** Mechanism-based inhibitors, in *Design of Enzyme Inhibitors as Drugs,* Sandler, M. and Smith, H. J., Eds., Oxford University Press, Oxford, 1989, 71.
136. **Silverman, R. B. and Hoffmann, J. S.,** The organic chemistry of mechanism based enzyme inhibition: a chemical approach to drug design, *Med. Res. Rev.,* 4, 415, 1984.
137. **Walsh, C. T.,** Suicide substrates, mechanism based enzyme inactivators: recent developments, *Annu. Rev. Biochem.,* 53, 493, 1984.
138. **Oritz de Montellano, P. R. and Reich, N. O.,** Inhibition of cytochrome P-450 enzymes, in *Cytochrome P-450—Structure, Mechanism, and Biochemistry,* Oritz de Montellano, P. R., Ed., Plenum Press, New York, 1986, 273.
139. **Ryder, N. S.,** Mode of action of allylamines, in *Sterol Biosynthesis Inhibitors—Pharmaceutical and Agrichemical Aspects,* Berg, D. and Plempel, M., Eds., Ellis Horwood, Chichester, 1988, 151.
140. **Balliano, G., Viola, F., Ceruti, M., and Cattel, L.,** Inhibition of sterol biosynthesis in *Saccharomyces cerevisiae* by *N,N*-diethylsqualene and derivatives, *Biochim. Biophys. Acta,* 959, 9, 1988.
141. **Hol, W. G.,** Protein crystallography and computer graphics—toward rational drug design, *Angew. Chem. Int. Ed. Engl.,* 25, 767, 1986.
142. **Marshall, G. R.,** Computer-aided drug design, *Annu. Rev. Pharmacol. Toxicol.,* 27, 193, 1987.
143. **Frühbeis, H., Klein, R., and Wallmeier, H.,** Computer-assisted molecular design, *Angew. Chem. Int. Ed. Engl.,* 26, 403, 1987.
144. **Marshall, G. R. and Cramer, R. D.,** Three-dimensional structure-activity relationships, *Trends Pharmacol. Sci.,* 9, 285, 1988.
145. **North, A. C. T.,** Potential of molecular graphics in the computer-aided design of inhibitors, in *Design of Enzyme Inhibitors as Drugs,* Sandler, M. and Smith, H. J., Eds., Oxford University Press, Oxford, 1989, 93.
146. **Scheraga, H. A.,** Some computational problems in the conformational analysis of polypeptides and proteins, in *Computer-Assisted Modeling of Receptor-Ligand Interactions—Theoretical Aspects and Applications to Drug Design,* Rein, R. and Golombeck, A., Eds., Alan R. Liss, New York, 1989, 3.
147. **Sali, A., Overington, J. P., Johnson, M. S., and Blundell, T. J.,** From comparisons of protein sequences and structures to protein modelling and design, *Trends Biochem. Sci.,* 15, 335, 1990.
148. **Cooke, R. M. and Campbell, I. D.,** Protein structure determination by nuclear magnetic resonance, *BioEssays,* 8, 52, 1988.
149. **Wolff, M. E. and McPherson, A.,** Antibody-directed drug discovery, *Nature,* 345, 365, 1990.
150. **Walsh, R. C. and Sisler, H. D.,** A mutant of *Ustilago maydis* deficient in sterol C-14 demethylation: characteristics and sensitivity to inhibitors of ergosterol biosynthesis, *Pestic. Biochem. Physiol.,* 18, 122, 1982.
151. **Yoshida, Y.,** Cytochronme P450 of fungi: primary target of azole antifungal agents, in *Current Topics in Medical Mycology,* Vol. 2, McGinnis, M. R., Ed., Springer-Verlag, New York, 1988, 388.
152. **Cooper, A. B., Wright, J. J., Ganguly, A. K., Desai, J., Loebenberg, D., and Parmegiani, R.,** Synthesis and antifungal properties of 14-aminomethyl-substituted lanosterol derivatives, *Ann. N.Y. Acad. Sci.,* 544, 109, 1988.
153. **Zollner, H.,** *Handbook of Enzyme Inhibitors,* VCH Verlagsgesellschaft, Weinheim, 1989.
154. **Montgomery, J. A. and Bennett, L. L.,** Inhibitors of purine biosynthesis, in *Design of Enzyme Inhibitors as Drugs,* Sandler, M. and Smith, H. J., Eds., Oxford University Press, Oxford, 1989, 402.

155. **Kishore, G. M. and Shah, D. M.,** Amino acid biosynthesis inhibitors as herbicides, *Annu. Rev. Biochem.*, 57, 627, 1988.
156. **Ray, T. B.,** Herbicides as inhibitors of amino acid biosynthesis, in *Target Sites of Herbicide Action*, Böger, P. and Sandmann, G., Eds., CRC Press, Boca Raton, FL, 1990, 105.
157. **Bode, R., Melo, C., and Birnbaum, D.,** Mode of action of glyphosate in *Candida malosa*, *Arch. Microbiol.*, 140, 83, 1984.
158. **Falco, S. C. and Dumas, K. S.,** Genetic analysis of mutants of *Saccharomyces cerevisiae* resistant to the herbicide sulfometuron methyl, *Genetics*, 109, 21, 1985.
159. **Yamaki, H., Yamaguchi, M., Imamaura, H., Suzuki, H., Nishimura, T., Saito, H., and Yamaguchi, H.,** The mechanism of antifungal action of (*S*)-2-amino-4-oxo-5-hydroxypentanoic acid, RI-331: the inhibition of homoserine dehydrogenase in *Saccharomyces cerevisiae*, *Biochem. Biophys. Res. Commun.*, 168, 837, 1990.
160. **Yamaguchi, M., Yamaki, H., Shinoda, T., Tago, Y., Suzuki, H., Nishimura, T., and Yamaguchi, H.,** The mode of action of (*S*)-2-amino-4-oxo-5-hydroxypentanoic acid, RI-331, *J. Antibiot.*, 43, 411, 1990.
161. **Gooday, G. W. and Gow, N. A. R.,** Enzymology of tip growth in fungi, in *Tip Growth in Plant and Fungal Cells*, Heath, B., Ed., Academic Press, New York, 1990, 31.
162. **Pedrazarreyes, M. and Lopezromero, E.,** Detection of nine chitinase species in germinating cells of *Mucor rouxii*, *Curr. Microbiol.*, 22, 43, 1991.
163. **Heby, O. and Persson, L.,** Molecular genetics of polyamine synthesis in eukaryotic cells, *Trends Biochem. Sci.*, 15, 153, 1990.
164. **Bitonti, A. J. and McCan, P. P.,** Inhibition of polyamine biosynthesis in microorganisms, in *Inhibition of Polyamine Metabolism*, McCann, P. P., Pegg, A. E., and Sjoerdsma, A., Eds., Academic Press, New York, 1987, 259.
165. **Bey, P., Danzin, C., and Jung, M.,** Inhibition of basic amino acid decarboxylases involved in polyamine biosynthesis, in *Inhibition of Polyamine Metabolism*, McCann, P. P., Pegg, A. E., and Sjoerdsma, A., Eds., Academic Press, New York, 1987, 1.
166. **Tyms, A. S., Williamson, J. D., and Bacchi, C. J.,** Polyamine inhibitors in antimicrobial chemotherapy, *J. Antimicrob. Chemother.*, 22, 403, 1988.
167. **Edman, J. C., Kovacs, J. A., Masure, H., Santi, D. V., Elwood, H. J., and Sogin, M. L.,** Ribosomal RNA sequence shows *Pneumocystis carinii* to be a member of the fungi, *Nature*, 334, 519, 1988.
168. **Slocum, R. D. and Galston, A. W.,** Inhibition of polyamine biosynthesis in plants and plant pathogenic fungi, in *Inhibition of Polyamine Metabolism*, McCann, P. P., Pegg, A. E., and Sjoerdsma, A., Eds., Academic Press, New York, 1987, 305.
169. **Galston, A. W. and Weistein, L. H.,** Control of phytopathogens by inhibition of polyamine biosynthesis, in *Progress in Polyamine Research: Novel Biochemical Pharmacological and Clinical Aspects*, Zappia, V. and Pegg, A. E., Eds., Plenum Press, New York, 1988, 589.
170. **Walters, D. R.,** Polyamines and plant disease, *Plants Today*, 2, 22, 1989.
171. **West, H. M. and Walters, D. R.,** Effects of polyamine biosynthesis inhibitors on growth of *Pyrenophora teres*, *Gaeumanomyces graminis*, *Fusarium culmorum* and *Septoria nodorum*, *Mycol. Res.*, 92, 453, 1989.
172. **Machatschke, S., Kamrowski, C., Moerschbacher, B. M., and Reisener, H.-J.,** Polyamine levels in stem rust infected wheat leaves and effects of alpha-difluoromethylornithine on fungal infection, *Physiol. Mol. Plant Pathol.*, 36, 451, 1990.
173. **Walters, D. R. and Kingham, G.,** Uptake and translocation of α-difluoromethylornithine, a polyamine biosynthesis inhibitor, by barley seedlings: effects on mildew infection, *New Phytol.*, 114, 659, 1990.
174. **Khan, A. J. and Minocha, S. C.,** Polyamine biosynthetic enzymes and the effect of their inhibition on the growth of some phytopathogenic fungi, *Plant Cell Physiol.*, 30, 655, 1989.
175. **Smith, T. A., Barker, J. H. A., and Jung, M.,** Effect of enzyme-activated inhibitors on ornithine decarboxylase and growth of *Botrytis cinerea*, *Phytochemistry*, 29, 1759, 1990.
176. **Elbein, A. D.,** The metabolism of α,α-trehalose, *Adv. Carbohydr. Chem. Biochem.*, 30, 227, 1974.
177. **Thevelein, J. M.,** Regulation of trehalose mobilization in fungi, *Microbiol. Rev.*, 48, 42, 1984.
178. **Van Laere, A.,** Trehalose, reserve and/or stress metabolite?, *FEMS Microbiol. Rev.*, 63, 201, 1989.
179. **Wakae, K. and Matsuura, K.,** Characteristics of validamycin as a fungicide for *Rhizoctonia* disease control, *Rev. Plant Prot. Res.*, 8, 81, 1975.
180. **Asano, N., Yamaguchi, T., Kameda, Y., and Matsui, K.,** Effect of validamycins on glycohydrolases of *Rhizoctonia solani*, *J. Antibiot.*, 40, 526, 1987.
181. **Shigemoto, R., Okuno, T., and Matsuura, K.,** Effect of validamycin A on the activity of trehalase of *Rhizoctonia solani* and several sclerotial fungi, *Ann. Phytopathol. Soc. Jpn.*, 55, 238, 1989.
182. **Kameda, Y., Asano, N., Yamaguchi, T., and Matsui, K.,** Validoxylamines as trehalase inhibitors, *J. Antibiot.*, 40, 563, 1987.
183. **Salleh, H. M. and Honek, J. F.,** Time-dependent inhibition of porcine kidney trehalase by aminosugars, *FEBS Lett.*, 262, 359, 1990.

184. **Kendall, E. J., Adams, R. P., and Kartha, K. K.**, Trehalase activity in plant tissue cultures, *Phytochemistry*, 29, 2525, 1990.
185. **Asano, N., Takeuchi, M., Kameda, Y., Matsui, K., and Kono, Y.**, Trehalase inhibitors, validamycin A and related compounds as insecticides, *J. Antibiot.*, 43, 722, 1990.
186. **Nishimura, K., Mori, K., Yoshikuni, Y., and Fujita, T.**, Inhibitors of trehalase prepared from american cockroaches and of their reproduction by 1-deoxyjirimycin and its derivatives, *J. Pestic.*, 15, 237, 1990.
187. **Thevelein, J. M.**, Regulation of trehalse activity by phosporylation-dephosphorylation during developmental transition in fungi, *Exp. Mycol.*, 12, 1, 1988.
188. **Sussman, R. M. and Surowy, T. K.**, Physiology and molecular biology of membrane ATPases, *Oxford Surv. Plant Mol. Cell Biol.*, 4, 47, 1987.
189. **Prasad, R.**, Nutrient transport in *Candida albicans*, a pathogenic yeast, *Yeast*, 3, 209, 1987.
190. **Serrano, R.**, Structure and function of proton translocating ATPase in plasma membranes of plants and fungi, *Biochim. Biophys. Acta*, 947, 1, 1988.
191. **Serrano, R.**, Structure and function of plasma membrane ATPase, *Annu. Rev. Plant Physiol. Plant Mol. Biol.*, 40, 61, 1989.
192. **Dixon, D. M., Szanislo, P. J., and Polak, A.**, Dihydroxynaphthalene (DHN) melanin and its relationship with virulence in the early stages of Phaeohyphomycosis, in *The Fungal Spore and Disease Initiation in Plants and Animals*, Cole, G. T. and Hoch, H. C., Eds., Plenum Press, New York, 1991, 297.
193. **Polak, A.**, Melanin as a virulence factor in pathogenic fungi, *Mycoses*, 33, 215, 1990.
194. **Aist, J. R.**, Cytology of penetration and infection—fungi, in *Physiological Plant Pathology*, Heitefuss, R. and Williams, P. H., Eds., Springer-Verlag, Berlin, 1976, 197.
195. **Dodman, R. L.**, How the defenses are breached, in *Plant Disease—An Advanced Treaties*, Vol. 4, Horsfall, J. G. and Cowling, E. B., Eds., Academic Press, New York, 1979, 135.
196. **Staples, R. C. and Hoch, H. C.**, Infection structures—form and function, *Exp. Mycol.*, 11, 163, 1987.
197. **Kolattukudy, P. E.**, Biochemistry and function of cutin and suberin, *Can. J. Bot.*, 62, 2918, 1985.
198. **Kolattukudy, P. E., Espelie, K. E., and Soliday, C. L.**, Hydrophobic layers attached to cell walls: cutin, suberin, and associated waxes, in *Plant Carbohydrates II—Extracellular Carbohydrates*, Tanner, W. and Loewus, F. A., Eds., Springer-Verlag, Berlin, 1981, 225.
199. **Kolattukudy, P. E. and Espelie, K. E.**, Biosynthesis of cutin, suberin, and associated waxes, in *Biosynthesis and Biodegradation of Wood Components*, Higuchi, M., Ed., Academic Press, New York, 1985, 161.
200. **Martin, J. T.**, Role of cuticle in the defense against plant disease, *Annu. Rev. Phytopathol.*, 2, 81, 1964.
201. **Kolattukudy, P. E. and Köller, W.**, Fungal penetration of the first line defensive barriers of plants, in *Biochemical Plant Pathology*, Callow, J. A., Ed., John Wiley & Sons, Chichester, 1983, 79.
202. **Kolattukudy, P. E.**, Enzymatic penetration of the plant cuticle by fungal pathogens, *Annu. Rev. Phytopathol.*, 23, 223, 1985.
203. **Kolattukudy, P. E. and Crawford, M. S.**, The role of polymer degrading compounds in fungal pathogenesis, in *Molecular Determinants in Plant Diseases*, Nishimura, S., Vance, C. P., and Doke, N., Eds., Springer-Verlag, Berlin, 1987, 75.
204. **Köller, W.**, The plant cuticle: a barrier to be overcome by plant pathogens, in *The Fungal Spore and Disease Initiation in Plants and Animals*, Cole, G. T. and Hoch, H. C., Eds., Plenum Press, New York, 1991, 219.
205. **Köller, W. and Kolattukudy, P. E.**, Mechanism of action of cutinase: chemical modification of the catalytic triad characteristic for serine esterases, *Biochemistry*, 21, 3083, 1982.
206. **Soliday, C. L., Flurkey, W. H., Okita, T. W., and Kolattukudy, P. E.**, Cloning and structure determination of cDNA for cutinase, an enzyme involved in fungal penetration of plants, *Proc. Natl. Acad. Sci. U.S.A.*, 81, 3939, 1984.
207. **Ettinger, W. F., Thukral, S. K., and Kolattukudy, P. E.**, Structure of cutinase gene, cDNA, and the derived amino acid sequence from phytopathogenic fungi, *Biochemistry*, 26, 7883, 1987.
208. **Soliday, C. L., Dickman, M. B., and Kolattukudy, P. E.**, Structure of the cutinase gene and detection of promoter activity in the 5′-flanking region by fungal transformation, *J. Bacteriol.*, 171, 1942, 1989.
209. **Shayk, M., Soliday, C. L., and Kolattukudy, P. E.**, Proof for the production of cutinase by *Fusarium solani* F. *pisi* during penetration into its host, *Plant Physiol.*, 60, 170, 1977.
210. **Maiti, I. B. and Kolattukudy, P. E.**, Prevention of fungal infection of plants by specific inhibition of cutinase, *Science*, 205, 507, 1979.
211. **Köller, W., Allan, C. R., and Kolattukudy, P. E.**, Role of cutinase and cell wall degrading enzymes in infection of *Pisum sativum* by *Fusarium solani* f. sp. *pisi*, *Physiol. Plant Pathol.*, 20, 47, 1982.
212. **Köller, W., Allan, C. R., and Kolattukudy, P. E.**, Protection of *Pisum sativum* from *Fusarium solani* f. sp. *pisi* by inhibition of cutinase with organophosphorous pesticides, *Phytopathology*, 72, 1425, 1982.
213. **Woloshuk, C. P. and Kolattukudy, P. E.**, Mechanism by which contact with plant cuticle triggers cutinase gene expression in the spores of *Fusarium solani* f. sp. *pisi*, *Proc. Natl. Acad. Sci. U.S.A.*, 83, 1704, 1986.

214. **Podila, G. K., Dickman, M. B., and Kolattukudy, P. E.,** Transcriptional activation of a cutinase gene in isolated fungal nuclei by plant cutin monomers, *Science*, 242, 922, 1988.
215. **Dickman, M. B., Podila, G. K., and Kolattukudy, P. E.,** Insertion of cutinase gene into a wound pathogen enables it to infect intact host, *Nature*, 342, 446, 1989.
216. **Kolattukudy, P. E., Podila, G. K., Roberts, E., and Dickman, M. D.,** Gene expression resulting from the early signals in plant-fungus interaction, in *Molecular Biology of plant Pathogen Interaction*, Staskavicz, B., Ahlquist, P., and Yoder, O., Eds., Alan R. Liss, New York, 1989, 87.
217. **Dickman, M. B. and Patil, S. S.,** The role of cutinase from *Colletotrichum gloeosporioides* in the penetration of papaya, in *Experimental and Conceptual Plant Pathology*, Vol. 2, Singh, R. S., Singh, U. S., Hess, W. M., and Weber, D. J., Eds., Gordon and Breach, New York, 1988, 175.
218. **Tanabe, K., Nishimura, S., and Kohmoto, K.,** Cutinase production by *Alternaria alternata* Japanese pear pathotype and its role in pathogenicity, *Ann. Phytopathol. Soc. Jpn.*, 54, 483, 1988.
219. **Tanabe, K., Nishimura, S., and Kohmoto, K.,** Pathogenicity of cutinase- and pectic enzymes deficient mutants of *Alternaria alternata* Japanese pear pathotype, *Ann. Phytopathol. Soc. Jpn.*, 54, 552, 1988.
220. **Köller, W., Parker, D. M., and Becker, C. M.,** Role of cutinase in the penetration of apple leaves by *Venturia inaequalis, Phytopathology*, in press.
221. **Howard, R. J. and Ferrari, M. A.,** Role of melanin in appressoria function, *Exp. Mycol.*, 13, 403, 1989.
222. **Wolkow, P. M., Sisler, H. D., and Vigil, E. L.,** Effect of inhibitors of melanin biosynthesis on structure and function of appressoria of *Colletotrichum lindemuthianum, Physiol. Plant Pathol.*, 23, 55, 1983.
223. **Bonner, A. M. and Hammerschmidt, R.,** Role of cutinolytic enzymes in infection of cucumber by *Colletotrichum lindemuthianum, Physiol. Mol. Plant Pathol.*, 35, 475, 1989.
224. **Salinas, J., Warnaar, F., and Verhoeff, K.,** Production of cutin hydrolyzing enzymes by *Botrytis cinerea, J. Phytopathol.*, 116, 299, 1986.
225. **Köller, W. and Parker, D. M.,** Purification and characterization of cutinase from *Venturia inaequalis, Phytopathology*, 79, 278, 1989.
226. **Trail, F. and Köller, W.,** Diversity of cutinases from plant pathogenic fungi: evidence for a relationship between enzyme properties and tissue specificity, *Physiol. Mol. Plant Pathol.*, 36, 495, 1990.
227. **Dickman, M. B., Patil, S. S., and Kolattukudy, P. E.,** Effects of organophosphorous pesticides on cutinase activity and infection of papayas by *Colletotrichum gloeosporioides, Phytopathology*, 73, 1209, 1983.
228. **Köller, W., Trail, F., and Parker, D. M.,** Ebelactones inhibit cutinases produced by fungal plant pathogens, *J. Antibiot.*, 43, 695, 1990.
229. **Maurer, F., Sommer, H., Köller, W., Brandes, W., and Reinecke, P.,** Preparation of pyridyl phosphate esters as agrochemical fungicides, *Chem. Abstr.*, 107, 231436n, 1987.
230. **Aldridge, W. N.,** Cholinesterase and esterase inhibitors and reactivation of organophosphorous inhibited esterases, in *Design of Enzyme Inhibitors as Drugs*, Sanders, M. and Smith, H. J., Eds., Oxford, Press, Oxford, 1989, 294.
231. **Powers, J. C. and Zimmerman, M.,** Inhibitors of elastases, chymases, and cathepsin G, in *Design of Enzyme Inhibitors as Drugs*, Sanders, M. and Smith, H. J., Eds., Oxford Press, Oxford, 1989, 596.
232. **Kraft, G. A. and Katzenellenbogen, J. A.,** Synthesis of haloenol lactones. Mechanism based inactivators of serin proteases, *J. Am. Chem. Soc.*, 103, 5459, 1981.
233. **Charcravarty, P. K., Krafft, G. A., and Katzenellenbogen, J. A.,** Haloenol lactones: enzyme-activated irreversible inhibitors for serine proteases, *J. Biol. Chem.*, 257, 610, 1982.
234. **Moorman, A. R. and Abeles, R. H.,** A new class of serine protease inactivators based on isatoic anhydrate, *J. Am. Chem. Soc.*, 104, 6785, 1984.
235. **Hammock, B. D., Wing, K. D., McLaughlin, J., Lovell, V. M., and Sparks, T. C.,** Trifluoromethylketones as possible transition state analog inhibitors of juvenile hormone esterase, *Pestic. Biochem. Physiol.*, 17, 76, 1982.
236. **Gelb, M. H., Svaren, J. P., and Abeles, R. H.,** Fluoro ketone inhibitors of hydrolytic enzymes, *Biochemistry*, 24, 1813, 1985.
237. **Lindermann, R. J., Leazer, J., Venkatesh, K., and Roe, R. M.,** The inhibition of insect juvenile hormone esterase by trifluoromethylketones: steric parameters at the active site, *Pestic. Biochem. Physiol.*, 29, 266, 1987.
238. **Allen, K. N. and Abeles, R. H.,** Inhibition of pig liver esterase by trifluoromethyl ketones: modulators of the catalytic reaction alter inhibition kinetics, *Biochemistry*, 28, 135, 1989.
239. **Sohl, J., Sutton, L. D., Burton, D. J., and Quinn, D. M.,** Haloketone transition state analog inhibitors of cholesterol esterase, *Biochem. Biophys. Res. Commun.*, 151, 554, 1988.
240. **Linderman, R. J., Leazer, J., Roe, R. M., Venkatesh, K., Selinsky, B. S., and London, R. E.,** ^{19}F-NMR spectral evidence that 3-octylthio-1,1,1-trifluoropropan-2-one, a potent inhibitor of insect juvenile hormone esterase, functions as a transition state analog inhibitor of acetylcholinesterase, *Pestic. Biochem. Physiol.*, 31, 187, 1988.

241. **Harper, J. W. and Powers, J. C.,** Reaction of serine proteases with substituted 3-alkoxy-4-chloroisocoumarines and 3-alkoxy-7-amino-4-chloroisocoumarins: new reactive mechanism-based inhibitors, *Biochemistry,* 24, 7200, 1985.
242. **Kam, C.-H., Fujikawa, K., and Powers, J. C.,** Mechanism-based isocoumarin inhibitors for trypsin and blood coagulation serine proteases: new anticoagulants, *Biochemistry,* 27, 2547, 1988.
243. **Powers, J. C., Kam, C.-M., Narasinham, L., Oleksyszyn, J., Hernandez, M. A., and Ueda, T.,** Mechanism-based isocoumarin inhibitors for serine proteases: use of active site structure and substrate specificity in inhibitor design, *J. Cell. Biochem.,* 39, 33, 1989.
244. **Weidmann, B. and Abeles, R. H.,** Mechanism of inactivation of chymotrypsin by 5-butyl-3*H*-1,3-oxazine-2,6-dione, *Biochemistry,* 23, 2373, 1984.
245. **Hemmi, K., Harper, J., and Powers, J. C.,** Inhibition of human leukocyte elastase, cathepsin G, chymotrypsin A, and porcine pancreatic elastase with substituted isobenzofuranones and benzopyrandiones, *Biochemistry,* 24, 1841, 1985.
246. **Spencer, R. W., Copp, L. J., Bona\entura, B., Tam, T. F., Liak, T. J., Billedeau, R. J., and Krantz, A.,** Inhibition of serine proteases by benzoxaziones: effects of electron withdrawal and 5-substitution, *Biochem. Biophys. Res. Commun.,* 140, 928, 1986.
247. **Bonney, R. J., Ashe, B., Maycock, A., Dellea, P., Hand, K., Osinga, D., Fletcher, D., Mumford, R., Davies, P., Frankenfield, D., Nolan, T., Schaeffer, L., Hagmann, W., Finke, P., Shah, S., Dorn, C., and Doherty, J.,** Pharmacological profile of the substituted beta-lactam L-659,286: a member of a new class of human PMN elastase inhibitors, *J. Cell. Biochem.,* 39, 47, 1989.
248. **Köller, W., Allan, C. R., and Kolattukudy, P. E.,** Inhibition of cutinase and prevention of fungal penetration into plants by benomyl—a possible protective mode of action, *Pestic. Biochem. Physiol.,* 18, 15, 1982.
249. **Odds, F. C.,** Potential for penetration of passive barriers to fungal invasion in humans, in *The Fungal Spore and Disease Initiation in Plants and Animals,* Cole, G. T. and Hoch, H. C., Eds., Plenum Press, New York, 1991, 287.
250. **Ross, I. K., De Bernadis, F., Emerson, G. W., Cassone, A., and Sullivan, P. A.,** The secreted aspartate proteinase of *Candida albicans*: physiology of secretion and virulence of a proteinase-deficient mutant, *J. Gen. Microbiol.,* 136, 687, 1990.
251. **Manocha, M. S. and Chen, Y.,** Specificity of attachment of fungal parasites to their host, *Can. J. Microbiol.,* 36, 69, 1990.
252. **Nicholson, R. L. and Epstein, L.,** Adhesion of fungi to the plant surface: prerequisite for pathogenesis, in *The Fungal Spore and Disease Initiation in Plants and Animals,* Cole, G. T. and Hoch, H. C., Eds., Plenum Press, New York, 1991, 3.
253. **Kennedy, M. J.,** Adhesion and association mechanisms of *Candida albicans,* in *Current Topics in Medical Mycology,* Vol. 2, McGinnis, Ed., Springer-Verlag, New York, 1988, 73.
254. **Kennedy, M. J.,** *Candida* blastospore adhesion, association and invasion of the gastrointestinal tract to of vertebrates, in *The Fungal Spore and Disease Initiation in Plants and Animals,* Cole, G. T. and Hoch, H. C., Eds., Plenum Press, New York, 1991, 157.
255. **Mehentee, J. F. and Hay, R. J.,** Effect of antifungal agents on the adherence of *Candida albicans* to murine gastrointestinal mucosal surfaces, *J. Antimicrob. Chemother.,* 25, 111, 1990.
256. **Shephard, M. G.,** Morphogenetic transformation of fungi, in *Current Topics in Medical Mycology,* Vol. 2, McGinnis, Ed., Springer-Verlag, New York, 1988, 278.
257. **Kurtz, M. B., Kirsch, D. R., and Kelly, R.,** The molecular genetics of *Candida albicans, Microbiol. Sci.,* 5, 58, 1988.
258. **Soll, D. R.,** Current status of the molecular basis of *Candida* pathogenicity, in *The Fungal Spore and Disease Initiation in Plants and Animals,* Cole, G. T. and Hoch, H. C., Eds., Plenum Press, New York, 1991, 503.
259. **Ernst, J. F.,** Molecular genetics of pathogenic fungi: some recent developments and perspectives, *Mycoses,* 33, 231, 1990.
260. **Leong, S. A.,** Recombinant DNA research in phytopathogenic fungi, *Adv. Plant Pathol.,* 6, 1, 1988.
261. **Garber, R. C.,** Molecular approaches to the analysis of pathogenicity genes from fungi causing plant diseases, in *The Fungal Spore and Disease Initiation in Plants and Animals,* Cole, G. T. and Hoch, H. C., Eds., Plenum Press, New York, 1991, 483.
262. **Leong, S. A. and Holden, D. W.,** Molecular genetic approaches to the study of fungal pathogenesis, *Annu. Rev. Phytopathol.,* 27, 463, 1989.
263. **Bhairi, S. M., Staples, R. C., Freve, P., and Yoder, O. C.,** Characterization of an infection structure-specific gene from the rust fungus *Uromyces appendiculatis, Gene,* 81, 237, 1989.
264. **Bhairi, S., Buckley, E. H., and Staples, R. C.,** Protein synthesis and gene expression during appressoria formation in *Glomerella magna, Exp. Mycol.,* 14, 207, 1990.
265. **Sewall, T. C., Mims, C. W., and Timberlake, W. E.,** *aba*A controls phialide differentiation in *Aspergillus nidulans, Plant Cell.,* 2, 731, 1990.

266. **Timberlake, W. E.**, Molecular genetics of Aspergillus development, *Annu. Rev. Genet.*, 24, 1, 1990.
267. **Smith, F. D. and Köller, W.**, Effect of flusilazole on germination of conidia: isolates of *Venturia inaequalis* with different sensitivities to sterol demethylation inhibitors, *Phytopathology*, 80, 1000, 1990.
268. **Nes, W. D. and Heupel, R. C.**, Physiological requirement for biosynthesis of multiple 24 β-methyl sterols in *Gibberella fujikuroi*, *Arch. Biochem. Biophys.*, 244, 211, 1986.
269. **Fukazawa, Y. and Kagaya, K.**, Host defense mechanisms against fungal infection, *Microbiol. Sci.*, 5, 124, 1988.
270. **Boller, T.**, Primary signals and second messengers in the reaction of plants to pathogens, in *Second Messengers in Plant Growth and Development*, Boss, W. F. and Morré, D. J., Eds., Alan R. Liss, New York, 1989, 227.
271. **Van Loon, L. C.**, Stress proteins in infected plants, in *Plant Microbe Interactions: Molecular and Genetic Perspectives*, Vol. 3, Kosuge, T. and Nester, E. W., Eds., McGraw-Hill, New York, 1989.
272. **Bird, P. M.**, The role of lignification in plant diseases, in *Experimental and Conceptual Plant Pathology*, Vol. 3, Singh, R. S., Singh, U. S., Hess, W. M., and Weber, D. J., Eds., Gordon and Breach, New York, 1988, 523.
273. **Aist, J. R. and Bushnell, W. R.**, Invasion of plant hosts by powdery mildew fungi and cellular mechanism of resistance, in *The Fungal Spore and Disease Initiation in Plants and Animals*, Cole, G. T. and Hoch, H. C., Eds., Plenum Press, New York, 1991, 321.
274. **Kolattukudy, P. E., Podila, G. K., and Mohan, R.**, Molecular basis of the early events in plant-fungus interaction, *Genome*, 31, 342, 1989.
275. **Scholtens-Thoma, I. M. J., Joosten, M. H. A. J., and De Wit, P. J. G. M.**, Appearance of pathogen-related proteins in plant hosts: relationships between compatible and incompatible interactions, in *The Fungal Spore and Disease Initiation in Plants and Animals*, Cole, G. T. and Hoch, H. C., Eds., Plenum Press, New York, 1991.
276. **Madamanchi, N. R. and Kuc, J.**, Induced systemic resistance in plants, in *The Fungal Spore and Disease Initiation in Plants and Animals*, Cole, G. T. and Hoch, H. C., Eds., Plenum Press, New York, 1991, 347.
277. **Cartwright, D., Langcake, P., Price, R. J., Leworthy, D., and Ride, J. P.**, Chemical activation of host defense mechanisms as a basis for crop protection, *Nature*, 267, 511, 1977.
278. **Emanuil, V. and Wood, R. K. S.**, Fungicides as inducers of the accumulation of fungitoxic substrates in plants, *Physiol. Plant Pathol.*, 22, 51, 1983.
279. **Smolka, S. and Wolf, G.**, Cytological studies on the mode of action of systemic fungicides on the host pathogen complex barley powdery mildew, *Pestic. Sci.*, 17, 249, 1986.
280. **Cohen, Y. and Coffey, M. D.**, Systemic fungicides and the control of oomycetes, *Annu. Rev. Phytopathol.*, 24, 311, 1986.
281. **Heath, M. C.**, Effect of fungal death or inhibition induced by oxcarboxin or polyoxin D on the interaction between resistant and susceptible bean cultivars and the bean rust fungus, *Phytopathology*, 78, 1454, 1988.
282. **Smillie, R., Grant, B. R., and Guest, D.**, The mode of action of phosphite: evidence for both direct and indirect modes of action on three *Phytophthora* spp. in plants, *Phytopathology*, 79, 921, 1989.
283. **Altman, J.**, Impact of herbicides on plant diseases, in *Ecology and Management of Soilborne Plant Pathogens*, Parker, C. A., Rovira, A. D., Moore, K. J., Wong, P. T. W., and Kollmorgen, J. F., Eds., APS Press, St. Paul, 1985, 227.
284. **Lazarovits, G.**, Induced resistance—xenobiotics, in *Experimental and Conceptual Plant Pathology*, Vol. 3, Singh, R. S., Singh, U. S., Hess, W. M., and Weber, D. J., Eds., Gordon and Breach, New York, 1988, 575.
285. **Grinstein, A., Lisker, N., Katan, J., and Eshel, Y.**, Herbicide-induced resistance to plant wilt diseases, *Physiol. Plant Pathol.*, 24, 347, 1984.
286. **Lyndon, J. and Duke, S. O.**, Pesticide effects on secondary metabolism of higher plants, *Pestic. Sci.*, 25, 361, 1989.
287. **Langcake, P., Cartwright, D. W., and Ride, J. P.**, The dichlorocyclopropanes and other fungicides with indirect mode of action, in *Systemische Fungizide und Antifungale Verindungen*, Lyr, H. and Polter, C., Eds., Akademie-Verlag, Berlin, 1983, 199.
288. **Dixon, R. A. and Harrison, M. J.**, Activation, structure, and organization of genes involved in microbial defense in plants, *Adv. Genet.*, 28, 165, 1990.
289. **Malamy, J., Carr, J. P., Klessig, D. F., and Raskin, I.**, Salicylic acid: a likely endogenous signal in the resistance response of tobacco to viral infection, *Science*, 250, 1002, 1990.
290. **Métraux, J. P., Signer, H., Ryals, J., Ward, E., Wyss-Benz, M., Gaudin, J., Raschdorf, K., Schmid, E., Blum, W., and Iverardi, B.**, Increase in salicylic acid at the onset of systemic acquired resistance in cucumber, *Science*, 250, 1004, 1990.
291. **Scholer, H. J. and Polak, A.**, Resistance to systemic antifungal agents, in *Antimicrobial Drug Resistance*, Academic Press, New York, 1984, 393.

292. **Kerridge, D., Fasoli, M., and Wayman, F. J.,** Drug resistance in *Candida albicans* and *Candida glabrata*, *Ann. N.Y. Acad. Sci.*, 544, 245, 1988.
293. **Skylakakis, G. and Hollomon, D. W.,** Epidemiology of fungicide resistance, in *Combating Resistance to Xenobiotics—Biological and Chemical Approaches*, Ford, M. G., Hollomon, D. W., Khambay, B. P. S., and Sawicki, R. M., Eds., Ellis Horwood, Chichester, 1987, 94.
294. **Smith, F. D., Parker, D. M., and Köller, W.,** Sensitivity distribution of *Venturia inaequalis* to the sterol demethylation inhibitor flusilazole: baseline sensitivity and implications for resistance monitoring, *Phytopathology*, 81, 392, 1991.
295. **Van Cutsem, J., Van Gerven, F., and Janssen, P. A. J.,** The in vitro and in vivo antifungal activity of itraconazole, in *Recent Developments in the Discovery, Development and Evaluation of Antifungal Agents*, Fromtling, R. A., Ed., J. R. Prous Science, Barcelona, 1987, 177.
296. **Fry, W. E. and Milgrom, M. G.,** Population biology and management of fungicide resistance, in *Managing Resistance to Agrochemicals—From Fundamental Research to Practical Strategies*, Green, M. B., LeBaron, H. M., and Moberg, W. K., Eds., American Chemical Society, Washington, D.C., 1990, 275.
297. **Krämer, W., Berg, D., and Köller, W.,** Chemical synthesis and fungicidal resistance, in *Combating Resistance to Xenobiotics—Biological and Chemical Approaches*, Ford, M. G., Hollomon, D. W., Khambay, B. P. S., and Sawicki, R. M., Eds., Ellis Horwood, Chichester, 1987, 291.
298. **Dekker, J.,** The development of resistance to fungicides, *Prog. Pestic. Biochem. Physiol.*, 4, 165, 1985.
299. **Uesugi, Y.,** Case study 3: Pyricularia oryzae of rice, in *Fungicide Resistance in Crop Protection*, Dekker, J. and Georgopoulos, S. G., Eds., Pudoc, Wageningen, 1982, 207.
300. **Orth, A. B., Henry, M. J., and Sisler, H. D.,** Mechanism of resistance to terbinafine in two isolates of *Ustilago maysid*, *Pestic. Biochem. Physiol.*, 37, 182, 1990.
301. **Orth, A. B. and Sisler, H. D.,** Mode of action of terbinafine in *Ustilago maydis* and characterization of resistant mutants, *Pestic. Biochem. Physiol.*, 37, 53, 1990.
302. **Aoyama, Y., Yoshida, Y., Nishino, T., Katsuki, H., Maitra, U. S., Mohan, V. P., and Sprinson, D. B.,** Isolation and characterization of an altered cytochrome P-450 from a yeast defective in lanosterol 14α-demethylation, *J. Biol. Chem.*, 262, 14260, 1987.
303. **Ishida, N., Aoyama, Y., Hatanaka, R., Oyama, Y., Imajo, S., Ishiguro, M., Oshima, T., Nakazato, H., Noguchi, T., Maitra, U. S., Mohan, V. P., Sprinson, D. B., and Yoshida, Y.,** A single amino acid substitution converts cytochrome $P\text{-}450_{14DM}$ to an inactive form, cytochrome $P450_{SG1}$: complete primary structures deduced from cloned DNAs, *Biochem. Biophys. Res. Commun.*, 155, 317, 1988.
304. **Sutter, T. R. and Loper, J. C.,** Disruption of the *Saccharomyces cerevisiae* gene for NADPH-cytochrome P450 reductase causes increased sensitivity to ketoconazole, *Biochem. Biophys. Res. Commun.*, 160, 1257, 1989.
305. **Vanden Bossche, H.,** Importance and role of sterols in fungal membranes, in *Biochemistry of Cell Walls and Membranes in Fungi*, Kuhn, P. J., Trinci, A. P. J., Jung, M. J., Goosey, M. W., and Copping, L. G., Eds., Springer-Verlag, Berlin, 1990, 135.
306. **Scott, J. G.,** Insecticide resistance in insects, in *CRC Handbook of Pest Management in Agriculture*, Vol. 2, 2nd ed., Pimentel, D., Ed., CRC Press, Boca Raton, FL, 1991, 663.
307. **Butt, T. R. and Ecker, D. J.,** Yeast metallothionein and application in biotechnology, *Microbiol. Rev.*, 51, 351, 1987.
308. **Barak, E. and Edgington, L. V.,** Glutathion synthesis in response to captan: a possible mechanism for resistance of *Botrytis cinerea* to the fungicide, *Pestic. Biochem. Physiol.*, 21, 412, 1984.
309. **VanEtten, H. D., Matthews, D. E., and Mackintosh, S. F.,** Adaptation of pathogenic fungi to toxic chemical barriers in plants: the pisatin demethylase of *Nectria haematococca* as an example, in *Molecular Strategies for Crop Protection*, Arntzen, C. J. and Ryan, C., Eds., Alan R. Liss, New York, 1987, 59.
310. **Uesugi, Y. and Katagiri, M.,** Metabolism of a phosphothiolate fungicide IBP by strains of *Pyricularia oryzae* with varied sensitivity, in *Pesticide, Human Welfare and the Environment*, Vol. 3, Miyamoto, Y. and Kearney, P. C., Eds., Pergamon Press, Elmsford, NY, 1983, 165.
311. **Nes, W. D. and Parish, E. J.,** Metabolism of 2 (*RS*), 3-epiminosqualene to 24(*RS*), 25-epiminolanosterol by *Gibberella fujikuroi*, *Lipids*, 23, 375, 1988.
312. **Nes, W. D., Xu, S., and Parish, E. J.,** Metabolism of 24 (*R,S*), 25-epiminolanosterol to 25-aminolanosterol and lanosterol by *Gibberella fujikuroi*, *Arch. Biochem. Biophys.*, 272, 323, 1989.
313. **Naider, F. and Becker, J. F.,** Peptide transport in *Candida albicans*: implications for the development of antifungal agents, in *Current Topics in Medical Mycology*, Vol. 2, Springer-Verlag, New York, 1988.
314. **Yadan, J.-C., Gonneau, M., Sarthou, P., and Le Goffic, F.,** Sensitivity to nikkomycin Z in *Candida albicans*: role of peptide permeases, *J. Bacteriol.*, 160, 884, 1984.
315. **De Waard, M. A. and Fuchs, A.,** Resistance to ergosterol biosynthesis inhibitors. II. Genetic and physiological aspects, in *Fungicide Resistance in Crop Protection*, Dekker, J. and Georgopoulos, S. G., Eds., Pudoc, Wageningen, 1982, 87.
316. **De Waard, M. A. and van Nistelrooy, J. G. M.,** Inhibitors of energy-dependent efflux of the fungicide fenarimol by *Aspergillus nidulans*, *Exp. Mycol.*, 11, 1, 1987.

317. **De Waard, M. A. and van Nistelrooy, J. G. M.,** Accumulation of SBI fungicides in wild type and fenarimol-resistant isolates of *Penicillium italicum, Pestic. Sci.*, 22, 371, 1988.
318. **Ryley, J. F., Wilson R. G., and Barrett-Bee, K. J.,** Azole resistance in *Candida albicans, J. Med. Vet. Mycol.*, 22, 53, 1984.
319. **Hippe, S.,** Combined application of low temperature preparation and electron microscopic autoradiography for the localization of systemic fungicides, *Histochemistry*, 87, 309, 1987.
320. **Rine, W., Hansen, E., Hademan, E., and Davis, R. W.,** Target selection of recombinant clones through gene dosage effects, *Proc. Natl. Acad. Sci. U.S.A.*, 80, 6750, 1983.
321. **Kalb, V. F., Woods, C. W., Turi, T. G., Dey, C. R., Sutter, T. R.,and Loper, J. C.,** Primary structure of the P450 lanosterol demethylase gene from *Saccharomyces cerevisiae, DNA*, 6, 529, 1987.
322. **Chen, C., Turi, T. G., Sanglard, D., and Loper, J. C.,** Primary structure of the *Candida tropicalis* gene for P450 lanosterol demethylase and its expression in *Saccharomyces cerevisiae, Biochem. Biophys. Res. Commun.*, 146, 1311, 1987.
323. **Pereira, R., Holmlund, C. E., and Whittacker, N.,** The effect of AY-9944 on yeast sterol and sterol ester metabolism, *Lipids*, 18, 545, 1983.
324. **Sobus, M. T., Holmlund, C. E., and Whittacker, N. F.,** Effects of the hypocholesteremic agent trifluperidol on the sterol, steryl esters, and fatty acid metabolism of *Saccharomyces cerevisiae, J. Bacteriol.*, 130, 1310, 1977.
325. **Smith, F. D. and Köller, W.,** The expression of resistance of *Ustilago avenae* to the sterol demethylation inhibitor triadimenol is an induced response, *Phytopathology*, 80, 584, 1990.
326. **Watson, P. F., Rose, M. E., Ellis, S. W., England, H., and Kelly, S. L.,** Defective sterol C5-6 desaturation and azole resistance: a new hypothesis for the mode of action of azole antifungals, *Biochem. Biophys. Res. Commun.*, 164, 1170, 1989.
327. **Henry, M. J. and Trivellas, A. E.,** Laboratory-induced fungicide resistance to benzimidazoles and azole fungicides in *Cercospora beticola, Pestic. Biochem. Physiol.*, 35, 89, 1989.
328. **Weete, J. D. and Wise, M. L.,** Effects of triazoles on fungi. V. Response by a naturally tolerant species, *Mucor rouxii, Exp. Mycol.*, 11, 214, 1987.
329. **Balzi, E., Chem, W., Ulaszewski, S., Capieaux, E., and Goffeau, A.,** The multidrug resistance gene PDR1 from *Saccharomyces cerevisiae, J. Biol. Chem.*, 25, 16871, 1987.
330. **Gisi, U. and Staehle-Csech,** Resistance risk evaluation of new candidates for disease control, in *Fungicide Resistance in North America*, Delp, C. J., Eds., APS-Press, St. Paul, 1988, 101.
331. **Brent, K. J., Hollomon, D. W., and Shaw, M. W.,** Predicting the evolution of fungicide resistance, in *Managing Resistance to Agrochemicals—From Fundamental Research to Practical Strategies*, Green, M. B., LeBaron, H. M., and Moberg, W. K., Eds., American Chemical Society, Washington, D.C., 1990, 303.
332. **Mazur, B. J. and Falco, S. C.,** The development of herbicide resistant crops, *Annu. Rev. Plant Physiol. Plant Mol. Biol.*, 40, 441, 1989.
333. **Hollomon, D. W.,** Molecular approaches to understanding the mechanisms of fungicide resistance, *Proc. Brighton Crop Prot. Conf. Pests Dis.*, 1990, 881, 1990.
334. **Gustafson, G. D., Waldron, C., and Davis, G. E.,** Isolation of a DNA fragment that transforms *Aspergillus nidulans* to resistance to the antifungal compound LY 214352 in Abstr. 4th Int. Mycol. Congr., IIF-282/1, 1990.
335. **Köller, W.,** Sterol demethylation inhibitors: mechanism of action and resistance, in *Fungicide Resistance in North America*, Delp, C. J., Ed., APS-Press, St. Louis, 1988, 79.
336. **Hollomon, D. W., Butters, J., and Hargreaves, J. A.,** Resistance to sterol biosynthesis-inhibiting fungicides: current status and biochemical basis, in *Managing Resistance to Agrochemicals—From Fundamental Research to Practical Strategies*, Green, M. B., LeBaron, H. M., and Moberg, W. K., Eds., American Chemical Society, Washington, D.C., 1990, 199.
337. **De Waard, M. A.,** Interactions of fungicide combinations, in *Fungicide Resistance in North America*, Delp, C. J., Eds., APS Press, St. Louis, 1988, 98.
338. **Köller, W. and Wubben, J. P.,** Variable resistance factors of fungicides acting as sterol demethylation inhibitors, *Pestic. Sci.*, 26, 133, 1989.
339. **Sambrook, J., Fritsch, E. F., and Maniatis, T.,** *Molecular Cloning—A Laboratory Manual*, Cold Spring Harbor Laboratory, Cold Spring Harbor, NY, 1989.
340. **Basson, M. E., Thorsness, M., Finer-Moore, J., Stroud, R. M., and Rine, J.,** Structural and functional conservation between yeast and human 3-hydroxy-3-methylglutaryl coenzyme A reductases, the rate-limiting enzyme of sterol biosynthesis, *Mol. Cell. Biol.*, 8, 3797, 1988.
341. **Learned, R. M. and Fink, G. R.,** 3-Hydroxy-3-methylglutaryl coenzyme A reductase from *Arabidopsis thaliana* is structurally distinct from the yeast and animal enzymes, *Proc. Natl. Acad. Sci. U.S.A.*, 86, 2779, 1989.
342. **Wallace, R. B. and Miyada, C. G.,** Oligonucleotide probes for the screening of recombinant DNA libraries, *Methods Enzymol.*, 152, 432, 1987.

343. **Campbell, W. H. and Gowri, G.,** Codon usage in higher plants, green algae, and cyanobacteria, *Plant Physiol.*, 92, 1, 1990.
344. **Fincham, J. R. S.,** Transformation of fungi, *Microbiol. Rev.*, 53, 148, 1989.
345. **Timberlake, W. E. and Marshall, M. A.,** Genetic engineering of filamentous fungi, *Science*, 244, 1313, 1989.
346. **Tsuge, T., Kobayashi, H., and Nishimura, S.,** Specific gene expression in the host-selective toxin producer, *Alternaria alternata* and its genetic transformation with high efficiency, in *Host-Specific Toxins and Specificity Factors in Plant Disease*, Kohmoto, K. and Durbin, R. D., Eds., Tottori University Press, Tottori, 1989, 153.
347. **Huang, D., Bhairi, S., and Staples, R. C.,** A transformation procedure for *Bortyotinia squamosa*, *Curr. Genet.*, 15, 411, 1989.
348. **Leung, H., Lehtinen, U., Karjalainen, R., Skinner, D., Tooley, P., Leong, S., and Ellingboe, A.,** Transformation of the rice blast fungus *Magnaporthe grisia* to hygromycin B resistance, *Curr. Genet.*, 17, 409, 1990.
349. **Langin, T., Daboussi, M. J., Gerlinger, C., and Brygoo, Y.,** Influence of biological parameters and gene transfer technique on transformation of *Fusarium oxysporum*, *Curr. Genet.*, 17, 313, 1990.
350. **Turgeon, B. G., Garber, R. C., and Yoder, O. C.,** Development of a fungal transformation system based on selection of sequences with promoter activity, *Mol. Cell. Biol.*, 7, 3297, 1987.
351. **Rodriguez, R. J. and Yoder, O. C.,** Selectable genes for transformation of the fungal plant pathogen *Glomerella cingulata* f.sp. *phaseoli*, *Gene*, 54, 73, 1987.
352. **Bej, A. K. and Perlin, M. H.,** A high efficiency transformation system for the basidiomycete *Ustilago violacea* employing hygromycin resistance and lithium-acetate treatment, *Gene*, 80, 171, 1989.
353. **Ward, M., Kodama, K. H., and Wilson, L. J.,** Transformation of *Aspergillus awamori* and *A. niger* by electroporation, *Exp. Mycol.*, 13, 289, 1989.
354. **Armaleo, D., Ye, G.-N., Klein, T. M., Shark, K. B., Sanford, J. C., and Johnston, S. A.,** Biolistic nuclear transformation of *Saccharomyces cerevisiae* and other fungi, *Curr. Genet.*, 17, 97, 1990.
355. **McClung, C. R., Phillips, J. D., Orbach, M. J., and Dunlap, J. C.,** New cloning vectors using benomyl resistance as a dominant marker for selection in *Neurospora crassa* and in other filamentous fungi, *Exp. Mycol.*, 13, 299, 1989.
356. **Poulsen, C. and Stougaard, P.,** Purification and properties of *Saccharomyces cerevisiae* acetolactate synthase from recombinant *Escherichia coli*, *Eur. J. Biochem.*, 185, 433, 1989.
357. **Hinnen, A., Meyhack, B., and Heim, J.,** Heterologous gene expression in yeast, in *Yeast Genetic Engineering*, Barr, P. J., Brake, A. J., and Valenzuela, P., Eds., Butterworths, Boston, 1989, 193.
358. **Saunders, G., Picknett, T. M., Tuite, M. F., and Ward, M.,** Heterologous gene expression in filamentous fungi, *Trends Biotechnol.*, 7, 283, 1989.
359. **Austin, B. and Tyler, B. M.,** Strategies for high-efficiency cotransformation of *Neurospora crassa*, *Exp. Mycol.*, 14, 9, 1990.
360. **Weber, J. M., Reiser, J., and Käppeli, O.,** Lanosterol 14α-demethylase-encoding gene: systematic analysis of homologous overexpression in *Saccharomyces cerevisiae* using strong yeast promoters, *Gene*, 87, 167, 1990.
361. **Zaworski, P. G. and Gill, G. S.,** Use of *Saccharomyces cerevisiae* expressing ß-galactosidase to screen for antimycotic agents directed against yeast cell wall biosynthesis and possible application to pathogenic fungi, *Antimicrobial Agents Chemother.*, 34, 660, 1990.
362. **Leatherbarrow, R. J. and Fersht, A. R.,** Use of protein engineering to study enzyme mechanisms, in *Enzyme Mechanisms*, Page, M. I. and Williams, A., Eds., Royal Society of Chemistry, London, 1987, 78.
363. **Shaw, W. V.,** Protein engineering—the design, synthesis and characterization of factitious proteins, *Biochem. J.*, 246, 1, 1987.
364. **Timberlake, W. E.,** Isolation of stage- and cell-specific genes from fungi, in *Biology and Molecular Biology of Plant-Pathogen Interactions*, Bailey, J., Eds., Springer-Verlag, 1986, 345.

Chapter 11

BIOTECHNOLOGY AND FUNGAL CONTROL

A. J. Poulose

TABLE OF CONTENTS

I. INTRODUCTION

Plants respond to fungal infection by producing inducible enzymes involved in either hydrolyzing the fungal cell wall polymers or catalyzing the synthesis of antimicrobial compounds.[1,2] However, certain pathogens can overcome the plant defense responses by producing enzymes to neutralize the antifungal compounds produced by the plant,[3,4] by inactivating the plant defense-related gene products, or by blocking the induction of plant defense response. It is also possible that the delay in recognizing the presence of a pathogen and turning on the defense system could result in inefficient levels of defense by the plant. After the entry of the fungus into the plant, the quantity of the antifungal defense compounds produced may not be sufficient to present the progress of infection. Therefore, externally applied chemical fungicides are needed to effectively control fungal diseases. Some of these chemicals are of general toxicity and thus are undesirable if alternative methods of fungal control are available.

In this chapter, an attempt is made to survey the studies done in the area of mycolytic and other enzymes and proteins that are fungicidal, and their direct use in fungal control. The results of several studies indicate the potential use of these enzymes in fungal control, but they are not widely used in commercial applications. In order to successfully commercialize biological fungicides, they have to be cost effective, stable in formulations, and active under the application conditions. Recent advances made in biotechnology will allow one to: (1) economically produce one or more of the fungal cell wall-degrading enzymes by the transfer and expression of genes encoding these enzymes into industrial production organisms: (2) modify the structural gene of these enzymes to engineer new properties such as temperature and storage stability, new pH optima, or increased catalytic efficiency, all of which would render enzymes more effective under the application conditions; (3) engineer these genes into microorganisms that colonize the root or the leaf so that they will be expressed in critical amounts to effectively control pathogenic fungi; and (4) engineer these genes into plants to be expressed in effective amounts before or during the early phases of fungal infection.

II. ENZYMES FOR FUNGAL CONTROL

A. FUNGAL CELL WALL-DEGRADING ENZYMES

Cell walls of fungi consist of a complex mixture of polymers which include polysaccharides, glycoproteins, and proteins. Even though the chemical composition of the cell walls from different fungi have been determined,[5,6] the knowledge of actual spacial organization of these polymers in fungal cell walls is limited. However, existence of α- and β-glucan, cellulose, and chitin in the outer cell walls of fungi has been described.[5,7-9] It is also well known that a complex group of covalently linked proteins and carbohydrates is part of the fungal cell walls.[10]

The cell wall plays a crucial role in the interaction of fungi with the environment and in cell recognition and interactions. Since fungal cell walls contain biodegradable polymers, enzymes can be used to disrupt cell wall structure by hydrolyzing the components from the cell wall. Mycolytic enzymes have been used for the hydrolysis of cell walls to prepare fungal protoplasts,[11] to extract fungal intracellular components,[12] and to elucidate the structure of fungal cell walls. Thus, mycolytic enzymes should have the potential to be effective biological fungicides of synergists to chemical fungicides.

Several bacteria and fungi produce mycolytic enzymes that degrade fungal cell wall polymers. *Arthrobacter*,[13] *Streptomyces*,[11] *Cytophaga*,[14] *Bacillus*,[15] *Rhizopus*,[16] and *Trichoderma*[17] are examples of some of the species that produce enzymes with fungal cell wall polymer-degrading activities. Another source of mycolytic enzymes is the gastric juice of the snail *Helix pomatia*.[18] Enzymes from these different sources consist of a mixture of glucanases, chitinases, and proteases.

Chitinase and β-1,3-glucanase from plants have been postulated to be involved in the protection of plants from pathogen attack by hydrolyzing the cell walls of invading pathogens.[19] These enzymes have been isolated from a variety of plants such as bean leaves,[19] wheat germ,[20] tomato stem,[21] barley and rye seedlings,[22] and soybean cotyledon.[23]

B. ENZYMES THAT INACTIVATE RIBOSOMES

Ribosome-inactivating proteins are enzymes that act on heterologous ribosomes to inhibit polypeptide chain elongation.[24] These enzymes are widely distributed in plants. Two types of these proteins are found in nature. One type, which is rare in nature, belongs to the group of the highly toxic plant proteins such as nicin. These proteins consist of two polypeptide chains, one which binds to the cell membrane and another which contains the enzymatic activity. A second type of protein, more commonly found in plants, contains only the chain that contains enzymatic activity. Since these proteins do not have the polypeptide that binds to the cell membrane, they are relatively nontoxic. This class of enzymes is found in seeds, roots, leaves, and sap of plants at very high concentration.[24-26] These enzymes, isolated from wheat, rye, barley, and corn, inhibited *T. reesei*.[27] These proteins, combined with fungal cell wall-degrading enzymes could be effective fungicidal combinations not toxic to mammalian cells.

III. APPLICATION OF ENZYMES IN FUNGAL CONTROL

Since the lytic enzymes have the capability to alter the structure of the cell wall, they could result in the inhibition of fungal growth due to (1) disruption of transport systems that control the uptake of nutrients, (2) release of reserve carbohydrates which might be needed in certain stages of the life cycle, (3) loss of cell wall components that might be important for cell recognition and interaction, or (4) loss of cellular integrity normally supported by the rigid cell wall matrix. Extensive degradation of the cell wall could affect the protoplast's osmotic protector function which would result in cell lysis. In addition to the inhibition of normal growth and metabolism of fungi, lytic enzymes could also facilitate the transport of other fungicidal compounds to their target site of action. Therefore, these enzymes could serve as fungicidal adjuvants or as fungicidal enzymes.

A. FUNGICIDAL ADJUVANT ENZYMES

Antifungal activity of amphotericin B has been enhanced when combined with lysozyme against *Candida albicans*.[28] Mycolytic enzymes isolated from *Physarum polycephalum* prolonged the survival of mice against *A. fumigatus* or *C. albicans* infection when injected along with amphotericin B.[29] Mycolases isolated from the culture filtrates of *Cytophaga* sp. grown on chitin potentiated the antifungal activity of amphotericin B against *C. pseudotropicalis*.[30] These studies indicate that the enzymes that hydrolyze fungal cell wall polymers enhance the antibiotic activity of amphotericin B.

The author has observed the synergistic effect of chitinase from *Serratia marcescens* on the inhibition of the growth of *Trichoderma reesei* and *Aspergillus nidulans* by Benomyl.[31] Similar effects are also observed with Novozyme 234, which contains a mixture of fungal cell wall-degrading enzymes.

Fungal lytic enzymes prepared from *Bacillus circulans* WL12, when used in combination with a nonionic detergent Emulgen 120, enhanced the fungicidal action of Polyoxin B and Kitazin P against *Pyricularia oryzae, Botrytis Cinerea,* and *Colletotrichum lagenarium*.[32] This enzyme system was not effective in decreasing the minimal inhibitory concentration of Polyoxin B and Kitazin P against *Glomerella cingulata* and *Alternaria kikuchiana* unless mannanase and protease were added from another source. This might be due to the differences in the cell wall structures between various fungi. However, the synergistic effect of nonionic detergents in the action of lytic enzymes is interesting.

Plant chitinases showed synergistic effects with antifungal compounds such as Amphotericin B, Polyoxin B, and Nikkomycin against *Candida albicans*.[33] The synergistic effects were more dramatic when a corn chitinase preparation containing another antifungal protein from corn was used. The mechanism of actin of this antifungal protein is not known. However, this protein enhanced the activity of Nikkomycin against *C. albicans* and *Rhodotorula*.[33] The author has observed that the minimum inhibitory concentration of Nikkomycin was considerably lower against *T. reesei, Aspergillus nidulans*, and *Phanerochaete chrysosporium* when this antifungal protein was used in conjunction with Nikkomycin.[34]

All of the above-mentioned studies clearly showed that enzymes from microbial and plant sources capable of degrading fungal cell wall polymers can synergistically improve the effect of several fungicides including Amphotericin B, Benomyl, Polyoxin B, Kitazin P, and Nikkomycin. Enzymes have the potential to increase the effectiveness of fungicides and, thus, to decrease the dosage level of chemical fungicides. When used in agriculture, this would have a positive environmental impact, since lesser amounts of toxic chemicals would be released into the environment. Since many of the antimycotic drugs are toxic to humans and animals, lowering the dosage by adding mycolytic enzymes for effective control of pathogenic fungi would be of significant advantage in medical applications.

B. FUNGICIDAL ENZYMES

Plants respond to fungal infections by producing mycolytic enzymes. It is possible that the mycolytic enzymes such as chitinase and β-1,3-glucanase could control pathogenic fungi by inhibiting their growth by the lysis of fungal hyphae.[35] Plant chitinases are also known to have higher antifungal activity than microbial chitinases. It has been shown that chitinases isolated from wheat, barley, and maize showed significantly more antifungal activity against *T. reesei* and *Phycomyces blakesleeanus* when compared to chitinase from *S. marcescens, Streptomyces griseus*, and *Pseudomonas stutzeri*.[36] The difference in activity is believed to be due to the fact that plant chitinase has more endochitinase activity, while microbial enzymes have more exochitinase activity.

In addition to being fungicidal, these enzymes generate oligosaccharide fragments from fungal cell walls, which induce defense responses in plants[37,38] resulting in the production of antifungal compound such as phytoalexins,[37] proteinase inhibitors,[39] and lipoxygenase.[40] Therefore, mycolytic enzymes could be effective in controlling fungal growth in plants by disrupting the organization of the fungal cell wall as well as by generating oligosaccharides for induction of plant antifungal compounds. There have been several studies showing the ability of microbes such as *S. marcescens, I. harzianum*, and others to control plant pathogenic fungi.[41-43] Fungal cell wall-degrading enzymes have been implicated to be responsible for the biological control activity of these microbes. However, very few studies have been done with isolated enzymes. In one such study, *S. marcescens* chitinase, when applied in irrigating water, was effective in controlling *S. rolfsii* and *R. solani* infection in bean seedlings grown under greenhouse conditions.[44] In a field study, protection of rice plants from *Pyricularia oryzae* was observed when rice plants were sprayed with a mixture of β-glucanase, chitinase, polyoxyethylene-lauryl ether, and isopropanol, before inoculation with the fungi.[45,46]

C. ENZYMES THAT ENHANCE THE UPTAKE OF FUNGICIDES INTO PLANTS

Major barriers for the entry of agricultural chemicals into plants are thought to include the surface wax and the polyester polymer cutin. Carbohydrate polymers pectin, cellulose, and hemicellulose could also be barriers for penetration of chemicals into the plant cell. The relative role of each of these barriers will depend on the structure and properties of the agricultural chemical.

It is possible to use enzymes such as cutinase, pectinase, and cellulase that will degrade plant polymers, to increase the uptake of agricultural chemicals into plants.[47] The author has shown that cutinase and a mixture of cellulase, pectinase, and hemicellulase increased the uptake of Benomyl into soybean plants. Use of enzymes to aid the penetration into plants could make fungicides more effective by increasing the amount present in the plant and thus prolonging the presence of fungicides in the leaf. This technology could also have a positive environmental impact by reducing the amount of chemicals that need to be applied in order to effectively control fungi. It might also be possible to discover new systemic fungicides which previously could not be taken up into the plant, but could be taken up in the presence of enzymes.

IV. ENGINEERING FUNGICIDAL ENZYMES INTO MICROBES AND PLANTS

The chitinase gene from *Serratia marcescens* has been cloned and expressed in *Escherichia coli*.[44,48,49] Under greenhouse conditions, *E. coli* expressing chitinase reduced the infection of bean plants by *Sclerotium rolfsii* and cotton plants by *Rhizoctonia solani*.[44] Chitinase was also cloned and expressed in *Pseudomonas fluorescens* and *P. putida*, both of which were isolated from the rhizosphere.[48]

Chitinase gene has also been introduced into tobacco plants[48] and active chitinase was expressed in the transformed plants. The rate of infection by *Alternaria alternata* of leaf disks of transformed tobacco plants expressing active chitinase was less than that of control plants. These studies show the feasibility of expressing antifungal enzymes in plants and microbes that inhabit the plants in order to increase the ability of plants to protect themselves from pathogenic fungi.

V. FUTURE PROSPECTS

It is evident from the literature and the author's work that enzymes can act as fungicidal adjuvants, fungicides, and as penetrating agents for fungicides into plants. However, applications of these technologies for effective control of fungi are not widely used. One of the major problems is the cost associated with enzyme production. For example, it is estimated that chitinase production costs are several hundred dollars per gram.[50] The cost of enzyme production can be lowered by biotechnological methods. Cloning and expression of the chitinase gene in an appropriate industrial production organism by genetic engineering could reduce the production cost of enzymes drastically. Availability of large amounts of enzymes will allow one to test these enzymes in field trials in order to assess the commercial feasibility. Thus, the low cost of fungicidal enzymes will enable the development of commercial fungicide formulations consisting of one or more enzymes. These novel fungicide formulations could then be used in crop protection by spraying the fields or coating the seeds. Antifungal enzymes effective in agriculture could also find applications in medicine for the control of fungal diseases.

Another approach in fungal control is to use organisms that inhabit the rizosphere or the phyllosphere which produce high levels of antifungal enzymes and proteins. One can isolate organisms from nature that contain antifungal proteins or use biotechnological methods to express the genes for antifungal proteins in other organisms. In turn, these organisms can be used to protect plants from fungal infection. Both approaches have been studied and the reports in the literature suggest the feasibility of these approaches in fungal control. Commercial products based on this approach, especially nonrecombinant microbes, can be expected to be developed in the near future.

The ultimate goal is to have plants that are resistant to the attack by fungal pathogens. This is an active area of research where several approaches are being investigated. One approach would be to engineer plants toward expression of antifungal enzymes and proteins. Constitutive expression of these enzymes may not be ideal for the plant because of the loss of energy that would be diverted to maintain constant high levels of these proteins. Furthermore, constant expression of these enzymes could facilitate the rapid development of pathogens that are resistant to the antifungal proteins. Therefore, the ideal situation might be to have these genes developmentally or environmentally regulated so that they are expressed at high levels during the growth stage when the plants are most susceptible to fungal pathogens, or upon environmental stimulus, such as invasion by pathogenic organisms.

ACKNOWLEDGMENTS

I would like to thank Dr. Roger O'Neill for the critical review of the manuscript and Ms. Jacqueline Aldinger for the preparation of the manuscript.

REFERENCES

1. **Boller, T.,** Induction of hydrolases as a defense reaction against pathogens, in *Cellular and Molecular Biology of Plant Stress,* Key, J. L. and Kosuge, T., Eds., Alan R. Liss, New York, 1985, 147.
2. **Bell, A. A.,** Biochemical mechanism of disease resistance, *Ann. Rev. Plant Physiol.*, 32, 21, 1981.
3. **Zhang, Y. and Smith, D. A.,** Concurrent metabolism of the phytoalexins Phaseolin Kievitone and Phaseolin Isoflavan by *Fusarium solani, Physiol. Plant Pathol.*, 23, 89, 1983.
4. **Weltring, k. M., Turgeon, B. G., Yoder, O. C., and Van Etten, H. D.,** Isolation of a phytoalexin-detoxification gene from the plant pathogenic fungus *Nectria haematocca* by detecting its expression in *Aspergillus nidulans, Gene,* 68, 335, 1988.
5. **Gander, J. E.,** Fungal cell wall glycoproteins and peptidopolysaccharides, *Ann. Rev. Microbiol.*, 28, 103, 1974.
6. **Bartnieke-Garcia, S.,** Cell wall chemistry and taxonomy of fungi, *Ann. Rev. Microbiol.*, 22, 87, 1968.
7. **Manners, D. J., Masson, A. J., and Patterson, J. C.,** The structure of a β-1, 3-D-glucan from yeast cell walls, *Biochem. J.*, 135, 19, 1970.
8. **Carbonell, L. M., Kanetsuna, F., and Gil, G.,** Chemical morphology of glucan and chitin in the cell wall of *Paracoccidioide brasiliensis, J. Bacteriol.*, 101, 636, 1970.
9. **Rundall, K. M. and Kenchington,** The chitin system, *Biol. Rev.*, 48, 597, 1973.
10. **Ballou, C. E.,** Structure and biosyntheses of the mannan component of the yeast cell envelope, *Adv. Microb. Physiol.*, 14, 93, 1976.
11. **Villanueva, J. R. and Garcia Acha, I.,** Production and the use of fungal protoplasts, in *Methods in Microbiology,* Vol. 4, Booth, C., Ed., Academic Press, New York, 1971, 665.
12. **Knorr, D., Shetty, K. J., and Kinsella, J. E.,** Enzymatic lysis of yeast cell walls, *Biotechnol. Bioeng.*, 21, 2011, 1979.
13. **Doi, K., Doi, A., and Fukui, T.,** Purification and properties of lytic β-glucanase from an *Arthrobacter* bacterium, *Agric. Biol. Chem.*, 37, 1619, 1973.
14. **Bacon, J. S. D., Gordon, A. H., Jones, D., Taylor, I. F., and Webley, D. M.,** The separation of β-glucanases produced by *Cytophaga johnsonii* and their role in the lysis of yeast cell walls, *Biochem. J.*, 120, 67, 1970.
15. **Tanaka, H. and Phaff, H. J.,** Enzymatic lysis of yeast cell walls. I. Isolation of wall decomposing organisms and separation and purification of lytic enzymes, *J. Bacteriol.*, 89, 1570, 1965.
16. **Yamamoto, S., Kobayashi, R., and Nagasaki, S.,** Physicochemical and enzymatic properties of a crystalline yeast cell lytic enzyme from a Rhizopus mold, *Agric. Biol. Chem.*, 38, 1563, 1974.
17. **de Vries, I. M. H. and Wessels, J. G. H.,** Release of protoplasts from *Schizophyllum commune* by a lytic enzyme preparation from *Trichoderma viride, J. Gen. Microbiol.*, 73, 13, 1972.
18. **Anderson, F. B. and Millbank, J. W.,** Protoplast formation and yeast cell wall structure. The action of the enzymes of the snail, *Biochem. J.*, 99, 628, 1966.
19. **Abels, F. G., Bosshart, R. P., Forrence, L. E., and Habig, W. H.,** Preparation and purification of glucanase and chitinase from bean leaves, *Plant Physiol.*, 47, 129, 1970.

20. **Molano, J., Polacheck, I., Duran, A., and Cabib, E.,** An endochitinase from wheat germ, *J. Biol. Chem.*, 254, 4901, 1979.
21. **Pegg, G. F. and Young, D. H.,** Purification and characterization of chitinase enzymes from healthy and *Verticillium albo-atrum-* infected tomato plants, and from *V. albo-atrum, Physiol. Plant Pathol.*, 21, 389, 1982.
22. **Ballance, G. M. and Manners, D. J.,** Partial purification and properties of an endo β-1, 3-D-glucanase from germinated rye, *Phytochemistry*, 18, 1539, 1978.
23. **Keen, N. T. and Yoshikawa, M.,** β-1,3-Endoglucanase from soybean releases elicitor-active carbohydrates from fungus cell walls, *Plant Physiol.*, 71, 460, 1983.
24. **Barbieri, L. and Stirpe, F.,** Ribosome-inactivating proteins from plants: properties and possible uses, *Cancer Surv.*, 1, 489, 1982.
25. **Reisbig, R. R. and Bruland, O.,** Dianthin 30 and Dianthin 32 from *Dianthus caryophyllus*. II. Inhibitors of plant protein synthesis and their tissue distribution, *Arch. Biochem. Biophys.*, 224, 700, 1983.
26. **Stirpe, F., Gasperi-Campani, A., Barnieri, L., Falasca, A., Abbondanza, A., and Stevens, W. A.,** Ribosome inactivating proteins from the seeds of *Saponaria-officinalis* (soapwort) of *Agrostemma-githago* (corn cockle) and of *Asparagus-Officinalis* (Asparagus) and from the latex of *Hura-crepitans* (sandbox tree), *Biochem. J.*, 216, 617, 1986.
27. **Roberts, W. K. and Selitrennikoff, C. P.,** Plant proteins that inactivate foreign ribosomes, *Biosci. Rep.*, 6, 19, 1986.
28. **Collins, M. S. and Pappagianis, D.,** Lysozyme enhanced killing of *Candida albicans* and *Coccidioides immitis* by amphotericin B, *Sabouraudia*, 12, 329, 1974.
29. **Pope, A. M. S. and Davies, D. A. L.,** The influence of carbohydrases on the growth of fungal pathogens *in vitro* and *in vivo, Postgrad. Med. J.*, 55, 674, 1979.
30. **Oranusi, N. A. and Trinei, A. P. J.,** Growth of bacteria on chitin, fungal cell walls and fungal biomass and the effect of extracellular enzymes produced by these cultures on the antifungal activity of amphotericin B, *Microbios*, 43, 17, 1985.
31. **Ashizawa, E. and Poulose, A. J.,** unpublished data, 1988.
32. **Watanabe, R., Ogasawara, N., Tanaka, H., and Uchiyama, T.,** Effects of fungal lytic enzymes and non-ionic detergents on the actions of some fungicides against *Pyricularia oryzae, Agric. Biol. Chem.*, 52, 895, 1988.
33. **Roberts, W. K., Laue, B. E., and Selitrennikoff, C. P.,** Antifungal proteins from plants, *Ann. N.Y. Acad. Sci.*, 544, 141, 1988.
34. **Lovett, M., Ashizawa, E., and Poulose, A. J.,** unpublished data, 1989.
35. **Pegg, G. G. and Vessey, J. C.,** Chitinase activity in *Lycopersicon esculentum* and its relationship to the *in vivo* lysis of *Verticillium albo-atrum* mycelium, *Physiol. Plant Pathol.*, 3, 207, 1973.
36. **Roberts, W. K. and Selitrennikoff, C. P.,** Plant and bacterial chitinases differ in antifungal activity, *J. Gen. Microbiol.*, 134, 169, 1988.
37. **Albersheim, P. and Valent, B.,** Host-pathogen interactions in plants. Plants, when exposed to oligosaccharides of fungal origin, defend themselves by accumulating antibiotics, *J. Cell Biol.*, 78, 627, 1978.
38. **Nichols, E. J., Beckman, J. M., and Hadwiger, L. A.,** Glycosidic enzyme activity in pea tissue and pea *Fusarium solani* interactions, *Plant Physiol.*, 66, 199, 1980.
39. **Ryan, C. A., Bishop, P. D., Graham, J. S., Broadway, R. M., and Duffey, S. S.,** Plant and fungal cell wall fragments activate expression of proteinase inhibitor genes for plant defense, *J. Chem. Ecol.*, 12, 1025, 1986.
40. **Rogers, K. K., Albert, F., and Anderson, A. J.,** Lipid peroxidation is a consequence of elicitor activity, *Plant Physiol.*, 86, 547, 1988.
41. **Chet, I.,** Trichoderma-application, mode of action and potential as a biocontrol agent of soil-borne plant pathogenic fungi, in *Innovative Approaches to Plant Disease Control*, Chet, I., Ed., John Wiley & Sons, New York, 1987, 137.
42. **Papavizas, G. C.,** Trichoderma and gliocladium: biology, ecology and the potential for biocontrol, *Annu. Rev. Phytopathol.*, 23, 23, 1985.
43. **Sneh, B.,** Use of rhizosphere chitinolytic bacteria for biological control, *Phytopathol. Z.*, 100, 251, 1981.
44. **Shapira, R., Ordentlick, A., Chet, I., and Oppenheim, A. B.,** Control of plant diseases by chitinase expressed from cloned DNA in *E. coli, Phytopathology*, 79, 1246, 1989.
45. **Tanaka, H., Ogasawara, N., Nakagima, T., and Tamari, K.,** Cell walls of *Piricularia oryzae*. I. Selective enzymolysis of *Piricularia oryzae* walls by wall-lytic enzymes of *Bacillus circulans* WL12, *J. Gen. Appl. Microbiol.*, 16, 36, 1970.
46. **Ogasawara, N., Tanaka, H., Yoshimaga, E., and Hirara, K.,** Japanese Patent 56,816, 1973.
47. **Poulose, A. J. and Kolattukudy, P. E.,** Enzymes as Agricultural Chemical Adjuvants, European Patent application, 87,310,146.3, 1987.
48. **Suslow, T. V. and Jones, J. D. G.,** Chitinase-Producing Bacteria and Plants, U.S. Patent 4,940,840, 1990.

49. **Jones, J. D. G., Grady, K. L., Suslow, T. V., and Bedbrook, J. R.,** Isolation and characterization of genes encoding two chitinase enzymes from *Serratia marcescens*—expression in *E. coli*, biological control agent, *EMBO J.*, 5, 467, 1986.
50. **Zikakis, J. P.,** Chitinolytic enzymes and their applications in *Biocatalysis in Agricultural Biotechnology*, Whitaker, J. R. and Sonnet, P. E., Eds., American Chemical Society, Washington, D.C., 1989, 116.

INDEX

A

B

C

F

G

H

I

K

L

M

N

O

P

R